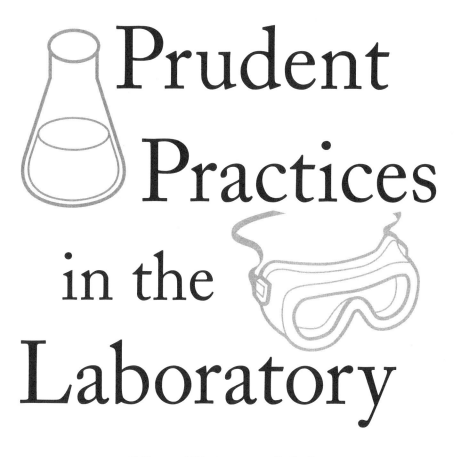

Prudent Practices in the Laboratory

Handling and Management of Chemical Hazards

Committee on Prudent Practices in the Laboratory: An Update

Board on Chemical Sciences and Technology

Division on Earth and Life Studies

NATIONAL RESEARCH COUNCIL
OF THE NATIONAL ACADEMIES

THE NATIONAL ACADEMIES PRESS
Washington, D.C.
www.nap.edu

THE NATIONAL ACADEMIES PRESS • 500 Fifth Street, N.W. • Washington, DC 20001

NOTICE: The project that is the subject of this report was approved by the Governing Board of the National Research Council, whose members are drawn from the councils of the National Academy of Sciences, the National Academy of Engineering, and the Institute of Medicine. The members of the committee responsible for the report were chosen for their special competences and with regard for appropriate balance.

This study was supported by the U.S. Department of Energy under grant number DE-FG02-08ER15932; the National Institutes of Health under contract number N01-OD-4-2139, TO #200; and the National Science Foundation under grant number CHE-0740356. Additional support was received from Air Products and Chemicals, Inc.; the American Chemical Society; E. I. du Pont de Nemours and Company; Eastman Chemical Company; the Howard Hughes Medical Institute; and PPG Industries.

Any opinions, findings, conclusions, or recommendations expressed in this publication are those of the authors and do not necessarily reflect the views of the organizations or agencies that provided support for the project.

Library of Congress Cataloging-in-Publication Data

Prudent practices in the laboratory : handling and management of chemical hazards / Committee on Prudent Practices in the Laboratory, Board on Chemical Sciences and Technology, Division on Earth and Life Studies. — Updated ed.
   p. cm.
 Includes bibliographical references and index.
 ISBN-13: 978-0-309-13864-2 (hardback)
 ISBN-10: 0-309-13864-7 (hardback)
 ISBN-13: 978-0-309-13865-9 (pdf)
 ISBN-10: 0-309-13865-5 (pdf)
 1. Hazardous substances. 2. Chemicals—Safety measures. 3. Hazardous wastes. I. National Research Council (U.S.). Committee on Prudent Practices in the Laboratory.
 T55.3.H3P78 2011
 660'.2804—dc22

          2010047731

Additional copies of this report are available from The National Academies Press, 500 Fifth Street, N.W., Lockbox 285, Washington, DC 20055; (800) 624-6242 or (202) 334-3313 (in the Washington metropolitan area); Internet, http://www.nap.edu.

THE NATIONAL ACADEMIES
Advisers to the Nation on Science, Engineering, and Medicine

The **National Academy of Sciences** is a private, nonprofit, self-perpetuating society of distinguished scholars engaged in scientific and engineering research, dedicated to the furtherance of science and technology and to their use for the general welfare. Upon the authority of the charter granted to it by the Congress in 1863, the Academy has a mandate that requires it to advise the federal government on scientific and technical matters. Dr. Ralph J. Cicerone is president of the National Academy of Sciences.

The **National Academy of Engineering** was established in 1964, under the charter of the National Academy of Sciences, as a parallel organization of outstanding engineers. It is autonomous in its administration and in the selection of its members, sharing with the National Academy of Sciences the responsibility for advising the federal government. The National Academy of Engineering also sponsors engineering programs aimed at meeting national needs, encourages education and research, and recognizes the superior achievements of engineers. Dr. Charles M. Vest is president of the National Academy of Engineering.

The **Institute of Medicine** was established in 1970 by the National Academy of Sciences to secure the services of eminent members of appropriate professions in the examination of policy matters pertaining to the health of the public. The Institute acts under the responsibility given to the National Academy of Sciences by its congressional charter to be an adviser to the federal government and, upon its own initiative, to identify issues of medical care, research, and education. Dr. Harvey V. Fineberg is president of the Institute of Medicine.

The **National Research Council** was organized by the National Academy of Sciences in 1916 to associate the broad community of science and technology with the Academy's purposes of furthering knowledge and advising the federal government. Functioning in accordance with general policies determined by the Academy, the Council has become the principal operating agency of both the National Academy of Sciences and the National Academy of Engineering in providing services to the government, the public, and the scientific and engineering communities. The Council is administered jointly by both Academies and the Institute of Medicine. Dr. Ralph J. Cicerone and Dr. Charles M. Vest are chair and vice chair, respectively, of the National Research Council.

www.national-academies.org

Preface

In the early 1980s, the National Research Council (NRC) produced two major reports on laboratory safety and laboratory waste disposal: *Prudent Practices for Handling Hazardous Chemicals in Laboratories* (1981) and *Prudent Practices for Disposal of Chemicals from Laboratories* (1983). In 1995, the NRC's Board on Chemical Sciences and Technology updated, combined, and revised the earlier studies in producing *Prudent Practices in the Laboratory: Handling and Disposal of Chemicals*. More than 10 years later, the Board on Chemical Sciences and Technology initiated an update and revision of the 1995 edition of *Prudent Practices*.

In 2007, the Department of Energy, the National Science Foundation, and the National Institutes of Health, with additional support from the American Chemical Society, Eastman Kodak Company, E. I. du Pont de Nemours and Company, Howard Hughes Medical Institute, Air Products and Chemicals, Inc., and PPG Industries, commissioned a study by NRC to "review and update the 1995 publication, *Prudent Practices in the Laboratory: Handling and Disposal of Chemicals*." The Committee on Prudent Practices in the Laboratory: An Update was charged to

- review and update the 1995 publication, *Prudent Practices in the Laboratory: Handling and Disposal of Chemicals*;
- modify the existing content and add content as required to reflect new fields and developments that have occurred since the previous publication;
- emphasize the concept of a "culture of safety" and how that culture can be established and nurtured;
- consider laboratory operations and the adverse impacts those operations might have on the surrounding environment and community.

The Committee on Prudent Practices in the Laboratory: An Update was established in June 2008. The first meeting was held in August 2008, and two subsequent meetings were held, one in October 2008 and the other in February 2009. All meetings were held in Washington, D.C.

The original motivation for drafting *Prudent Practices 1981* and *Prudent Practices 1983* was to provide an authoritative reference on the handling and disposal of chemicals at the laboratory level. These volumes not only served as a guide to laboratory workers, but also offered prudent guidelines for the development of regulatory policy by government agencies concerned with safety in the workplace and protection of the environment.

Pertinent health-related parts of *Prudent Practices 1981* are incorporated in a non-mandatory section of the Occupational Safety and Health Administration (OSHA) Laboratory Standard (29 CFR § 1910.1450, "Occupational Exposure to Hazardous Chemicals in Laboratories," reprinted in this edition as Appendix A). OSHA's purpose was to provide guidance for developing and implementing its required Chemical Hygiene Plan. Since their original publication in the early 1980s, these reports have been distributed widely both nationally and internationally. In 1992, the International Union of Pure and Applied Chemistry and the World Health Organization published *Chemical Safety Matters*, a document based on *Prudent Practices 1981* and *Prudent Practices 1983*, for wide international use.

The next volume (*Prudent Practices 1995*) responded to societal and technical developments that were driving significant change in the laboratory culture and laboratory operations relative to safety, health, and environmental protection.

The major drivers for this new culture of laboratory safety included an increase in regulations regarding laboratory practice, technical advances in hazard and risk evaluation, and an improvement in the understanding of the elements necessary for an effective culture of safety.

Building on this history, the updated (2011) edition of *Prudent Practices in the Laboratory* also considers technical, regulatory, and societal changes that have occurred since the last publication. As a reflection of some of those changes, it provides information on new topics, including

- emergency planning,
- laboratory security,
- handling of nanomaterials, and
- an expanded discussion of environment, health, and safety management systems.

Throughout the development of this book, the committee engaged in discussions with subject matter experts and industrial and academic researchers and teachers. The goal of these discussions was to determine what the various constituencies considered to be *prudent practices* for laboratory operations.

Public support for the laboratory use of chemicals depends on compliance with regulatory laws, respect by organizations and individuals of the concerns of the public, and the open acknowledgment and management of the risks to personnel who work in laboratory environments. Addressing these issues is the joint responsibility of everyone who handles or makes decisions about chemicals, from shipping and receiving clerks to laboratory personnel and managers, environmental health and safety staff, and institutional administrators.

The writers of the preface to the 1995 edition stated that, "This shared responsibility is now a fact of laboratory work as inexorable as the properties of the chemicals that are being handled," and we restate that sentiment here. Organizations and institutions must create environments where safe laboratory practice is standard practice. Each individual influences the "culture of safety" in the laboratory. All of us should recognize that the safety of each of us depends on teamwork and personal responsibility as well as the knowledge of chemistry. Faculty, research advisors, and teachers should note that a vital component of chemical education is teaching students how to identify the risks and hazards in a laboratory. Such education serves scientists well in their ultimate careers in government, industry, academe, and the health sciences.

The promotion of a "culture of safety" has come a long way since 1995; however, in some ways, the "culture of chemistry" is still at odds with that of safety. Some of us may have witnessed unsafe behavior or minor accidents, and yet, rather than viewing these incidents with concern and as opportunities to modify practices and behavior, we often have failed to act upon these "teachable moments." Ironically, however, we shudder when, even today, we hear of accidents—some fatal—that might have been our near misses.

Rigorous practitioners argue that, in principle, all accidental injuries are preventable if systems and attitudes are in place to prevent them. Even in these days of technological advancements, tracking of near misses and adaptation of systems to eradicate them is inconsistent across the enterprise. Within the research and teaching communities, less rigorous practitioners seem to accept different safety tolerances for different environments. It is common during a discussion of laboratory safety to hear the statement, "Industry is much stricter on safety than academia. Things happen in academic research labs that would never be allowed where I work." This is often accompanied by a "when I was a student . . ." story. The path to failure illustrated by this colloquy should be obvious and unacceptable. To fully implement a culture of safety, even with improved technology, everyone who is associated with the laboratory must be mindful of maintaining a safe environment.

Prudent Practices (1995) has been used worldwide and has served as a leading reference book for laboratory practice. The committee hopes that this new edition of the book will expand upon that tradition, and that this edition will assist the readers to provide a safe and healthy laboratory environment in which to teach, learn, and conduct research.

Acknowledgments

Many technical experts provided input to this book. Their involvement, by speaking to the committee or by providing technical reviews of material prepared by the committee, greatly enhanced this work. The Committee on Prudent Practices in the Laboratory: An Update thanks the following people for their contributions to this revision of *Prudent Practices*.

Chyree Batton, Spelman College
Kevin Charbonneau, Yale University
Jasmaine Coleman, Spelman College
Dennis Deziel, Department of Homeland Security
Michael Ellenbecker, Toxic Use Reduction Institute, University of
    Massachusetts–Lowell
Drew Endy, Stanford University
Dennis Fantin, California Polytechnic State University
Charles Geraci, National Institute for Occupational Safety and Health
Lawrence Gibbs, Stanford University
Laura Hodson, National Institute for Occupational Safety and Health
Barbara Karn, Environmental Protection Agency
Cathleen King, Yale University
Robert Klein, Yale University
Stanley K. Lengerich, Eli Lilly & Company
Thomas J. Lentz, National Institute for Occupational Safety and Health
Clyde Miller, BASF Corporation
John Miller, Department of Energy
Richard W. Niemeier, National Institute for Occupational Safety and Health
Todd Pagano, Rochester Institute of Technology
Tammy Stemen, Yale University
Cary Supalo, Pennsylvania State University
Candice Tsai, Toxic Use Reduction Institute, University of Massachusetts–Lowell
Bryana Williams, Spelman College

Acknowledgment of Reviewers

This report has been reviewed in draft form by persons chosen for their diverse perspectives and technical expertise in accordance with procedures approved by the National Research Council's Report Review Committee. The purpose of this independent review is to provide candid and critical comments that will assist the institution in making the published report as sound as possible and to ensure that it meets institutional standards of objectivity, evidence, and responsiveness to the study charge. The review comments and draft manuscript remain confidential to protect the integrity of the deliberative process. We wish to thank the following for their review of this report:

Robert J. Alaimo, Proctor & Gamble Pharmaceuticals, retired
Bruce Backus, Washington University
Janet Baum, Independent Consultant
L. Casey Chosewood, Centers for Disease Control and Prevention
Rick L. Danheiser, Massachusetts Institute of Technology
Louis J. DiBerardinis, Massachusetts Institute of Technology
Charles L. Geraci, Centers for Disease Control and Prevention, National Institute
    for Occupational Safety and Health
Lawrence M. Gibbs, Stanford University
Stephanie Graham-Sims, West Virginia University
Scott C. Jackson, E. I. du Pont de Nemours & Company
Donald Lucas, Lawrence Berkeley National Laboratory
Edward H. Rau, National Institutes of Health
Robin D. Rogers, University of Alabama
Timothy J. Scott, The Dow Chemical Company
Robert W. Shaw, U.S. Army Research Laboratory, retired
Erik A. Talley, Cornell University
William C. Trogler, University of California, San Diego
Douglas B. Walters, KPC, Inc.

Although the reviewers listed above provided many constructive comments and suggestions, they were not asked to endorse the conclusions or recommendations, nor did they see the final draft of the report before its release. The review of this report was overseen by **Stanley Pine, University of California, San Diego**. Appointed by the National Research Council, he was responsible for making certain that an independent examination of this report was carried out in accordance with institutional procedures and that all review comments were carefully considered. Responsibility for the final content of this report rests entirely with the authors and the institution.

Contents

1 The Culture of Laboratory Safety   1

1.A   Introduction, 2
1.B   The Culture of Laboratory Safety, 2
1.C   Responsibility and Accountability for Laboratory Safety, 2
1.D   Special Safety Considerations in Academic Laboratories, 3
1.E   The Safety Culture in Industrial and Governmental Laboratories, 4
1.F   Other Factors That Influence Laboratory Safety Programs, 5
1.G   Laboratory Security, 7
1.H   Structure of the Book, 7
1.I   Summary, 7

2 Environmental Health and Safety Management System   9

2.A   Introduction, 10
2.B   Chemical Hygiene Plan, 14
2.C   Safety Rules and Policies, 15
2.D   Chemical Management Program, 20
2.E   Laboratory Inspection Program, 23
2.F   Emergency Procedures, 27
2.G   Employee Safety Training Program, 29

3 Emergency Planning   31

3.A   Introduction, 33
3.B   Preplanning, 33
3.C   Leadership and Priorities, 37
3.D   Communication During an Emergency, 38
3.E   Evacuations, 39
3.F   Shelter in Place, 30
3.G   Loss of Power, 40
3.H   Institutional or Building Closure, 41
3.I   Emergency Affecting the Community, 42
3.J   Fire or Loss of Laboratory, 42
3.K   Drills and Exercises, 43
3.L   Outside Responders and Resources, 43

4 Evaluating Hazards and Assessing Risks in the Laboratory   45

4.A   Introduction, 47
4.B   Sources of Information, 47
4.C   Toxic Effects of Laboratory Chemicals, 53
4.D   Flammable, Reactive, and Explosive Hazards, 65
4.E   Physical Hazards, 74
4.F   Nanomaterials, 77
4.G   Biohazards, 79
4.H   Hazards from Radioactivity, 79

5 Management of Chemicals   83

5.A   Introduction, 84
5.B   Green Chemistry for Every Laboratory, 84
5.C   Acquisition of Chemicals, 88

5.D   Inventory and Tracking of Chemicals, 90
5.E   Storage of Chemicals in Stockrooms and Laboratories, 94
5.F   Transfer, Transport, and Shipment of Chemicals, 101

6   Working with Chemicals   105
6.A   Introduction, 107
6.B   Prudent Planning, 107
6.C   General Procedures for Working with Hazardous Chemicals, 108
6.D   Working with Substances of High Toxicity, 122
6.E   Working with Biohazardous and Radioactive Materials, 126
6.F   Working with Flammable Chemicals, 127
6.G   Working with Highly Reactive or Explosive Chemicals, 130
6.H   Working with Compressed Gases, 140
6.I   Working with Microwave Ovens, 141
6.J   Working with Nanoparticles, 141

7   Working with Laboratory Equipment   147
7.A   Introduction, 149
7.B   Working with Water-Cooled Equipment, 149
7.C   Working with Electrically Powered Laboratory Equipment, 149
7.D   Working with Compressed Gases, 164
7.E   Working with High or Low Pressures and Temperatures, 170
7.F   Using Personal Protective, Safety, and Emergency Equipment, 175
7.G   Emergency Procedures, 181

8   Management of Waste   183
8.A   Introduction, 185
8.B   Chemical Hazardous Waste, 186
8.C   Multihazardous Waste, 201
8.D   Procedures for the Laboratory-Scale Treatment of Surplus and
      Waste Chemicals, 209

9   Laboratory Facilities   211
9.A   Introduction, 213
9.B   General Laboratory Design Considerations, 213
9.C   Laboratory Ventilation, 219
9.D   Room Pressure Control Systems, 242
9.E   Special Systems, 243
9.F   Maintenance of Ventilation Systems, 248
9.G   Ventilation System Management Program, 249
9.H   Safety and Sustainability, 250
9.I   Laboratory Decommissioning, 253

10 Laboratory Security   255
10.A Introduction, 256
10.B Security Basics, 256
10.C Systems Integration, 259
10.D Dual-Use Hazard of Laboratory Materials, 259
10.E Laboratory Security Requirements, 260
10.F Security Vulnerability Assessment, 261
10.G Dual-Use Security, 262
10.H Security Plans, 262

11 Safety Laws and Standards Pertinent to Laboratories   265
11.A Introduction, 267
11.B Regulation of Laboratory Design and Construction, 272
11.C Regulation of Chemicals Used in Laboratories, 273
11.D Regulation of Biohazards and Radioactive Materials Used in Laboratories, 276
11.E Environmental Regulations Pertaining to Laboratories, 276
11.F Shipping, Export, and Import of Laboratory Materials, 278
11.G Laboratory Accidents, Spills, Releases, and Incidents, 281

Bibliography   283

Appendixes
A OSHA Laboratory Standard   289
B Statement of Task   307
C Committee Member Biographies   309

Index

Supplemental Materials on CD
1. Sample Inspection Checklist
2. ACS Security and Vulnerability Checklist for Academic and Small Chemical Laboratory Facilities
3. Chemical Compatibility Storage Guide
4. Chemical Compatibility Storage Codes
5. Sample Incident Report Form
6. Laboratory Closeout Checklist
7. Laboratory Emergency Information Poster
8. Laboratory Hazard Assessment Checklist
9. Environmental Protection Agency (40 CFR Parts 261 and 262) Standards Applicable to Generators of Hazardous Waste; Alternative Requirements for Hazardous Waste Determination and Accumulation of Unwanted Material at Laboratories Owned by Colleges and Universities and Other Eligible Academic Entities Formally Affiliated With Colleges and Universities; Final Rule
10. Laboratory Chemical Safety Summaries
11 Blank Form for Laboratory Chemical Safety Summaries
12. Procedures for the Laboratory Scale Treatment of Surplus and Waste Chemicals
13. Electronic Copy of *Prudent Practices in the Laboratory: Handling and Management of Chemical Hazards*

Tables, Figures, Boxes, and Vignettes

FIGURES

2.1 Overview of environmental health and safety management system, 11
2.2 Accident report form, 30

3.1 Impact/Likelihood of occurrence mapping, 34

4.1 GHS placards for labeling containers of hazardous chemicals, 49
4.2 A simple representation of possible dose-response curves, 56
4.3 The fire triangle, 67
4.4 National Fire Protection Association (NFPA) system for classification of hazards, 68
4.5 U.S. Department of Energy graded exposure risk for nanomaterials, 78

5.1 Compatible storage group classification system, 97
5.2 Recommended inner packaging label for on-site transfer of nanomaterials, 103

6.1 U.S. Department of Energy graded exposure risk for nanomaterials, 143

7.1 Representative design for a three-wire grounded outlet, 150
7.2 Standard wiring convention for 110-V electric power to equipment, 151
7.3 Schematic diagram of a properly wired variable autotransformer, 155
7.4 Example of a column purification system, 160

8.1 Flowchart for categorizing unknown chemicals for waste disposal, 188
8.2 Example of Uniform Hazardous Waste Manifest, 200

9.1 Open versus closed laboratory design, 215
9.2 Specifications for barrier-free safety showers and eyewash units, 218
9.3 Laminar versus turbulent velocity profile, 226
9.4 Effect of baffles on face velocity profile in a laboratory chemical hood, 227
9.5 Effect of sash placement on airflow in a nonbypass laboratory chemical hood, 229
9.6 Effect of sash placement on airflow in a bypass laboratory chemical hood, 230
9.7 Diagram of a typical benchtop laboratory chemical hood, 232
9.8 Diagram of a typical distillation hood, 233
9.9 Diagram of a typical walk-in laboratory chemical hood, 234
9.10 Schematic of a typical laboratory chemical hood scrubber, 235
9.11 Fume extractor or snorkel, 238
9.12 Diagrams of typical slot hoods, 238
9.13 Example of a Class II biosafety cabinet, 246
9.14 Examples of postings for laboratory chemical hoods, 250
9.15 Carbon inventory of a research university campus, 251

10.1 Concentric circles of physical protection, 257

TABLES

4.1   Acute Toxicity Hazard Level, 59
4.2   Probable Lethal Dose for Humans, 60
4.3   Examples of Compounds with a High Level of Acute Toxicity, 60
4.4   NFPA Fire Hazard Ratings, Flash Points (FP), Boiling Points (bp), Ignition Temperatures, and Flammable Limits of Some Common Laboratory Chemicals, 67
4.5   Additional Symbols Seen in the NFPA Diamond, 69
4.6   Examples of Oxidants, 69
4.7   Functional Groups in Some Explosive Compounds, 71
4.8   Classes of Chemicals That Can Form Peroxides, 72
4.9   Types of Compounds Known to Autooxidize to Form Peroxides, 72
4.10 Examples of β Emitters, 81
4.11 Radiation Quality Factors, 81
4.12 U.S. Nuclear Regulatory Commission Dose Limits, 82

5.1   Examples of Compatible Storage Groups, 96
5.2   Storage Limits for Flammable and Combustible Liquids for Laboratories with Sprinkler System, 99
5.3   Container Size for Storage of Flammable and Combustible Liquids, 99

7.1   Summary of Magnetic Field Effects, 163

8.1   Assignment of Tasks for Waste Handling, 194
8.2   Classes and Functional Groupings of Organic Chemicals for Which There Are Existing Treatment Methods, 210
8.3   Classes and Functional Groupings of Inorganic Chemicals for Which There Are Existing Treatment Methods, 210

9.1   Some Activities, Equipment, or Materials That May Require Separation from the Main Laboratory, 215
9.2   Examples of Equipment That Can Be Shared Between Researchers and Research Groups, 216
9.3   Laboratory Engineering Controls for Personal Protection, 220
9.4   US FED STD 209E Clean Room Classification, 243
9.5   ISO Classification of Air Cleanliness for Clean Rooms, 244
9.6   Comparison of Biosafety Cabinet Characteristics, 247

10.1 Security Features for Security Level 1, 263
10.2 Security Features for Security Level 2, 264
10.3 Security Features for Security Level 3, 264

11.1 Federal Safety Laws and Regulations That Pertain to Laboratories, 270
11.2 Chemicals Covered by Specific OSHA Standards, 273

BOXES

1.1   Tips for Encouraging a Culture of Safety Within an Academic Laboratory, 5

2.1   Chemical Hygiene Responsibilities in a Typical Academic Institution, 16
2.2   Chemical Hygiene Responsibilities in a Typical Industry Research Facility, 18

2.3   Chemical Hygiene Responsibilities in a Typical Governmental Laboratory, 20

2.4   Excerpt from an Inspection Checklist, 27

3.1   Continuity of Laboratory Operations Checklist, 41

4.1   Quick Guide for Toxicity Risk Assessment of Chemicals, 55

4.2   Quick Guide to Risk Assessment for Physical, Flammable, Explosive, and Reactive Hazards in the Laboratory, 66

4.3   Quick Guide to Risk Assessment for Biological Hazards in the Laboratory, 80

6.1   A Simple Qualitative Method to Verify Adequate Laboratory Chemical Hood Ventilation, 109

9.1   Quick Guide for Maximizing Efficiency of Laboratory Chemical Hoods, 223

9.2   Quick Guide for Working in Environmental Rooms, 245

VIGNETTES

3.1   Preplanning reduces the impact of a fire on continuity of operations, 43

5.1   Pollution prevention reduces solvent waste, 85

6.1   Finger laceration from broken tubing connector, 115

6.2   Runaway reaction during scale-up, 115

6.3   Solvent fire, 128

6.4   Fluorine inhalation, 137

7.1   Oil bath fire as a result of a loose temperature sensor, 156

7.2   Muffle furnace fire, 157

7.3   Centrifuge explosion from use of improper rotor, 161

7.4   Hydrogen leak from jammed cylinder cap, 166

7.5   Injury while working on equipment under pressure, 171

9.1   Appropriate use of personal protective equipment in shared spaces, 214

9.2   Sustainability considerations in laboratory ventilation design, 251

The Culture of Laboratory Safety

1.A   INTRODUCTION   2

1.B   THE CULTURE OF LABORATORY SAFETY   2

1.C   RESPONSIBILITY AND ACCOUNTABILITY FOR LABORATORY
        SAFETY   2

1.D   SPECIAL SAFETY CONSIDERATIONS IN ACADEMIC
        LABORATORIES   3
        1.D.1   High School Teaching Laboratories   3
        1.D.2   Undergraduate Teaching Laboratories   3
        1.D.3   Academic Research Laboratories   4

1.E   THE SAFETY CULTURE IN INDUSTRIAL AND GOVERNMENTAL
        LABORATORIES   4

1.F   OTHER FACTORS THAT INFLUENCE LABORATORY SAFETY
        PROGRAMS   5
        1.F.1   Advances in Technology   5
        1.F.2   Environmental Impact   5
        1.F.3   Changes in the Legal and Regulatory Requirements   5
        1.F.4   Accessibility for Scientists with Disabilities   5

1.G   LABORATORY SECURITY   7

1.H   STRUCTURE OF THE BOOK   7

1.I   SUMMARY   7

1.A INTRODUCTION

Over the past century, chemistry has increased our understanding of the physical and biological world as well as our ability to manipulate it. As a result, most of the items we take for granted in modern life involve synthetic or natural chemical processing.

We acquire that understanding, carry out those manipulations, and develop those items in the chemical laboratory; consequently, we also must monitor and control thousands of chemicals in routine use. Since the age of alchemy, laboratory chemicals have demonstrated dramatic and dangerous properties. Some are insidious poisons.

During the "heroic age" of chemistry, martyrdom for the sake of science was acceptable, according to an 1890 address by the great chemist August Kekulé: "If you want to become a chemist, so Liebig told me, when I worked in his laboratory, you have to ruin your health. Who does not ruin his health by his studies, nowadays will not get anywhere in Chemistry" (as quoted in Purchase, 1994).

Today that attitude seems as ancient as alchemy. Over the years, we have developed special techniques for handling chemicals safely. Institutions that sponsor chemical laboratories hold themselves accountable for providing safe working environments. Local, state, and federal regulations codify this accountability.

Beyond regulation, employers and scientists also hold themselves responsible for the well-being of building occupants and the general public. Development of a "culture of safety"—with accountability up and down the managerial (or administrative) and scientific ladders—has resulted in laboratories that are, in fact, safe and healthy environments in which to teach, learn, and work. Injury, never mind martyrdom, is out of style.

1.B THE CULTURE OF LABORATORY SAFETY

As a result of the promulgation of the Occupational Safety and Health Administration (OSHA) Laboratory Standard (29 CFR § 1910.1450), a culture of safety consciousness, accountability, organization, and education has developed in industrial, governmental, and academic laboratories. Safety and training programs, often coordinated through an office of environment, health, and safety (EHS), have been implemented to monitor the handling of chemicals from the moment they are ordered until their departure for ultimate disposal and to train laboratory personnel in safe practices.[1]

Laboratory personnel realize that the welfare and safety of each individual depends on clearly defined attitudes of teamwork and personal responsibility and that laboratory safety is not simply a matter of materials and equipment but also of processes and behaviors. Learning to participate in this culture of habitual risk assessment, experiment planning, and consideration of worst-case possibilities—for oneself and one's fellow workers—is as much part of a scientific education as learning the theoretical background of experiments or the step-by-step protocols for doing them in a professional manner.[2]

Accordingly, a crucial component of chemical education at every level is to nurture basic attitudes and habits of prudent behavior so that safety is a valued and inseparable part of all laboratory activities. In this way, a culture of laboratory safety becomes an internalized attitude, not just an external expectation driven by institutional rules. This process must be included in each person's chemical education throughout his or her scientific career.

1.C RESPONSIBILITY AND ACCOUNTABILITY FOR LABORATORY SAFETY

Ensuring a safe laboratory environment is the combined responsibility of laboratory personnel, EHS personnel, and the management of an organization, though the primary responsibility lies with the individual performing the work. Of course, federal, state, and local laws and regulations make safety in the laboratory a legal requirement and an economic necessity. Laboratory safety, although altruistic, is not a purely voluntary function; it requires mandatory safety rules and programs and an ongoing commitment to them. A sound safety organization that is respected by all requires the participation and support of laboratory administrators, employees, and students.

The ultimate responsibility for creating a safe environment and for encouraging a culture of safety rests with the head of the organization and its operating units. Leadership by those in charge ensures that an effective safety program is embraced by all. Even a well-conceived safety program will be treated casually by workers if it is neglected by top management.

Direct responsibility for the management of the laboratory safety program typically rests with the chemical

---

[1]Throughout this book, the committee uses the word *training* in its usual sense of "making proficient through specialized instruction" with no direct reference to regulatory language.

[2]With regard to safe use of chemicals, the committee distinguishes between hazard, which is an inherent danger in a material or system, and the risk that is assumed by using it in various ways. *Hazards* are dangers intrinsic to a substance or operation; *risk* refers to the probability of injury associated with working with a substance or carrying out a particular laboratory operation. For a given chemical, risk can be reduced; hazard cannot.

hygiene officer (CHO) or safety director; responsibility for working safely, however, lies with those scientists, technicians, faculty, students, and others who actually do the work. A detailed organizational chart with regard to each individual's responsibility for chemical hygiene can be a valuable addition to the Chemical Hygiene Plan (CHP). (See Chapter 2, section 2.B.)

In course work, laboratory instructors carry direct responsibility for actions taken by students. Instructors are responsible for promoting a culture of safety as well as for teaching the requisite skills needed to handle chemicals safely.

As federal, state, and local regulations became more stringent, institutions developed infrastructures to oversee compliance. Most industrial, governmental, and academic organizations that maintain laboratory operations have an EHS office staffed with credentialed professionals. These individuals have a collective expertise in chemical safety, industrial hygiene, engineering, biological safety, environmental health, environmental management (air, water, waste), occupational medicine, health physics, fire safety, and toxicology.

EHS offices consult on or manage hazardous waste issues, accident reviews, inspections and audits, compliance monitoring, training, record keeping, and emergency response. They assist laboratory management in establishing policies and promoting high standards of laboratory safety. To be most effective, they should partner with department chairpersons, safety directors, CHOs, principal investigators or managers, and laboratory personnel to design safety programs that provide technical guidance and training support that are relevant to the operations of the laboratory, are practical to carry out, and comply with existing codes and regulations.

In view of the importance of these offices, safety directors should be highly knowledgeable in the field and given responsibility for the development of a unified safety program, which will be vetted by institutional authorities and implemented by all. As a result, EHS directors should also have direct access, when necessary, to those senior authorities in the institution who are ultimately accountable to the public.

1.D SPECIAL SAFETY CONSIDERATIONS IN ACADEMIC LABORATORIES

Academic laboratories, like industrial and governmental laboratories, are concerned with meeting the fundamental safety goals of minimizing accidents and injuries, but there are differences. Forming the foundation for a lifelong attitude of safety consciousness, risk assessment, and prudent laboratory practice is an integral part of every stage of scientific education—from classroom to laboratory and from primary school through postdoctoral training. Teaching and academic institutions must accept this unique responsibility for attitude development.

Resources are limited and administration must provide support for teachers who are not subject matter experts. The manifold requirements for record keeping and waste handling can be especially burdensome for overworked teachers in high school or college laboratories. Institutions with graduate programs teach, but they also conduct research activities that often involve unpredictable hazards. The safety goals and the allocation of resources to achieve them are sufficiently different for high school, undergraduate, and graduate teaching laboratories that they are discussed separately here.

1.D.1 High School Teaching Laboratories

Laboratory safety involves recognizing and evaluating hazards, assessing risks, selecting appropriate personal protective equipment, and performing the experimental work in a safe manner. Training must start early in a chemist's career. Even a student's first chemical experiments should cover the proper approach to understanding and dealing with the hazardous properties of chemicals (e.g., flammability, reactivity, corrosiveness, and toxicity) as an introduction to laboratory safety and should also teach sound environmental practice when managing chemical waste. Advanced high school chemistry courses should assume the same responsibilities for developing professional attitudes toward safety and waste management as are expected of college and university courses.

1.D.2 Undergraduate Teaching Laboratories

Undergraduate chemistry courses are faced with the problem of introducing inexperienced people to the culture of laboratory safety. Although some students enroll in their first undergraduate course with good preparation from their high school science courses, many others bring little or no experience in the laboratory. They must learn to evaluate the wide range of hazards in laboratories and learn risk management techniques that are designed to eliminate various potential dangers in the laboratory.

Undergraduate laboratory instruction is often assigned to graduate—and in some cases undergraduate—teaching assistants, who have widely different backgrounds and communication skills. Supervising and supporting teaching assistants is a special departmental responsibility that is needed to ensure the safe operation of the undergraduate laboratories in the department. The assistants are teaching chemistry while they are trying to learn it and teaching safety when they may not be prepared to do so. However, they are in a position to act as role models of safe laboratory practice for the students in the laboratory, and adequate support and training are required for them to fill that role appropriately.

To this end, a manual designed and written specifically for teaching assistants in undergraduate laboratories is an extremely effective training tool. The manual can include sections on principles of laboratory safety; laboratory facilities; teaching assistant duties during the laboratory session; chemical management; applicable safety rules; teaching assistant and student apparel; teaching assistant and student personal protective equipment; departmental policy on pregnant students in laboratories; and emergency preparedness in the event of a fire, chemical spill, or injury in the laboratory.

There should be resolute commitment by the entire faculty to the departmental safety program to minimize exposure to hazardous materials and unsafe work practices in the laboratory. Teaching safety and safe work practices in the laboratory should be a top priority for faculty as they prepare students for careers in industrial, governmental, academic, and health sciences laboratories. By promoting safety during the undergraduate and graduate years, the faculty will have a significant impact not just on their students but also on everyone who will share their future work environments.

1.D.3 Academic Research Laboratories

Advanced training in safety should be mandatory for students engaged in research, and hands-on training is recommended whenever possible. Unlike laboratory course work, where training comes primarily from repeating well-established procedures, research often involves making new materials by new methods, which may pose unknown hazards. As a result, workers in academic research laboratories do not always operate from a deep experience base.

Thus, faculty is expected to provide a safe environment for research via careful oversight of the student's work. Responsibility for the promotion of safe laboratory practices extends beyond the EHS department, and all senior researchers—faculty, postdoctoral, and experienced students—should endeavor to teach the principles and set a good example for their associates. The ability to maintain a safe laboratory environment is necessary for a chemist entering the workforce, and students who are not adequately trained in safety are placed at a professional disadvantage when compared with their peers. To underscore the importance of maintaining a safe and healthy laboratory environment, many chemistry departments provide laboratory

safety training and seminars for incoming graduate students. However, in many cases these sessions are designed to prepare graduate students for their work as teaching assistants rather than for their work as research scientists.

Formal safety education for advanced students and laboratory personnel should be made as relevant to their work activities as possible. Training conducted simply to satisfy regulatory requirements may seem like compliance, and researchers may sense that the training does not have the leader's full support. EHS offices and researchers can work together to address such concerns and to design training sessions that fulfill regulatory requirements, provide training perceived as directly relevant to the researchers' work, and provide hands-on experience with safety practices whenever possible.

Safety training is an ongoing process, integral to the daily activities of laboratory personnel. As a new laboratory technique is formally taught or used, relevant safe practices should be included; however, informal training through collegial interactions is a good way to exchange safety information, provide guidance, and reinforce good work habits.

Although principal investigators and project managers are legally accountable for the maintenance of safety in laboratories under their direction, this activity, like much of the research effort, is distributable. Well-organized academic research groups develop hierarchical structures of experienced postdoctoral research associates, graduate students at different levels, undergraduates, and technicians, which can be highly effective in transmitting the importance of safe, prudent laboratory operations. Box 1.1 provides some examples of how to encourage a culture of safety within an academic laboratory.

When each principal investigator offers leadership that demonstrates a deep concern for safety, fewer people get hurt. If any principal investigator projects an attitude that appears to be cavalier or hostile to the university safety program, that research group and others can mirror the poor example and exhibit behavior that sets the stage for potential accidents, loss of institutional property, and costly litigation.

1.E THE SAFETY CULTURE IN INDUSTRIAL AND GOVERNMENTAL LABORATORIES

The degree of commitment to EHS programs varies widely among companies and governmental laboratories, as well. Many chemical companies recognize both their moral responsibility and their own self-interest in developing the best possible safety programs, extending them not just to employees but also to contractors. Others do little more than is absolutely required by law and regulations. Unfortunately, bad publicity from a serious accident in one careless operation tarnishes the credibility of all committed supervisors and employees. Fortunately, chemical companies that excel in safety are becoming more common, and safety is often recognized as equal in importance to productivity, quality, profitability, and efficiency.

The industrial or governmental laboratory environment provides strong corporate structure and discipline for maintaining a well-organized safety program where the culture of safety is thoroughly understood, respected, and enforced from the highest level of management down. New employees coming from academic research laboratories are often surprised to discover the detailed planning and extensive safety checks that are required before running experiments. In return for their efforts, they learn the sense of personal security that goes with high professional standards.

1.F   OTHER FACTORS THAT INFLUENCE LABORATORY SAFETY PROGRAMS

Several key factors continue to affect the evolution of laboratory safety programs in industry, government, and academe. These factors include advances in technology, environmental impact, and changes in legal and regulatory requirements.

1.F.1   Advances in Technology

In response to the increasingly high cost of chemical management, from procurement to waste disposal, a steady movement toward miniaturizing chemical operations exists in both teaching and research laboratories. This trend has had a significant effect on laboratory design and has also reduced the costs associated with procurement, handling, and disposal of chemicals. Another trend—motivated at least partially by safety concerns—is the simulation of laboratory experiments by computer. Such programs are a valuable conceptual adjunct to laboratory training but are by no means a substitute for hands-on experimental work. Only students who have been carefully educated through a series of hands-on experiments in the laboratory have the confidence and expertise needed to handle real laboratory procedures safely as they move on to advanced courses, research work, and eventually to their careers in industry, academe, health sciences, or government laboratories.

1.F.2   Environmental Impact

If a laboratory operation produces less waste, there is less waste to dispose of and less impact on the environment. A frequent, but not universal, corollary is that costs are also reduced. The terms "waste reduction," "waste minimization," and "source reduction" are often used interchangeably with "pollution prevention." In most cases the distinction is not important. However, the term "source reduction" may be used in a narrower sense than the other terms, and the limited definition has been suggested as a regulatory approach that mandates pollution prevention. The narrow definition of source reduction includes only procedural and process changes that actually use less material and produce less waste. The definition does not include recycling or treatment to reduce the hazard of a waste. For example, changing to microscale techniques is considered source reduction, but recycling a solvent waste is not.

Many advantages are gained by taking an active pollution prevention approach to laboratory work, and these are well documented throughout this book. Some potential drawbacks do exist, and these are discussed as well and should be kept in mind when planning activities. For example, dramatically reducing the quantity of chemicals used in teaching laboratories may leave the student with an unrealistic appreciation of his or her behavior when using them on a larger scale. Also, certain types of pollution prevention activities, such as solvent recycling, may cost far more in dollars and time than the potential value of recovered solvent. For more information about solvent recycling, see Chapter 5, section 5.D.3.2. Before embarking on any pollution prevention program, it is worthwhile to review the options thoroughly with local EHS program managers and to review other organizations' programs to become fully aware of the relative merits of those options.

Perhaps the most significant impediment to comprehensive waste reduction in laboratories is the element of scale. Techniques that are practical and cost-effective on a 55-gal or tank-car quantity of material may be highly unrealistic when applied to a 50-g (or milligram) quantity, or vice versa. Evaluating the costs of both equipment and time becomes especially important when dealing with very small quantities.

1.F.3 Changes in the Legal and Regulatory Requirements

Changes in the legal and regulatory requirements over the past several decades have greatly affected laboratory operations. Because of increased regulations, the collection and disposal of laboratory waste constitute major budget items in the operation of every chemical laboratory. The cost of accidents in terms of time and money spent on fines for regulatory violations and on litigation are significant. Of course, protection of students and research personnel from toxic materials is not only an economic necessity but an ethical obligation. Laboratory accidents have resulted in serious, debilitating injuries and death, and the personal impact of such events cannot be forgotten.

In 1990, OSHA issued the Laboratory Standard (29 CFR § 1910.1450), a performance-based rule that serves the community well. In line with some of the developments in laboratory practice, the committee recommends that OSHA review the standard in current context. In particular, the section on CHPs, 1910.1450(e), does not currently include emergency preparedness, emergency response, and consideration of physical hazards as well as chemical hazards. In addition, this book provides guidance that could be a basis for strengthening the employee information and training section, 1910.1450(f). Finally, the nonmandatory Appendix A of the Laboratory Standard was based on the original edition of *Prudent Practices in the Laboratory*, published in 1981 and currently out of print. The committee recommends that the appendix be updated to reflect the changes in the current edition in both content and reference.

The Laboratory Standard requires that every workplace conducting research or training where hazardous chemicals are used develop a CHP. This requirement has generated a greater awareness of safety issues at all educational science and technology departments and research institutions. Although the priority assigned to safety varies widely among personnel within academic departments and divisions, increasing pressure comes from several other directions in addition to the regulatory agencies and to the potential for accident litigation. In some cases, significant fines have been imposed on principal investigators who received citations for safety violations. These actions serve to increase the faculty's concern for laboratory safety. Boards of trustees or regents of educational institutions often include prominent industrial leaders who are aware of the increasing national concern with safety and environmental issues and are particularly sensitive to the possibility of institutional liability as a result of laboratory accidents. Academic and government laboratories can be the targets of expensive lawsuits. The trustees assist academic officers both by helping to develop an appropriate institutional safety system with an effective EHS office and by supporting departmental requests for modifications of facilities to comply with safety regulations.

Federal granting agencies recognize the importance of sound laboratory practices and active laboratory safety programs in academe. Some require documentation of the institution's safety program as part of the grant proposal. When negligent or cavalier treatment of laboratory safety regulations jeopardizes everybody's ability to obtain funding, a powerful incentive is created to improve laboratory safety.

1.F.4 Accessibility for Scientists with Disabilities

Over the years, chemical manufacturers have modernized their views of safety. Approaches to safety for all—including scientists with disabilities—have largely changed in laboratories as well. In the past, full mobility and full eyesight and hearing capabilities were considered necessary for safe laboratory operations. Now, encouraged legally by the adoption of the Americans with Disabilities Act of 1990 (ADA) and the ADA Amendments Act of 2008, leaders in laboratory design and management realize that a nimble mind is more difficult to come by than modified space or instrumentation.

As a result, assistive technologies now exist to circumvent almost any inaccessibility, and laboratories can be equipped to take advantage of them. Many of the modifications to laboratory space and fixtures have benefits for all. Consider, as a single example, the assistance of ramps and an automatic door opener to all lab personnel moving a large cart or carrying two heavy containers.

It is a logical extension of the culture of safety to include a culture of accessibility. For information about compliance with the ADA in the laboratory, see Chapter 9, section 9.B.8.

1.G LABORATORY SECURITY

Laboratory security is an issue that has grown in prominence in recent years and is complementary to laboratory safety. In short, a laboratory safety program should be designed to protect people and chemicals from accidental misuse of materials; the laboratory security program should be designed to protect workers from intentional misuse or misappropriation of materials. Security procedures and programs will no doubt be familiar to some readers, but others may have encountered it only in the context of locking the laboratory door. However, in the coming years, a working awareness of security will likely become a common requirement for anyone working in a chemical laboratory. Risks to laboratory security include theft or diversion of high-value equipment, theft of chemicals to commit criminal acts, intentional release of hazardous materials, or loss or release of sensitive information, and will vary with the organization and the work performed. Chapter 10 of this book provides a broad introduction to laboratory security, including discussions of the elements of a security program, performing a security vulnerability assessment, dual-use hazards of laboratory materials, and regulations that affect security requirements. The chapter is not intended to provide all the details needed to create a security program, but rather to acquaint laboratory personnel with the rationale behind developing such a program and to provide the basic tools needed to begin identifying and addressing concerns within their own laboratories.

1.H STRUCTURE OF THE BOOK

This edition of *Prudent Practices in the Laboratory* builds on the work provided in previous editions. Among other changes, it has two new chapters, one on Emergency Planning and one on Laboratory Security, described above, and the discussion of EHS management systems has been extensively revised. Chapters 2, 3, and 10 cover administrative and organizational concerns that affect the laboratory environment; Chapters 4–8 discuss practical concerns when working in a laboratory; Chapter 9 discusses laboratory facilities; and Chapter 11 provides an overview of federal regulations that affect laboratory activities. Acknowledging the stronger regulatory environment that exists today, this edition provides more references to relevant codes, standards, and regulations than the prior versions. This is not intended to imply that safety has become a matter of regulation rather than of good practice; it is a reflection of laboratory practice today and is intended to provide a resource for personnel who must remain in compliance with these regulations or face legal consequences.

1.I SUMMARY

A strong culture of safety within an organization creates a solid foundation upon which a successful laboratory health and safety program can be built. As part of that culture, all levels of the organization (i.e., administrative personnel, scientists, laboratory technicians) should understand the importance of minimizing the risk of exposure to hazardous materials in the laboratory and should work together toward this end. In particular, laboratory personnel should consider the health, physical, and environmental hazards of the chemicals that will be used when planning a new experiment and perform their work in a prudent manner. However, the ability to accurately identify and assess hazards in the laboratory is not a skill that comes naturally, and it must be taught and encouraged through training and ongoing organizational support. A successful health and safety program requires a daily commitment from everyone in the organization, and setting a good example is the best method of demonstrating commitment.

2 Environmental Health and Safety Management System

2.A	INTRODUCTION	10
	2.A.1 Environmental Health and Safety Policy	10
	2.A.2 Management Commitment	10
	2.A.3 Planning	10
	2.A.4 Implementation	12
	2.A.5 Performance Measurement and Change Management	12
	2.A.6 Management Review of EHS Management System	13
	2.A.7 Example Management System: Department of Energy Integrated Safety Management System	13
2.B	CHEMICAL HYGIENE PLAN	14
2.C	SAFETY RULES AND POLICIES	15
	2.C.1 General Safety Rules	15
	2.C.2 Working Alone in the Laboratory	17
	2.C.3 How to Avoid Routine Exposure to Hazardous Chemicals	18
	2.C.4 General Housekeeping Practices in the Laboratory	19
2.D	CHEMICAL MANAGEMENT PROGRAM	20
	2.D.1 Chemical Procurement	20
	2.D.2 Chemical Storage	21
	2.D.3 Chemical Handling	22
	2.D.4 Chemical Inventory	22
	2.D.5 Transporting, Transferring, and Shipping Chemicals	23
	2.D.6 Chemical Waste	23
2.E	LABORATORY INSPECTION PROGRAM	23
	2.E.1 Types of Inspection Programs: Who Conducts Them and What They Offer	24
	2.E.1.1 Routine Inspections	24
	2.E.1.2 Self-Audits	24
	2.E.1.3 Program Audits	24
	2.E.1.4 Peer Inspections	24
	2.E.1.5 Environmental Health and Safety Inspections	24
	2.E.1.6 Inspections by External Entities	25
	2.E.2 Elements of an Inspection	25
	2.E.2.1 Preparing for an Inspection	25
	2.E.2.2 Inspection Checklists	25
	2.E.2.3 Conducting the Inspection	26
	2.E.2.4 Inspection Report	26
	2.E.2.5 Corrective Actions	26
	2.E.3 Items to Include in an Inspection Program	26
2.F	EMERGENCY PROCEDURES	27
	2.F.1 Fire Alarm Policy	27
	2.F.2 Emergency Safety Equipment	27
	2.F.3 Chemical Spill Policy	28
	2.F.4 Accident Procedures	29
2.G	EMPLOYEE SAFETY TRAINING PROGRAM	29

2.A INTRODUCTION

Many people are interested in an organization's approach to laboratory environmental health and safety (EHS) management including laboratory personnel; customers, clients, and students (if applicable); suppliers; the community; shareholders; contractors; insurers; and regulatory agencies. More and more organizations attach the same importance to high standards in EHS management as they do to other key aspects of their activities. High standards demand a structured approach to the identification of hazards and the evaluation and control of work-related risks.

A comprehensive legal framework already exists for laboratory EHS management. This framework requires organizations to manage their activities in order to anticipate and prevent circumstances that might result in occupational injury, ill health, or adverse environmental impact. This chapter seeks to improve the EHS performance of organizations by providing guidance on EHS to integrate EHS management with other aspects of the organization.

Many features of effective EHS management are identical to management practices advocated by proponents of quality assurance and business excellence. The guidelines presented here are based on general principles of good management and are designed to integrate EHS management within an overall management system.[1] By establishing an EHS management system, EHS risks are controlled in a systematic proactive manner.

Within many organizations, some elements of EHS management are already in place, such as policy and risk assessment records, but other aspects need to be developed. It is important that all the elements described here are incorporated into the EHS management system. The manner and extent to which individual elements are applied, however, depend on factors such as the size of the organization, the nature of its activities, the hazards, and the conditions in which it operates. An initial status review should be carried out in all organizations that do not have an established EHS management system. This initial status review will provide information on the scope, adequacy, and implementation of the current management system. Where no formal management system exists, or if the organization is newly established, the initial status review should indicate where the organization stands with respect to managing risks.

Figure 2.1 illustrates the major elements of an EHS management system.

---

[1]A general definition of a management system is "a series of elements for establishing policy, objectives, and processes for implementation, review, and continual improvement."

2.A.1 Environmental Health and Safety Policy

Top management should set in place procedures to define, document, and endorse a formal EHS policy for an organization. The policy should clearly outline the roles and expectations for the organization, faculty, EHS personnel, and individual employees or students. It should be developed in communication with laboratory personnel to ensure that all major concerns are adequately addressed.

The EHS policy should state intent to

- prevent or mitigate both human and economic losses arising from accidents, adverse occupational exposures, and environmental events;
- build EHS considerations into all phases of the operations, including laboratory discovery and development environments;
- achieve and maintain compliance with laws and regulations; and
- continually improve EHS performance.

The EHS policy and policy statement should be reviewed, revalidated, and where necessary, revised by top management as often as necessary. It should be communicated and made readily accessible to all employees and made available to relevant interested parties, as appropriate.

2.A.2 Management Commitment

Management commitment to EHS performance is widely recognized as one of the elements most critical to EHS program success and to the development of a strong culture of safety within an organization. Therefore, the management system document establishes management commitment with a formal statement of intent, which defines examples of how performance goals are supported. Examples of how this commitment is supported include the following:

- Establish methods to use energy more efficiently, reduce waste, and prevent accidents.
- Comply with laws, regulations, and organizational requirements applicable to their operations.
- Improve EHS performance continually.
- Conduct periodic assessments to verify and validate EHS performance.

2.A.3 Planning

Planning is an integral part of all elements of the management system and to be effective involves the design and development of suitable processes and

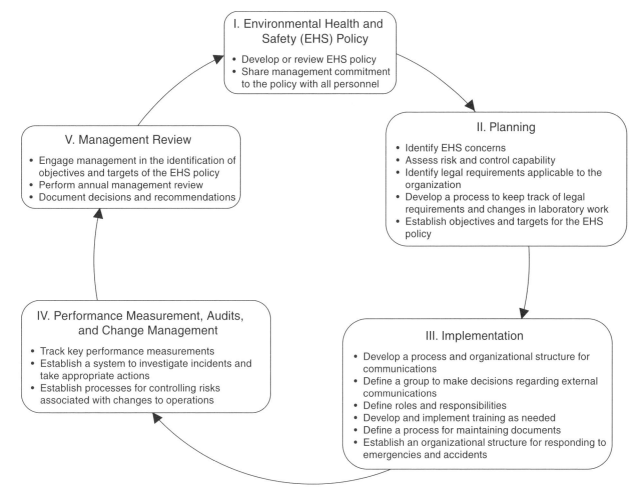

FIGURE 2.1 Overview of environmental health and safety management system.

organizational structure to manage EHS aspects and their associated risk control systems proportionately to the needs, hazards, and risks of the organization. Planning is equally important to deal with health risks that might only become apparent after a long latency period. It also establishes objectives that define the criteria for judging success or failure of the management system. Objectives are identified on the basis of either the results of the initial status review, subsequent periodic reviews, or other available data.

Various sources of information are used to identify applicable EHS aspects and to assess the risk associated with each. Examples include, but are not limited to, information obtained from the following:

- hazard/exposure assessment,
- risk assessment,
- inspections,
- permits,
- event investigations (injury and illness investiga-

tions, environmental incident investigations, root-cause analysis, trend analysis),
- internal audits and/or external agency audits,
- fire and building codes,
- employee feedback concerning unsafe work conditions or situations,
- emerging issues,
- corporate/institution goals, and
- emergency management.

Once applicable EHS aspects are identified, a risk-based evaluation is performed to determine the potential impact and adequacy of existing control measures. If additional controls or corrective actions are needed to reduce risks to acceptable levels, they are integrated into business planning. Categorizing each item in this manner allows gaps that are identified to be prioritized and incorporated, based on level of importance and available resources.

Care should be taken when developing and disseminating new controls and corrective actions. If requirements are perceived by laboratory personnel as unnecessarily onerous, there is potential for lower compliance within the organization and a loss of credibility on the part of EHS personnel. While understanding that some individuals will never be convinced of the need for new controls, it is important to provide clear, supported justifications for changes to existing protocols to encourage adoption of the new policies and procedures.

2.A.4 Implementation

The design of management arrangements should reflect the organization's business needs and the nature of their risks. However, there should be appropriate activity across all elements of the model (policy; planning; implementation; performance measurement, audits, and change management; and management review).

Specifically the organization should make arrangements to cover the following key areas:

- overall plans and objectives, including employees and resources, for the organization to implement its policy;
- operational plans to implement arrangements to control the risks identified;
- contingency plans for foreseeable emergencies and to mitigate their effects (e.g., prevention, preparedness, and response procedures);
- plans covering the management of change of either a permanent or a temporary nature (e.g., associated with new processes or plant working procedures, production fluctuations, legal requirements, and organizational and staffing changes);
- plans covering interactions with other interested parties (e.g., control, selection, and management of contractors; liaison with emergency services; visitor control);
- performance measures, audits, and status reviews;
- corrective action implementation;
- plans for assisting recovery and return to work of any staff member who is injured or becomes ill through work activities;
- communication networks to management, employees, and the public;
- clear performance and measurement criteria defining what is to be done, who is responsible, when it is to be done, and the desired outcome;
- education and training requirements associated with EHS;
- document control system; and
- contractors should have written safety plans and qualified staff whose qualifications are thoroughly

reviewed before a contract is awarded. All contractor personnel should be required to comply with the sponsoring organization's safety policies and plans.

Though it is the responsibility of each individual researcher to ensure that work is performed in a prudent and safe manner, achieving a safe laboratory environment is a cooperative endeavor between management, EHS personnel, and laboratory personnel. Regulations, policies, and plans will never cover every contingency, and it is important for these different groups to communicate with each other to ensure that new situations can be handled appropriately. One way to ensure that the needs of all groups are being met is by creating safety committees consisting of representatives from each part of an organization. In this forum, safety concerns can be raised, information can be distributed to affected parties, and a rough sense of the efficacy of policies and programs can be gained.

2.A.5 Performance Measurement and Change Management

The primary purpose of measuring EHS performance is to judge the implementation and effectiveness of the processes established for controlling risk. Performance measurement provides information on the progress and current status of the arrangements (strategies, processes, and activities) used by an organization to control risks to EHS. Measurement information includes data to judge the management system by

- gathering information on how the system operates in practice,
- identifying areas where corrective action is necessary, and
- providing a basis for continual improvement.

All of the components of the EHS management system should be adequately inspected, evaluated, maintained, and monitored to ensure continued effective operation. Risk assessment and risk control should be reviewed in the light of modifications or technological developments. Results of evaluation activities are used as part of the planning process and management review, to improve performance and correct deficiencies over time.

Periodic audits that enable a deeper and more critical appraisal of all of the elements of the EHS management system (see Figure 2.1) should be scheduled and should reflect the nature of the organization's hazards and risks. To maximize benefits, competent persons independent of the area or activity should conduct the audits. The use of external, impartial auditors should be considered to assist in evaluation of the EHS management system. When performing these reviews, it is important that the organization have a plan for following up on the results of the audit to ensure that problems are addressed and that recognition is given where it is deserved.

The concept of change management in the laboratory environment varies markedly from methods typically prescribed, for example, in manufacturing operations. By its very nature, the business of conducting experiments is constantly changing. Therefore, it is a part of everyday activities to evaluate modifications and/or technological developments in experimental and scale-up processes. As such, a number of standard practices are used to identify appropriate handling practices, containment methods, and required procedures for conducting laboratory work in a safe manner. Several examples of these practices include

- identification of molecules as particularly hazardous substances (PHSs),[2] which specifies certain handling and containment requirements and the use of personal protective equipment (PPE);
- approval and training for new radioisotope users;
- completion of biosafety risk assessments for the use of infectious agents; and
- Material Safety Data Sheet (MSDS) review of chemicals being used.

2.A.6 Management Review of EHS Management System

Top management should review the organization's EHS management system at regular intervals to ensure its continuing suitability, adequacy, and effectiveness. This review includes assessing opportunities for improvement and the need for changes in the management system, including the EHS policy and objectives. The results of the management review should be documented.

Among other information, a management review should include the following:

- results of EHS management system audits,
- results from any external audits,
- communications from interested parties,
- extent to which objectives have been met,
- status of corrective and preventive actions,
- follow-up actions from previous management reviews, and
- recommendations for improvement based on changing circumstances.

---

[2]The term "particularly hazardous substances" is used by Occupational Safety and Health Administration (OSHA) and defined in the Laboratory Standard 29 CFR § 1910.1450. (For more information see Chapter 4, section 4.C.3.)

The outputs from management review should include any decisions and actions related to possible change to EHS policy, objectives, and other elements of the management system, consistent with the commitment to continual improvement.

The management system review ensures a regular process that evaluates the EHS management system in order to identify deficiencies and modify them. Systemic gaps, evidence that targets are not being met, or compliance issues that are discovered during compliance or risk assessments indicate a possible need for revision to the management system or its implementation.

2.A.7 Example Management System: Department of Energy Integrated Safety Management System

One example of a common EHS management system is that used by the Department of Energy (DOE). The agency's Integrated Safety Management (ISM) system, adopted in 1996, is used at all DOE facilities, and has been used as a model for other agencies and institutions. The system consists of six guiding principles and five core management safety functions. The principles and functions in DOE Policy DOE P 450.4 (DOE, 1994), outlined below, require planning, identification of hazards and controls before work begins, and for work to be performed within these defined and planned methods.

Principles:

- **Line management responsibility for safety.** Line management is directly responsible for the protection of the public, the workers, and the environment. As a complement to line management, the Department's Office of Environment, Safety, and Health provides safety policy, enforcement, and independent oversight functions.
- **Clear roles and responsibilities.** Clear and unambiguous lines of authority and responsibility for ensuring safety shall be established and maintained at all organizational levels within the Department and its contractors.
- **Competence commensurate with responsibilities.** Personnel shall possess the experience, knowledge, skills, and abilities that are necessary to discharge their responsibilities.
- **Balanced priorities.** Resources shall be effectively allocated to address safety, programmatic, and operational considerations. Protecting the public, the workers, and the environment shall be a priority whenever activities are planned and performed.
- **Identification of safety standards and requirements.** Before work is performed, the associated

hazards shall be evaluated and an agreed-upon set of safety standards and requirements shall be established which, if properly implemented, will provide adequate assurance that the public, the workers, and the environment are protected from adverse consequences.

- **Hazard controls tailored to work being performed.** Administrative and engineering controls to prevent and mitigate hazards shall be tailored to the work being performed and associated hazards.
- **Operations authorization.** The conditions and requirements to be satisfied for operations to be initiated and conducted shall be clearly established and agreed upon.

Functions:

- **Define the scope of work.** Missions are translated into work, expectations are set, tasks are identified and prioritized, and resources are allocated.
- **Analyze the hazards.** Hazards associated with the work are identified, analyzed, and categorized.
- **Develop and implement hazard controls.** Applicable standards and requirements are identified and agreed upon, controls to prevent/mitigate hazards are identified, the safety envelope is established, and controls are implemented.
- **Perform work within controls.** Readiness is confirmed and work is performed safely.
- **Provide feedback and continuous improvement.** Feedback information on the adequacy of controls is gathered, opportunities for improving the definition and planning of work are identified and implemented, line and independent oversight is conducted, and, if necessary, regulatory enforcement actions occur.

In addition, in 2006, and in recognition of a gap within the management system, DOE identified four supplemental safety culture elements. These, as described in DOE Manual DOE M 450.4-1 (DOE, 2006), are as follows:

- **Individual attitude and responsibility for safety.** Every individual accepts responsibility for safe mission performance. Individuals demonstrate a questioning attitude by challenging assumptions, investigating anomalies, and considering potential adverse consequences of planned actions. All employees are mindful of work conditions that may impact safety, and assist each other in preventing unsafe acts or behaviors.
- **Operational excellence.** Organizations achieve sustained, high levels of operational performance, encompassing all DOE and contractor activities to meet mission, safety, productivity, quality, environmental, and other objectives. High reliability is achieved through a focus on operations, conservative decision making, open communications, deference to expertise, and systematic approaches to eliminate or mitigate error-likely situations.
- **Oversight for performance assurance.** Competent, robust, periodic, and independent oversight is an essential source of feedback that verifies expectations are being met and identifies opportunities for improvement. Performance assurance activities verify whether standards and requirements are being met. Performance assurance through conscious, directed, independent previews at all levels brings fresh insights and observations to be considered for safety and performance improvement.
- **Organizational learning for performance improvement.** The organization demonstrates excellence in performance monitoring, problem analysis, solution planning, and solution implementation. The organization encourages openness and trust, and cultivates a continuous learning environment.

More information about the DOE ISM system can be found at www.directives.doe.gov.

The DOE ISM system is only one example of an EHS management system, and many others exist. It is important that each organization develop a management system to meet the needs of the organization. Small organizations or those that do not handle particularly hazardous materials should not be tempted to "over-engineer" the system. If the burden of organizational oversight and management of the ESH program is not appropriately tied to the organizational risk, then the safety program may lose credibility in the eyes of the people it supports.

2.B CHEMICAL HYGIENE PLAN

The foundation of all management system approaches is the identification of EHS concerns, which if not adequately controlled, can result in employee injury or illness, adverse effects on the environment, and regulatory action. One of the most critical EHS aspects for laboratories is the requirement for chemical safety, which in the United States is specifically regulated by OSHA Laboratory Standard, 29 CFR § 1910.1450, *Occupational Exposure to Hazardous Chemicals in Laboratories*. This standard was created to minimize employee exposure to hazardous chemicals in the laboratory and sets forth guidelines for employers and trained laboratory personnel engaged in the use of hazardous chemicals.[3]

---

[3]29 CFR § 1910.1450 (1990), http://www.osha.gov/.

The OSHA Laboratory Standard defines a Chemical Hygiene Plan (CHP) as "a written program developed and implemented by the employer which sets forth procedures, equipment, personal protective equipment and work practices that are capable of protecting employees from the health hazards presented by hazardous chemicals used in that particular workplace." "Where hazardous chemicals as defined by this standard are used in the workplace, the employer shall develop and carry out the provisions of a written Chemical Hygiene Plan." The CHP is the foundation of the laboratory safety program and should be reviewed and updated, as needed, on an annual basis to reflect changes in policies and personnel. A CHP that is facility specific can assist in promoting a culture of safety to protect employees from exposure to hazardous materials.

Topics included in a CHP are

1. individual responsibilities for chemical hygiene within the organization (see Boxes 2.1, 2.2, and 2.3),
2. emergency preparedness and facility security issues,
3. personal apparel and PPE,
4. chemical management,
5. laboratory housekeeping,
6. standard operating procedures,
7. emergency action plan (EAP) for accidents and spills,
8. safety equipment,
9. chemical waste policies,
10. required training,
11. safety rules and regulations,
12. facility design and laboratory ventilation,
13. medical and environmental monitoring,
14. compressed gas safety,
15. laboratory equipment,
16. biological safety, and
17. radiation safety.

Determining what belongs in the CHP for a given laboratory should be the result of conversations between the Chemical Hygiene Officer (CHO), the director of the laboratory, and laboratory personnel. The laboratory director and the individuals performing the research are responsible for following safe practices, and they are the people most familiar with the work being performed. However, they are less likely to be familiar with all relevant regulations, standards, and codes than the CHO, and they may benefit from assistance in identification and assessment of hazards within the laboratory. Thus there must be communication across the groups to ensure that the CHP is complete and that it contains no irrelevant information

(e.g., information on biological safety in a laboratory that only works with inorganic materials).

2.C SAFETY RULES AND POLICIES

Safety rules and regulations are created to protect laboratory personnel from unsafe work practices and exposure to hazardous materials. Consistently following and enforcing the safety rules in order to create a safe and healthful laboratory environment in which to work will help encourage a culture of safety within the workplace. What follows is a description of laboratory safety rules, but these will not cover every contingency. Part of the culture of safety is communication and discussion about safety hazards within the laboratory, so that new concerns can be addressed as quickly as possible.

2.C.1 General Safety Rules

Below are some basic guidelines for maintaining a safe laboratory environment.

1. To ensure that help is available if needed, do not work alone if using hazardous materials or performing hazardous procedures.
2. To ensure that help is available in case of emergencies, laboratory personnel should not deviate from the assigned work schedule without prior authorization from the laboratory supervisor.
3. Do not perform unauthorized experiments.
4. Plan appropriate protective procedures and the positioning of all equipment before beginning any operation. Follow the appropriate standard operating procedures at all times in the laboratory.
5. Always read the MSDS and the label before using a chemical in the laboratory.
6. Wear appropriate PPE, including a laboratory apron or coat, at all times in the laboratory. Everyone, including visitors, must wear appropriate eye protection in areas where laboratory chemicals are used or stored.
7. Wear appropriate gloves when handling hazardous materials. Inspect all gloves for holes and defects before using.
8. Use appropriate ventilation such as laboratory chemical hoods when working with hazardous chemicals.
9. Contact the CHO or the EHS office if you have questions about the adequacy of the safety equipment available or chemical handling procedures.
10. Know the location and proper use of the safety equipment (i.e., eyewash unit, safety shower, fire extinguisher, first-aid kit, fire blanket, emergency telephone, and fire alarm pulls).

BOX 2.1
Chemical Hygiene Responsibilities in a Typical Academic Institution

Chemical Hygiene Officer (CHO)

The duties of the CHO vary widely from one institution to another but may include the following:

- Establish, maintain, and revise the Chemical Hygiene Plan (CHP).
- Create and revise safety rules and regulations.
- Serve on appropriate safety committees.
- Monitor procurement, use, storage, and disposal of chemicals.
- Conduct regular inspections of the laboratories, preparations rooms, and chemical storage rooms, and submit detailed laboratory inspection reports to administration.
- Maintain inspection, personnel training, and inventory records.
- Oversee chemical inventory updates.
- Assist laboratory supervisors in developing and maintaining adequate facilities.
- Know current legal requirements concerning regulated substances.
- Seek ways to improve the chemical hygiene program.
- Attend CHO training that is conducted by the institution.
- Encourage laboratory personnel to attend specialized training that is provided by the institution (i.e., first-aid training, fire extinguisher training, and gas cylinder training).
- Notify employees of the availability of medical attention under the following circumstances:
  - Whenever an employee develops signs or symptoms associated with a hazardous chemical to which the employee may have been exposed in the laboratory;
  - Where exposure monitoring reveals an exposure level routinely above the action level for an Occupational Safety and Health Administration–regulated substance for which there are exposure monitoring and medical surveillance requirements;
  - Whenever a spill, leak, explosion, or other occurrence resulting in the likelihood of a hazardous exposure occurs, a medical consultation to ascertain if a medical examination is warranted.

Department Chairperson or Director

- Assumes responsibility for personnel engaged in the laboratory use of hazardous chemicals.
- Provides the CHO with the support necessary to implement and maintain the CHP.
- After receipt of laboratory inspection report from the CHO, meets with laboratory supervisors to discuss cited violations and to ensure timely actions to protect trained laboratory personnel and facilities and to ensure that the department remains in compliance with all applicable federal, state, university, local, and departmental codes and regulations.
- Provides budgetary arrangements to ensure the health and safety of the departmental personnel, visitors, and students.
- Serves as chair of the departmental safety committee

---

11. Maintain situational awareness. Be aware of the hazards posed by the work of others in the laboratory and any additional hazards that may result from contact between materials and chemicals from different work areas.
12. Make others in the laboratory aware of any special hazards associated with your work.
13. Notify supervisors of any chemical sensitivities or allergies.
14. Report all injuries, accidents, incidents, and near misses as directed by the organization's policy.
15. For liability, safety, and security reasons, do not allow unauthorized persons in the laboratory.
16. Report any unsafe conditions to the laboratory supervisor or CHO.
17. Properly dispose of all chemical wastes. Follow organizational policies for drain and trash disposal of chemicals.

Visitors, including children, are permitted in laboratories where hazardous substances are stored or are in use or hazardous activities are in progress as long as they are properly protected. If minors are expected in a laboratory (e.g., as part of an educational or classroom activity), ensure that they are under the direct supervision of qualified adults at all times. The institution should have a review process regarding minors in the laboratory, and prior to their arrival, scheduled activities should be approved. Other laboratory personnel in the area should be made aware that minors will be present.

No pets are permitted in laboratories. Note that service animals are not pets. They are highly trained and may be present in a laboratory. However, a clean, safe area should be provided where the animal can wait.

To prevent some common laboratory accidents:

1. Always protect hands with appropriate gloves

and appoints a faculty member and a graduate student member, if applicable, to serve on the departmental safety committee.

Departmental Safety Committee

- Reviews accident reports and makes appropriate recommendations to the department chairperson regarding proposed changes in the laboratory procedures.
- Performs laboratory inspections on an annual basis, or as needed. Prepares a detailed inspection report to be submitted to each faculty member/laboratory supervisor.

Laboratory Supervisor

- Ensures that laboratory personnel comply with the departmental CHP and do not operate equipment or handle hazardous chemicals without proper training and authorization.
- Always wears personal protective equipment (PPE) that is compatible to the degree of hazard of the chemical.
- Follows all pertinent safety rules when working in the laboratory to set an example.
- Reviews laboratory procedures for potential safety problems before assigning to other laboratory personnel.
- Ensures that visitors follow the laboratory rules and assumes responsibility for laboratory visitors
- Ensures that PPE is available and properly used by each laboratory employee and visitor.
- Maintains and implements safe laboratory practices.
- Monitors the facilities and the chemical fume hoods to ensure that they are maintained and function properly.

Contacts the appropriate person, as designated by the department chairperson, to report problems with the facilities or the chemical fume hoods.

Laboratory Personnel

- Reads, understands, and follows all safety rules and regulations that apply to the work area.
- Plans and conducts each operation, laboratory class, or research project in accordance with the departmental and institutional CHP.
- Promotes good housekeeping practices in the laboratory or work area.
- Communicates appropriate portions of the CHP to students in the work area.
- Notifies the supervisor of any hazardous conditions or unsafe work practices in the work area.
- Uses PPE as appropriate for each procedure that involves hazardous chemicals.
- Immediately reports any job-related illness or injury to the supervisor.

Facilities, Maintenance, and Custodial Service Personnel Assigned to Laboratories and Laboratory Areas

- Completes training on CHP awareness.
- Understands how to read and use MSDSs.
- Knows hazards of the chemicals being used.
- Obtains information about hazards in the work area from lab personnel before work is started.

---

when cutting glass tubing. To avoid breakage, do not attempt to dry glassware by inserting a glass rod wrapped with paper towels. Always lubricate glassware with soap or glycerin before inserting rods, tubing, or thermometers into stoppers.
2. To reduce the chances of injuries from projectiles, when heating a test tube or other apparatus, never point the apparatus toward yourself or others.
3. Be sure that glassware has cooled before touching it. Hot glass looks just like cold glass.
4. Dilute concentrated acids and bases by slowly pouring the acid or base into the water while stirring.

2.C.2 Working Alone in the Laboratory

It is not prudent to work alone in a laboratory. The American Chemical Society states that one should, "[n]ever work alone in the laboratory" (ACS, 2003). In

Alaimo (2001) it states that "[w]ork should be absolutely forbidden unless there are at least two people present". The OSHA Laboratory Standard states "Avoid working alone in a building; do not work alone in a laboratory if the procedures being conducted are hazardous." Accidents are unexpected by definition, and if a person is working alone when one occurs, his or her ability to respond appropriately could be severely impaired, which could result in personal injury or death and catastrophic facility damage. Thus it is imperative that, whenever working in the laboratory, others are actively aware of your activities. If faced with a situation where you feel it is necessary to work alone in a laboratory:

1. Reconsider the need. Are the increased risks to your health and safety really outweighed by the return?
2. Reconsider the timing and setup of the work. Is there any way to accomplish the required tasks during a time when others will be present?

BOX 2.2
Chemical Hygiene Responsibilities in a Typical Industry Research Facility

Chemical Hygiene Officer (CHO)

- Qualified by training or experience to provide technical guidance in the development and implementation of the provisions of the Laboratory Standard.
- Oversees implementation and communication of the Laboratory Standard.
- Conducts appropriate audits.
- Facilitates continuous improvement of safety policies and practice.
- Acts as liaison between Environmental Health and Safety, Environmental Affairs, and laboratory management.
- Chairs and schedules meetings of chemical hygiene coordination committee (CHCC).
- Ensures that Chemical Hygiene Plan (CHP) and training courses are reviewed annually;
- Revises the CHP, per CHCC instructions.
- Facilitates revisions of the CHP and requisite training courses.
- Approves/reviews all prior-approval chemical uses.
- Maintains minutes and other documents for CHP implementation and makes available for CHCC.

Chemical Hygiene Coordination Committee

The committee includes the CHO and representatives from business area (appointed by line management). These representatives are referred to as site chemical hygiene contacts.

- Provides opportunity for communication between plant sites.
- Facilitates continuous improvement of safety policies and practices.
- Meets at least annually to review CHP and training.

Line Management

- Completes CHP training.
- Implements the CHP, including but not limited to practices and procedures for the following:
  o particularly hazardous materials,
  o prior-approval process,
  o laboratory inspections.
- Supervises CHP training by all laboratory personnel.
- Supports and enforces laboratory standards program.
- Provides ongoing ownership.
- Ensures that local safety procedures are written, training is provided, and procedures are followed.
- Designates a local chemical hygiene contact (CHC) (by site, building, division, or department).

Site/Area Chemical Hygiene Contact

- Completes CHP training.
- Serves as contact person for CHO and laboratory personnel for site/area CHP implementation.
- Represents site/area on CHCC.

---

3. If the timing of the task cannot be changed and you still feel it must be accomplished during a period when the laboratory is empty, is there any other person trained in laboratory procedures who can accompany you while you work?
4. If not, is there anyone else within the building who could act as a "buddy" to check on you periodically during the time that you feel you must work alone?
5. If no one can accompany you and you cannot find a "buddy," **do not proceed with the work**. The situation is unsafe. Speak to your supervisor or the organizational safety office to make arrangements to complete the work in a safe manner.

2.C.3   How to Avoid Routine Exposure to Hazardous Chemicals

Many chemicals and solutions routinely used in laboratories present a significant health risk when handled improperly. The Swiss physician and alchemist Theophrastus Phillippus Aureolus Bombastus von Hohenheim (1493–1541), who took the name Paracelsus later in life in homage to Celsus, a Roman physician, is known as "the father of toxicology." Paracelsus is famous for his quote, "What is it that is not poison? All things are poison and nothing is without poison. It is the dose alone that makes a thing not a poison" (Dillon, 1994). Today, in that same spirit, trained laboratory personnel are encouraged to reduce personal risk by minimizing exposure to hazardous chemicals and by eliminating unsafe work practices in the laboratory.

The OSHA Laboratory Standard defines a hazardous chemical as one "for which there is statistically significant evidence based on at least one study conducted in accordance with established scientific principles that acute or chronic health effects may occur in exposed persons." Note that this definition is not limited to toxic chemicals and includes corrosives, explosives, and other hazard classes. Routes of exposure to hazardous materials include contact with skin and eyes, inhala-

- Presents CHP compliance or implementation concerns/issues to the site/area management.
- Meets and reviews laboratory inspection issues.
- Ensures availability of the CHP.
- Maintains site/area documentation for CHP.
- Coordinates prior-approval process of chemicals subject to Occupational Safety and Health Administration substance-specific standards.
- Coordinates and reviews annual laboratory inspections.
- Coordinates written reports of laboratory inspections and distributes to management.
- Communicates with site safety office regarding site/area compliance issues.
- Serves as technical resource for CHP questions and interpretations.
- Ensures that staff follow the controls identified in the research safety summaries.
- Shall be knowledgeable of the requirements for procurement, use, and disposal of the chemicals used in their projects.
- Suggests and implements ways to minimize all chemical exposures.
- Promptly reports spills, accidents, and employee exposures to the appropriate person.
- Assists during investigations.
- Ensures that an inventory of hazardous materials is maintained and updated in accordance with the requirements of the Maintenance of Hazardous Materials Inventory procedure, facility safety basis documents, and facility use agreements.

Staff

- Plans and conducts all research in accordance with Work Control, the CHP, supplemental CHPs, and relevant subject areas.
- Ensures that hazardous chemicals are procured, labeled, bar coded, handled, inventoried, stored, used, and properly disposed of in accordance with laboratory procedures and the applicable CHP.
- Shall be knowledgeable of hazards in their work area and the proper practices and procedures to minimize all chemical exposure. Consults available information sources such as the Material Safety Data Sheet database, training courses, Web searches, the supervisor, and health and safety personnel to identify hazards.
- Wears the appropriate personal protective equipment and follows the work controls that have been identified.
- Ensures that new hazardous chemicals are bar-coded and that the inventory is updated as required in the Maintenance of Hazardous Materials Inventory procedure.
- Develops good personal chemical hygiene habits.
- Promptly reports spills, accidents, or abnormal events to immediate line manager.
- Provides feedback to principal investigator, laboratory space manager, and immediate line manager identifying changes that may have introduced new hazards for determination of the need for reevaluation of the research safety summaries and ensuring continuous improvement.

tion, ingestion, and injection. Acute exposure is defined as short durations of exposure to high concentrations of hazardous materials in the workplace. Chronic exposure is defined as continual exposure over a long period of time to low concentrations of hazardous materials in the workplace. Overexposure to chemicals, whether a result of a single episode or long-term exposure, can result in adverse health effects. These effects are categorized as acute or chronic. Acute health effects appear rapidly after only one exposure and symptoms include rashes, dizziness, coughing, and burns. Chronic health effects may take months or years before they are diagnosed. Symptoms of chronic health effects include joint paint, neurological disorders, and tumors. (For more information on toxicity of laboratory chemicals, see Chapter 4, section 4.C.)

In addition to the hazards associated with the chemicals themselves, flammable, reactive, explosive, and physical hazards may be present in the laboratory. Reactive hazards include pyrophorics and incompatible chemicals; explosive hazards include peroxide formers and powders; and physical hazards include cryogenic liquids, electrical equipment, lasers, compressed gas cylinders and reactions that involve high pressure or vacuum lines. (For more information about these hazards within a laboratory, see Chapter 4, sections 4.D and 4.E.)

An array of controls exists to protect laboratory personnel from the hazards listed above. Engineering controls (e.g., laboratory chemical hoods and gloveboxes), administrative controls (e.g., safety rules, CHPs, and standard operating procedures), and PPE (e.g., gloves, laboratory coats, and chemical splash goggles) are all designed to minimize the risks posed by these hazards.

Work practices to minimize exposure to hazardous chemicals can be found in Chapter 6.

2.C.4 General Housekeeping Practices in the Laboratory

Good housekeeping practices in the laboratory has a number of benefits. For example, in terms of safety,

---

BOX 2.3
Chemical Hygiene Responsibilities in a Typical Governmental Laboratory

Chemical Hygiene Officer (CHO)

- Is given authority by the Director of Safety Services Division to provide technical guidance in the development and implementation of the provisions of the Chemical Hygiene Plan (CHP).
- Interfaces with safety, radiological protection, quality, health, and other organizations, as requested, on chemical hygiene matters.
- Knows current policies, procedures, and legal requirements concerning use and handling of chemicals in laboratories (such as Occupational Safety and Health Administration Laboratory Standard 29 CFR § 1910.1450, *Occupational Exposure to Hazardous Chemicals in Laboratories*).
- Ensures that the CHP is reviewed annually.

Divisional Chemical Hygiene Officer(s) (DCHO)

- Is appointed by division directors (line management) to assist in the development and implementation of the CHP and supplemental CHP within their respective division.
- Works with division management in implementing appropriate chemical hygiene practices. The role of the DCHO may be performed by division/facility safety officer/representatives.
- Develops appropriate addendums to the CHP, as necessary, to address division-specific hazards or to meet division-specific requirements.
- Should be knowledgeable in chemical handling, use, and disposal techniques and requirements.

Laboratory Space Manager

- Maintains a laboratory-space posting for each laboratory.
- Implements laboratory access controls and monitors compliance.

- Communicates hazard and safety information, on a continuing basis, to researchers and staff using the laboratory space.
- Communicates issues concerning the laboratory space to the laboratory space group leader.
- Periodically conducts walk-throughs of the laboratory space according to the division or directorate annual performance plan.
- Acts as a mentor for laboratory occupants, encouraging safe chemical hygiene practices.
- Assists in preparation of research safety summaries (RSSs) relevant to his or her assigned laboratory.

Line Management

- Ensures that the staff knows and follows the rules from the CHP and that it is fully implemented.
- Ensures that laboratory personnel and guests under their supervision receive appropriate site-specific information and training on the hazards of the chemicals in the workplace at the time of their initial assignment.
- Ensures updates are made to RSSs and training when new chemical hazards are introduced into the workplace.
- Ensures that hazards associated with reaction intermediates and products that will be synthesized are identified and analyzed.
- Ensures that appropriate controls are established and documented in the RSS. Subject matter experts may provide assistance.
- Ensures that the staff has adequate facilities, personal protective equipment, equipment, and training to handle and use the chemicals that are currently on inventory in the laboratories.
- Ensures that the required personal protective equipment (PPE) is available and in working order and that appropriate training on the use and limitations of PPE has been provided and documented.

---

it can reduce the number of chemical hazards (health, physical, reactive, etc.) in the laboratory and help control the risks from hazards that cannot be eliminated. Practices that encourage the appropriate labeling and storage of chemicals can reduce the risks of mixing of incompatible chemicals and assist with regulatory compliance. From a security standpoint, order in the laboratory makes it easier to identify items out of place or missing. And finally, good housekeeping can help reduce scientific error by, for example, reducing

the chances of samples becoming confused or contaminated and keeping equipment clean and in good working order. More information about housekeeping practices can be found in Chapter 6, section 6.C.3.

2.D CHEMICAL MANAGEMENT PROGRAM

One of the most important components of a laboratory safety program is chemical management. Prudent chemical management includes the following processes.

2.D.1 Chemical Procurement

According to the nonmandatory OSHA Laboratory Standard (Appendix A, section D.2(a), *Chemical Procurement, Distribution, and Storage*), "Before a substance is received, information on proper handling, storage, and disposal should be known to those who will be involved." The standard further states that "No container should be accepted without an adequate identifying label. Preferably, all substances should be received in a central location." These procedures are strongly recommended. Personnel should be trained to identify signs of breakage (e.g., rattling) and leakage (e.g., wet spot or stain) on shipments and such shipments should be refused or opened in a hood by laboratory staff.

Some organizations have specific purchasing policies to prohibit unauthorized purchases of chemicals and other hazardous materials. The purchaser must assume responsibility for ownership of the chemical. Because of the possibility of a chemical leak or release and subsequent exposure, chemical shipments should only be received by trained personnel in a laboratory or central receiving area with proper ventilation. Neither administrative offices nor the mail room is appropriate for receipt or opening of chemical shipments.

When preparing to order a chemical for an experiment, several questions should be asked:

- What is the minimum amount of this chemical that is needed to perform the experiment? Is it available elsewhere in the facility? Remember, when ordering chemicals, less is always best. Prudent purchasing methods will save storage space, money, and disposal costs. Larger containers require more storage space and will incur additional disposal costs if the chemical is not used.
- Has the purchase been reviewed by the CHO to ensure that any special requirements can be met?
- Is the proper PPE available in the laboratory to handle this chemical?
- What are the special handling precautions?
- Where will the chemical be stored in the laboratory?
- Does the laboratory chemical hood provide proper ventilation?
- Are there special containment considerations in the event of a spill, fire, or flood?
- Will there be additional costs or considerations related to the disposal of this chemical?

2.D.2 Chemical Storage

To lessen risk of exposure to hazardous chemicals, trained laboratory personnel should separate and store all chemicals according to hazard category and compatibility. In the event of an accident involving a broken container or a chemical spill, incompatible chemicals that are stored in close proximity can mix to produce fires, hazardous fumes, and explosions. Laboratory personnel should read the MSDS and heed the precautions regarding the storage requirements of the chemicals in the laboratory. A detailed chemical compatibility table is included in Chapter 5, section 5.E.2, Table 5.1.

To avoid accidents, all chemical containers must be properly labeled with the full chemical name, not abbreviations, and using a permanent marker. All transfer vessels should have the following label information:

- chemical name,
- hazard warnings,
- name of manufacturer,
- name of researcher in charge, and
- date of transfer to the vessel.

Incoming chemical shipments should be dated promptly upon receipt, and chemical stock should be rotated to ensure use of older chemicals. It is good practice to date peroxide formers upon receipt and date again when the container is opened so that the user can dispose of the material according to the recommendations on the MSDS. Peroxide formers should be stored away from heat and light in sealed airtight containers with tight-fitting, nonmetal lids. Test regularly for peroxides and discard the material prior to the expiration date. (For more information about storage and handling of peroxides, see Chapter 4, section 4.D.3.2, and Chapter 6, section 6.G.3.)

When storing chemicals on open shelves, always use sturdy shelves that are secured to the wall and contain ¾-in. lips. Do not store liquid chemicals higher than 5 ft on open shelves. Do not store chemicals within 18 in. of sprinkler heads in the laboratory. Use secondary containment devices (i.e., chemical-resistant trays) where appropriate. Do not store chemicals in the laboratory chemical hood, on the floor, in the aisles, in hallways, in areas of egress, or on the benchtop. Chemicals should be stored away from heat and direct sunlight.

Only laboratory-grade explosion-proof refrigerators and freezers should be used to store properly sealed and labeled chemicals that require cool storage in the laboratory. Periodically clean and defrost the refrigerator and freezer to ensure maximum efficiency. Domestic refrigerators and freezers should not be used to store chemicals; they possess ignition sources and can cause dangerous and costly laboratory fires and explosions. Do not store food or beverages in the laboratory refrigerator. (For more information, see Chapter 7, section 7.C.3.)

Highly hazardous chemicals must be stored in a well-ventilated secure area that is designated for this purpose. Cyanides must be stored in a tightly closed container that is securely locked in a cool dry cabinet to which access is restricted. Protect cyanide containers against physical damage and separate them from incompatibles. When handling cyanides, follow good hygiene practices and regularly inspect your PPE. Use proper disposal techniques.

Flammable liquids should be stored in approved flammable-liquid containers and storage cabinets. Observe National Fire Protection Association, International Building Code, International Fire Code, and other local code requirements that limit the quantity of flammables per cabinet, laboratory space, and building. Consult the local fire marshal for assistance, if needed. Store odiferous materials in ventilated cabinets. Chemical storage cabinets may be used for long-term storage of limited amounts of chemicals.

Rooms that are used specifically for chemical storage and handling (i.e., preparation rooms, storerooms, waste collection rooms, and laboratories) should be controlled-access areas that are identified with appropriate signage. Chemical storage rooms should be designed to provide proper ventilation, two means of access/egress, vents and intakes at both ceiling and floor levels, a diked floor, and a fire suppression system. If flammable chemicals are stored in the room, the chemical storage area must be a spark-free environment and only spark-free tools should be used within the room. Special grounding and bonding must be installed to prevent static charge while dispensing solvents.

2.D.3 Chemical Handling

Important information about handling chemicals can be found in the MSDS. A comprehensive file of MSDSs must be kept in the laboratory or be readily accessible online to all employees during all work shifts. Trained laboratory personnel should always *read* and *heed* the label and the MSDS before using a chemical for the first time. Laboratory personnel should be familiar with the types of PPE that must be worn when handling the chemical. Ensure that the ventilation will be adequate to handle the chemicals in the laboratory. One should be familiar with the institutional CHP and EAP so that appropriate actions are taken in the event of a chemical spill, fire, or explosion.

2.D.4 Chemical Inventory

The OSHA Laboratory Standard, Appendix A, section D.2(b) (Chemical Procurement, Distribution, and Storage), states, "Stored chemicals should be examined periodically (at least annually) for replacement,

deterioration, and container integrity." Section D.2(d) states, "Periodic inventories should be conducted, with unneeded items being discarded or returned to the storeroom/stockroom." Though Appendix A is not mandatory, compliance with the standard is an element of good laboratory management. On a basic level, you cannot safely manage something if you do not know that you have it on-site. Thus, a system for maintaining an accurate inventory of the laboratory chemicals on campus or within an organization is essential for compliance with local and state regulations and any building codes that apply.

There are many benefits of performing annual physical chemical inventory updates:

- ensures that chemicals are stored according to compatibility tables,
- eliminates unneeded or outdated chemicals,
- increases ability to locate and share chemicals in emergency situations,
- updates the hazard warning signage on the laboratory door,
- promotes more efficient use of laboratory space,
- checks expiration dates of peroxide formers,
- ensures integrity of shelving and storage cabinets,
- encourages laboratory supervisors to make "executive decisions" about discarding dusty bottles of chemicals,
- repairs/replaces torn or missing labels and broken caps on bottles,
- ensures compliance with all federal, state, and local record-keeping regulations,
- promotes good relations and a sense of trust with the community and the emergency responders,
- reduces the risk of exposure to hazardous materials and ensures a clean and healthful laboratory environment, and
- may reduce costs by making staff aware of chemicals available within the organization.

Every laboratory should maintain an up-to-date chemical inventory. A physical chemical inventory should be performed at least annually, or as requested by the CHO. Although the software that is used to maintain the inventory and the method of performing the chemical inventory will vary from one institution to another, ultimately, the chemical inventory should include the following information:

- chemical name,
- Chemical Abstract Service number,
- manufacturer,
- owner,
- room number, and
- location of chemical within the room.

Note that the chemical name should be listed with its synonyms. This will allow for cross-indexing for tracking of chemicals and help reduce unnecessary inventory.

Important safety issues to consider when performing a chemical inventory are:

- Wear appropriate PPE and have extra gloves available.
- Use a chemical cart with side rails and secondary containment.
- Use a laboratory step stool to reach chemicals on high shelves.
- Read the EAP and be familiar with the institution's safety equipment.
- If necessary cease all other work in the laboratory while performing the inventory.

Once the inventory is complete, use suitable security precautions regarding the accessibility of the information in the chemical inventory. For example, precautions should be taken when the database shows the location of Department of Homeland Security (DHS) Chemicals of Interest in excess of DHS threshold quantities. (For more information about laboratory security, see Chapter 10.)

2.D.5 Transporting, Transferring, and Shipping Chemicals

It is prudent practice to use a secondary containment device (i.e., rubber pail) when transporting chemicals from the storeroom to the laboratory or even short distances within the laboratory. When transporting several containers, use carts with attached side rails and trays of single piece construction at least 2 in. deep to contain a spill that may occur. Bottles of liquids should be separated to avoid breakage and spills. Avoid high-traffic areas when moving chemicals within the building. When possible, use freight elevators when transporting chemicals and do not allow other passengers. If you must use a general traffic elevator, ask other passengers to wait until you have delivered the chemicals.

Always ground and bond the drum and receiving vessel when transferring flammable liquids from a drum to prevent static charge buildup. Use a properly operating chemical fume hood, local exhaust, or adequate ventilation, as verified by monitoring, when transferring PHSs.

All outgoing domestic and international chemical shipments must be authorized and handled by the institutional shipper. The shipper must be trained in U.S. Department of Transportation (DOT) regulations for ground shipments and must receive mandatory International Air Transport Association training for air shipments. DOT oversees the shipment of hazardous materials and has the authority to impose citations and fines in the event of noncompliance. (For more detailed information on the shipment of chemicals, see Chapter 5, section 5.F.)

2.D.6 Chemical Waste

All chemical waste must be stored and disposed of in compliance with applicable federal, state, local, and institutional regulatory requirements. Waste containers should be properly labeled and should be the minimum size that is required. There should be at least 2 in. of headspace in the liquid waste container to avoid a buildup of gas that could cause an explosion or a container rupture. (For more information about handling of hazardous waste, see Chapter 8.)

2.E LABORATORY INSPECTION PROGRAM

A program of periodic laboratory inspections helps keep laboratory facilities and equipment in a safe operating condition. Inspections safeguard the quality of the institution's laboratory safety program. A variety of inspection protocols may be used, and the organization's management should select and participate in the design of the inspection program appropriate for that institution's unique needs. The program should embrace the following goals:

- Maintain laboratory facilities and equipment in a safe, code-compliant operating condition.
- Provide a comfortable and safe working environment for all personnel and the public.
- Ensure that all laboratory activities are conducted in a manner to avoid employee exposure to hazardous chemicals.
- Ensure that trained laboratory personnel follow institutional CHPs.

Approach these goals with a degree of flexibility. Consider the different types of inspection, the frequency with which they are conducted, and who conducts them. A discussion of items to inspect and several possible inspection protocols follows, but is not all-inclusive.

Laboratory inspections are performed by EHS staff, the CHO, the safety director, laboratory staff, a safety committee, or an outside entity with the requisite qualifications and experience. The inspection checklist can include sections on chemical storage, chemical waste, housekeeping, PPE, laboratory chemical hoods, gas cylinder storage, emergency safety equipment, signs and labels, and facility issues.

Following each inspection, a detailed report is sent to the laboratory supervisor and appropriate administration. Photographs taken during the inspection process can emphasize the critical nature of a violation. Consider giving special recognition to laboratories demonstrating good laboratory practice and those that have demonstrated significant improvements in safety.

2.E.1 Types of Inspection Programs: Who Conducts Them and What They Offer

There are several types of inspection programs, each providing a different perspective and function. A comprehensive laboratory inspection program includes a combination of some or all of these programs.

2.E.1.1   Routine Inspections

Trained laboratory personnel and supervisors should complete general equipment and facility inspections on a regular basis. For certain types of equipment in constant use, such as gas chromatographs, daily inspections may be appropriate. Other types of equipment may need only weekly or monthly inspection or inspection prior to use if operated infrequently. Keep a record of inspection attached to the equipment or in a visible area. The challenge for any inspection program is to keep laboratory personnel continuously vigilant. They need positive encouragement to develop the habit of inspection and to adopt the philosophy that good housekeeping and maintenance for their workspace protect them and may help them produce better research results.

2.E.1.2   Self-Audits

To supplement an inspection program, some institutions promote self-inspections within the laboratories. Laboratory personnel may conduct their own inspections for their own benefit or management may ask them to self-audit and report their findings, using the routine inspections as a check on the self-inspections. This approach can be mutually beneficial, raising awareness, promoting the institutional safety culture, and easing the burden on management.

2.E.1.3   Program Audits

A program audit includes both a physical inspection and a review of the operations and the facilities. This type of audit is generally conducted by a team, which includes the laboratory supervisor, senior management, and laboratory safety representatives, and presents an excellent opportunity to promote a culture of safety and prudence within an organization. The su-

pervisor and senior management have the opportunity to take a close look at the facilities and operations. They can discuss with individual workers issues of interest or concern that may fall outside the scope of the actual inspection. A constructive and positive approach to observed problems and issues fosters an attitude of cooperation and leadership with regard to safety and helps build and reinforce a culture of teamwork and cooperation that has benefits far beyond protecting personnel and the physical facilities.

The audit begins with a discussion of the safety program and culture, and a review of operations, written programs, training records, and pertinent policies and procedures and how they are implemented in the laboratory. A laboratory inspection that includes interviews with laboratory personnel follows to determine the level of safety awareness. An open discussion with key personnel can ascertain how personnel, supervisors, managers, and safety officers can better support each other.

This type of audit provides a much more comprehensive view of the laboratory than a routine inspection.

2.E.1.4   Peer Inspections

One of the most effective safety tools a facility can use is periodic peer-level inspections. Usually, the people who fulfill this role work in the organization they serve, but not in the area being surveyed. Personnel may participate on an ad hoc basis, or the institution may select specific individuals to be part of a more formal, ongoing inspection team. A peer inspection program has the intrinsic advantage of being perceived as less threatening than other forms of surveys or audits.

Peer inspections depend heavily on the knowledge and commitment of the people who conduct them. Individuals who volunteer or are selected to perform inspections for only a brief time may not learn enough about an operation or procedure to observe and comment constructively. People who receive involuntary appointments or who serve too long may not maintain the desired level of diligence.

A high-quality peer-level inspection program reduces the need for frequent inspections by supervisory personnel. However, peer inspections should not replace other inspections completely. Walk-throughs by the organization's leadership demonstrate commitment to the safety programs, which is key to their continuing success.

2.E.1.5   Environmental Health and Safety Inspections

The organization's EHS staff, the safety committee, or an equivalent group may also conduct laboratory inspections on a routine basis. These inspections may be comprehensive, targeted to certain operations or experiments, focused on a particular type of inspection such as safety equipment and systems, or audits to check the work of other inspectors.

Safety staff are not the only nonlaboratory personnel who may conduct safety inspections. Facility engineers or maintenance personnel may add considerable value to safety inspection programs. They are also given the opportunity to gain a better perspective on laboratory work. It is advisable to have a representative from facilities engineering present during inspections so physical deficiencies can be appropriately and clearly noted and understood and priorities set for correction.

2.E.1.6 Inspections by External Entities

Many types of elective inspections or audits are conducted by outside experts, regulatory agencies, emergency responders, or other organizations. They may inspect a particular facility, equipment, or procedure either during the preexperiment design phase or during operations. As a matter of safety and security, if someone requests entry to a laboratory for the purpose of an audit without a recognized escort, ask to see his or her credentials and contact the EHS office or other appropriate parties.

Tours, walk-throughs, and inspections by regulatory or municipal organizations offer the opportunity to build relationships with governmental agencies and the public. For example, an annual visit by the fire department serving a particular facility will acquaint personnel with the operations and the location of particular hazards. If these individuals are ever called into the facility to handle an emergency, their familiarity with it will make them more effective. During their walk-through, they may offer comments and suggestions for improvements. A relationship built over time helps make this input positive and constructive.

If a pending operation or facility change may raise public attention and concern, an invitation targeted to specific people or groups may prevent problems. Holding public open houses from time to time helps build a spirit of support and trust. Many opportunities exist to apply this type of open approach to dealing with the public. An organization only needs to consider when to use it and what potential benefits may accrue.

Inspections and audits by outside consultants or peer institutions are especially helpful to identify both best practices and vulnerabilities. Many times, the inspectors bring with them experiences and examples from other laboratories that prove useful. When choosing a consultant, best practice is to find one with experience conducting similar audits of peer institutions. More and more often, health and safety experts, facilities staff, and laboratory personnel from peer institutions form inspection teams that conduct inspections of each other's laboratories. Such an arrangement can be beneficial and economical.

Many regulatory agencies promote institutions conducting self-audits, by either consultants or peer auditors, and reporting the findings to the agency. As an incentive, any violations noted in the self-audit may result in reduced or waived fines and fewer visits from the agency inspectors. It is important to fully understand the regulatory agency's self-reporting policy before implementing this option. In some cases, the institution must commit to remediating identified deficiencies within a specific time period.

Finally, regulatory agencies may conduct announced or unannounced inspections on a routine or sporadic basis. Laboratories and institutions should keep their programs and records up-to-date at all times to be prepared for such inspections. Any significant incident or accident within a facility may trigger one or more inspections or investigations by outside agencies. Evidence that the underlying safety programs are sound may help limit negative findings and potential penalties.

2.E.2 Elements of an Inspection

2.E.2.1 Preparing for an Inspection

Whether an inspection is announced or unannounced depends on the objective. There are many advantages to announcing an inspection ahead of time. By announcing and scheduling inspections, the inspectors are more likely to interact with the laboratory personnel and the supervisors. The inspection can be a good learning experience for all and will feel less like a safety-police action and more like a value-added service, with the right attitude and approach. However, if the objective is to observe real-time conditions in preparation for a regulatory inspection, an unannounced targeted inspection might be appropriate.

Before the inspection, have a checklist of inspection items, along with the criteria and the basis for each issue. The criteria may be based on regulations, institutional policies, or recommended practices. Sharing the checklist with laboratory personnel prior to the inspection helps them perform their own inspections before and periodically after the inspection.

Bring a camera. A photograph is much more effective than a long explanation in convincing a manager that something needs attention.

2.E.2.2   Inspection Checklists

Inspection checklists take a variety of formats and vary in length depending on the type and focus of the inspection. Although most inspection forms are paper, some are computer based. Make each inspection item a YES or NO question. Pose the issue so that a positive outcome is a YES, making it easy to spot problems. Always leave room for comments.

There are a number of commercial products on the market offering Web-based applications that work on a laptop or notebook computer. Checklist programs are available for handheld digital devices. Some may download into spreadsheets or word-processing programs. Others automatically create reports that can be e-mailed to recipients. All are intended to streamline the record-keeping and reporting process.

2.E.2.3   Conducting the Inspection

When conducting an inspection, interacting with the individuals in the laboratory is important. Even if inspectors are mainly looking at equipment and conditions, laboratory personnel can provide a great deal of information and the conversation itself may foster positive relationships between laboratory personnel and the group conducting the inspection. Speaking with laboratory personnel also helps gauge how well training programs are working and provides feedback for possible improvements to the laboratory safety program.

Take notes and make comments on the inspection form to be able to recall the details and describe any problems in the report. Where possible, take photographs of issues that need particular attention.

Point out problems as they are found and show laboratory personnel how to fix them. If the problem is corrected during the inspection, make a note that it was resolved.

2.E.2.4   Inspection Report

As soon as possible after an inspection, prepare a report for the laboratory supervisor and others, as appropriate. This may include the CHO, the chair or manager of the department, line supervisors, and directors. Depending on the type and focus of the inspection, it may be helpful to hold a meeting with the key individuals to review the findings.

The report should include all problems noted during the inspection, along with the criteria for correcting them. If photographs were taken, include them in the report. The report should also note any best practices and any improvements since the last inspection.

Include a reasonable time line for corrective actions.

Be sure to follow up with the laboratory to ensure that recommended corrections are made.

2.E.2.5   Corrective Actions

In most cases, laboratory personnel will take the appropriate corrective actions once they have been made aware of an issue. If the laboratory supervisor is not supportive and the necessary changes are not made, the inspectors and EHS and other appropriate individuals in the organization will have to decide whether the infractions are serious enough to put either the health or safety of laboratory personnel at risk or the institution at risk for violation of a regulation or code.

The organization must decide what steps to take for those individuals or laboratory groups that are using unsafe work practices or are not in compliance with institutional policies or external regulations.

2.E.3   Items to Include in an Inspection Program

The following list is representative, not exhaustive:

- Required PPE is available and used consistently and correctly (e.g., laboratory coat, gloves, safety glasses, chemical splash goggles, face shield).
- Compressed gas cylinders are secured correctly, cylinders are capped if not connected for use, and proper regulators are used.
- Limitations on where food and drink storage and eating and drinking are allowed are observed.
- Electrical cords are off surfaces where spills of flammable materials are likely, and cords are in good condition, not displaying signs of excessive wear (fraying, cords are not pinched). Equipment not meeting National Electrical Safety Code Division 1, Group C and D explosion-resistance specifications are electrically inspected prior to use in the laboratory. (See Chapter 7, section 7.C.)
- Laboratory chemical hoods have been tested and are operated with inspection information visible, hoods are used properly, work is conducted inside 6 in. from hood face, airflow is not significantly impeded by large pieces of equipment.
- Vacuum glassware is inspected and maintained in good condition, pressure reaction vessels with pressure relief and temperature/pressure measuring capability are used for high-pressure reactions.
- Health classification of materials is conducted (particularly for unknown compounds), and associated work practices and containment based on hazard/risk classification of the material are followed (e.g., low hazard, hazardous, particularly

hazardous materials and associated requirements for use of ventilated enclosures, disposal of waste, labeling of areas where work with high-hazard materials is conducted, decontamination of work surfaces).

- Access to emergency equipment is unobstructed (e.g., safety showers, eyewash units, fire extinguishers), and equipment is maintained in good working order. Aisles are unobstructed and minimum egress is maintained. Minimum clearance to sprinkler heads, as required by local building and fire codes, is maintained.
- Chemicals are properly stored and segregated (e.g., flammables, strong acids, strong bases, peroxides).
- Personnel demonstrate ability to access MSDSs or other chemical safety references and knowledge of handling requirements for various classifications of materials.
- Rotating machinery and high-temperature devices have appropriate guards. Safety switches and emergency stops are working.
- Associated egress corridors are unobstructed and minimum egress as required by building and fire codes is maintained. Combustible and surplus materials and equipment are removed from exit passageways.

Depending on the laboratory and the type of work conducted in it, other items may also be targeted for inspection (Box 2.4).

2.F EMERGENCY PROCEDURES

2.F.1 Fire Alarm Policy

When a fire alarm sounds in the facility, evacuate the laboratory immediately via the nearest exit. Extinguish all Bunsen burner and equipment flames. If the fire originates in your laboratory, follow all institutional policies regarding firefighting and suppression. Check restrooms and other areas with possible limited audio or visual notification of an alarm before exiting the facility. Where necessary, provide assistance to persons with disabilities to ensure they are able to exit the facility.

2.F.2 Emergency Safety Equipment

The following is a guide to safety equipment found in a laboratory.

1. A written EAP has been developed and communicated to all personnel in the unit. The plan

---

BOX 2.4
Excerpt from an
Inspection Checklist[a]

Department/Group/Laboratory:
Inspector:
Date:
Building and Room:
Laboratory Supervisor:

LABORATORY ENVIRONMENT

Work areas illuminated	Y N NA	
Storage of combustible materials minimized	Y N NA	
Aisles and passageways are clear and unobstructed	Y N NA	
Trash is removed promptly	Y N NA	
No evidence of food or drink in active laboratory areas	Y N NA	
Wet surfaces are covered with nonslip materials	Y N NA	
Exits are illuminated and unobstructed	Y N NA	
Proper management of hazardous materials and waste	Y N NA	

COMMENTS:

Other elements of the checklist can include:

- Emergency equipment and planning
- Personal protective equipment
- Signs, labels, plans, and postings
- Electrical hazards
- Storage
- Compressed gases and cryogenics
- Pressure and vacuum systems
- Laboratory hoods and ventilation
- Security
- Training/awareness

[a]For a full checklist, see the CD that accompanies this book.

---

includes procedures for evacuation, ventilation failure, first aid, and incident reporting.
2. Fire extinguishers are available in the laboratory and tested on a regular basis. If a fire extinguisher is activated for any reason, make an immediate report of the activity to the CHO, fire marshal, or appropriate individual responsible for fire

safety equipment so that the fire extinguisher is replaced in a timely manner.

3. Eyewash units are available, inspected, and tested on a regular basis.
4. Safety showers are available and tested routinely.
5. Fire blankets are available in the laboratory, as required. Fire blankets can be used to wrap a burn victim to douse flames as well as to cover a shock victim and to provide a privacy shield when treating a victim under a safety shower in the event of a chemical spill.

NOTE: Laboratory personnel should be taught that fire blankets can be dangerous if used incorrectly. Wrapping a fire blanket around a person on fire can result in a chimney-like effect that intensifies, rather than extinguishes, the fire. Fire blankets should never be used on a person when they are standing. (See Chapter 7, section 7.F.2.3 for more information on responding to fires.)

6. First-aid equipment is accessible, whether through a kit available in the laboratory or by request through the organization.
7. Fire alarms and telephones are available and accessible for emergency use.
8. Pathways to fire extinguishers, eyewash units, fire blankets, first-aid kits, and safety showers are clear.

2.F.3 Chemical Spill Policy

Laboratory personnel should be familiar with the chemical, physical, and toxicological properties of each hazardous substance in the laboratory. Consult the label and the MSDS prior to the initial use of each hazardous substance. Always use the minimal amount of the chemical and use caution when transporting the chemical. In the event of an accidental chemical release or spill, personnel should refer to the following general guidelines.

Most laboratory workers should be able to clean up incidental spills of the materials they use. Large spills, for example, 4 L or more, may require materials, protective equipment, and special handling that make it unsafe for cleanup by laboratory workers themselves. Lab workers should be instructed to contact EHS personnel to evaluate how to proceed with spill cleanup.

In the event that the spill material has been released to the environment, notify EHS personnel immediately. A release to the environment includes spills directly into a drain or waterway or onto land, such as grass or dirt.

Low-flammability and low-toxicity materials that are not volatile (e.g., inorganic acids and caustic bases)

1. Decontaminate any victim at the nearest safety shower or eyewash unit. Take other appropriate action as described in the MSDS.
2. Notify appropriate personnel immediately.[4]
3. Limit or restrict access to the area as necessary.
4. Wear PPE that is appropriate to the degree of hazard of the spilled substance.
5. Use chemical spill kits that contain an inert absorbent to clean up the affected area if this action can be accomplished without risk of additional injury or contamination to personnel. If the spill is located on the laboratory floor, be aware that some absorbents can create a slipping hazard.
6. Dispose of contaminated materials according to institutional policy.
7. Complete an incident report and submit it to the appropriate office or individual.
8. Label all phones with emergency phone numbers.

Flammable solvents of low toxicity (e.g., diethyl ether and tetrahydrofuran)

1. Decontaminate any victims at the nearest safety shower or eyewash unit. Take other appropriate action as described in the MSDS.
2. Alert all other personnel in the laboratory and the general vicinity of the spill.
3. Extinguish all flames and turn off any spark-producing equipment. If necessary, turn off power to the laboratory at the circuit breaker. The ventilation system must remain operational.
4. Immediately notify appropriate personnel.[4]
5. Limit or restrict access to the area as necessary.
6. Wear PPE that is appropriate to the degree of hazard of the spilled substance.
7. Use spill pillows or spill absorbent and nonsparking tools to soak up the solvent as quickly as possible. Be sure to soak up chemicals that have seeped under equipment and other objects in the laboratory. If the spill is located on the laboratory floor, be aware that some absorbents can create a slipping hazard.
8. Dispose of contaminated materials according to institutional policy.
9. Complete an incident report and submit it to the appropriate office or individual.

---

[4]The person to notify in case of an incident in the laboratory varies by organization. It may be the CHO, the safety director, on-site security, or another party. Check with the organization to determine the appropriate individual or office.

Highly toxic materials (e.g., dimethylmercury)

1. Alert all trained laboratory personnel in the laboratory and the general vicinity of the spill and immediately evacuate the area.
2. Decontaminate any victims at a safety shower or eyewash unit in a safe location. Take other appropriate decontamination action as described in the MSDS.
3. Immediately notify appropriate personnel.[4]
4. Limit or restrict access to the area as necessary.
5. Do not attempt to clean up the spill. EHS personnel will evaluate the hazards that are involved with the spill and will take the appropriate actions.
6. Only EHS personnel and appropriate outside industrial hygienists are authorized to decontaminate the area and dispose of the contaminated waste.
7. Complete an incident report and submit it to the appropriate office or individual.

2.F.4 Accident Procedures

In the event of an accident, follow all institutional policies for emergency response and notify the internal point of contact for laboratory safety and local emergency responders. All accidents involving personal injury, however slight, must be immediately reported according to your institution's procedure. Provide a copy of the appropriate MSDS to the attending physician, as needed. Complete an accident report (Figure 2.2) and submit it to the appropriate office or individual within 24 hours of the incident.

2.G EMPLOYEE SAFETY TRAINING PROGRAM

Newly hired employees or students working in a laboratory should be required to attend basic safety training prior to their first day. Additional training should be provided to laboratory personnel as they advance in their laboratory duties or when they are required to handle a chemical or use equipment for the first time.

Safety training should be viewed as a vital component of the laboratory safety program within the organization. The organization should provide ongoing safety activities that serve to promote a culture of safety in the workplace that will begin when the person begins work and will continue for the length of their tenure. Personnel should be encouraged to suggest or request training if they feel it would be beneficial. The training should be recorded and related documents maintained in accordance with organizational requirements.

Training sessions may be provided in-house by professional trainers or may be provided via online training courses. Hands-on, scenario-based training should be incorporated whenever possible. Safety training topics that may prove to be helpful to laboratory personnel include

- use of CHPs and MSDSs,
- chemical segregation,
- PPE,
- safety showers and eyewash units,
- first aid and cardiopulmonary resuscitation,
- chemical management,
- gas cylinder use,
- fire extinguisher training,
- laser safety, and
- emergency procedures.

Personal Data

Employee/Student Name		Case No.
Employee/Student Phone No.		
Employee/Student Dept.		Investigation Date
Employee Supervisor		Investigator Name

Event Details

Employee/Student Statement (Description of event—before, during, and after)

_____

_____

_____

_____

_____

_____

_____

_____

_____

_____

_____

_____

_____

Work Related?	Yes ☐ No ☐	Body Part Injured	
Event Date/Time	/	Event Location	[lab, corridor, stairs, outside, etc.]
Reported Injury Date/Time	/	Specific Location	[building, floor, room, column]

Injury Severity	☐ Observation/Near Miss	☐ First Aid	☐ MTBFA (OSHA)
	☐ Work Restrictions	☐ Lost Time Restriction	

Accident Type	☐ Allergen Exposure	☐ Bitten By	☐ Car/Truck/Motorized Vehicle
	☐ Caught In/Between	☐ Contact with Chemical	☐ Contact with Hot Surface
	☐ Environmental Exposure	☐ Ergonomic	☐ Needle Stick
	☐ Pushing/Pulling	☐ Slip/Trip/Fall	☐ Struck Against
	☐ Struck By	☐ Twist/Turn	☐ Other

	Device Type	**Device Brand**
Contaminated Sharp Involved		
Needle Stick		

FIGURE 2.2 Accident report form.

3 Emergency Planning

3.A	INTRODUCTION	33
3.B	PREPLANNING	33
	3.B.1 Vulnerability Assessment	33
	3.B.1.1 Fire	34
	3.B.1.2 Flood	34
	3.B.1.3 Severe Weather	34
	3.B.1.4 Seismic Activity	35
	3.B.1.5 Extensive Absences Due to Illness	35
	3.B.1.6 Hazardous Material Spill or Release	35
	3.B.1.7 High-Profile Visitors	35
	3.B.1.8 Political or Controversial Researchers or Research	35
	3.B.1.9 Intentional Acts of Violence or Theft	35
	3.B.1.10 Loss of Laboratory Materials or Equipment	36
	3.B.1.11 Loss of Data or Computer Systems	36
	3.B.1.12 Loss of Mission-Critical Equipment	36
	3.B.1.13 Loss of High-Value or Difficult-to-Replace Equipment	36
	3.B.2 What Every Laboratory Should Know and Have	36
	3.B.2.1 Survival Kit	36
	3.B.2.2 Training	37
3.C	LEADERSHIP AND PRIORITIES	37
	3.C.1 Decision Makers, with Succession	37
	3.C.2 Laboratory Priorities	37
	3.C.3 Essential Personnel	37
3.D	COMMUNICATION DURING AN EMERGENCY	38
	3.D.1 Contact List	38
	3.D.2 Communication Plan	38
	3.D.2.1 Telephone	38
	3.D.2.2 Text Messages	38
	3.D.2.3 E-Mail	38
	3.D.2.4 Internet and Blogs	38
	3.D.2.5 Emergency Contacts	39
	3.D.2.6 Media and Community Relations	39
	3.D.3 Assembly Point	39
3.E	EVACUATIONS	39
	3.E.1 Shutdown Procedures	39
	3.E.1.1 Processes Requiring Special Shutdown Procedures	39
	3.E.1.2 Experiments Running Unattended	39
	3.E.2 Assembly Points and Evacuation Routes	39
3.F	SHELTER IN PLACE	39

3.G   LOSS OF POWER   40
   3.G.1   Short-Term Power Loss   40
      3.G.1.1   Potential Effects   40
      3.G.1.2   Laboratory Procedures   40
   3.G.2   Long-Term Power Loss   40
      3.G.2.1   Security Issues   40
      3.G.2.2   Environmental and Storage Conditions   40
      3.G.2.3   Discontinuation of Experiments   40
   3.G.3   Preplanning   40
      3.G.3.1   Generator Power   41
      3.G.3.2   Uninterruptible Power Supply (UPS)   41
      3.G.3.3   Dry Ice   41
      3.G.3.4   Other   41

3.H   INSTITUTIONAL OR BUILDING CLOSURE   41
   3.H.1   Short-Term Closure   41
   3.H.2   Long-Term Closure   42
   3.H.3   Alternative Laboratory Facilities   42

3.I   EMERGENCY AFFECTING THE COMMUNITY   42
   3.I.1   Disruption of Deliveries of Goods and Services   42
   3.I.2   Laboratory Staff Shortage   42

3.J   FIRE OR LOSS OF LABORATORY   42
   3.J.1   Records for Replacement of Laboratory Equipment   42
   3.J.2   Alternative Laboratories to Continue Operations   43
      3.J.2.1   Preplanning and Prevention   43

3.K   DRILLS AND EXERCISES   43

3.L   OUTSIDE RESPONDERS AND RESOURCES   43

3.A INTRODUCTION

Although most laboratory personnel are prepared to handle incidental spills or minor chemical exposures, many other types of emergencies can affect a laboratory, ranging from power outages to floods or intentional malicious acts. Some may have long-term consequences and may severely affect the continuity of laboratory operations. Although these issues must be considered on an organizational level, laboratory personnel should be trained in how to respond to large-scale emergencies. Laboratory security can play a role in reducing the likelihood of some emergencies and assisting in preparation and response for others. (For more information about laboratory security, see Chapter 10.)

There are four major phases to managing an emergency: mitigation, preparedness, response, and recovery.

The **mitigation phase** includes efforts to minimize the likelihood that an incident will occur and to limit the effects of an incident that does occur. Mitigation efforts may be procedural, such as safe storage of materials, or physical, such as a sprinkler system.

The **preparedness phase** is the process of developing plans for managing an emergency and taking action to ensure that the laboratory is ready to handle an emergency. This phase might include ensuring that adequate supplies are available, training personnel, and preparing a communication plan.

The **response phase** involves efforts to manage the emergency as it occurs and may include outside responders as well as laboratory staff. The response is more effective and efficient when those involved in it understand their roles, have the training to perform their duties, and have the supplies they need on hand.

The **recovery phase** encompasses the actions taken to restore the laboratory and affected areas to a point where the functions of the laboratory can be carried out safely. Usually, these actions restore the laboratory to its previous condition; however, this stage provides an opportunity for improvement.

The four phases are interconnected. Effective mitigation efforts reduce the impact of the emergency and ease the response and recovery stages. Lessons learned during an emergency may lead to further mitigation and preparedness efforts during the recovery phase. Good planning in the preparedness stage makes the response and recovery less complicated. **However, a plan is not a substitute for thinking.** It offers guidance and helps prepare for emergencies. It is not intended to replace analyzing the situation and formulating the best response based on the resources and situation at hand.

3.B PREPLANNING

Every institution, department, and individual laboratory should have an emergency preparedness plan. The level of detail of the plan will vary depending on the function of the group and institutional planning efforts already in place.

Planning proceeds in several steps. First, determine what types of incidents are most likely to occur to determine the type and magnitude of planning required. This will require input from multiple levels of the organization, and discussions with laboratory personnel should be integral to the process. Next, decide who the decision makers and stakeholders are and how to handle communications. Then, do the actual plan for the types of emergencies identified in the first step. Finally, train staff in the procedures outlined in the plan.

Emergency planning is a dynamic process. As personnel, operations, and events change, plans need updating and modification.

It is not possible to account for every emergency. When handling an emergency, do not use the plan as a recipe; use it as a list of ingredients and guidance.

3.B.1 Vulnerability Assessment

To determine the type and level of emergency planning needed, laboratory personnel need to perform a vulnerability assessment. What kinds of emergencies are most likely? What is the possible effect on laboratory operations?

For every potential emergency, the group should consider the history of occurrence in their laboratory or institution and at institutions with similar circumstances. The group should evaluate how the emergency would affect the laboratory, for example, damage to critical equipment, staffing limitations, loss of data, and the severity of the resulting conditions on laboratory operations. Making a list of available emergency response equipment and the location of that equipment assists in this task.

When planning, especially when determining where to spend time and resources, use impact/occurrence mapping (Figure 3.1). Where time and/or resources are limited, focus attention on events that would have higher impact and higher likelihood, and less attention on issues that are unlikely to occur or would have little impact.

The types of incidents and emergencies to consider vary depending on the type of laboratory, geographical location, and other factors that are unique to an institution or laboratory. The following sections cover most common issues faced in laboratories.

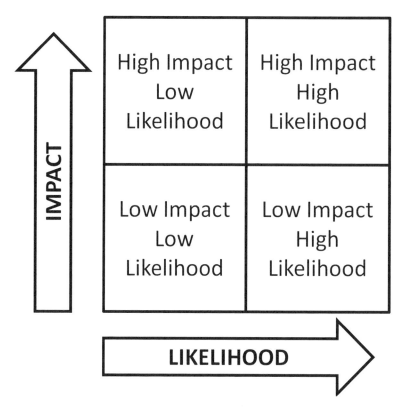

FIGURE 3.1 Impact/Likelihood ofoccurrence mapping.

3.B.1.1 Fire

A fire could occur in any laboratory but is more likely where chemicals such as flammable liquids, oxidizers, and pyrophoric compounds are stored and used. Consider the amount of combustible materials in the laboratory, potential ignition sources, and any other factors that increase the potential for fire. Some equipment is vulnerable even to minor smoke damage, such as laser optics, and plates used for semiconductor work.

Assess the impact of a fire. Does the laboratory contain mission-critical equipment that could be damaged by fire or smoke? Are there smoke detectors in place? Are there detectors for hazardous vapors and gas? Is there a sprinkler system or other automatic extinguishing system? Are the correct type and number of fire extinguishers present and do people know how to use them? If possible, fire extinguisher training should involve practice putting out fires.

3.B.1.2 Flood

Floods could be due to rain, rising levels of bodies of water, water pipe breaks, or accidental or deliberate acts. Some areas are more prone to floods than others. Laboratories on the basement or ground level are more likely to be flooded in a storm than those on

higher floors. Safety showers and eyewash stations that are not properly plumbed or do not have floor drains nearby may also be a source of flooding. Consider the likelihood of flooding and its impact. Also consider whether the laboratory contains equipment that is very sensitive to water damage. If flooding occurs, could it affect the space below the flood? If so, is the floor sealed appropriately? Are there overhead pipes?

3.B.1.3 Severe Weather

Storms and flooding can disrupt power, cause damage to buildings, and result in impassable roads. In severe cases, a local state of emergency in response to weather could close roads to all but essential travel. In areas that are prone to tornadoes or hurricanes, consideration should be given to the adequate protection of critical hazardous operations. For example, in some areas, hydrogen cylinders and liquid nitrogen tanks are located outside the building. It may be sensible in some areas to locate lab space away from outside windows.

If there are travel restrictions, would anyone be able to reach the laboratory? If so, is there a means of communication to inform the individual(s) able to travel that they need to do so? Are there operations that run unattended? What possible problems could arise if no one is able to come to the laboratory for a day, a few days, or longer? Have rally points been identified and shelter-in-place protocols distributed in areas where tornadoes occur?

3.B.1.4 Seismic Activity

Laboratories in areas where seismic activity is common should take special precautions to secure and restrain equipment and chemicals within a laboratory. Consideration should be given to the damage that could be caused by falling equipment. An earthquake can render a building unusable for days or months or cause it to be condemned. Note that the earthquake may cause secondary hazards such as gas leaks, fires, chemical spills, electrical hazards, broken glass, reduced structural integrity of buildings, and flooding from broken water pipes.

Is all freestanding equipment that may shift or fall during an earthquake secured appropriately? Are plumbed connections to that equipment, including gas and water lines, flexible to allow for movement? Are items stored on open shelves appropriately organized (e.g., heavier items below, appropriate lips on the shelving, restraints where necessary)? If multiple containers fall and are damaged during a quake, is there potential for incompatible chemicals to come in contact? Are all compressed gas cylinders secured in accordance with the guidance in Chapter 5, section 5.E.6? Also consider the likelihood of other sensitive equipment, such as computers and analytical equipment, falling to the ground. If possible, secure those items to the desk or benchtop. How will continuity of operations be maintained in the event that the laboratory is inaccessible for a significant period of time?

3.B.1.5 Extensive Absences Due to Illness

Although pandemic planning is something that all institutions should complete, other circumstances, such as foodborne illnesses or communicable diseases could result in a large percentage of laboratory personnel unable to come to the laboratory for a short or extended period.

Some experts have estimated that in the event of pandemic influenza, an institution or laboratory may experience a 50% reduction in workforce for a period of 4 to 8 weeks. How might this affect laboratory operations? Are there experiments that cannot be suspended? Have laboratory personnel been cross-trained to be able to fill in for a person who is absent?

3.B.1.6 Hazardous Material Spill or Release

Incidental spills may happen at any time. Most are easily managed by laboratory personnel. Are there

large stores of chemicals in the laboratory or building? What are the most hazardous materials and what are the consequences of a release of those materials? Does the laboratory have sufficient spill control materials to handle any spill?

Some spills may be too large or too hazardous for laboratory personnel to clean up safely. What plans are in place in the event assistance is needed?

Consider the likelihood of an environmental release to the ground, air, or sewer. What procedures are in place to report such an incident and remediate it?

Are there unusually hazardous gases or materials that should be continuously monitored to detect a spill or leak? If such monitoring is in place, does it sound an alarm or send a signal? Where does that signal go (e.g., to security personnel, local only)? Are staff fully trained in how to respond and who to contact?

3.B.1.7 High-Profile Visitors

Visiting dignitaries and other individuals with some level of fame or notoriety can attract negative attention from protesters, paparazzi, and others who want to make their opinion known. In some cases, acts of civil disobedience may occur that impede access to the building and interrupt operations. Consider the security risks and how your institution handles such matters.

3.B.1.8 Political or Controversial Researchers or Research

Certain types of research and outspoken researchers with controversial views may engender protests, hate mail, and other concerns. There is the possibility that protestors may engage in civil disobedience in response. The vulnerabilities may vary from nuisance issues to more serious matters. Consider the level of security in and around the lab, mail handling, and other factors. Ensure procedures are in place to deal with these situations effectively, such as screening of incoming mail with irradiation procedures if deemed necessary.

3.B.1.9 Intentional Acts of Violence or Theft

Planning for and preventing intentional acts (including theft, sabotage, or terrorism) are difficult, especially if they are conducted by individuals within the laboratory or organization. The scale of the event will determine the extent of the disruption for a laboratory. If the act only affects one experiment or material, disruption will likely be minimal. However, acts of violence or theft that target a laboratory or building could cause significant disruption of laboratory operations. High-profile, destructive acts of terrorism can also have an effect on laboratory operations, even if they occur in another locale. In such cases, lockdowns of buildings, instructions to shelter in place or to evacuate, and simple distraction of laboratory personnel from their work can affect the laboratory environment. Any activity that causes significant damage to a building, such as an explosion, can have an effect similar to that of seismic activity.

Consider the history of such events at the institution, at similar institutions, and in the geographical area. Is there a known cause for concern? Are laboratory personnel prepared and trained in case of a shelter-in-place emergency? Are all emergency contact numbers posted in a high-visibility area? Has a rally or gather point been designated in case of evacuation of the laboratory?

3.B.1.10 Loss of Laboratory Materials or Equipment

Equipment, chemicals, samples, or other materials in the laboratory could be lost due to theft, sabotage, fire, flood, or other events. Think about the materials and equipment in the laboratory and consider the impact of their loss.

Planning for loss of such equipment is prudent. Keep purchasing and other records that would be helpful for an insurance claim. If equipment can be replaced, make note of where to find that equipment and the specifications needed. For custom-made equipment, keep the plans that show how to rebuild it.

Even with good planning, several days or longer may elapse before equipment is in place or usable. Make note of other laboratories or institutions with similar equipment or functions. Make arrangements, if possible, to use such facilities as a backup, if needed.

3.B.1.11 Loss of Data or Computer Systems

Because many laboratories store data in a digital format and rely on computerized systems, loss of critical data or systems poses serious problems.

Every laboratory and all laboratory personnel should have a backup plan for their digital data. A plan may include the following items:

- Data that should be stored off-site or in special storage and how to back this up using USB drives, external hard drives, or other external storage device;
- Whether networked computers are backed up automatically on a schedule;

- Resources that are available in the event there are problems with a computer system; and
- Backup or other procedures that can be used to continue operations in the event that a system is not available.

3.B.1.12 Loss of Mission-Critical Equipment

Some equipment may be so mission-critical that its loss will shut down operations until it is replaced. Ensure that this equipment has all the necessary protection (e.g., security, fire protection) and plan what to do if it is not available.

3.B.1.13 Loss of High-Value or Difficult-to-Replace Equipment

Some equipment is impossible or very difficult to replace. When it is lost, the laboratory may not be able to complete this function for an extended period of time. Very expensive equipment may take longer to process through insurance or may not be able to be replaced immediately.

3.B.2 What Every Laboratory Should Know and Have

3.B.2.1 Survival Kit

Every laboratory and all laboratory personnel should consider the possibility of having to stay at work for an extended time or under unusual conditions, such as a power loss. Consider keeping the following on hand:

For the laboratory:

- emergency contact information,
- flashlight,
- radio and batteries,
- first-aid kit, and
- safety glasses and gloves.

For individuals:

- change of clothing and shoes,
- medications,
- contact lens solution,
- nonperishable snacks,
- water, and
- blanket, jacket, or fleece.

This list is not complete. Organizations such as the Red Cross and the U.S. Department of Homeland Security have comprehensive Web pages that describe materials to have on hand in case of an emergency that requires personnel to shelter in place.

3.B.2.2 Training

In addition to laboratory safety issues, laboratory personnel should be familiar with what to do in an emergency. Topics may include

- evacuation procedures;
- emergency shutdown procedures—equipment shutdown and materials that should be stored safely;
- communications during an emergency—what to expect, where to call or look for information;
- how and when to use a fire extinguisher;
- security issues;
- protocol for absences due to travel restrictions or illness;
- safe practices for power outage;
- shelter in place;
- handling suspicious mail or phone calls;
- laboratory-specific protocols relating to emergency planning and response;
- handling violent behavior in the workplace;
- first-aid and CPR training, including automated external defibrillator training if available.

Periodic drills to assist in training and evaluation of the emergency plan are recommended as part of the training program.

3.C LEADERSHIP AND PRIORITIES

In an emergency situation, even with good planning, a number of factors tend to create a chaotic environment. Emotions may run high, uncertainties may exist regarding how long the conditions will last, and the general routine of the laboratory environment is disrupted.

Decisions need to be made, priorities set, and plans put in motion. Having a clear succession of leadership and priorities ahead of time can help provide clarity to the situation.

3.C.1 Decision Makers, with Succession

Determine who will provide leadership for the institution, department, group, or laboratory. Make a list of individuals authorized to make decisions, including financial commitments. Assume that there will be absences and include a succession. Keep in mind that in an emergency situation, the most practical leadership

succession does not always follow the organizational chart. Ensure that the people on that list know their designation and understand their responsibilities.

3.C.2 Laboratory Priorities

In the event of a reduction of staff, a limited amount of freezer space for sample storage, or other circumstances that place limitations on laboratory operations, experiments may need to be suspended or laboratory materials allowed to deteriorate. Consider laboratory priorities ahead of time, to reduce the decision-making burden during an emergency. Examples of priorities include securing pathogenic microbe libraries; securing toxic, flammable, or unstable compounds; and securing compounds that could be precursors to pharmaceuticals.

Review the operations and materials in the laboratory and formulate a hierarchy. Although each laboratory has unique needs, the following is one example:

- Priority 1: Protect human life.
- Priority 2: Protect research animals:
  - Grant-funded research animals,
  - Thesis-related research animals,
  - Other research animals.
- Priority 3: Protect property and the environment:
  - Mission-critical property,
  - High-value equipment,
  - Difficult to replace materials.
- Priority 4: Maintain integrity of research:
  - Grant-funded research,
  - Thesis-related research,
  - Other research.

3.C.3 Essential Personnel

In an emergency, there may be a facility closure and/or travel bans in place that would restrict personnel in their ability to report to work. If the laboratory must remain at least partially operational and personnel must report to work, it is important that these individuals be recognized as "essential personnel." There are human resources and payroll issues that may factor into this designation, as well as institutional policies.

In an emergency, the duties and responsibilities of the individuals reporting to work may be different than their responsibilities under normal conditions. Ensure that personnel understand and accept these responsibilities.

When there is a travel ban due to a state of emergency, those who must travel by car will need documentation from the institution stating that they are essential personnel. These individuals should keep such documentation in their vehicle to provide to a law enforcement officer. It remains the decision of that law enforcement officer whether to allow the travel.

3.D COMMUNICATION DURING AN EMERGENCY

Communication is key during an unexpected incident. Depending on the circumstances, some regular means of communication may be compromised: telephones may not work; a power loss may affect access to computers.

Among the most important elements of emergency preparedness is the communications plan. Laboratory personnel should know how to find information, how to contact people, and what to expect in terms of communications.

3.D.1 Contact List

Institutions should have extended contact information, including home, office, and cell phone numbers, for key personnel, including individuals familiar with the operations of the laboratories. In an emergency, particularly when outside emergency responders, such as police and ambulance attendants, are on-site, being able to speak with someone who can describe what is behind the laboratory doors can sometimes mean the difference between a reasonably appropriate response to the situation at hand and an overresponse that could tie up resources for an extended period.

Within the laboratories, laboratory managers, principal investigators, or others assigned leadership responsibility for emergencies should have up-to-date contact lists for all laboratory personnel. Such lists should be accessible from both the laboratory and from home.

Consider collecting information regarding an individual's ability to get to the laboratory during an emergency. Know who is within walking distance, who has access to a vehicle that can travel in all types of weather, or who has commitments that would preclude them from coming to the laboratory.

To aid emergency responders, many laboratories also post contact information on the laboratory door, as well as information about the hazards within the laboratory. An example emergency response poster can be found on the CD that accompanies this book.

3.D.2 Communication Plan

There are numerous ways to communicate during an emergency. Each institution, department, and laboratory group should have a communication plan that details which means of communication may be implemented. Laboratory personnel should be aware of the plan and should know what to expect and what is expected of them.

When an emergency affects a large population, telephone systems may quickly become overloaded, and local or institutional police, security, or public safety officials may be bombarded with calls. Instruct laboratory personnel to limit their use of phones during such times and use text messaging, e-mail, and the internet as primary means of communication.

3.D.2.1 Telephone

The telephone is often the most direct way to contact people. Some institutions have implemented mass notification systems that send voice messages to several phone numbers simultaneously. For a department or laboratory, a telephone chain may be an effective means of sharing information.

In an emergency that affects a large population, telephone systems may quickly become overloaded. Other circumstances may render telephones unusable. Do not rely only on telephones for communication of important instructions or information.

Hotlines with recorded messages are also helpful. For a laboratory, the number could be used for this purpose. In an emergency, the person in charge could leave a message with instructions on the main telephone that is available to anyone who calls.

3.D.2.2 Text Messages

Text messaging utilizes cellular phone service but can be more reliable. Even when cellular service is too weak or overloaded for calls, text messaging is often available. Text messages can be sent via cell phone or through e-mail. Check with the individual's service provider to determine the domain name to send text messages via e-mail. For example, for a Verizon Wireless customer with the phone number 123-456-7890, sending an e-mail to 1234567890@vtext.com would deliver the message as a text message. Most text message services have a limit of 120 characters per message.

3.D.2.3 E-Mail

E-mail can be a reliable way of sharing information. In the event that the institution's computer system is affected, it is prudent to have an alternative e-mail address for each person. Consider preparing a Listserve or e-mail list for use during an emergency.

3.D.2.4 Internet and Blogs

Posting updates on the institution or laboratory Web site is an easy way to reach multiple people. Instruct individuals to visit the site in the event of an emergency.

If using the laboratory Web site for this purpose is not practical, consider using a blog. Many internet providers and search engine sites offer free blog services. Blogs allow the users to post information easily without the use of a Web-page editor.

3.D.2.5 Emergency Contacts

Having the name and contact information for at least one friend or family member of laboratory personnel is prudent. The information would be useful if a person cannot be reached or in an emergency involving the laboratory person.

3.D.2.6 Media and Community Relations

If an incident has caught the attention of the media, whether local, national, or even a school or facility newspaper, ensure that the institution's spokesperson is involved in any conversations with reporters. Media inquiries should go through the person or group that is used to working with the media, because it is very easy for facts or issues to be misconstrued or presented in an inflammatory manner. All involved should be instructed to forward calls and interviewers to the media relations group.

When an incident command system has been instituted, a press officer will be appointed. All inquiries and statements go through this individual or group.

3.D.3 Assembly Point

Consider establishing an assembly point for laboratory personnel. In an emergency, essential personnel would be expected to report to that assembly point whether or not they have received specific instructions. This plan is especially helpful when communications are limited.

3.E EVACUATIONS

Fires, spills, and other emergencies may require evacuation of the building or the laboratory. All laboratory personnel should be aware of the evacuation procedures for the building and laboratory.

3.E.1 Shutdown Procedures

Some laboratories may have operations, materials, or equipment that could pose a hazard if simply abandoned and left unattended for an extended period. If a building is evacuated for an emergency, hours may elapse before personnel are allowed back inside. Consider the hazards in the laboratory and establish procedures to follow during an evacuation.

In the event that processes, experiments, or equipment were not shut down prior to evacuation and may pose a risk to health, the environment, or property, inform emergency responders of the situation. Emergency responders may escort a person into the laboratory to shut down the process, or they may ask for advice on how to do so themselves.

3.E.1.1 Processes Requiring Special Shutdown Procedures

Make a list of processes that need to be shut down prior to evacuation. Post the procedures in a conspicuous place and ensure that all laboratory personnel are aware of them. Posting a list at the exits may be helpful as a reminder to laboratory personnel as they leave.

3.E.1.2 Experiments Running Unattended

Note the hazards of experiments left unattended for an extended period. For routine procedures that fit into this category, establish procedures for safely terminating the procedure prior to evacuation.

3.E.2 Assembly Points and Evacuation Routes

Each building, section of a building, or group should have a designated assembly point to which individuals evacuate. At the assembly point, the emergency coordinator will account for individuals who should have evacuated, to advise emergency responders on the probability of individuals left in the building.

Main and alternative evacuation routes should be posted. Supervisors should ensure that all laboratory personnel are familiar with the safest way to evacuate the building and where to assemble. In case of evacuation, sign-in/sign-out boards or other check-in methods can be used as an aid to determine whether employees are in the building.

3.F SHELTER IN PLACE

For certain emergency situations, rather than evacuation, emergency responders may advise that people shelter in place, meaning that they remain inside the building. Such circumstances may include hazardous material releases outdoors; weather emergencies, such as hurricanes or tornadoes, or suspects wielding weapons.

When directed to shelter in place, take the following actions:

- Go or stay inside the building.
- Do not use elevators.
- Close and lock doors and windows.

- If possible, go to a location within the building that has no exterior doors or windows.
- As possible, monitor the situation via a radio, the internet, or a telephone.

Ensure that the laboratory is prepared by having a radio and flashlight on hand and provide an overview of shelter-in-place procedures to laboratory personnel as part of their orientation.

3.G   LOSS OF POWER

Most laboratory buildings experience occasional brief periods of power loss. Such instances may be minor disturbances or could damage equipment or ruin experimentation. Longer term power outages may cause significant disruption and loss. It is prudent to consider the effects of long-term and short-term power loss and implement plans to minimize negative outcomes.

3.G.1   Short-Term Power Loss

3.G.1.1   Potential Effects

Consider what can happen in the event of short-term power loss. If the outcome may be more than just an inconvenience, implement steps to reduce the impact. For example, if temperature is regulated by a heating mantle and loss of heat for even a few minutes could create an unacceptable variation, the result may be loss of that particular experimental run.

When developing a plan for handling a short-term power loss, consideration should be given as to what "state" a piece of equipment goes to during a loss of power or a resumption of power. Equipment should enter a fail-safe state and it should be tested for this state by purposely shutting off power to it and then reenergizing the circuit. Any interlocks (e.g., against high temperatures on heating mantels) should be rechecked after a loss of power. Some equipment must be restarted manually after a shutdown, resulting in longer term power loss even when power is restored. Uninterruptable power supplies and automatic generators should be considered for freezers and refrigerators that are used to store unstable compounds.

3.G.1.2   Laboratory Procedures

If laboratory personnel are present when power is lost, and power is not restored immediately, consider the following actions:

- *Turn off equipment*, particularly if leaving before power is restored. Some equipment can be dam-

aged if turned on abruptly once power comes back online. If no one is in the laboratory when the power is restored, equipment that does turn on will be running unattended.
- Discontinue operations requiring local ventilation, such as laboratory chemical hoods. The building ventilation system may not be on emergency power.
- Close laboratory chemical hood sashes.

3.G.2   Long-Term Power Loss

Damaged power distribution systems and other conditions may result in power loss that lasts hours or days. This has implications for security, safety, and experimental work that go well beyond those for a short-term power loss.

3.G.2.1   Security Issues

For laboratories with specialized security systems, such as card readers or electronic locks, know if the locks are locked or unlocked in the event of power failure. Develop a backup plan for laboratory security in the absence of such systems.

3.G.2.2   Environmental and Storage Conditions

The most common problem during a power outage is storage of materials that require specialized environmental conditions, such as refrigeration and humidity controls.

For example, sub-80 °C freezers, may hold their temperature for a few hours after a power loss but will eventually warm. This warming may lead to loss of samples or, for materials that become unstable when warmed, to more hazardous conditions, including fire, overpressurization, or release.

3.G.2.3   Discontinuation of Experiments

Experiments that rely on power may need to be discontinued and disassembled. Leaving the materials in place may not be prudent. Someone should be assigned responsibility for walking through the laboratory to identify problems and ensure that materials are safely stored.

3.G.3   Preplanning

There are many options for minimizing the effects of a power loss, including alternative energy sources and, when that is not practical, prioritizing experimental needs, consolidating, and using dry ice.

3.G.3.1 Generator Power

The laboratory building may be connected to a generator. If so, know what will continue to run during a power loss. In some buildings, for example, the generator may only run emergency lighting and security systems. In others, the ventilation system, all or in part, may be connected to the generator. Some buildings may have specially marked outlets that are connected to the generator.

One potentially negative aspect of a generator is that there is usually a slight delay, up to several seconds, from the time the power is lost to the time that the power load is taken up by the generator. Equipment that is sensitive to a minor power disruption may be affected and a generator may not be the right solution.

Know what will continue to operate during a power loss. Determine how long the laboratory can rely on the generator. If there is equipment that would benefit from connection to the generator, inquire about the possibility of such a connection being made.

3.G.3.2 Uninterruptible Power Supply (UPS)

When generator power is not available or if equipment is sensitive to the slight power delay, UPS systems may be the right choice for continued power. UPS systems are composed of large rechargeable batteries that immediately provide emergency power when the main supply is interrupted.

UPS systems come in a variety of types and sizes. The three basic types are offline, line interactive, and online. The differences among the three are related to the level and type of surge protection, with the offline providing the least amount of surge protection and the online providing the most sophisticated protection. Size varies based on power needs. When purchasing an UPS for equipment other than a computer, consult with the equipment manufacturer to help choose the right solution.

All UPS systems require some degree of maintenance. The battery needs to be replaced at an interval specified by the manufacturer. Batteries may be expensive and should be figured into the cost of the system.

3.G.3.3 Dry Ice

Dry ice may be helpful in maintaining temperatures in refrigerators or freezers. Because demand for dry ice increases significantly during a power loss, have a list of alternative vendors in case the regular vendors are unable to provide supplies.

To preserve resources, researchers should prioritize their experimental materials needing refrigeration and combine them as much as possible.

3.G.3.4 Other

As with any crisis, cooperation among laboratory groups and individuals results in the best outcome. Creative problem solving is something at which most researchers are skilled and it should never be overlooked in an emergency.

3.H INSTITUTIONAL OR BUILDING CLOSURE

Weather emergencies, fire, or other circumstances may require closure of an institution or building. Laboratories may be inaccessible or special permission may be required to enter or work in the laboratory. Whereas interruption in research or teaching may be a nuisance, other conditions may pose a hazard or a risk of significant loss of research.

Ensure that personnel expected to report to work even when there is a closure are aware of their responsibilities and have been designated as essential personnel. (See section 3.C.3 for more information.)

3.H.1 Short-Term Closure

For laboratory closures lasting a day or less, the main concerns include experiments running unattended and security. Whether the closure is planned or unexpected, it is important to consider how it will affect laboratory operations, and how critical operations can be maintained. See Box 3.1 for a checklist of things to consider while planning for a closure.

BOX 3.1
Continuity of Laboratory Operations Checklist

☐ List of high-priority operations
☐ List of personnel who can perform these operations
☐ Communication plan
☐ Data backup plan
☐ Leadership succession
☐ Key dependencies within the organization (e.g., essential goods and services that other departments or groups provide) and alternatives
☐ Key dependencies outside the organization, with alternative vendors
☐ List of essential equipment, purchase records, and information on how to replace it permanently or temporarily
☐ Restoration plan and priorities

If the closure is unexpected, experiments may be left running unattended. Depending on the experimental procedure, problems may occur with temperature regulation, integrity of containers, evaporation of solutions, concentration of solutions, and numerous other possibilities.

3.H.2 Long-Term Closure

A fire or other event that causes serious building damage, police activity, and communicable disease outbreaks are just a few examples of incidents that could result in a building or institution closing for several days, weeks, or months. It may be necessary to place the laboratory into a state of inactivity or hibernation during an emergency that causes serious staffing disruptions. Plans for making the transition from active to suspended laboratory operations should be a part of the organization's emergency response policy.

Consider the impact of laboratory closure on research, services provided to outside entities, and other groups. Communicate disruptions in services to those that rely on them.

3.H.3 Alternative Laboratory Facilities

If the laboratory will be inaccessible, it may be possible to share another laboratory at or outside the institution or to set up a temporary laboratory in another space. Preplanning for such an event reduces the amount of downtime.

Make a list of what is essential for an alternative facility:

- equipment and materials needed to perform priority tasks,
- space,
- environmental controls (e.g., temperature, humidity),
- security requirements, and
- ventilation requirements.

3.I EMERGENCY AFFECTING THE COMMUNITY

When an emergency affects only the institution, building, or laboratory, community resources, emergency responders, and external services are generally available for assistance or for business continuity. However, when an emergency affects the local community or a larger area, resuming normal operations may take longer.

A laboratory may be indirectly affected by a community emergency when goods and services are unavailable.

3.I.1 Disruption of Deliveries of Goods and Services

Many laboratories rely on just-in-time delivery of chemicals and supplies because stockpiling chemicals poses its own risks and should be avoided. Excessive storage of other supplies may result in an increased fire risk from combustible materials.

As part of the preplanning process, consider the following:

- Prepare a list of alternative vendors and service providers in the event that the primary vendor is unavailable. Add them to the vendor list for centralized purchasing or have a contract on hand if necessary.
- Ensure that primary vendors have up-to-date business continuity plans.
- Ensure that the institution or laboratory is a priority for your primary vendors and service providers.

3.I.2 Laboratory Staff Shortage

Staff may not be able to report to the laboratory. For continuity of laboratory operations, ensure that personnel are cross-trained to be able to fill in for a person who is absent. Have a succession plan that clarifies who is responsible when supervisors are not available.

3.J FIRE OR LOSS OF LABORATORY

A fire can be devastating. Even when fire does not damage the laboratory directly, it may result in disruption of services or limited access to the laboratory, or damage may be caused by smoke, water, or fire-extinguishing materials.

First, assess the vulnerabilities within the laboratory; then take action to prevent fire. Maintaining safe chemical storage, minimizing combustible materials, and controlling ignition sources are just a few examples of fire prevention. Next, ensure that there is an adequate level of detection and, where possible, extinguishing systems, and take additional steps to limit the impact of a fire. Finally, consider how the laboratory would manage after a fire and implement plans for facilitating continuity of operations. (See Vignette 3.1.)

3.J.1 Records for Replacement of Laboratory Equipment

Keep records for both existing equipment and replacement equipment. Having purchasing records readily available can make a difference in how long it takes for insurance claims to be processed. Because

VIGNETTE 3.1
Preplanning reduces the impact of a fire on continuity of operations

A DNA sequencing/synthesis laboratory that provided services to several other laboratories at the institution experienced a fire that caused a total loss of all equipment due to smoke damage. The source of the fire was a fault in the power supply of a computer. The laboratory did not have a sprinkler system, which would have reduced the magnitude of the damage.

However, the laboratory had planned for such an occurrence and the immediate availability of purchasing records facilitated the insurance claim. The laboratory manager had backup plans and had a temporary but fully functional laboratory operating in 3 days. It took more than a year to renovate the burned laboratory but the services were disrupted for less than 1 week.

older equipment may not be replaceable, knowing what alternatives are available and where to get them may speed up resumption of laboratory activities.

3.J.2 Alternative Laboratories to Continue Operations

It may be necessary to use existing laboratories or furnish a temporary laboratory in order to continue operations. (See section 3.H.3 for more information.)

3.J.2.1 *Preplanning and Prevention*

To help prevent a fire or limit the effect of a fire, ensure that the laboratory has the following in place:

- appropriate types and number of fire extinguishers and individuals trained to use them,
- sprinkler systems or other automatic extinguishing systems for sensitive areas or equipment,
- fire-safe storage of data and mission-critical samples, and
- good chemical storage practices.

3.K DRILLS AND EXERCISES

We all hope we never have to implement emergency plans. However, because the plans are rarely implemented, it is even more important to have drills and/or exercises that allow laboratory personnel to simulate their response.

Test alarm systems regularly, at least once a year. Prudent practice coordinates the testing of the system with a drill that exercises the appropriate response. Fire drills are a classic example. By sounding the alarm and expecting everyone to evacuate, one can uncover problems with the planning.

Drills and exercises may be full scale, where individuals are expected to carry out the responsibilities and procedures; tabletop exercises, where individuals discuss their response but do not physically respond; or a combination of both.

3.L OUTSIDE RESPONDERS AND RESOURCES

Some emergencies require response by police, fire, ambulance, or other outside responders. Prudent practice establishes good communication with these responders before they are expected to respond to an emergency. You can facilitate this by

- inviting responders to the facility for a tour of the areas of most concern;
- providing information about areas of higher risk for a fire, spill, or other emergency;
- providing maps and other tools to help them navigate the facility and familiarize themselves with the location of laboratory buildings or special facilities;
- informing emergency responders and local hospitals of the use of chemicals that present unusual hazards; and
- having chemical inventories accessible remotely through a password-protected system or file, which allows emergency responders to have an idea of what could be in the laboratory or building before they enter.

4 Evaluating Hazards and Assessing Risks in the Laboratory

4.A   INTRODUCTION   47

4.B   SOURCES OF INFORMATION   47
      4.B.1   Chemical Hygiene Plan (CHP)   47
      4.B.2   Material Safety Data Sheets (MSDSs)   47
      4.B.3   Globally Harmonized System for Hazard Communication   49
      4.B.4   Laboratory Chemical Safety Summaries   50
      4.B.5   Labels   51
      4.B.6   Additional Sources of Information   51
      4.B.7   Computer Services   52
              4.B.7.1   The National Library of Medicine Databases   53
              4.B.7.2   Chemical Abstracts Databases   53
              4.B.7.3   Informal Forums   53
      4.B.8   Training   53

4.C   TOXIC EFFECTS OF LABORATORY CHEMICALS   53
      4.C.1   Basic Principles   53
              4.C.1.1   Dose-Response Relationships   54
              4.C.1.2   Duration and Frequency of Exposure   56
              4.C.1.3   Routes of Exposure   57
      4.C.2   Assessing Risks of Exposure to Toxic Laboratory Chemicals   58
              4.C.2.1   Acute Toxicants   59
      4.C.3   Types of Toxins   60
              4.C.3.1   Irritants, Corrosive Substances, Allergens, and
                        Sensitizers   60
              4.C.3.2   Asphyxiants   62
              4.C.3.3   Neurotoxins   62
              4.C.3.4   Reproductive and Developmental Toxins   62
              4.C.3.5   Toxins Affecting Other Target Organs   63
              4.C.3.6   Carcinogens   63
              4.C.3.7   Control Banding   64

4.D   FLAMMABLE, REACTIVE, AND EXPLOSIVE HAZARDS   65
      4.D.1   Flammable Hazards   65
              4.D.1.1   Flammable Substances   65
              4.D.1.2   Flammability Characteristics   65
              4.D.1.3   Classes of Flammability   68
              4.D.1.4   Causes of Ignition   69
              4.D.1.5   Special Hazards   69
      4.D.2   Reactive Hazards   70
              4.D.2.1   Water Reactives   70
              4.D.2.2   Pyrophorics   70
              4.D.2.3   Incompatible Chemicals   70
      4.D.3   Explosive Hazards   70
              4.D.3.1   Explosives   70
              4.D.3.2   Azos, Peroxides, and Peroxidizables   72

4.D.3.3   Other Oxidizers   73
4.D.3.4   Powders and Dusts   73
4.D.3.5   Explosive Boiling   73
4.D.3.6   Other Considerations   73

4.E   PHYSICAL HAZARDS   74
4.E.1   Compressed Gases   74
4.E.2   Nonflammable Cryogens   74
4.E.3   High-Pressure Reactions   74
4.E.4   Vacuum Work   75
4.E.5   Ultraviolet, Visible, and Near-Infrared Radiation   75
4.E.6   Radio Frequency and Microwave Hazards   75
4.E.7   Electrical Hazards   76
4.E.8   Magnetic Fields   76
4.E.9   Sharp Edges   76
4.E.10   Slips, Trips, and Falls   77
4.E.11   Ergonomic Hazards in the Laboratory   77

4.F   NANOMATERIALS   77

4.G   BIOHAZARDS   79

4.H   HAZARDS FROM RADIOACTIVITY   79

4.A INTRODUCTION

A key element of planning an experiment is assessing the hazards and potential risks associated with the chemicals and laboratory operations to be used. This chapter provides a practical guide for the trained laboratory personnel engaged in these activities. Section 4.B introduces the sources of information for data on toxic, flammable, reactive, and explosive chemical substances. Section 4.C discusses the toxic effects of laboratory chemicals by first presenting the basic principles that form the foundation for evaluating hazards for toxic substances. The remainder of this section describes how trained laboratory personnel can use this understanding and the sources of information to assess the risks associated with potential hazards of chemical substances and then to select the appropriate level of laboratory practice as discussed in Chapter 4. Sections 4.D and 4.E present guidelines for evaluating hazards associated with the use of flammable, reactive, and explosive substances and physical hazards, respectively. Finally, nanomaterials, biohazards, and radioactivity hazards are discussed briefly in sections 4.F and 4.G, respectively.

The primary responsibility for proper hazard evaluations and risk assessments lies with the person performing the experiment. That being said, the responsibility is shared by the laboratory supervisor. The actual evaluations and assessments may be performed by trained laboratory personnel, but these should be checked and authorized by the supervisor. The supervisor is also responsible for ensuring that everyone involved in an experiment and those nearby understand the evaluations and assessments. For example, depending on the level of training and experience, the immediate laboratory supervisor may be involved in the experimental work itself. In addition, some organizations have environmental health and safety (EHS) offices, with industrial hygiene specialists to advise trained laboratory personnel and their supervisors in risk assessment. When required by federal regulation, Chemical Hygiene Officers (CHOs) play similar roles in many organizations. As part of a culture of safety, all of these groups work cooperatively to create a safe environment and to ensure that hazards are appropriately identified and assessed prior to beginning work.

4.B SOURCES OF INFORMATION

4.B.1 Chemical Hygiene Plan (CHP)

Beginning in 1991, every laboratory in which hazardous chemicals are used has been required by federal regulations (Occupational Safety and Health Administration [OSHA] Occupational Exposure to Hazard-

ous Chemicals in Laboratories, 29 CFR § 1910.1450) to have a written CHP, which includes provisions capable of protecting personnel from the "health hazards presented by hazardous chemicals used in that particular workplace." All laboratory personnel should be familiar with and have ready access to their institution's CHP. In some laboratories, CHPs include standard operating procedures for work with specific chemical substances, and the CHP may be sufficient as the primary source of information used for risk assessment and experiment planning. However, most CHPs provide only general procedures for handling chemicals, and prudent experiment planning requires that laboratory personnel consult additional sources for information on the properties of the substances that will be encountered in the proposed experiment. Many laboratories require documentation of specific hazards and controls for a proposed experiment.

4.B.2 Material Safety Data Sheets (MSDSs)

Federal regulations (OSHA Hazard Communication Standard 29 CFR § 1910.1200) require that manufacturers and distributors of hazardous chemicals provide users with material safety data sheets (MSDSs),[1] which are designed to provide the information needed to protect users from any hazards that may be associated with the product. MSDSs have become the primary vehicle through which the potential hazards of materials obtained from commercial sources are communicated to trained laboratory personnel. Institutions are required by law (OSHA Hazard Communication Standard) to retain and make readily available the MSDSs provided by chemical suppliers. The MSDSs themselves may be electronic or on paper, as long as employees have unrestricted access to the documents. Be aware that some laboratories have been asked by local emergency personnel to print paper copies in the event of an emergency.

As the first step in risk assessment, trained laboratory personnel should examine any plan for a proposed experiment and identify the chemicals with toxicological properties they are not familiar with from previous experience. The MSDS for each unfamiliar chemical should be examined. Procedures for accessing MSDS files vary from institution to institution. In some cases, MSDS files are present in each laboratory, but often complete files of MSDSs are maintained only in a central location, such as the institution's EHS office. Many laboratories are able to access MSDSs electroni-

---

[1]In the Globally Harmonized System for Hazard Communication, the term "material safety data sheet" has been shortened to "safety data sheet (SDS)." This book will continue to use the term MSDS as it is more recognizable at the time of writing than SDS.

cally, either from CD-ROM disks, via the internet, or from other computer networks. Laboratory personnel can always contact the chemical supplier directly and request that an MSDS be sent by mail.

MSDSs are technical documents, several pages long, typically beginning with a compilation of data on the physical, chemical, and toxicological properties of the substance and providing concise suggestions for handling, storage, and disposal. Finally, emergency and first-aid procedures are usually outlined. At present, there is no required format for an MSDS; however, OSHA recommends the general 16-part format created by the American National Standards Institute (ANSI Z400.1). The information typically found in an MSDS follows:

1. **Supplier (with address and phone number) and date MSDS was prepared or revised.** Toxicity data and exposure limits sometimes undergo revision, and for this reason MSDSs should be reviewed periodically to check that they contain up-to-date information. Phone numbers are provided so that, if necessary, users can contact the supplier to obtain additional information on hazards and emergency procedures.
2. **Chemical.** For products that are mixtures, this section may include the identity of most but not every ingredient. Hazardous chemicals must be identified. Common synonyms are usually listed.
3. **Physical and chemical properties.** Data such as melting point, boiling point, and molecular weight are included here.
4. **Physical hazards.** This section provides data related to flammability, reactivity, and explosion hazards.
5. **Toxicity data.** OSHA, the National Institute for Occupational Safety and Health (NIOSH), and the American Conference of Governmental Industrial Hygienists (ACGIH) exposure limits (as discussed in section 4.C.2.1) are listed. Many MSDSs provide lengthy and comprehensive compilations of toxicity data and even references to applicable federal standards and regulations.
6. **Health hazards.** Acute and chronic health hazards are listed, together with the signs and symptoms of exposure. The primary routes of entry of the substance into the body are also described. In addition, potential carcinogens are explicitly identified. In some MSDSs, this list of toxic effects is quite lengthy and includes every possible harmful effect the substance has under the conditions of every conceivable use.
7. **Storage and handling procedures.** This section usually consists of a list of precautions to be taken in handling and storing the material. Particular attention is devoted to listing appropriate control measures, such as the use of engineering controls and personal protective equipment necessary to prevent harmful exposures. Because an MSDS is written to address the largest scale at which the material could conceivably be used, the procedures recommended may involve more stringent precautions than are necessary in the context of laboratory use.
8. **Emergency and first-aid procedures.** This section usually includes recommendations for firefighting procedures, first-aid treatment, and steps to be taken if the material is released or spilled. Again, the measures outlined here are chosen to encompass worst-case scenarios, including accidents on a larger scale than are likely to occur in a laboratory.
9. **Disposal considerations.** Some MSDSs provide guidelines for the proper disposal of waste material. Others direct the users to dispose of the material in accordance with federal, state, and local guidelines.
10. **Transportation information.** This chapter only evaluates the hazards and assesses the risks associated with chemicals *in the context of laboratory use.* MSDSs, in contrast, must address the hazards associated with chemicals in all possible situations, including industrial manufacturing operations and large-scale transportation accidents. For this reason, some of the information in an MSDS may not be relevant to the handling and use of that chemical in a laboratory. For example, most MSDSs stipulate that self-contained breathing apparatus and heavy rubber gloves and boots be worn in cleaning up spills, even of relatively nontoxic materials such as acetone. Such precautions, however, might be unnecessary in laboratory-scale spills of acetone and other substances of low toxicity.

Originally, the principal audience for MSDSs was constituted of health and safety professionals (who are responsible for formulating safe workplace practices), medical personnel (who direct medical surveillance programs and treat exposed workers), and emergency responders (e.g., fire department personnel). With the promulgation of federal regulations such as the OSHA Hazard Communication Standard (29 CFR § 1910.1200) and the OSHA Laboratory Standard (29 CFR § 1910.1450), the audience for MSDSs has expanded to include trained laboratory personnel in industrial and academic laboratories. However, not all MSDSs are written to meet the requirements of this new audience effectively.

In summary, among the currently available resources, MSDSs remain the best single source of information for the purpose of evaluating the hazards and assessing the risks of chemical substances. However, laboratory personnel should recognize the limitations of MSDSs as applied to laboratory-scale operations. If MSDSs are not adequate, specific laboratory operating procedures should be available for the specific laboratory manipulations to be employed:

1. The quality of MSDSs produced by different chemical suppliers varies widely. The utility of some MSDSs is compromised by vague and unqualified generalizations and internal inconsistencies.
2. Unique morphology of solid hazardous chemicals may not be addressed in MSDSs; for example, an MSDS for nano-size titanium dioxide may not present the unique toxicity considerations for these ultrafine particulates.
3. MSDSs must describe control measures and precautions for work on a variety of scales, ranging from microscale laboratory experiments to large manufacturing operations. Some procedures outlined in an MSDS may therefore be unnecessary or inappropriate for laboratory-scale work. An unfortunate consequence of this problem is that it tends to breed a lack of confidence in the relevance of the MSDS to laboratory-scale work.
4. Many MSDSs comprehensively list all conceivable health hazards associated with a substance without differentiating which are most significant and which are most likely to actually be encountered. As a result, trained laboratory personnel may not distinguish highly hazardous materials from moderately hazardous and relatively harmless ones.

4.B.3 Globally Harmonized System (GHS) for Hazard Communication

The GHS of Classification and Labeling of Chemicals is an internationally recognized system for hazard classification and communication. (Available at http://www.unece.org.) It was developed with support from the International Labour Organization (ILO), the Organisation for Economic Co-operation and Development, and the United Nations Sub-Committee of Experts on the Transport of Dangerous Goods with the goal of standardizing hazard communication to improve the safety of international trade and commerce. Within the United States, the responsibility for implementing the GHS falls to four agencies: OSHA,

the Department of Transportation, the EPA, and the Consumer Product Safety Commission. At the time this book was written, the agencies had not yet provided final guidance on use of GHS. The revised Hazard Communication Standard (29 CFR § 1910.1200) is expected to be issued by OSHA in the near future.

GHS classifies substances by the physical, health, and environmental hazards that they pose, and provides signal words (e.g., Danger), hazard statements (e.g., may cause fire or explosion), and standard pictogram-based labels to indicate the hazards and their severity. When transporting hazardous chemicals, use the pictograms specified in the UN Recommendations on the Transport of Dangerous Goods, Model Regulations. For other purposes, the pictograms in Figure 4.1 should be used. Container labels should have a product identifier with hazardous ingredient disclosure, supplier information, a hazard pictogram, a signal word, a hazard statement, first-aid information, and supplemental information. Three of these elements—the pictograms, signal word, and hazard statements—are standardized under GHS. The signal words, either "Danger" or "Warning," reflect the severity of hazard posed. Hazard statements are standard phrases that describe the nature of the hazard posed by the material (e.g., heating may cause explosion).

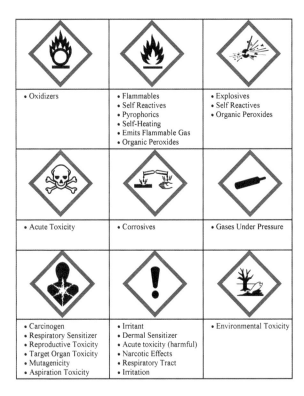

FIGURE 4.1 GHS placards for labeling containers of hazardous chemicals.

GHS recognizes 16 types of physical hazards, 10 types of health hazard, and an environmental hazard.

Physical hazards include

- explosives;
- flammable gases;
- flammable aerosols;
- oxidizing gases;
- gases under pressure;
- flammable liquids;
- flammable solids;
- self-reactive substances;
- pyrophoric liquids;
- pyrophoric solids;
- self-heating substances;
- substances which, in contact with water, emit flammable gases;
- oxidizing liquids;
- oxidizing solids;
- organic peroxides; and
- corrosive to metals.

Health hazards include

- acute toxicity,
- skin corrosion or irritation,
- serious eye damage or eye irritation,
- respiratory or skin sensitization,
- germ cell mutagenicity,
- carcinogenicity,
- reproductive toxicology,
- target organ systemic toxicity—single exposure,
- target organ systemic toxicity—repeated exposure, and
- aspiration hazard.

Environmental hazard includes

- Hazardous to the aquatic environment:
  - acute aquatic toxicity or
  - chronic aquatic toxicity with
    - bioaccumulation potential
    - rapid degradability.

In addition to the labeling requirements, GHS requires a standard format for Safety Data Sheets (SDS) that accompany hazardous chemicals. Note the change in terminology from MSDS. SDSs must contain a minimum of 16 elements:

1. identification,
2. hazard(s) identification,
3. composition/information on ingredients,
4. first-aid measures,
5. firefighting measures,
6. accidental release measures,
7. handling and storage,
8. exposure controls/personal protection,
9. physical and chemical properties,
10. stability and reactivity,
11. toxicological information,
12. ecological information,
13. disposal considerations,
14. transport information,
15. regulatory information, and
16. other information.

As with current MSDSs, these sheets are intended to inform employers and personnel of the hazards associated with the chemicals they are handling, and to act as a resource for management of the chemicals. Trained personnel should evaluate the information and use it to develop safety and emergency response policies, protocols, and procedures that are tailored to the workplace or laboratory.

4.B.4 Laboratory Chemical Safety Summaries (LCSSs)

As discussed above, although MSDSs are invaluable resources, they suffer some limitations as applied to risk assessment in the specific context of the laboratory. Committee-generated LCSSs, which are tailored to trained laboratory personnel, are on the CD accompanying this book. As indicated in their name, LCSSs provide information on chemicals in the context of laboratory use. These documents are summaries and are not intended to be comprehensive or to fulfill the needs of all conceivable users of a chemical. In conjunction with the guidelines described in this chapter, the LCSS gives essential information required to assess the risks associated with the use of a particular chemical in the laboratory.

The format, organization, and contents of LCSSs are described in detail in the introduction on the CD. Included in an LCSS are the key physical, chemical, and toxicological data necessary to evaluate the relative degree of hazard posed by a substance. LCSSs also contain a concise critical discussion, presented in a style readily understandable to trained laboratory personnel, of the toxicity, flammability, reactivity, and explosivity of the chemical; recommendations for the handling, storage, and disposal of the title substance; and first-aid and emergency response procedures.

The CD contains LCSSs for 91 chemical substances. Several criteria were used in selecting these chemicals, the most important consideration being whether the substance is commonly used in laboratories. Preference was also given to materials that pose relatively serious hazards. Finally, an effort was made to select chemicals representing a variety of classes of substances, so as to provide models for the future development of additional LCSSs. A blank copy of the form is provided for development of laboratory-specific LCSSs.

4.B.5 Labels

Commercial suppliers are required by law (OSHA Hazard Communication Standard) to provide their chemicals in containers with precautionary labels. Labels usually present concise and nontechnical summaries of the principal hazards associated with their contents. Note that precautionary labels do not replace MSDSs and LCSSs as the primary sources of information for risk assessment in the laboratory. However, labels serve as valuable reminders of the key hazards associated with the substance. As with the MSDS, the quality of information presented on a label can be inconsistent. Additionally, labeling is not always required for chemicals transferred between laboratories within the same building.

4.B.6 Additional Sources of Information

The resources described above provide the foundation for risk assessment of chemicals in the laboratory. This section highlights the sources that should be consulted for additional information on specific harmful effects of chemical substances. Although MSDSs and LCSSs include information on toxic effects, in some situations laboratory personnel should seek additional more detailed information. This step is particularly important when laboratory personnel are planning to use chemicals that have a high degree of acute or chronic toxicity or when it is anticipated that work will be conducted with a particular toxic substance frequently or over an extended period of time. Institutional CHPs include the requirement for CHOs, who are capable of providing information on hazards and controls. CHOs can assist laboratory personnel in obtaining and interpreting hazard information and in ensuring the availability of training and information for all laboratory personnel.

Sections 4.B.2 and 4.B.4 of this chapter provide explicit guidelines on how laboratory personnel use the information in an MSDS or LCSS, respectively, to recognize when it is necessary to seek such additional information.

The following annotated list provides references on the hazardous properties of chemicals and which are useful for assessing risks in the laboratory.

1. *International Chemical Safety Cards* from the International Programme on Chemical Safety (IPCS, 2009). The IPCS is a joint activity of the ILO, the United Nations Environment Programme, and the World Health Organization. The cards contain hazard and exposure information from recognized sources and undergo international peer review. They are designed to be understandable to employers and employees in factories, agriculture, industrial shops, and other areas, and can be considered complements to MSDSs. They are available in 18 languages and can be found online through the NIOSH Web site, www.cdc.gov/niosh, or through the ILO Web site, www.ilo.org.

2. *NIOSH Pocket Guide to Chemical Hazards* (HHS/CDC/NIOSH, 2007). This volume is updated regularly and is found on the NIOSH Web site (http://www.cdc.gov/niosh). These charts are quick guides to chemical properties, reactivities, exposure routes and limits, and first-aid measures.

3. *A Comprehensive Guide to the Hazardous Properties of Chemical Substances*, 3rd edition (Patnaik, 2007). This particularly valuable guide is written at a level appropriate for typical laboratory personnel. It covers more than 1,500 substances; sections in each entry include uses and exposure risk, physical properties, health hazards, exposure limits, fire and explosion hazards, and disposal or destruction. Entries are organized into chapters according to functional group classes, and each chapter begins with a general discussion of the properties and hazards of the class.

4. *2009 TLVs and BEIs: Based on the Documentation of the Threshold Limit Values for Chemical Substances and Physical Agents and Biological Exposure Indices* (ACGIH, 2009). A handy booklet listing ACGIH threshold limit values (TLVs) and short-term exposure limits (STELs). These values are under continuous review, and this booklet is updated annually. The multivolume publication *Documentation of the Threshold Limit Values and Biological Exposure Indices* (ACGIH, 2008b) reviews the data (with reference to literature sources) that were used to establish the TLVs. (For more information about TLVs, see section 4.C.2.1 of this chapter.)

5. *Fire Protection for Laboratories Using Chemicals* (NFPA, 2004). This is the national fire safety code pertaining to laboratory use of chemicals. It describes the basic requirements for fire protection of life and property in the laboratory. For example, the document outlines technical

requirements for equipment such as fire suppression systems and ventilation systems for flammables and defines the maximum allowable quantities for flammable materials within the laboratory.

6. *Fire Protection Guide to Hazardous Materials*, 13th edition (NFPA, 2001). This resource contains hazard data on hundreds of chemicals and guidance on handling and storage of, and emergency procedures for, those chemicals.

7. *Bretherick's Handbook of Reactive Chemical Hazards*, 7th edition (Urben, 2007). This handbook is a comprehensive compilation of examples of violent reactions, fires, and explosions due to unstable chemicals, as well as reports on known incompatibility between reactive chemicals.

8. *Hazardous Chemicals Handbook*, 2nd edition (Carson and Mumford, 2002). This book is geared toward an industrial audience. It provides basic information about chemical hazards and synthesizes technical guidance from a number of authorities in chemical safety. The chapters are organized by hazard (e.g., "Toxic Chemicals," "Reactive Chemicals," and "Cryogens").

9. *Sax's Dangerous Properties of Industrial Materials*, 11th edition, three volumes (Lewis, 2004). Also available on CD, this compilation of data for more than 26,000 chemical substances contains much of the information found in a typical MSDS, including physical and chemical properties; data on toxicity, flammability, reactivity, and explosivity; and a concise safety profile describing symptoms of exposure. It also contains immediately dangerous to life or health (IDLH) levels for approximately 1,000 chemicals, and for laboratory personnel it is a useful reference for checking the accuracy of an MSDS and a valuable resource in preparing a laboratory's own LCSSs.

10. *Patty's Industrial Toxicology*, 5th edition (Bingham et al., 2001). Also available on CD, this authoritative reference on the toxicology of different classes of organic and inorganic compounds focuses on health effects; hazards due to flammability, reactivity, and explosivity are not covered.

11. *Proctor and Hughes' Chemical Hazards of the Workplace*, 5th edition (Hathaway and Proctor, 2004). This resource provides an excellent summary of the toxicology of more than 600 chemicals. Most entries are one to two pages and include signs and symptoms of exposure with reference to specific clinical reports.

12. *Sittig's Handbook of Toxic and Hazardous Chemicals and Carcinogens*, 5th edition, two volumes (Pohanish, 2008). This very good reference, which is written with the industrial hygienists and first responder in mind, covers 2,100 substances.

13. *Clinical Toxicology*, 1st edition (Ford et al., 2001). This book is designed for clinicians and other health care providers. It describes the symptoms and treatment of poisoning from various sources.

14. *Casarett and Doull's Toxicology: The Basic Science of Poisons*, 7th edition (Klaassen, 2007). This complete and readable overview of toxicology is a good textbook but is not arranged as a ready reference for handling laboratory emergencies.

15. *Catalog of Teratogenic Agents*, 11th edition (Shepard and Lemire, 2004). This catalog is one of the best references available on the subject of reproductive and developmental toxins.

16. *Wiley Guide to Chemical Incompatibilities*, 2nd edition (Pohanish and Greene, 2003). Simple-to-use reference listing the incompatibilities of more than 11,000 chemicals. Includes information about chemical incompatibility, conditions that favor undesirable reactions, and corrosivity data.

17. *Occupational Health Guidelines for Chemical Hazards* (HHS/CDC/NIOSH, 1981) and a supplement (HHS/CDC/NIOSH, 1995). The guidelines currently cover more than 400 substances and are based on the information assembled under the Standards Completion Program, which served as the basis for the promulgation of federal occupational health regulations ("substance-specific standards"). Typically five pages long and written clearly at a level readily understood by trained laboratory personnel, each set of guidelines includes information on physical, chemical, and toxicological properties; signs and symptoms of exposure; and considerable detail on control measures, medical surveillance practices, and emergency first-aid procedures. However, some guidelines date back to 1978 and may not be current, particularly with regard to chronic toxic effects. These guidelines are available on the NIOSH Web site (http://www.cdc.gov/niosh/).

A number of Web-based resources also exist. Some of these are NIOSH Databases and Information Resources (www.cdc.gov/niosh) and TOXNET through the National Library of Medicine (NLM; www.nlm.nih.gov).

4.B.7 Computer Services

In addition to computerized MSDSs, a number of computer databases are available that supply data for creating or supplementing MSDSs, for example, the NLM and Chemical Abstracts (CA) databases. These and other such databases are accessible through various online computer data services; also, most of this information is available as CD and computer updates. Many of these services can be accessed for up-to-date toxicity information.

Governmental sources of EHS information include

- NIOSH (www.cdc.gov/niosh),
- OSHA (www.osha.gov),
- Environmental Protection Agency (EPA; www.epa.gov).

4.B.7.1 The National Library of Medicine Databases

The databases supplied by NLM are easy to use and free to access via the Web. TOXNET is an online collection of toxicological and environmental health databases. TOXLINE, for example, is an online database that accesses journals and other resources for current toxicological information on drugs and chemicals. It covers data published from 1900 to the present. Databases accessible through TOXNET include the Hazardous Substance Data Base (HSDB) Carcinogenic Potency Database (CPDB), the Developmental and Reproductive Toxicology Database (DART), the Genetic Toxicology Data Bank (GENE-TOX), the Integrated Risk Information System (IRIS), the Chemical Carcinogenesis Research Information System (CCRIS), and the International Toxicity Estimates for Risk (ITER). Other databases supplied by NLM that provide access to toxicological information are PubMed, which includes access to MEDLINE, PubChem, and ChemIDPlus. Free text searching is available on most of the databases.

4.B.7.2 Chemical Abstracts Databases

Another source of toxicity data is Chemical Abstracts Service (CAS). In addition to the NLM, several services provide CAS, including DIALOG, ORBIT, STN, and SciFinder. Searching procedures for CAS depend on the various services supplying the database. Searching costs are considerably higher than for NLM databases because CAS royalties must be paid. Telephone numbers for the above suppliers are as follows:

DIALOG, 800-334-2564;
Questel, 800-456-7248;
STN, 800-734-4227;
SciFinder, 800-753-4227.

Additional information can be found on the CAS Web site, www.cas.org.

Specialized databases also exist. One example is the ECOTOX database from EPA (www.epa.gov/ecotox).

This database provides information on toxicity of chemicals to aquatic life, terrestrial plants, and wildlife.

Searching any database listed above is best done using the CAS registry number for the particular chemical.

4.B.7.3 Informal Forums

The "Letters to the Editor" column of *Chemical & Engineering News* (C&EN), published weekly by the American Chemical Society (ACS), was for many years an informal but widely accepted forum for reporting anecdotal information on chemical reactivity hazards and other safety-related information. Although less frequently updated, the ACS maintains an archive of all safety-related letters submitted to C&EN on the Web site of the Division of Chemical Health and Safety (CHAS) of ACS. CHAS also publishes the *Journal of Chemical Health and Safety*. Additional resources include the annual safety editorial called "Safety Notables: Information from the Literature" in the *Organic Process Research and Development* and community Listservs relating to laboratory safety.

4.B.8 Training

One important source of information for laboratory personnel is training sessions, and the critical place it holds in creating a safe environment should not be underestimated. Facts are only as useful as one's ability to interpret and apply them to a given problem, and training provides context for their use. Hands-on, scenario-based training is ideal because it provides the participants with the chance to practice activities and behaviors in a safe way. Such training is especially useful for learning emergency response procedures. Another effective tool, particularly when trying to build awareness of a given safety concern, is case studies. Prior to beginning any laboratory activity, it is important to ensure that personnel have enough training to safely perform required tasks. If new equipment, materials, or techniques are to be used, a risk assessment should be performed, and any knowledge gaps should be filled before beginning work. (More information about training programs can be found in Chapter 2, section 2.G.)

4.C TOXIC EFFECTS OF LABORATORY CHEMICALS

4.C.1 Basic Principles

The chemicals encountered in the laboratory have a broad spectrum of physical, chemical, and toxicological properties and physiological effects. The risks associated with chemicals must be well understood prior to their use in an experiment. The risk of toxic effects is related to both the extent of exposure and the inherent toxicity of a chemical. As discussed in detail below, extent of exposure is determined by the dose, the duration and frequency of exposure, and the route of exposure. Exposure to even large doses of chemicals with little inherent toxicity, such as phosphate buffer, presents low risk. In contrast, even small quantities of chemicals with high inherent toxicity or corrosivity may cause significant adverse effects. The duration and frequency of exposure are also critical factors in determining whether a chemical will produce harmful effects. A single exposure to some chemicals is sufficient to produce an adverse health effect; for other chemicals repeated exposure is required to produce toxic effects. For most substances, the route of exposure (through the skin, the eyes, the gastrointestinal tract, or the respiratory tract) is also an important consideration in risk assessment. For chemicals that are systemic toxicants, the internal dose to the target organ is a critical factor. Exposure to acute toxicants can be guided by well-defined toxicity parameters based on animal studies and often human exposure from accidental poisoning. The analogous quantitative data needed to make decisions about the neurotoxicity and immunogenicity of various chemicals is often unavailable.

When considering possible toxicity hazards while planning an experiment, recognizing that *the combination of the toxic effects of two substances may be significantly greater than the toxic effect of either substance alone* is important. Because most chemical reactions produce mixtures of substances with combined toxicities that have never been evaluated, it is prudent to assume that mixtures of different substances (i.e., chemical reaction mixtures) will be more toxic than their most toxic ingredient. Furthermore, chemical reactions involving two or more substances may form reaction products that are significantly more toxic than the starting reactants. This possibility of generating toxic reaction products may not be anticipated by trained laboratory personnel in cases where the reactants are mixed unintentionally. For example, inadvertent mixing of formaldehyde (a common tissue fixative) and hydrogen chloride results in the generation of bis(chloromethyl)ether, a potent human carcinogen.

All laboratory personnel must understand certain basic principles of toxicology and recognize the major classes of toxic and corrosive chemicals. The next sections of this chapter summarize the key concepts involved in assessing the risks associated with the use of toxic chemicals in the laboratory. (Also see Chapter 6, section 6.D.) Box 4.1 provides a quick guide for performing a toxicity-based risk assessment for laboratory chemicals.

4.C.1.1 Dose-Response Relationships

Toxicology is the study of the adverse effects of chemicals on living systems. The basic tenets of toxicology are that no substance is entirely safe and that all chemicals result in some toxic effects if a high enough amount (dose) of the substance comes in contact with a living system. As mentioned in Chapter 2, Paracelsus noted that the dose makes the poison and is perhaps the most important concept for all trained laboratory personnel to know. For example, water, a vital substance for life, results in death if a sufficiently large amount (i.e., gallons) is ingested at one time. On the other hand, sodium cyanide, a highly lethal chemical, produces no permanent (acute) effects if a living system is exposed to a sufficiently low dose. The single most important factor that determines whether a substance is harmful (or, conversely, safe) to an individual is the relationship between the amount (and concentration) of the chemical reaching the target organ, and the toxic effect it produces. For all chemicals, there is a range of concentrations that result in a graded effect between the extremes of no effect and death. In toxicology, this range is referred to as the dose-response relationship for the chemical. The dose is the amount of the chemical and the response is the effect of the chemical. This relationship is unique for each chemical, although for similar types of chemicals, the dose-response relationships are often similar. (See Figure 4.2.) Among the thousands of laboratory chemicals, a wide spectrum of doses exists that are required to produce toxic effects and even death. For most chemicals, a threshold dose has been established (by rule or by consensus) below which a chemical is not considered to be harmful to most individuals.

In these curves, dosage is plotted against the percent of the population affected by the dosage. Curve A represents a compound that has an effect on some percent of the population even at small doses. Curve B represents a compound that has an effect only above a dosage threshold.

Some chemicals (e.g., dioxin) produce death in laboratory animals exposed to microgram doses and therefore are extremely toxic. Other substances, however, have no harmful effects following doses in excess of several grams. One way to evaluate the acute toxicity (i.e., the toxicity occurring after a single exposure) of laboratory chemicals involves their lethal dose 50 (LD_{50}) or lethal concentration 50 (LC_{50}) value. The LD_{50} is defined as the amount of a chemical that when ingested, injected, or applied to the skin of a test animal under controlled laboratory conditions kills one-half (50%) of the animals. The LD_{50} is usually expressed in milligrams or grams per kilogram of body weight. For volatile chemicals (i.e., chemicals with sufficient

---

BOX 4.1
Quick Guide for Toxicity Risk Assessment of Chemicals

The following outline provides a summary of the steps discussed in this chapter that trained laboratory personnel should use to assess the risks of handling toxic chemicals. Note that if a laboratory chemical safety summary (LCSS) is not already available, this enables a worker to prepare his or her own LCSS.

1. **Identify chemicals to be used and circumstances of use.** Identify the chemicals involved in the proposed experiment and determine the amounts that will be used. Is the experiment to be done once, or will the chemicals be handled repeatedly? Will the experiment be conducted in an open laboratory, in an enclosed apparatus, or in a chemical fume hood? Is it possible that new or unknown substances will be generated in the experiment? Are any of the trained laboratory personnel involved in the experiment pregnant or likely to become pregnant? Do they have any known sensitivities to specific chemicals?

2. **Consult sources of information.** Consult an up-to-date LCSS for each chemical involved in the planned experiment or examine an up-to-date material safety data sheet (MSDS) if an LCSS is not available. In cases where substances with significant or unusual potential hazards are involved, consult more detailed references such as Patnaik (2007), Bingham et al. (2001), and other sources discussed in section 4.B. Depending on the laboratory personnel's level of experience and the degree of potential hazard associated with the proposed experiment, obtain the assistance of supervisors and safety professionals before proceeding with risk assessment.

3. **Evaluate type of toxicity.** Use the above sources of information to determine the type of toxicity associated with each chemical involved in the proposed experiment. Are any of the chemicals to be used acutely toxic or corrosive? Are any of the chemicals to be used irritants or sensitizers? Will any select carcinogens or possibly carcinogenic substances be encountered? Consult the listings of the resources described in section C.4.6 of this chapter to identify chemical similarities to known carcinogens. Are any chemicals involved in the proposed experiment suspected to be reproductive or developmental toxins or neurotoxins?

4. **Consider possible routes of exposure.** Determine the potential routes of exposure for each chemical. Are the chemicals gases, or are they volatile enough to present a significant risk of exposure through inhalation? If liquid, can the substances be absorbed through the skin? Is it possible that dusts or aerosols will be formed in the experiment? Does the experiment involve a significant risk of inadvertent ingestion or injection of chemicals?

5. **Evaluate quantitative information on toxicity.** Consult the information sources to determine the LD_{50}, discussed in section 4.C.1.1 of this chapter, for each chemical via the relevant routes of exposure. Determine the acute toxicity hazard level for each substance, classifying each chemical as highly toxic, moderately toxic, slightly toxic, and so forth. For substances that pose inhalation hazards, take note of the threshold limit value–time weighted average (TLV-TWA), short-term exposure limit, and permissible exposure limit values. (See section 4.C.2.1.)

6. **Select appropriate procedures to minimize exposure.** Use the basic prudent practices for handling chemicals, which are discussed in Chapter 6, section 6.C for all work with chemicals in the laboratory. In addition, determine whether any of the chemicals to be handled in the planned experiment meet the definition of a particularly hazardous substance due to high acute toxicity, carcinogenicity, and/or reproductive toxicity. If so, consider the total amount of the substance that will be used, the expected frequency of use, the chemical's routes of exposure, and the circumstances of its use in the proposed experiment. As discussed in this chapter, use this information to determine whether it is appropriate to apply the additional procedures for work with highly toxic substances and whether additional consultation with safety professionals is warranted (see Chapter 6, section 6.D).

7. **Prepare for contingencies.** Note the signs and symptoms of exposure to the chemicals to be used in the proposed experiment. Note appropriate measures to be taken in the event of exposure or accidental release of any of the chemicals, including first aid or containment actions.

Note: See Box 4.2 for a quick guide for assessing the physical, flammable, explosive, and reactive hazards in the laboratory and Box 4.3 for a quick guide for assessing biological hazards in the laboratory.

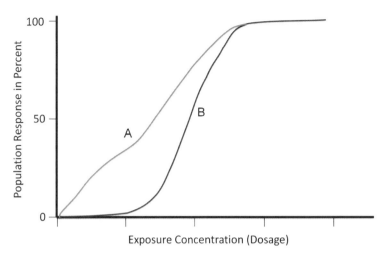

FIGURE 4.2 A simple representation of possible dose-response curves.

vapor pressure that inhalation is an important route of chemical entry into the body), the LC_{50} is often reported instead of the LD_{50}. The LC_{50} is the concentration of the chemical in air that will kill 50% of the test animals exposed to it. The LC_{50} is given in parts per million, milligrams per liter, or milligrams per cubic meter. Also reported are LC_{LO} and LD_{LO} values, which are defined as the lowest concentration or dose that causes the death of test animals. In general, the larger the LD_{50} or LC_{50}, the more chemical it takes to kill the test animals and, therefore, the lower the toxicity of the chemical. Although lethal dose values may vary among animal species and between animals and humans, chemicals that are highly toxic to animals are generally highly toxic to humans.

4.C.1.2 Duration and Frequency of Exposure

Toxic effects of chemicals occur after single (acute), intermittent (repeated), or long-term repeated (chronic) exposure. An acutely toxic substance causes damage as the result of a single short-duration exposure. Hydrogen cyanide, hydrogen sulfide, and nitrogen dioxide are examples of acute toxins. In contrast, a chronically toxic substance causes damage after repeated or long-duration exposure or causes damage that becomes evident only after a long latency period. Chronic toxins include all carcinogens, reproductive toxins, and certain heavy metals and their compounds. Many chronic toxins are extremely dangerous because of their long latency periods: the cumulative effect of low exposures to such substances may not become apparent for many years. Many chemicals may be hazardous both acutely and chronically depending on exposure level and duration.

In a general sense, the longer the duration of ex-posure, that is, the longer the body (or tissues in the body) is in contact with a chemical, the greater the opportunity for toxic effects to occur. Frequency of exposure also has an important influence on the nature and extent of toxicity. The total amount of a chemical required to produce a toxic effect is generally less for a single exposure than for intermittent or repeated exposures because many chemicals are eliminated from the body over time, because injuries are often repaired, and because tissues may adapt in response to repeated low-dose exposures. Some toxic effects occur only after long-term exposure because sufficient amounts of chemical cannot be attained in the tissue by a single exposure. Sometimes a chemical has to be present in a tissue for a considerable time to produce injury. For example, the neurotoxic and carcinogenic effects from exposure to heavy metals usually require long-term, repeated exposure.

The time between exposure to a chemical and onset of toxic effects varies depending on the chemical and the exposure. For example, the toxic effects of carbon monoxide, sodium cyanide, and carbon disulfide are evident within minutes. The chemical reaches the target organ rapidly and the organ responds rapidly. For many chemicals, the toxic effect is most severe between one and a few days after exposure. However, some chemicals produce delayed toxicity; in fact, the neurotoxicity produced by some chemicals is not observed until a few weeks after exposure. Delayed toxic effects are produced by chemical carcinogens and some organ toxins that produce progressive diseases such as pulmonary fibrosis and emphysema: in humans, it usually takes 10 to 30 years between exposure to a known human carcinogen and the detection of a tumor, and pulmonary fibrosis may take 10 or more years to result in symptoms.

4.C.1.3   Routes of Exposure

Exposure to chemicals in the laboratory occurs by several routes: (1) inhalation, (2) contact with skin or eyes, (3) ingestion, and (4) injection. Important features of these different pathways are detailed below.

4.C.1.3.1   Inhalation

Toxic materials that enter the body via inhalation include gases, the vapors of volatile liquids, mists and sprays of both volatile and nonvolatile liquid substances, and solid chemicals in the form of particles, fibers, and dusts. Inhalation of toxic gases and vapors produces poisoning by absorption through the mucous membranes of the mouth, throat, and lungs and also damages these tissues seriously by local action. Inhaled gases and vapors pass into the capillaries of the lungs and are carried into the circulatory system, where absorption is extremely rapid. Because of the large surface area of the lungs in humans (approximately 75 m^2), they are the main site for absorption of many toxic materials.

The factors governing the absorption of gases and vapors from the respiratory tract differ significantly from those that govern the absorption of particulate substances. Factors controlling the absorption of inhaled gases and vapors include the solubility of the gas in body fluids and the reactivity of the gas with tissues and the fluid lining the respiratory tract. Gases or vapors that are highly water soluble, such as methanol, acetone, hydrogen chloride, and ammonia, dissolve predominantly in the lining of the nose and windpipe (trachea) and therefore tend to be absorbed from those regions. These sites of absorption are also potential sites of toxicity. Formaldehyde is an example of a reactive highly water-soluble vapor for which the nose is a major site of deposition. In contrast to water-soluble gases, reactive gases with low water solubility, such as ozone, phosgene, and nitrogen dioxide, penetrate farther into the respiratory tract and thus come into contact with the smaller tubes of the airways. Gases and vapors that are not water soluble but are more fat soluble, such as benzene, methylene chloride, and trichloroethylene, are not completely removed by interaction with the surfaces of the nose, trachea, and small airways. As a result, these gases penetrate the airways down into the deep lung, where they can diffuse across the thin alveoli lung tissue into the blood. The more soluble a gas is in the blood, the more it will be dissolved and transported to other organs.

For inhaled solid chemicals, an important factor in determining if and where a particle will be deposited in the respiratory tract is its size. One generalization is that the largest particles (>5 μm) are deposited primarily in the nose, smaller particles (1 to 5 μm) in the trachea and small airways, and the smallest particles in the alveoli region of the lungs. Thus, depending on the size of an inhaled particle, it will be deposited in different sections of the respiratory tract, and the location affects the local toxicity and the absorption of the material. In general, particles that are water soluble dissolve within minutes or days, and chemicals that are not water soluble but have a moderate degree of fat solubility also clear rapidly into the blood. Those that are not water soluble or highly fat soluble do not dissolve and are retained in the lungs for long periods of time. Metal oxides, asbestos, fiberglass, and silica are examples of water-insoluble inorganic particles that are retained in the lungs for years.

A number of factors affect the airborne concentrations of chemicals, but vapor pressure (the tendency of molecules to escape from the liquid or solid phase into the gaseous phase) is the most important characteristic. The higher the vapor pressure is, the greater the potential concentration of the chemical in the air. For example, acetone (with a vapor pressure of 180 mmHg at 20 °C) reaches an equilibrium concentration in air of 240,000 ppm, or approximately 24%. Fortunately, the ventilation system in most laboratories prevents an equilibrium concentration from developing in the breathing zone of laboratory personnel.

Even very low vapor pressure chemicals are dangerous if the material is highly toxic. A classic example is elemental mercury. Although the vapor pressure of mercury at room temperature is only 0.0012 mmHg, the resulting equilibrium concentration of mercury vapor is 1.58 ppm, or approximately 13 mg/m^3. The TLV for mercury is 0.05 mg/m^3, more than two orders of magnitude lower.

The vapor pressure of a chemical increases with temperature; therefore, heating solvents or reaction mixtures increases the potential for high airborne concentrations. Also, a spilled volatile chemical evaporates very quickly because of its large surface area, creating a significant exposure potential. Clearly, careful handling of volatile chemicals is very important; keeping containers tightly closed or covered and using volatiles in laboratory chemical hoods help avoid unnecessary exposure to inhaled chemicals.

Certain types of particulate materials also present potential for airborne exposure. If a material has a very low density or a very small particle size, it tends to remain airborne for a considerable time. For example, the very fine dust cloud generated by emptying a low-density particulate (e.g., vermiculite or nanomaterials) into a transfer vessel takes a long time to settle. Such operations should therefore be carried out in a laboratory chemical hood or in a glovebox.

Operations that generate aerosols (suspensions of microscopic droplets in air), such as vigorous boiling, high-speed blending, or bubbling gas through a liquid, increase the potential for exposure via inhalation. Consequently, these and other such operations on toxic chemicals should also be carried out in a laboratory chemical hood.

4.C.1.3.2   Contact with Skin or Eyes

Chemical contact with the skin is a frequent mode of injury in the laboratory. Many chemicals injure the skin directly by causing skin irritation and allergic skin reactions. Corrosive chemicals cause severe burns. In addition to causing local toxic effects, many chemicals are absorbed through the skin in sufficient quantity to produce systemic toxicity. The main avenues by which chemicals enter the body through the skin are the hair follicles, sebaceous glands, sweat glands, and cuts or abrasions of the outer layer. Absorption of chemicals through the skin depends on a number of factors, including chemical concentration, chemical reactivity, and the solubility of the chemical in fat and water. Absorption is also dependent on the condition of the skin, the part of the body exposed, and duration of contact. Differences in skin structure affect the degree to which chemicals are absorbed. In general, toxicants cross membranes and thin skin (e.g., scrotum) much more easily than thick skin (e.g., palms). Although an acid burn on the skin is felt immediately, an alkaline burn takes time to be felt and its damage goes deeper than the acid. When skin is damaged, penetration of chemicals increases. Acids and alkalis injure the skin and increase its permeability. Burns and skin diseases are the most common examples of skin damage that increase penetration. Also, hydrated skin absorbs chemicals better than dehydrated skin. Some chemicals such as dimethyl sulfoxide actually increase the penetration of other chemicals through the skin by increasing its permeability.

Contact of chemicals with the eyes is of particular concern because the eyes are sensitive to irritants. Few substances are innocuous in contact with the eyes; most are painful and irritating, and a considerable number are capable of causing burns and loss of vision. Alkaline materials, phenols, and acids are particularly corrosive and can cause permanent loss of vision. Because the eyes contain many blood vessels, they also are a route for the rapid absorption of many chemicals.

4.C.1.3.3   Ingestion

Many of the chemicals used in the laboratory are extremely hazardous if they enter the mouth and are swallowed. The gastrointestinal tract, which consists of the mouth, esophagus, stomach, and small and large intestines, can be thought of as a tube of variable diameter (approximately 5 m long) with a large surface area (approximately 200 m^2) for absorption. Toxicants that enter the gastrointestinal tract must be absorbed into the blood to produce a systemic injury, although some chemicals are caustic or irritating to the gastrointestinal tract tissue itself. Absorption of toxicants takes place along the entire gastrointestinal tract, even in the mouth, and depends on many factors, including the physical properties of the chemical and the speed at which it dissolves. Absorption increases with surface area, permeability, and residence time in various segments of the tract. Some chemicals increase intestinal permeability and thus increase the rate of absorption. More chemical will be absorbed if the chemical remains in the intestine for a long time. If a chemical is in a relatively insoluble solid form, it will have limited contact with gastrointestinal tissue, and its rate of absorption will be low. If it is an organic acid or base, it will be absorbed in that part of the gastrointestinal tract where it is most fat soluble. Fat-soluble chemicals are absorbed more rapidly and extensively than water-soluble chemicals.

4.C.1.3.4   Injection

Exposure to toxic chemicals by injection does not occur frequently in the laboratory, but it occurs inadvertently through mechanical injury from sharp objects such as glass or metal contaminated with chemicals or syringes used for handling chemicals. The intravenous route of administration is especially dangerous because it introduces the toxicant directly into the bloodstream, eliminating the process of absorption. Nonlaboratory personnel, such as custodial workers or waste handlers, must be protected from exposure by placing sharp objects in special trash containers and not ordinary scrap baskets. Hypodermic needles with blunt ends are available for laboratory use.

4.C.2   Assessing Risks of Exposure to Toxic Laboratory Chemicals

Exposure to a harmful chemical results in local toxic effects, systemic toxic effects, or both. Local effects involve injury at the site of first contact; the eyes, the skin, the nose and lungs, and the digestive tract are typical sites of local reactions. Examples of local effects include (1) inhalation of hazardous materials causing toxic effects in the nose and lungs; (2) contact with harmful materials on the skin or eyes leading to effects ranging from mild irritation to severe tissue damage; and (3) ingestion of caustic substances causing burns and ulcers in the mouth, esophagus, stomach, and intestines. Systemic effects, by contrast, occur after the toxicant has been absorbed from the site of contact into the bloodstream and distributed throughout the body. Some chemicals produce adverse effects on all tissues of the body, but others tend to selectively injure a particular tissue or organ without affecting others. The affected organ (e.g., liver, lungs, kidney, and central nervous system) is referred to as the target organ of toxicity, although it is not necessarily the organ where the highest concentration of the chemical is found. Hundreds of systemic toxic effects of chemicals are known; they result from single (acute) exposures or from repeated or long-duration (chronic) exposures that become evident only after a long latency period.

Toxic effects are classified as either reversible or irreversible. Reversible toxicity is possible when tissues have the capacity to repair toxic damage, and the damage disappears after cessation of exposure. Irreversible damage, in contrast, persists after cessation of exposure. Recovery from a burn is a good example of reversible toxicity; cancer is considered irreversible, although appropriate treatment may reduce the effects in this case.

Laboratory chemicals are grouped into several classes of toxic substances, and many chemicals display more than one type of toxicity. The first step in assessing the risks associated with a planned laboratory experiment involves identifying which chemicals in the proposed experiment are potentially hazardous substances. The OSHA Laboratory Standard (29 CFR § 1910.1450) defines a hazardous substance as a chemical for which there is statistically significant evidence based on at least one study conducted in accordance with established scientific principles that acute or chronic health effects may occur in exposed employees. The term "health hazard" includes chemicals that are carcinogens, toxic or highly toxic agents, reproductive toxins, irritants, corrosives, sensitizers, hepatotoxins, nephrotoxins, neurotoxins, agents that act on the hematopoietic systems, and agents that damage the lungs, skin, eyes, or mucous membranes.

The OSHA Laboratory Standard further requires that certain chemicals be identified as particularly hazardous substances (commonly known as PHSs) and handled using special additional procedures. PHSs include chemicals that are select carcinogens (those strongly implicated as a potential cause of cancer in humans), reproductive toxins, and compounds with a high degree of acute toxicity. When working with these substances for the first time, it is prudent to consult with a safety professional prior to beginning work. This will provide a second set of trained eyes to review the safety protocols in place and will help ensure that any special emergency response requirements can be met in the event of exposure of personnel to the material or accidental release.

Highly flammable and explosive substances make up another category of hazardous compounds, and the assessment of risk for these classes of chemicals is discussed in section 4.D. This section considers the assessment of risks associated with specific classes of toxic chemicals, including those that pose hazards due to acute toxicity and chronic toxicity.

The following are the most common classes of toxic substances encountered in laboratories.

4.C.2.1 Acute Toxicants

Acute toxicity is the ability of a chemical to cause a harmful effect after a single exposure. Acutely toxic agents cause local toxic effects, systemic toxic effects, or both, and this class of toxicants includes corrosive chemicals, irritants, and allergens (sensitizers).

In assessing the risks associated with acute toxicants, it is useful to classify a substance according to the acute toxicity hazard level as shown in Table 4.1. LD_{50} values can be found in the LCSS or MSDS for a given substance, and in references such as *Sax's Dangerous Properties of Industrial Materials* (Lewis, 2004), *A Comprehensive Guide to the Hazardous Properties of Chemical Substances*, 3rd Edition (Patnaik, 2007), and the *Registry of Toxic Effects of Chemical Substances* (RTECS) (NIOSH). Table 4.2 relates test animal LD_{50} values expressed as milligrams or grams per kilogram of body weight to the probable human lethal dose, expressed in easily understood units, for a 70-kg person.

Special attention is given to any substance classified according to the above criteria as having a high level of acute toxicity hazard. Chemicals with a high level of acute toxicity make up one of the categories of PHSs defined by the OSHA Laboratory Standard. Any compound rated as highly toxic in Table 4.1 meets the OSHA criteria for handling as a PHS.

Table 4.3 lists some of the most common chemicals with a high level of acute toxicity that are encountered in the laboratory. These compounds are handled using the additional procedures outlined in Chapter 6,

TABLE 4.1 Acute Toxicity Hazard Level

Hazard Level	Toxicity Rating	Oral LD_{50} (rats, per kg)	Skin Contact LD_{50} (rabbits, per kg)	Inhalation LC_{50} (rats, ppm for 1 h)	Inhalation LC_{50} (rats, mg/m³ for 1 h)
High	Highly toxic	<50 mg	<200 mg	<200	<2,000
Medium	Moderately toxic	50 to 500 mg	200 mg to 1 g	200 to 2,000	2,000 to 20,000
Low	Slightly toxic	500 mg to 5 g	1 to 5 g	2,000 to 20,000	20,000 to 200,000

TABLE 4.2 Probable Lethal Dose for Humans

Toxicity Rating	Animal LD$_{50}$ (per kg)	Lethal Dose When Ingested by 70-kg (150-lb) Human
Extremely toxic	<5 mg	A taste (<7 drops)
Highly toxic	5 to 50 mg	Between 7 drops and 1 tsp
Moderately toxic	50 to 500 mg	Between 1 tsp and 1 oz
Slightly toxic	500 mg to 5 g	Between 1 oz and 1 pint
Practically nontoxic	>5 g	>1 pint

SOURCE: Modified, by permission, from Gosselin et al. (1984); reprinted by permission from Lippincott Williams and Wilkins, http://lww.com.

section 6.D. In some circumstances, all these special precautions may not be necessary, such as when the total amount of an acutely toxic substance is a small fraction of the harmful dose. An essential part of prudent experiment planning is to determine whether a chemical with a high degree of acute toxicity should be treated as a PHS in the context of a specific planned use. This determination not only involves consideration of the total amount of the substance to be used but also requires a review of the physical properties of the substance (e.g., Is it volatile? Does it tend to form dusts?), its potential routes of exposure (e.g., Is it readily absorbed through the skin?), and the circumstances of its use in the proposed experiment (e.g., Will the substance be heated? Is there likelihood that aerosols may be generated?). Depending on the laboratory personnel's level of experience and the degree of potential hazard, this determination may require consultation with supervisors and safety professionals.

Because the greatest risk of exposure to many laboratory chemicals is by inhalation, trained laboratory personnel must understand the use of exposure limits that have been established by agencies such as OSHA and NIOSH and by an organization such as ACGIH.

The TLV assigned by the ACGIH, defines the concentration of a chemical in air to which nearly all individuals can be exposed without adverse effects. These limits reflect a view of an informed scientific community and are not legal standards. They are designed to be an aid to industrial hygienists. The TLV time-weighted average (TWA) refers to the concentration safe for exposure during an entire 8-hour workday; the TLV-STEL is a

TABLE 4.3 Examples of Compounds with a High Level of Acute Toxicity

Acrolein	Methyl fluorosulfonate
Arsine	Nickel carbonyl
Chlorine	Nitrogen dioxide
Diazomethane	Osmium tetroxide
Diborane (gas)	Ozone
Dimethyl mercury	Phosgene
Hydrogen cyanide	Sodium azide
Hydrogen fluoride	Sodium cyanide (and other cyanide salts)

higher concentration to which workers may be exposed safely for a 15-minute period up to four times during an 8-hour shift and at least 60 minutes between these periods. TLVs are intended for use by professionals after they have read and understood the documentation of the TLV for the chemical or physical agent under study.

OSHA defines the permissible exposure limit (PEL) analogously to the ACGIH values, with corresponding 8-hour TWA and ceiling limits based on either a 15-minute TWA or an instantaneous reading, whichever is possible. In some cases, OSHA also defines a maximum peak concentration that cannot be exceeded beyond a given duration. Compliance with PELs is required, and the limits are enforceable by OSHA. PEL values allow trained laboratory personnel to quickly determine the relative inhalation hazards of chemicals. In general, substances with 8-hour TWA PELs of less than 50 ppm should be handled in a laboratory chemical hood. Comparison of these values to the odor threshold for a given substance often indicates whether the odor of the chemical provides sufficient warning of possible hazard. However, individual differences in ability to detect some odors as well as anosmia for ethylene oxide or olfactory fatigue for hydrogen sulfide can limit the usefulness of odors as warning signs of overexposure. LCSSs contain information on odor threshold ranges and whether a substance is known to cause olfactory fatigue.

Recommended exposure limits (RELs) are occupational exposure limits recommended by NIOSH to protect the health and safety of individuals over a working lifetime. Compliance with RELs is not required by law. RELs may also be expressed as a ceiling limit that should never be exceeded over a given time period, but the limit is usually expressed as a TWA exposure for up to 10 hours per day during a 40-hour workweek. As with TLVs, RELs are also expressed as STELs. One should not exceed the STEL for longer than 15 minutes at anytime throughout a workday.

A variety of devices are available for measuring the concentration of chemicals in laboratory air, so that the degree of hazard associated with the use of a chemical is assessed directly. Industrial hygiene offices of many institutions assist trained laboratory personnel in measuring the air concentrations of chemicals.

4.C.3 Types of Toxins

4.C.3.1 Irritants, Corrosive Substances, Allergens, and Sensitizers

Lethal dose and other quantitative toxicological parameters generally provide little guidance in assessing the risks associated with corrosives, irritants, allergens, and sensitizers because these toxic substances exert their harmful effects locally. It would be very useful for the chemical research community if a quantitative measure for such effects were developed. When planning an experiment that involves corrosive substances, basic prudent handling practices should be reviewed to ensure that the skin, face, and eyes are protected adequately by the proper choice of corrosion-resistant gloves and protective clothing and eyewear, including, in some cases, face shields. Similarly, LD_{50} and LC_{50} data are not indicators of the irritant effects of chemicals, and therefore special attention should be paid to the identification of irritant chemicals by consulting LCSSs, MSDSs, and other sources of information. Allergens and sensitizers are another class of acute toxicants with effects that are not included in LD_{50} or LC_{50} data.

4.C.3.1.1 Irritants

Irritants are noncorrosive chemicals that cause reversible inflammatory effects (swelling and redness) on living tissue by chemical action at the site of contact. A wide variety of organic and inorganic chemicals are irritants, and consequently, skin and eye contact with all reagent chemicals in the laboratory should be minimized. Examples include formaldehyde, iodine, and benzoyl chloride.

4.C.3.1.2 Corrosive Substances

Corrosive substances are those that cause destruction of living tissue by chemical action at the site of contact and are solids, liquids, or gases. Corrosive effects occur not only on the skin and eyes but also in the respiratory tract and, in the case of ingestion, in the gastrointestinal tract as well. Corrosive materials are probably the most common toxic substances encountered in the laboratory. Corrosive liquids are especially dangerous because their effect on tissue is rapid. Bromine, sulfuric acid, aqueous sodium hydroxide solution, and hydrogen peroxide are examples of highly corrosive liquids. Corrosive gases are also frequently encountered. Gases such as chlorine, ammonia, chloramine, and nitrogen dioxide damage the lining of the lungs, leading, after a delay of several hours, to the fatal buildup of fluid known as pulmonary edema. Finally, a number of solid chemicals have corrosive effects on living tissue. Examples of common corrosive solids include sodium hydroxide, phosphorus, and phenol. If dust from corrosive solids is inhaled, it causes serious damage to the respiratory tract.

There are several major classes of corrosive substances. Strong acids such as nitric, sulfuric, and hydrochloric acid cause serious damage to the skin and eyes. Hydrofluoric acid is particularly dangerous and produces slow-healing painful burns (see Chapter 6, section 6.G.6). Strong bases, such as metal hydroxides and ammonia, are another class of corrosive chemicals. Strong dehydrating agents, such as phosphorus pentoxide and calcium oxide, have a powerful affinity for water and cause serious burns on contact with the skin. Finally, strong oxidizing agents, such as concentrated solutions of hydrogen peroxide, also have serious corrosive effects and should never come into contact with the skin or eyes.

4.C.3.1.3 Allergens and Sensitizers

A chemical allergy is an adverse reaction by the immune system to a chemical. Such allergic reactions result from previous sensitization to that chemical or a structurally similar chemical. Once sensitization occurs, allergic reactions result from exposure to extremely low doses of the chemical. Some allergic reactions are immediate, occurring within a few minutes after exposure. Anaphylactic shock is a severe immediate allergic reaction that results in death if not treated quickly. Delayed allergic reactions take hours or even days to develop, the skin is the usual site of such delayed reactions, becoming red, swollen, and itchy. Delayed chemical allergy occurs even after the chemical has been removed; contact with poison ivy is a familiar example of an exposure that causes a delayed allergic reaction due to uroshiol. Also, just as people vary widely in their susceptibility to sensitization by environmental allergens such as dust and pollen, individuals also exhibit wide differences in their sensitivity to laboratory chemicals.

Because individuals differ widely in their tendency to become sensitized to allergens, compounds with a proven ability to cause sensitization should be classified as highly toxic agents within the institution's CHP. When working with chemicals known to cause allergic sensitization, follow institutional policy on handling and containment of allergens and highly toxic agents. Once a person has become sensitized to an allergen, subsequent contact often leads to immediate or delayed allergic reactions.

Because an allergic response is triggered in a sensitized individual by an extremely small quantity of the allergen, it may occur despite personal protection measures that are adequate to protect against the acute effects of chemicals. Laboratory personnel should be alert for signs of allergic responses to chemicals. Examples of chemical substances that cause allergic reactions in some individuals include diazomethane; dicyclohexylcarbodiimide; formaldehyde and phenol derivatives; various isocyanates (e.g., methylene diphenyl diisocyanate (MDI) or toluene diisocyanate (TDI), used in adhesives, elastomers, and coatings); benzylic and allylic halides; metals including nickel, beryllium, platinum, cobalt, tin, and chromium; and acid anhydrides such as acetic anhydrides.

4.C.3.2 Asphyxiants

Asphyxiants are substances that interfere with the transport of an adequate supply of oxygen to vital organs of the body. The brain is the organ most easily affected by oxygen starvation, and exposure to asphyxiants leads to rapid collapse and death. Simple asphyxiants are substances that displace oxygen from the air being breathed to such an extent that adverse effects result. Acetylene, carbon dioxide, argon, helium, ethane, nitrogen, and methane are common asphyxiants. Certain other chemicals have the ability to combine with hemoglobin, thus reducing the capacity of the blood to transport oxygen. Carbon monoxide, hydrogen cyanide, and certain organic and inorganic cyanides are examples of such substances.

4.C.3.3 Neurotoxins

Neurotoxic chemicals induce an adverse effect on the structure or function of the central or peripheral nervous system, which can be permanent or reversible. The detection of neurotoxic effects may require specialized laboratory techniques, but often they are inferred from behavior such as slurred speech and staggered gait. Many neurotoxins are chronically toxic substances with adverse effects that are not immediately apparent. Some chemical neurotoxins that may be found in the laboratory are mercury (inorganic and organic), organophosphate pesticides, carbon disulfide, xylene, tricholoroethylene, and *n*-hexane. (For information about reducing the presence of mercury in laboratories, see Chapter 5, section 5.B.8.)

4.C.3.4 Reproductive and Developmental Toxins

Reproductive toxins are defined by the OSHA Laboratory Standard as substances that cause chromosomal damage (mutagens) and substances with lethal or teratogenic (malformation) effects on fetuses. These substances have adverse effects on various aspects of reproduction, including fertility, gestation, lactation, and general reproductive performance, and can affect both men and women. Many reproductive toxins are chronic toxins that cause damage after repeated or long-duration exposures with effects that become evident only after long latency periods. Developmental toxins act during pregnancy and cause adverse effects on the fetus; these effects include embryo lethality (death of the fertilized egg, embryo, or fetus), teratogenic effects, and postnatal functional defects. Male reproductive toxins in some cases lead to sterility.

When a pregnant woman is exposed to a chemical, generally the fetus is exposed as well because the placenta is an extremely poor barrier to chemicals. Embryotoxins have the greatest impact during the first trimester of pregnancy. *Because a woman often does not know that she is pregnant during this period of high susceptibility, women of childbearing potential are advised to be especially cautious when working with chemicals, especially those rapidly absorbed through the skin (e.g., formamide).* Pregnant women and women intending to become pregnant should seek advice from knowledgeable sources before working with substances that are suspected to be reproductive toxins. As minimal precautions, the general procedures outlined in Chapter 6, section 6.D, should be followed, though in some cases it will be appropriate to handle the compounds as PHSs.

For example, among the numerous reproductive hazards to female laboratory scientists, gestational exposure to organic solvents should be of concern (HHS/CDC/NIOSH, 1999; Khattak et al., 1999). Some common solvents in high doses have been shown to be teratogenic in laboratory animals, resulting in developmental defects. Although retrospective studies of the teratogenic risk in women of childbearing age of occupational exposure to common solvents have reached mixed conclusions, at least one such study of exposure during pregnancy to multiple solvents detected increased fetal malformations. Thus, inhalation exposure to organic solvents should be minimized during pregnancy. Also, exposure to lead or to anticancer drugs, such as methotrexate, or to ionizing radiation can cause infertility, miscarriage, birth defects, and low birth weight. Certain ethylene glycol ethers such as 2-ethoxyethanol and 2-methoxyethanol can cause miscarriages. Carbon disulfide can cause menstrual cycle changes. One cannot assume that any given substance is safe if no data on gestational exposure are available.

Specific hazards of chemical exposure are associated with the male reproductive system, including suppression of sperm production and survival, alteration in sperm shape and motility, and changes in sexual drive and performance. Various reproductive hazards have been noted in males following exposure to halogenated hydrocarbons, nitro aromatics, arylamines, ethylene glycol derivatives, mercury, bromine, carbon disulfide, and other chemical reagents (HHS/CDC/NIOSH, 1996).

Information on reproductive toxins can be obtained from LCSSs, MSDSs, and by consulting safety professionals in the environmental safety department, industrial hygiene office, or medical department. Literature sources of information on reproductive and developmental toxins include the *Catalog of Teratogenic Agents* (Shepard and Lemire, 2007), *Reproductively Active Chemicals: A Reference Guide* (Lewis, 1991), and "What Every Chemist Should Know About Teratogens" in the *Journal of Chemical Education* (Beyler and Meyers, 1982). The State of California maintains a list of chemicals it considers reproductive toxins, and additional information can be found through the NLM TOXNET system. The study of reproductive toxins is an active area of research, and laboratory personnel should consult resources that are updated regularly for information.

4.C.3.5   Toxins Affecting Other Target Organs

Target organs outside the reproductive and neurological systems are also affected by toxic substances in the laboratory. Most of the chlorinated hydrocarbons, benzene, other aromatic hydrocarbons, some metals, carbon monoxide, and cyanides, among others, produce one or more effects in target organs. Such an effect may be the most probable result of exposure to the particular chemical. Although this chapter does not include specific sections on liver, kidney, lung, or blood toxins, many of the LCSSs mention those effects in the toxicology section.

4.C.3.6   Carcinogens

A carcinogen is a substance capable of causing cancer. Cancer, in the simplest sense, is the uncontrolled growth of cells and can occur in any organ. The mechanism by which cancer develops is not well understood, but the current thinking is that some chemicals interact directly with DNA, the genetic material in all cells, to result in permanent alterations. Other chemical carcinogens modify DNA indirectly by changing the way cells grow. Carcinogens are chronically toxic substances; that is, they cause damage after repeated or long-duration exposure, and their effects may become evident only after a long latency period. Carcinogens are particularly insidious toxins because they may have no immediate apparent harmful effects.

Because cancer is a widespread cause of human mortality, and because exposure to chemicals may play a significant role in the onset of cancer, a great deal of attention has been focused on evaluation of the carcinogenic potential of chemicals. However, a vast majority of substances involved in research, especially in laboratories concerned primarily with the synthesis of novel compounds, have not been tested for carcinogenicity. Compounds that are known to pose the greatest carcinogenic hazard are referred to as select carcinogens, and they constitute another category of substances that must be handled as PHSs according to the OSHA Laboratory Standard. A select carcinogen is defined in the OSHA Laboratory Standard as a substance that meets one of the following criteria:

1. It is regulated by OSHA as a carcinogen.
2. It is listed as known to be a carcinogen in the latest *Annual Report on Carcinogens* issued by the National Toxicology Program (NTP) (HHS/CDC/NTP, 2005).
3. It is listed under Group 1 (carcinogenic to humans) by the International Agency for Research on Cancer (IARC).
4. It is listed under IARC Group 2A (probably carcinogenic to humans) or 2B (possibly carcinogenic to humans), or under the category "reasonably anticipated to be a carcinogen by the NTP," *and* causes statistically significant tumor incidence in experimental animals in accordance with any of the following criteria: (a) after inhalation exposure of 6 to 7 hours per day, 5 days per week, for a significant portion of a lifetime to dosages of less than 10 mg/m^3; (b) after repeated skin application of less than 300 mg/kg of body weight per week; or (c) after oral dosages of less than 50 mg/kg of body weight per day.

Chemicals that meet the criteria of a select carcinogen are classified as PHSs and should be handled using the basic prudent practices given in Chapter 6, section 6.C, supplemented by the additional special practices outlined in Chapter 6, section 6.D. Work with compounds that are *possible* human carcinogens may or may not require the additional precautions given in section 6.D. For these compounds, the LCSS should indicate whether the substance meets the additional criteria listed in category 4 and must therefore be treated as a select carcinogen. If an LCSS is not available, consultation with a safety professional such as a CHO may be necessary to determine whether a substance should be classified as PHS. Lists of known human carcinogens and compounds that are "reasonably anticipated to be carcinogens" based on animal tests can be found in the *11th Report on Carcinogens* (HHS/CDC/NTP, 2005). This report is updated periodically. Check the NTP Web site (ntp.niehs.nih.gov) for the most recent edition. Additional information can be found on the OSHA and IARC Web sites (www.osha.gov and www.iarc.fr).

In the laboratory many chemical substances are encountered for which there is no animal test or human epidemiological data on carcinogenicity. In these cases, trained laboratory personnel must evaluate the potential risk that the chemical in question is a carcinogenic substance. This determination is sometimes made on the basis of knowledge of the specific classes of compounds and functional group types that have previously been correlated with carcinogenic activity. For example, chloromethyl methyl ether is a known human carcinogen and therefore is regarded as an OSHA select carcinogen requiring the handling procedures outlined in section 6.D. On the other hand, the carcinogenicity of ethyl chloromethyl ether and certain other alkyl chloromethyl ethers is not established, and these substances do not necessarily have to be treated as select carcinogens. However, because of the chemical similarity of these compounds to chloromethyl methyl ether, these substances may have comparable carcinogenicity, and it is prudent to regard them as select carcinogens requiring the special handling procedures outlined in section 6.D.

Whether a suspected carcinogenic chemical is treated as a PHS in the context of a specific laboratory use is affected by the scale and circumstances associated with the intended experiment. Trained laboratory personnel must decide whether the amount and frequency of use, as well as other circumstances, require additional precautions beyond the basic prudent practices of section 6.C. For example, the large-scale or recurring use of such a chemical might suggest that the special precautions of section 6.D be followed to control exposure, whereas adequate protection from a single use of a small amount of such a substance may be obtained through the use of the basic procedures in section 6.C.

When evaluating the carcinogenic potential of chemicals, note that exposure to certain combinations of compounds (not necessarily simultaneously) causes cancer even at exposure levels where neither of the individual compounds would have been carcinogenic. 1,8,9-Trihydroxyanthracene and certain phorbol esters are examples of tumor promoters that are not carcinogenic themselves but dramatically amplify the carcinogenicity of other compounds. Understand that the response of an organism to a toxicant typically increases with the dose given, but the relationship is not always a linear one. Some carcinogenic alkylating agents exhibit a dose threshold above which the tendency to cause mutations increases markedly. At lower doses, natural protective systems prevent genetic damage, but when the capacity of these systems is overwhelmed, the organism becomes much more sensitive to the toxicant. However, individuals have differences in the levels of protection against genetic damage as well as in other defense systems. These differences are determined in part by genetic factors and in part by the aggregate exposure of the individual to all chemicals within and outside the laboratory.

4.C.3.7 Control Banding

Control banding is a qualitative risk assessment and management approach to assist in determining the appropriate handling of materials without occupational exposure limits (OELs) and to minimize the exposure of personnel to hazardous material.[2] It is not intended to be a replacement for OELs but as an additional tool. The system uses a range of exposure and hazard

"bands" that, when mapped for a given material and application, help the user determine the appropriate safety controls that should be in place. The approach is built on two major premises: (1) there are a limited number of control approaches and (2) that many problems have been encountered and solved before. Control banding uses the solutions that experts have developed previously to control occupational chemical exposures and applies those solutions to other tasks with similar exposure concerns.

By considering the physical and chemical characteristics and hazards posed by the material (e.g., toxicity), the quantity used, the intended use or application, and the mode of exposure (e.g., inhalation), a graduated scale of controls can be applied, from general ventilation requirements to requiring containment of the material to recommending that the user seek expert advice. Because this approach is expected to provide simplified guidance for assessing hazards and applying controls, it is anticipated that control banding will have utility for small- and medium-size nonchemical businesses; however, larger companies may also find it useful for prioritizing chemical hazards and hazard communication.

Note that a number of control banding models exist, each with its own level of complexity and applicability to a variety of scenarios. Within the United States, questions about the utility of control banding for workplaces initiated a review by NIOSH on the critical issues and potential applications of the system. The resulting report, *Qualitative Risk Characterization and Management of Occupational Hazards: Control Banding (CB)* (HHS/CDC/NIOSH, 2009b), can be found on the NIOSH Web site. It provides an overview of the major concepts and methodologies and presents a critical analysis of control banding.

Control banding is of interest internationally, and variations on the methodology can be found in many countries. More information about control banding can be found by consulting these Web sites and articles.

- (UK Health and Safety Executive) Control of Substances Hazardous to Health Regulations, www.coshh-essentials.org.uk/
- ILO Programme on Safety and Health at Work and the Environment (SafeWork), www.ilo.org/
- NIOSH, www.cdc.gov/niosh/
- "Training Health and Safety Committees to Use Control Banding: Lessons Learned and Opportunities for the United States" (Bracker et al., 2009)
- "Evaluation of COSHH Essentials: Methylene Chloride, Isopropanol, and Acetone Exposures in a Small Printing Plant" (Lee et al., 2009)

---

[2]For information on how OELs are determined, see Alaimo (2001).

- "Application of a Pilot Control Banding Tool for Risk Level Assessment and Control of Nanoparticle Exposures" (Paik et al., 2008)
- "'Stoffenmanager,' a Web-Based Control Banding Tool Using an Exposure Process Model" (Marquart et al., 2008)
- "History and Evolution of Control Banding: A Review" (Zalk and Nelson, 2008)
- *"Control Banding: Issues and Opportunities." A Report of the ACGIH Exposure Control Banding Task Force* (ACGIH, 2008a)
- "Evaluation of the Control Banding Method—Comparison with Measurement-Based Comprehensive Risk Assessment" (Hashimoto et al., 2007)
- *Guidance for Conducting Control Banding Analyses* (American Industrial Hygiene Association, 2008)

4.D FLAMMABLE, REACTIVE, AND EXPLOSIVE HAZARDS

In addition to the hazards due to the toxic effects of chemicals, hazards due to flammability, explosivity, and reactivity need to be considered in risk assessment. These hazards are described in detail in the following sections. Further information can be found in *Bretherick's Handbook of Reactive Chemical Hazards* (Urben, 2007), an extensive compendium that is the basis for lists of incompatible chemicals included in other reference works. The handbook describes computational protocols that consider thermodynamic and kinetic parameters of a system to arrive at quantitative measures such as the reaction hazard index. Reactive hazards arise when the release of energy from a chemical reaction occurs in quantities or at rates too great for the energy to be absorbed by the immediate environment of the reacting system, and material damage results. An additional resource is the *Hazardous Chemical Handbook* (Carson and Mumford, 2002). The book is geared toward an industrial audience and contains basic descriptions of chemical hazards along with technical guidance.

Box 4.2 is a quick guide for assisting in the assessment of the physical, flammable, explosive, and reactive hazards in the laboratory.

4.D.1 Flammable Hazards

4.D.1.1 Flammable Substances

Flammable substances, those that readily catch fire and burn in air, may be solid, liquid, or gaseous. The most common fire hazard in the laboratory is a flammable liquid or the vapor produced from such a liquid. An additional hazard is that a compound can enflame so rapidly that it produces an explosion. Proper use of substances that cause fire requires knowledge of their tendencies to vaporize, ignite, or burn under the variety of conditions in the laboratory.

For a fire to occur, three conditions must exist simultaneously: an atmosphere containing oxygen, usually air; a fuel, such as a concentration of flammable gas or vapor that is within the flammable limits of the substance; and a source of ignition (see Figure 4.3). Prevention of the coexistence of flammable vapors and an ignition source is the optimal way to deal with the hazard. When the vapors of a flammable liquid cannot always be controlled, strict control of ignition sources is the principal approach to reduce the risk of flammability. The rates at which different liquids produce flammable vapors depend on their vapor pressures, which increase with increasing temperature. The degree of fire hazard of a substance depends also on its ability to form combustible or explosive mixtures with air and on the ease of ignition of these mixtures. Also important are the relative density and solubility of a liquid with respect to water and of a gas with respect to air. These characteristics can be evaluated and compared in terms of the following specific properties.

4.D.1.2 Flammability Characteristics

4.D.1.2.1 Flash Point

The flash point is the lowest temperature at which a liquid has a sufficient vapor pressure to form an ignitable mixture with air near the surface of the liquid. Note that many common organic liquids have a flash point below room temperature: for example, acetone (–18 °C), benzene (–11.1 °C), diethyl ether (–45 °C), and methyl alcohol (11.1 °C). The degree of hazard associated with a flammable liquid also depends on other properties, such as its ignition point and boiling point. Commercially obtained chemicals are clearly labeled as to flammability and flash point. Consider the example of acetone given in section 4.C.1.3.1. At ambient pressure and temperature, an acetone spill produces a concentration as high as 23.7% acetone in air. Although it is not particularly toxic, with a flash point of –18 °C and upper and lower flammable limits of 2.6% and 12.8% acetone in air, respectively (see Table 4.4), clearly an acetone spill produces an extreme fire hazard. Thus the major hazard given for acetone in the LCSS is flammability.

4.D.1.2.2 Ignition Temperature

The ignition temperature (autoignition temperature) of a substance, whether solid, liquid, or gaseous, is the minimum temperature required to initiate or cause self-sustained combustion independent of the heat source. The lower the ignition temperature, the greater the potential for a fire started by typical laboratory

---

BOX 4.2
Quick Guide to Risk Assessment for Physical, Flammable,
Explosive, and Reactive Hazards in the Laboratory

The following outline provides a summary of the steps discussed in this chapter that laboratory personnel should use to assess the risks of managing physical hazards in the laboratory.

1. **Identify chemicals to be used and circumstances of use.** Identify the chemicals involved in the proposed experiment and determine the amounts that will be used. Is the experiment to be done once, or will the chemicals be handled repeatedly? Will the experiment be conducted in an open laboratory, in an enclosed apparatus, or in a chemical fume hood? Is it possible that new or unknown substances will be generated in the experiment?

2. **Consult sources of information.** Consult an up-to-date laboratory chemical safety summary, material safety data sheet, or NIOSH *Pocket Guide to Chemical Hazards* (HHS/CDC/NIOSH, 2007). In cases where substances with significant or unusual potential physical hazards are involved, consult more detailed references such as Urben (2007) and other sources discussed in section 4.B of this chapter. Depending on the laboratory personnel's level of experience and the degree of potential hazard associated with the proposed experiment, obtain the assistance of supervisors and safety professionals before proceeding with risk assessment.

3. **Evaluate type of physical, flammable, explosive, or reactive hazard(s) posed by the chemicals.** Use the above sources of information to determine the hazards associated with each chemical involved in the proposed experiment. Examples of questions that could be asked are: Are any of the chemicals to be used corrosive? Are any of the chemicals to be used flammable or explosive? Are any of the chemicals strong oxidizers? Will the reaction result in a large release of gases? Are any cryogens required?

4. **Evaluate the hazards posed by chemical changes over the course of the experiment.** Consider the changes in pressure, heat, flammability, and explosivity during the experiment. If the

experiment is a scale-up of a reaction, consider how changes in gas production and temperature may change the engineering controls required for safety.

5. **Evaluate type of physical hazard(s) posed by the equipment required.** Review procedures for correct use of all equipment. Identify heat and ignition sources and consider appropriate placement for them in the laboratory. Are any pieces of equipment going to be under high or low pressure (e.g., a vacuum line)? Is the glassware free of cracks and chips?

6. **Select appropriate procedures to minimize risk.** Use the basic prudent practices for handling chemicals, which are discussed in Chapter 6, section 6.C for all work with chemicals in the laboratory and the guidance on using laboratory equipment in Chapter 7. Determine if the chemicals or equipment in use pose physical hazards to laboratory personnel. If so, consider methods to minimize the risk posed by the hazard through substitution with another chemical, if possible, or use of engineering controls (e.g., chemical fume hoods, grounding cables, inert atmospheres) and personal protective equipment if not. Use information in this chapter and Chapter 7 to determine whether it is appropriate to apply the additional procedures before continuing work and whether additional consultation with safety professionals is warranted (see section 4.D).

7. **Prepare for contingencies.** Be aware of institutional procedures in the event of emergencies and accidents. Some questions to consider are: What are appropriate measures to take in the event of fire, explosion, or injury? What are the appropriate first-aid procedures in the event of heat, cold, or chemical burns? What is the procedure in the event of an emergency or non-emergency spill of the chemicals?

NOTE: For a quick guide for assessing the toxicity hazards associated with laboratory chemicals, see Box 4.1. For a quick guide for assessing biological hazards in the laboratory, see Box 4.3.

---

equipment. A spark is not necessary for ignition when the flammable vapor reaches its autoignition temperature. For instance, carbon disulfide has an ignition temperature of 90 °C, and it can be set off by a steam line or a glowing light bulb. Diethyl ether has an ignition temperature of 160 °C and can be ignited by a hot plate.

4.D.1.2.3 Limits of Flammability

Each flammable gas and liquid (as a vapor) has two fairly definite limits of flammability defining the range of concentrations in mixtures with air that will propagate a flame and cause an explosion. At the low extreme, the mixture is oxygen rich but contains insufficient fuel. The lower flammable limit (lower explosive limit [LEL]) is the minimum concentration (percent by volume) of the fuel (vapor) in air at which a flame is propagated when an ignition source is present. The upper flammable limit (upper explosive limit [UEL]) is the maximum concentration (percent by volume) of the vapor in air above which a flame is not propagated. The flammable range (explosive range) consists of all concentrations between the LEL and the UEL. This range becomes wider with increasing temperature and in oxygen-rich atmospheres and also changes depending on the presence of other components. The limitations of the flammability range, however, provide little margin of safety from the practical point of view because, when a solvent is spilled in the presence of an energy source, the LEL is reached very quickly and a fire or explosion ensues before the UEL is reached.

4.D.1.3 Classes of Flammability

Several systems are in use for classifying the flammability of materials. Some (e.g., Class I—flammable

FIGURE 4.3 The fire triangle.

TABLE 4.4 NFPA Fire Hazard Ratings, Flash Points (FP), Boiling Points (bp), Ignition Temperatures, and Flammable Limits of Some Common Laboratory Chemicals

	NFPA Flammability Rating[a]	Flash Point (°C)	Boiling Point (°C)	Ignition Temperature (°C)	Flammable Limits (% by volume)	
					Lower	Upper
Acetaldehyde	4	−39	21	175	4	60
Acetic acid (glacial)	2	39	118	463	4	19.9
Acetone	3	−20	56	465	2.5	12.8
Acetonitrile	3	6	82	524	3	16
Carbon disulfide	4	−30	46	90	1.3	50
Cyclohexane	3	−20	82	245	1.3	8
Diethylamine	3	−23	57	312	1.8	10.1
Diethyl ether	4	−45	35	180	1.9	36
Dimethyl sulfoxide	2	95	189	215	2.6	42
Ethyl alcohol	3	13	78	363	3.3	19
Heptane	3	−4	98	204	1.05	6.7
Hexane	3	−22	69	225	1.1	7.5
Hydrogen	4		−252	500	4	75
Isopropyl alcohol	3	12	83	399	2	12.7 @ 200 (93)
Methyl alcohol	3	11	64	464	6	36
Methyl ethyl ketone	3	−9	80	404	1.4 @ 200 (93)	11.4 @ 200 (93)
Pentane	4	<−40	36	260	1.5	7.8
Styrene	3	31	146	490	0.9	6.8
Tetrahydrofuran	3	−14	66	321	2	11.8
Toluene	3	4	111	480	1.1	7.1
p-Xylene	3	25	138	528	1.1	7

[a]0, will not burn under typical fire conditions; 1, must be preheated to burn, liquids with FP ≥ 93.4 °C (200 °F); 2, ignitable when moderately heated, liquids with FP between 37.8 °C (100 °F) and 93.4 °C (200 °F); 3, ignitable at ambient temperature, liquids with FP < 22.8 °C (73 °F), bp ≥ 37.8 °C (100 °F) or FP between 22.8 °C and 37.8 °C (100 °F); 4, extremely flammable, readily dispersed in air, and burns readily, liquids with FP < 22.8 °C (73 °F), bp < 37.8 °C (100 °F).
SOURCE: Adapted with permission from *Fire Guide to Hazardous Materials* (13th Edition), Copyright © 2001, National Fire Protection Association.

liquid, see Chapter 5, section 5.E.5, Table 5.2) apply to storage or transportation considerations. Another (Class A, B, C—paper, liquid, electrical fire) specifies the type of fire extinguisher to be used (see Chapter 7, section 7.F.2 on emergency equipment). To assess risk quickly, the most direct indicator is the NFPA system, which classifies flammables according to the severity of the fire hazard with numbers 0 to 4 in order of increasing hazard: 0, will not burn; 1, must be preheated to burn; 2, ignites when moderately heated; 3, ignites at normal temperature; 4, extremely flammable (Figure 4.4). Substances rated 3 or 4 under this system require particularly careful handling and storage in the laboratory. Some vendors include the NFPA hazard diamond on the labels of chemicals. The *Fire Protection Guide on Hazardous Materials* (NFPA, 2001) contains a comprehensive listing of flammability data and ratings. Note that other symbols may be found in the Special Hazard quadrant of the diamond. These symbols (see Table 4.5) are not endorsed by NFPA.

The NFPA fire hazard ratings, flash points, boiling points, ignition temperatures, and flammability limits of a number of common laboratory chemicals are given in Table 4.4 and in the LCSSs (see accompanying CD). The data illustrate the range of flammability for liquids commonly used in laboratories. Dimethyl sulfoxide

and glacial acetic acid (NFPA fire hazard ratings of 1 and 2, respectively) are handled in the laboratory without great concern about their fire hazards. By contrast, both acetone (NFPA rating 3) and diethyl ether (NFPA rating 4) have flash points well below room temperature.

Note that tabulations of properties of flammable substances are based on standard test methods, which have very different conditions from those encountered in practical laboratory use. Large safety factors should be applied. For example, the published flammability limits of vapors are for uniform mixtures with air. In a real situation, local concentrations that are much higher than the average may exist. Thus, it is good practice to set the maximum allowable concentration for safe working conditions at some fraction of the tabulated LEL; 10% is a commonly accepted value.

Among the most hazardous liquids are those that have flash points near or below 38 °C (100 °F) according to OSHA (29 CFR § 1910.106) and below 60.5 °C (140.9 °F) according to the U.S. Department of Transportation (49 CFR § 173.120). These materials can be hazardous in the common laboratory environment. There is particular risk if their range of flammability is broad. Note that some commonly used substances are potentially very hazardous, even under relatively

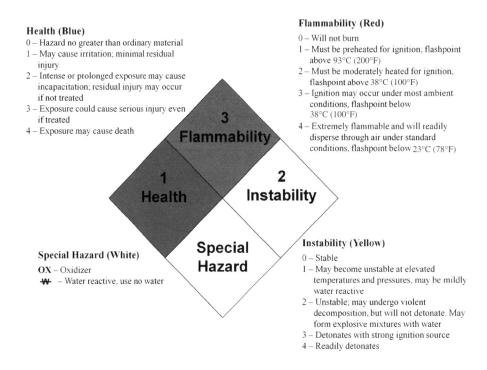

Health (Blue)
0 – Hazard no greater than ordinary material
1 – May cause irritation; minimal residual injury
2 – Intense or prolonged exposure may cause incapacitation; residual injury may occur if not treated
3 – Exposure could cause serious injury even if treated
4 – Exposure may cause death

Flammability (Red)
0 – Will not burn
1 – Must be preheated for ignition; flashpoint above 93°C (200°F)
2 – Must be moderately heated for ignition, flashpoint above 38°C (100°F)
3 – Ignition may occur under most ambient conditions, flashpoint below 38°C (100°F)
4 – Extremely flammable and will readily disperse through air under standard conditions, flashpoint below 23°C (78°F)

3 Flammability

1 Health

2 Instability

Special Hazard

Special Hazard (White)
OX – Oxidizer
W – Water reactive, use no water

Instability (Yellow)
0 – Stable
1 – May become unstable at elevated temperatures and pressures, may be mildly water reactive
2 – Unstable; may undergo violent decomposition, but will not detonate. May form explosive mixtures with water
3 – Detonates with strong ignition source
4 – Readily detonates

FIGURE 4.4 National Fire Protection Association (NFPA) system for classification of hazards.
SOURCE: Reproduced with permission from NFPA 704-2007. *System for the Identification of the Hazards of Materials for Emergency Response*, Copyright © 2007 National Fire Protection Association. This symbol is for illustrative purposes only. The NFPA does not classify individual chemicals and is not responsible in any way for the numerical values assigned to any chemical.

TABLE 4.5 Additional Symbols Seen in the NFPA Diamond

Symbol	Meaning
ACID	Acid
ALK	Alkali
CORR	Corrosive
☢	Radioactive

cool conditions (see Table 4.4). Some flammable liquids maintain their flammability even at concentrations of 10% by weight in water. Methanol and isopropyl alcohol have flash points below 38 °C (100 °F) at concentrations as low as 30% by weight in water. High-performance liquid chromatography users generate acetonitrile–water mixtures that contain from 15–30% acetonitrile in water, a waste that is considered toxic and flammable and thus cannot be added to a sewer.

Because of its extreme flammability and tendency for peroxide formation, diethyl ether is available for laboratory use only in metal containers. Carbon disulfide is almost as hazardous.

4.D.1.4 Causes of Ignition

4.D.1.4.1 Spontaneous Combustion

Spontaneous ignition (autoignition) or combustion takes place when a substance reaches its ignition temperature without the application of external heat. The possibility of spontaneous combustion should always be considered, especially when storing or disposing of materials. Examples of materials susceptible to spontaneous combustion include oily rags, dust accumulations, organic materials mixed with strong oxidizing agents (e.g., nitric acid, chlorates, permanganates, peroxides, and persulfates), alkali metals (e.g., sodium and potassium), finely divided pyrophoric metals, and phosphorus.

4.D.1.4.2 Ignition Sources

Potential ignition sources in the laboratory include the obvious torch and Bunsen burner, as well as a number of less obvious electrically powered sources ranging from refrigerators, stirring motors, and heat guns to microwave ovens (see Chapter 7, section 7.C). Whenever possible, open flames should be replaced by electrical heating. Because the vapors of most flammable liquids are heavier than air and capable of traveling considerable distances, special note should be taken of ignition sources situated at a lower level than that at which the substance is being used. Flammable vapors from massive sources such as spills have been

known to descend into stairwells and elevator shafts and ignite on a lower story. If the path of vapor within the flammable range is continuous, as along a floor or benchtop, the flame propagates itself from the point of ignition back to its source.

Metal lines and vessels discharging flammable substances should be bonded and grounded properly to discharge static electricity. There are many sources of static electricity, particularly in cold dry atmospheres, and caution should be exercised.

4.D.1.4.3 Oxidants Other Than Oxygen

The most familiar fire involves a combustible material burning in air. However, the oxidant driving a fire or explosion need not be oxygen itself, depending on the nature of the reducing agent. All oxidants have the ability to accept electrons, and fuels are reducing agents or electron donors [see Young (1991)].

Examples of nonoxygen oxidants are shown in Table 4.6. When potassium ignites on addition to water, the metal is the reducing agent and water is the oxidant. If the hydrogen produced ignites, it becomes the fuel for a conventional fire, with oxygen as the oxidant. In ammonium nitrate explosions, the ammonium cation is oxidized by the nitrate anion. These hazardous combinations are treated further in section 4.D.2. (See Chapter 6, section 6.F, for a more detailed discussion on flammable substances.)

4.D.1.5 Special Hazards

Compressed or liquefied gases present fire hazards because the heat causes the pressure to increase and the container may rupture (Yaws and Braker, 2001). Leakage or escape of flammable gases produces an explosive atmosphere in the laboratory; acetylene, hydrogen, ammonia, hydrogen sulfide, propane, and carbon monoxide are especially hazardous.

Even if not under pressure, a liquefied gas is more concentrated than in the vapor phase and evaporates rapidly. Oxygen is an extreme hazard and liquefied air is almost as dangerous because nitrogen boils away first, leaving an increasing concentration of oxygen. Liquid nitrogen standing for a period of time may have condensed enough oxygen to require careful handling. When a liquefied gas is used in a closed system, pres-

TABLE 4.6 Examples of Oxidants

Gases: chlorine, fluorine, nitrous oxide, oxygen, ozone, steam

Liquids: bromide, hydrogen peroxide, nitric acid, perchloric acid, sulfuric acid

Solids: bromates, chlorates, chlorites, chromates, dichromates, hypochlorites, iodates, nitrates, nitrites, perchlorates, peroxides, permanganates, picrates

sure may build up and adequate venting is required. If the liquid is flammable (e.g., hydrogen and methane), explosive concentrations may develop without warning unless an odorant has been added. Flammability, toxicity, and pressure buildup become more serious on exposure of gases to heat.

(Also see Chapter 6, section 6.G.2.5, for more information.)

4.D.2 Reactive Hazards

4.D.2.1 Water Reactives

Water-reactive materials are those that react violently with water. Alkali metals (e.g., lithium, sodium, and potassium), many organometallic compounds, and some hydrides react with water to produce heat and flammable hydrogen gas, which ignites or combines explosively with atmospheric oxygen. Some anhydrous metal halides (e.g., aluminum bromide), oxides (e.g., calcium oxide), and nonmetal oxides (e.g., sulfur trioxide), and halides (e.g., phosphorus pentachloride) react exothermically with water, resulting in a violent reaction if there is insufficient coolant water to dissipate the heat produced.

(See Chapter 6, section 6.G, for further information.)

4.D.2.2 Pyrophorics

For pyrophoric materials, oxidation of the compound by oxygen or moisture in air proceeds so rapidly that ignition occurs. Many finely divided metals are pyrophoric, and their degree of reactivity depends on particle size, as well as factors such as the presence of moisture and the thermodynamics of metal oxide or metal nitride formation. Other reducing agents, such as metal hydrides, alloys of reactive metals, low-valent metal salts, and iron sulfides, are also pyrophoric.

4.D.2.3 Incompatible Chemicals

Accidental contact of incompatible substances results in a serious explosion or the formation of substances that are highly toxic or flammable or both. Although trained laboratory personnel question the necessity of following storage compatibility guidelines, the reasons for such guidelines are obvious after reading descriptions of laboratories following California earthquakes in recent decades [see Pine (1994)]. Those who do not live in seismically active zones should take these accounts to heart, as well. Other natural disasters and chemical explosions themselves can set off shock waves that empty chemical shelves and result in inadvertent mixing of chemicals.

Some compounds pose either a reactive or a toxic hazard, depending on the conditions. Thus, hydro-

cyanic acid (HCN), when used as a pure liquid or gas in industrial applications, is incompatible with bases because it is stabilized against (violent) polymerization by the addition of acid inhibitor. HCN can also be formed when cyanide salt is mixed with an acid. In this case, the toxicity of HCN gas, rather than the instability of the liquid, is the characteristic of concern.

Some general guidelines lessen the risks involved with these substances. Concentrated oxidizing agents are incompatible with concentrated reducing agents. Indeed, either may pose a reactive hazard even with chemicals that are not strongly oxidizing or reducing. For example, sodium or potassium, strong reducing agents frequently used to dry organic solvents, are extremely reactive toward halocarbon solvents (which are not strong oxidizing agents). Strong oxidizing agents are frequently used to clean glassware, but they should be used only on the last traces of contaminating material. Because the magnitude of risk depends on quantities, chemical incompatibilities will not usually pose much, if any, risk if the quantity of the substance is small (a solution in an NMR tube or a microscale synthesis). However, storage of commercially obtained chemicals (e.g., in 500-g jars or 1-L bottles) should be carefully managed from the standpoint of chemical compatibility.

(For more information about compatible and incompatible chemicals, see Chapter 5, section 5.E.2.)

4.D.3 Explosive Hazards

4.D.3.1 Explosives

An explosive is any chemical compound or mechanical mixture that, when subjected to heat, impact, friction, detonation, or other suitable initiation, undergoes rapid chemical change, evolving large volumes of gases that exert pressure on the surrounding medium. The term applies to materials that either detonate or deflagrate. Heat, light, mechanical shock, and certain catalysts initiate explosive reactions. Hydrogen and chlorine react explosively in the presence of light. Acids, bases, and other substances catalyze the explosive polymerization of acrolein, and many metal ions can catalyze the violent decomposition of hydrogen peroxide. Shock-sensitive materials include acetylides, azides, nitrogen triiodide, organic nitrates, nitro compounds, perchlorate salts (especially those of heavy metals such as ruthenium and osmium), many organic peroxides, and compounds containing diazo, halamine, nitroso, and ozonide functional groups.

Table 4.7 lists a number of explosive compounds. Some are set off by the action of a metal spatula on the solid; some are so sensitive that they are set off by the action of their own crystal formation. Diazomethane (CH_2N_2) and organic azides, for example, may decom-

TABLE 4.7 Functional Groups in Some Explosive Compounds

Structural Feature	Compound	Structural Feature	Compound
— C ≡ C —	Acetylenic compounds	\C—N=N—S—N=N—C/	Bis-arenediazo sulfides
— C ≡ C —Metal	Metal acetylides	—C—N=N—N—C R	Trizazenes (R = H, —CN, –OH, —NO)
— C ≡ C —X	Haloacetylene derivatives	N=N—N=N	High-nitrogen compounds, tetrazoles
N=N C	Diazirines	—C—O—O—H	Alkylhydroperoxides
CN₂	Diazo compounds	—C—CO—COOH	Peroxyacids
—C—N=O	Nitroso compounds	—C—O—O—C—	Peroxides (cyclic, diacyl, dialkyl)
—C—NO₂	Nitroalkanes, C-nitro and polynitroaryl compounds	—C—CO—COOR	Peroxyesters
NO₂ C NO₂	Polynitroalkyl compounds	—O—O—Metal	Metal peroxides, peroxoacid salts
—C—O—N=O	Acyl or alkyl nitrites	—O—O—Non-metal	Peroxoacids
—C—O—NO₂	Acyl or alkyl nitrates	N→Cr—O₂	Aminechromium peroxocomplexes
O C—C	1,2-Epoxides	—N₃	Azides (acyl, halogen, nonmetal, organic)
C=N—O—Metal	Metal fulminates or aci-nitro salts	—C—N₂⁺S⁻	Diazoniumsulfides and derivatives, "xanthates"
NO₂ —C—F NO₂	Fluorodinitromethyl compounds	N⁺–HZ⁻	Hydrazinium salts, oxosalts of nitrogenous bases
N—Metal	N-Metal derivatives	—N⁺–OH Z⁻	Hydroxylammonium salts
N—N=O	N-Nitroso compounds	—C—N₂⁺Z⁻	Diazonium carboxylates or salts
N—NO₂	N-Nitro compounds	(N-Metal)⁺Z⁻	Aminemetal oxosalts
—C—N=N—C—	Azo compounds	Ar—Metal—X X—Ar—Metal	Halo-arylmetals
—C—N=N—O—C—	Arenediazoates	N—X	Halogen azides, N-halogen compounds, N-haloimides
—C—N=N—S—C—	Arenediazo aryl sulfides	—NF₂	Difluoroamino compounds
—C—N=N—O—N=N—C—	Bis-arenediazo oxides	—O—X	Alkyl perchlorates, chlorite salts, halogen oxides, hypohalites, perchloric acid, perchloryl compounds

pose explosively when exposed to a ground glass joint or other sharp surfaces (*Organic Syntheses*, 1973, 1961).

4.D.3.2 Azos, Peroxides, and Peroxidizables

Organic azo compounds and peroxides are among the most hazardous substances handled in the chemical laboratory but are also common reagents that often are used as free radical sources and oxidants. They are generally low-power explosives that are sensitive to shock, sparks, or other accidental ignition. They are far more shock sensitive than most primary explosives such as TNT. Inventories of these chemicals should be limited and subject to routine inspection. Many require refrigerated storage. Liquids or solutions of these compounds should not be cooled to the point at which the material freezes or crystallizes from solution, however, because this significantly increases the risk of explosion. Refrigerators and freezers storing such compounds should have a backup power supply in the event of electricity loss. Users should be familiar with the hazards of these materials and trained in their proper handling.

Certain common laboratory chemicals form peroxides on exposure to oxygen in air (see Tables 4.8 and 4.9). Over time, some chemicals continue to build peroxides to potentially dangerous levels, whereas others accumulate a relatively low equilibrium concentration of peroxide, which becomes dangerous only after being concentrated by evaporation or distillation. (See Chapter 6, section 6.G.3.) The peroxide becomes concentrated because it is less volatile than the parent chemical. A related class of compounds includes inhibitor-free monomers prone to free radical polymerization that on exposure to air can form peroxides or other free radical sources capable of initiating violent polymerization. Note that care must be taken when storing and using these monomers—most of the inhibitors used to stabilize these compounds require the presence of oxygen to function properly, as described below. Always refer to the MSDS and supplier instructions for proper use and storage of polymerizable monomers.

Essentially all compounds containing C—H bonds pose the risk of peroxide formation if contaminated with various radical initiators, photosensitizers, or catalysts. For instance, secondary alcohols such as isopropanol form peroxides when exposed to normal fluorescent lighting and contaminated with photosensitizers, such as benzophenone. Acetaldehyde, under normal conditions, autoxidizes to form acetic acid. Although this autoxidation proceeds through a peroxy acid intermediate, the steady-state concentrations of that intermediate are extremely low and pose no hazard. However, in the presence of catalysts (Co^{2+})

TABLE 4.8 Classes of Chemicals That Can Form Peroxides

Class A: Chemicals that form explosive levels of peroxides without concentration

Isopropyl ether	Sodium amide (sodamide)
Butadiene	Tetrafluoroethylene
Chlorobutadiene (chloroprene)	Divinyl acetylene
Potassium amide	Vinylidene chloride
Potassium metal	

Class B: These chemicals are a peroxide hazard on concentration (distillation/evaporation). A test for peroxide should be performed if concentration is intended or suspected.* (See Chapter 6, section 6.C.3)

Acetal	Dioxane (*p*-dioxane)
Cumene	Ethylene glycol dimethyl
Cyclohexene	ether (glyme)
Cyclooctene	Furan
Cyclopentene	Methyl acetylene
Diaacetylene	Methyl cyclopentane
Dicyclopentadiene	Methyl-isobutyl ketone
Diethylene glycol dimethyl	Tetrahydrofuran
ether (diglyme)	Tetrahydronaphthalene
Diethyl ether	Vinyl ethers

Class C: Unsaturated monomers that may autopolymerize as a result of peroxide accumulation if inhibitors have been removed or are depleted*[a]*

Acrylic acid	Styrene
Butadiene	Vinyl acetate
Chlorotrifluoroethylene	Vinyl chloride
Ethyl acrylate	Vinyl pyridine
Methyl methacrylate	

*These lists are illustrative, not comprehensive.
SOURCES: Jackson et al. (1970) and Kelly (1996).

and under the proper conditions of ultraviolet light, temperature, and oxygen concentration, high concentrations of an explosive peroxide can be formed. The chemicals described in Table 4.9 represent only those materials that form peroxides in the absence of such contaminants or otherwise atypical circumstances.

Although not a requirement, it is prudent to discard old samples of organic compounds of unknown origin or history, or those prone to peroxidation if contaminated; secondary alcohols are a specific example.

Class A compounds are especially dangerous when

TABLE 4.9 Types of Compounds Known to Autoxidize to Form Peroxides

Ethers containing primary and secondary alkyl groups (never distill an ether before it has been shown to be free of peroxide)
Compounds containing benzylic hydrogens
Compounds containing allylic hydrogens (C=C—CH)
Compounds containing a tertiary C—H group (e.g., decalin and 2,5-dimethylhexane
Compounds containing conjugated, polyunsaturated alkenes and alkynes (e.g., 1,3-butadiene, vinyl acetylene)
Compounds containing secondary or tertiary C—H groups adjacent to an amide (e.g., 1-methyl-2-pyrrolidinone)

peroxidized and should not be stored for long periods in the laboratory. Good practice requires they be discarded within 3 months of receipt. Inventories of Class B and C materials should be kept to a minimum and managed on a first-in, first-out basis. Class B and C materials should be stored in dark locations. If stored in glass bottles, the glass should be amber. Containers should be marked with their opening date and inspected every 6 months thereafter.

Class B materials are often sold with autoxidation inhibitors. If the inhibitor is removed, or if inhibitor-free material is purchased, particular care must be taken in their long-term storage because of the enhanced probability of peroxide formation. Purging the container headspace with nitrogen is recommended. Several procedures, including test strips, are available to check Class B materials for peroxide contamination. (For information about testing for peroxides, see Chapter 6, section 6.G.3.2.) No special disposal precautions are required for peroxide-contaminated Class B materials.

In most cases, commercial samples of Class C materials are provided with polymerization inhibitors that **require** the presence of oxygen to function and therefore are not to be stored under inert atmosphere. Inhibitor-free samples of Class C compounds (i.e., the compound has been synthesized in the laboratory or the inhibitor has been removed from the commercial sample) should be kept in the smallest quantities required and under inert atmosphere. Unused material should be properly disposed of immediately, or if long-term storage is necessary, an appropriate inhibitor should be added.

(For more information about handling of peroxides, see Chapter 6, section 6.G.3.)

4.D.3.3 Other Oxidizers

Oxidizing agents may react violently when they come into contact with reducing materials and sometimes with ordinary combustibles. Such oxidizing agents include halogens, oxyhalogens and organic peroxyhalogens, chromates, and persulfates as well as peroxides. Inorganic peroxides are generally stable. However, they may generate organic peroxides and hydroperoxides in contact with organic compounds, react violently with water (alkali metal peroxides), and form superoxides and ozonides (alkali metal peroxides). Perchloric acid is a powerful oxidizing agent with organic compounds and other reducing agents. Perchlorate salts are explosive and should be treated as potentially hazardous compounds.

Baths to clean glassware generally contain strong oxidizers and should be handled with care. For many years, sulfuric acid–dichromate mixtures were used to clean glassware. These solutions are corrosive and toxic and present difficulties for disposal. Their use should be avoided if at all possible. A common substitute is a sulfuric acid–peroxydisulfate solution, and commercial cleaning solutions that contain no chromium are readily available. Confusion about appropriate cleaning bath solutions has led to explosions due to mixing of incompatible chemicals such as potassium permanganate with sulfuric acid or nitric acid with alcohols. For information about how to clean glassware appropriately, consider contacting the manufacturer of the equipment.

4.D.3.4 Powders and Dusts

Suspensions of oxidizable particles (e.g., flour, coal dust, magnesium powder, zinc dust, carbon powder, and flowers of sulfur) in the air constitute a powerful explosive mixture. These materials should be used with adequate ventilation and should not be exposed to ignition sources. Some solid materials, when finely divided, spontaneously combust if allowed to dry while exposed to air. These materials include zirconium, titanium, Raney nickel, finely divided lead (such as prepared by pyrolysis of lead tartrate), and catalysts such as activated carbon containing active metals and hydrogen.

4.D.3.5 Explosive Boiling

Not all explosions result from chemical reactions; some are caused physically. A dangerous explosion can occur if a hot liquid or a collection of very hot particles comes into sudden contact with a lower boiling-point material. Sudden boiling eruptions occur when a nucleating agent (e.g., charcoal, "boiling chips") is added to a liquid heated above its boiling point. Even if the material does not explode directly, the sudden formation of a mass of explosive or flammable vapor can be very dangerous.

4.D.3.6 Other Considerations

The hazards of running a new reaction should be considered especially carefully if the chemical species involved contain functional groups associated with explosions or are unstable near the reaction or work-up temperature, if the reaction is subject to an induction period, or if gases are byproducts. Modern analytical techniques (see Chapter 6, section 6.G) can be used to determine reaction exothermicity under suitable conditions.

Even a small sample may be dangerous. Furthermore, the hazard is associated not with the total energy released but with the remarkably high rate of a detonation reaction. A high-order explosion of even milligram quantities can drive small fragments of glass or other matter deep into the body; therefore, use minimum amounts of these hazardous materials with adequate shielding and personal protection. A compound is apt to be explosive if its heat of formation is more than 100 cal/g less than the sum of the heats of formation of its products. In making this calculation, a reasonable reaction should be used to yield the most exothermic products.

Scaling up reactions introduces several hazards. Unfortunately, the current use of microscale teaching methods in undergraduate laboratories increases the likelihood that graduate students and others are unprepared for problems that arise when a reaction is run on a larger scale. These problems include heat buildup and the serious hazard of explosion from incompatible materials. The rate of heat input and production must be weighed against that of heat removal. Bumping the solution or a runaway reaction can result when heat builds up too rapidly.

Exothermic reactions can run away if the heat evolved is not dissipated. When scaling up experiments, sufficient cooling and surface for heat exchange should be provided, and mixing and stirring rates should be considered. Detailed guidelines for circumstances that require a systematic hazard evaluation and thermal analysis are given in Chapter 6, section 6.G.

Another situation that can lead to problems is a reaction susceptible to an induction period; particular care must be given to the rate of reagent addition versus its rate of consumption. Finally, the hazards of exothermic reactions or unstable or reactive chemicals are exacerbated under extreme conditions, such as high temperature or high pressure used for hydrogenations, oxygenations, or work with supercritical fluids.

4.E PHYSICAL HAZARDS

4.E.1 Compressed Gases

Compressed gases can expose the trained laboratory personnel to both mechanical and chemical hazards, depending on the gas. Hazards can result from the flammability, reactivity, or toxicity of the gas; from the possibility of asphyxiation; and from the gas compression itself, which could lead to a rupture of the tank or valve. (See Chapter 7, section 7.D.)

4.E.2 Nonflammable Cryogens

Nonflammable cryogens (chiefly liquid nitrogen) can cause tissue damage from extreme cold because of contact with either liquid or boil-off gases. In poorly ventilated areas, inhalation of gas due to boil off or spills can result in asphyxiation. Another hazard is explosion from liquid oxygen condensation in vacuum traps or from ice plug formation or lack of functioning vent valves in storage Dewars. Because 1 volume of liquid nitrogen at atmospheric pressure vaporizes to 694 volumes of nitrogen gas at 20 °C, the warming of such a cryogenic liquid in a sealed container produces enormous pressure, which can rupture the vessel. (See Chapter 6, section 6.G.4, and Chapter 7, section 7.E.2, for detailed discussion.)

4.E.3 High-Pressure Reactions

Experiments that generate high pressures or are carried out at pressures above 1 atm can lead to explosion from equipment failure. For example, hydrogenation reactions are frequently carried out at elevated pressures, and a potential hazard is the formation of explosive O_2/H_2 mixtures and the reactivity/pyrophoricity of the catalyst (see section 6.G.5). High pressures can also be associated with the use of supercritical fluids.

When evaluating whether a reaction generates high pressures, it is important to consider not just the initial reaction conditions, but the kinetics and thermodynamics of the reaction as a whole. Is any stage of the reaction exothermic? What are the characteristics of the reactants, products, intermediates, and synthetic byproducts (explosive, gaseous, etc.)? What are the temperature and pressure requirements for equipment used during the reaction? If scaling up a reaction, carefully calculate the expected temperatures and pressures that will be generated and the rates at which any pressures will be generated. Be sure to choose laboratory equipment that is appropriate for every stage of the reaction, and consult with the manufacturer if there are any questions or concerns about whether a given reactor or piece of equipment is appropriate for high-pressure work. (For more information about using high-pressure equipment, see Chapter 7, section 7.E.)

In many cases, barricading is not necessary if the appropriate reaction vessel, fittings, and other equipment are used. However, the laboratory environment must be designed to accommodate the failure of the equipment: ventilation must be adequate to handle discharge from a high-pressure reaction to prevent asphyxiation, laboratory personnel may require hearing protection to guard against the sound of a rupture disc failure, and barricades are necessary if catastrophic failure could result in injury or death of laboratory personnel. For specific information regarding barricade design, see Porter et al. (1956); Smith (1964); and the *Handbook of Chemical Health and Safety* (Alaimo, 2001).

4.E.4 Vacuum Work

Precautions to be taken when working with vacuum lines and other glassware used at subambient pressure are mainly concerned with the substantial danger of injury in the event of glass breakage. The degree of hazard does not depend significantly on the magnitude of the vacuum because the external pressure leading to implosion is always 1 atmosphere. Thus, evacuated systems using aspirators merit as much respect as high-vacuum systems. Injury due to flying glass is not the only hazard in vacuum work. Additional dangers can result from the possible toxicity of the chemicals contained in the vacuum system, as well as from fire following breakage of a flask (e.g., of a solvent stored over sodium or potassium). (For more information about working with equipment under vacuum, see Chapter 7, section 7.E.)

Because vacuum lines typically require cold traps (generally liquid nitrogen) between the pumps and the vacuum line, precautions regarding the use of cryogens should be observed also. Health hazards associated with vacuum gauges have been reviewed (Peacock, 1993). The hazards include the toxicity of mercury used in manometers and McLeod gauges, overpressure and underpressure situations arising with thermal conductivity gauges, electric shock with hot cathode ionization systems, and the radioactivity of the thorium dioxide used in some cathodes. (For information about reducing the presence of mercury in laboratories, see Chapter 5, section 5.B.8.)

4.E.5 Ultraviolet, Visible, and Near-Infrared Radiation

Ultraviolet, visible, and infrared radiation from lamps and lasers in the laboratory can produce a number of hazards. Medium-pressure Hanovia 450 Hg lamps are commonly used for ultraviolet irradiation in photochemical experiments. Ultraviolet lights used in biosafety cabinets, as decontamination devices, or in light boxes to visualize DNA can cause serious skin and corneal burns. Powerful arc lamps can cause eye damage and blindness within seconds. Some compounds (e.g., chlorine dioxide) are explosively photosensitive.

When incorrectly used, the light from lasers poses a hazard to the eyes of the operators and other people present in the room and is also a potential fire hazard. Depending on the type of laser, the associated hazards can include mutagenic, carcinogenic, or otherwise toxic laser dyes and solvents; flammable solvents; ultraviolet or visible radiation from the pump lamps; and electric shock from lamp power supplies.

At the time of this publication, two systems for classifying lasers are in use. Before 2002, lasers were classified as I, II, IIIA, IIIB, and IV. From 2002 forward, a revised system is being phased in which classifies lasers as 1, 1M, 2, 2M, 3R, 3B, and 4. Although they have different designations, both systems classify lasers based on their ability to cause damage to individuals. The older designation is given in the text with the new designation in parentheses. Class I (1) lasers are either completely enclosed or have such a low output of power that even a direct beam in the eye could not cause damage. Class II (2) lasers, can be a hazard if a person stares into the beam and resists the natural reaction to blink or turn away. Class IIIA (1M, 2M, or 3R, depending on power output) lasers can present an eye hazard if a person stares into the beam and resists the natural reaction to blink or turn away or views the beam with focusing optical instruments. Class IIIB (3B) lasers can produce eye injuries instantly from both direct and specularly reflected beams, although diffuse reflections are not hazardous. The highest class of lasers, Class IV (4), presents all the hazards of Class III (3B) lasers but because of their higher power output may also produce eye or skin damage from diffuse scattered light. In addition to these skin and eye hazards, Class IV (4) lasers are a potential fire hazard.

Select protective eyewear with the proper optical density for the specific type of laser in use. Dark lenses can be hazardous because of the risk of looking over the top of the glasses. Leave laser safety glasses in a bin outside the laboratory so that people entering use the appropriate laser safety glasses. When operating or adjusting a laser, remove or cover any reflective objects on hands and wrists to reduce the chance of reflections. Consider using beam blocks and containment walls to reduce the chance of stray reflections in the laboratory. When using a laser-based microscope, consider using a camera and computer display to view the sample rather than direct viewing through the eyepiece. Anyone who is not the authorized operator of a laser system should never enter a posted laser-controlled laboratory if the laser is in use. Visitors may be present when a laser is in use, but they must be authorized by the laboratory supervisor. Visitors must not operate the equipment and should be under the direct supervision of an approved operator.

4.E.6 Radio Frequency and Microwave Hazards

Radio frequency (rf) and microwaves occur within the range 10 kHz to 300,000 MHz and are used in rf ovens and furnaces, induction heaters, and microwave ovens. Extreme overexposure to microwaves can result in the development of cataracts or sterility or both. Microwave ovens are increasingly being used in laboratories for organic synthesis and digestion of analytical samples. Only microwave ovens designed for laboratory or industrial use should be used in a laboratory. Use of metal in microwave ovens can result in arcing and, if a flammable solvent is present, in fire or explosion. Superheating of liquids can occur. Capping of vials and other containers used in the oven can result in explosion from pressure buildup within the vial. Inappropriately selected plastic containers may melt.

4.E.7 Electrical Hazards

The electrocution hazards of electrically powered instruments, tools, and other equipment are almost eliminated by taking reasonable precautions, and the presence of electrically powered equipment in the laboratory need not pose a significant risk. Many electrically powered devices are used in homes and workplaces in the United States, often with little awareness of the safety features incorporated in their design and construction. But, in the laboratory these safety features should not be defeated by thoughtless or ill-informed modification. The possibility of serious injury or death by electrocution is very real if careful attention is not paid to engineering, maintenance, and personal work practices. Equipment malfunctions can lead to electrical fires. If there is a need to build, repair, or modify electrical equipment, the work should ideally be performed or, at a minimum, inspected by a trained and licensed electrician or electrical expert. All laboratory personnel should know the location of electrical shutoff switches and circuit breaker switches and should know how to turn off power to burning equipment by using these switches. Laboratory equipment should be correctly bonded and grounded to reduce the chances of electric shock if a fault occurs.

Some special concerns arise in laboratory settings. The insulation on wires can be eroded by corrosive chemicals, organic solvent vapors, or ozone (from ultraviolet lights, copying machines, and so forth). Eroded insulation on electrical equipment in wet locations such as cold rooms or cooling baths must be repaired immediately. In addition, sparks from electrical equipment can serve as an ignition source in the presence of flammable vapor. Operation of certain equipment (e.g., lasers, electrophoresis equipment) may involve high voltages and stored electrical energy. The large capacitors used in many flash lamps and other systems are capable of storing lethal amounts of electrical energy and should be regarded as live even if the power source has been disconnected.

Loss of electrical power can produce extremely hazardous situations. Flammable or toxic vapors may be released from freezers and refrigerators as chemicals stored there warm up; certain reactive materials may decompose energetically on warming. Laboratory chemical hoods may cease to function. Stirring (motor or magnetic) required for safe reagent mixing may cease. Return of power to an area containing flammable vapors may ignite them.

4.E.8 Magnetic Fields

Increasingly, instruments that generate large static magnetic fields (e.g., NMR spectrometers) are present in research laboratories. Such magnets typically have fields of 14,000 to 235,000 G (1.4 to 23.5 T), far above that of Earth's magnetic field, which is approximately 0.5 G. The magnitude of these large static magnetic fields falls off rapidly with distance. Many instruments now have internal shielding, which reduces the strength of the magnetic field outside of the instrument (see Chapter 7, Table 7.1). Strong attraction occurs when the magnetic field is greater than 50 to 100 G and increases by the seventh power as the separation is reduced. However, this highly nonlinear falloff of magnetic field with distance results in an insidious hazard. Objects made of ferromagnetic materials such as ordinary steel may be scarcely affected beyond a certain distance, but at a slightly shorter distance may experience a significant attraction to the field. If the object is able to move closer, the attraction force increases rapidly, and the object can become a projectile aimed at the magnet. Objects ranging from scissors, knives, wrenches, and other tools, keys, steel gas cylinders, buffing machines, and wheelchairs have been pulled from a considerable distance to the magnet itself.

Superconducting magnets use liquid nitrogen and liquid helium coolants. Thus, the hazards associated with cryogenic liquids (see section 4.E.2) are of concern, as well.

The health effects of exposure to static magnetic fields is an area of active research. Currently, there is no clear evidence of a negative health impact from exposure to static magnetic fields, although biological effects have been observed (Schenck, 2000), and recently, guidelines on limits of exposure to static magnetic fields have been issued by the International Commission on Non-ionizing Radiation (ICNIRP, 2009), which is a collaborating organization with the World Health Organization's International Electromagnetic Field Project.

(For more information about magnetic fields, see Chapter 7, section 7.C.8.4.1.)

4.E.9 Sharp Edges

Among the most common injuries in laboratories are cuts from broken glass. Cuts can be minimized by the use of correct procedures (e.g., the procedure for inserting glass tubing into rubber stoppers and tubing, which is taught in introductory laboratories), through the appropriate use of protective equipment, and by careful attention to manipulation. Glassware should always be checked for chips and cracks before use and discarded if any are found. Never dispose of glass in the general laboratory trash. It should only be placed in specific glassware disposal bins. This will reduce the chance of anyone changing the trash receiving a cut.

Other cut hazards include razors, box cutters, knives, wire cutters, and any other sharp-edged tool. When working with these tools, it is important to wear appropriate eye protection and cut-resistant gloves. Follow basic safety procedures when using a cutting tool:

- Inspect the tool prior to use. Do not use it if it is damaged.
- When cutting, always use a tool with a sharp edge. Dull edges are more likely to slip and cause harm.
- Keep hands out of the line of the cut.
- Stand off-line from the direction of the cut.
- If using a box cutter or other tool with a mounted blade, ensure that the blade is well seated before use.
- Never use a cutting tool for a task for which it was not designed, for example, as a screwdriver or lever for opening a container.
- Never submerge a sharp object in soapy or dirty water. It can be difficult to see and poses a risk to the dishwasher.

4.E.10 Slips, Trips, and Falls

Other common injuries in the laboratory arise from slipping, tripping, or improper lifting. Spills resulting from dropping chemicals not stored in protective rubber buckets or laboratory carts can be serious because the laboratory worker can fall or slip into the spilled chemical, thereby risking injury from both the fall and exposure to the chemical. Chemical spills resulting from tripping over bottles of chemicals stored on laboratory floors are part of a general pattern of bad housekeeping that can also lead to serious accidents. Wet floors around ice, dry ice, or liquid nitrogen dispensers can be slippery if the areas are not carpeted and if drops or small puddles are not wiped up as soon as they form.

Attempts to retrieve 5-gallon bottles of distilled water, jars of bulk chemicals, and rarely used equipment stored on high shelves often lead to back injuries in laboratory environments. Careful planning of where to store difficult-to-handle equipment and containers (because of weight, shape, or overall size) reduces the incidence of back injuries.

4.E.11 Ergonomic Hazards in the Laboratory

General workplace hazards also apply in the laboratory. For example, laboratory personnel are often involved in actions such as pipetting and computer work that can result in repetitive-motion injuries. Working at a bench or at a microscope without considering posture can result in back strain, and some instruments require additional in-room ventilation that may raise the background noise level to uncomfortable or hazardous levels. With these and other issues such as high or low room temperatures and exposure to vibrations, it is important to be aware of and to control such issues to reduce occupational injuries. For example, microscope users may find that using a camera to view images on a screen, rather than direct viewing through the eyepiece, reduces back and eye strain.

The Centers for Disease Control and Prevention (CDC) and the National Institutes of Health have information on their Web sites (www.cdc.gov and www.nih.gov, respectively) describing specific ergonomic concerns for laboratories and proposed solutions. The CDC provides a downloadable self-assessment form to aid in evaluating these hazards. NIOSH (www.cdc.gov/niosh) and OSHA (www.osha.gov) provide information about vibration, noise levels, and other workplace hazards.

4.F NANOMATERIALS

Nanoscale materials are of considerable scientific interest because some chemical and physical properties can change at this scale. (See definition of engineered nanomaterials below.) These changes challenge the researcher's, manager's, and safety professional's understanding of hazards, and their ability to anticipate, recognize, evaluate, and control potential health, safety, and environmental risks. Essentially any solid may be formed in the nano size range, and in general, the term "nanomaterials" has been broadly accepted as including a number of nanometer-scale objects, including: nanoplates, nanofibers (including nanotubes); and nanoparticles. In addition to the conventional hazards posed by the material, hazard properties may also change.

Nanoparticles are dispersible particles that are between 1 and 100 nm in size that may or may not exhibit a size-related intensive property. The U.S. Department of Energy (DOE, 2008, 2009) states that *engineered nanomaterials* are intentionally created, in contrast with natural or incidentally formed, and engineered to be between 1 and 100 nm. This definition

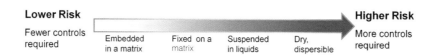

FIGURE 4.5 U.S. Department of Energy graded exposure risk for nanomaterials. This figure assumes that no disruptive force (e.g., sonication, grinding, burning) is applied to the matrix. SOURCE: Adapted from Karn (2008).

excludes biomolecules (proteins, nucleic acids, and carbohydrates).[3] Incidentally formed nanoparticles are often called "ultrafine" particles.

As with hazardous chemicals, exposures to these materials may occur through inhalation, dermal contact, accidental injection, and ingestion, and the risk increases with duration of exposure and the concentration of nanoparticles in the sample or air. Inhalation presents the greatest exposure hazard. Nanomaterials suspended in a solution or slurry pose a lesser hazard, but because the solutions can dry into a powder, they should be handled with care. Nanomaterials suspended in a solution or slurry present a hazard whenever mechanical energy is imparted to the suspension of slurry. Sonication, shaking, stirring, pouring, or spraying of a suspended nanomaterial can result in an inhalation exposure. Suspensions also represent a dermal exposure potential. Nanoparticles that are fixed within a matrix pose the least hazard as long as no mechanical disruption, such as grinding, cutting, or burning, occurs. (See Figure 4.5.)

Nanoparticles can enter the laboratory in a variety of ways. For example, the materials may be imported into the lab for characterizations or be incorporated into a study. Alternatively, they could be created (synthesized) in the lab as part of an experiment. In either case, it is important for laboratory personnel to know about the presence and physical state of the nanomaterial (i.e., powder, in solution, on a solid matrix, or in solid matrix) so they can manage the hazards accordingly.

Nanoparticles have significantly greater relative surface areas than larger particles of an equivalent mass, and animal studies have demonstrated a correlation between biological effects (toxic response) and surface

area. Thus, nanoparticles represent a greater toxic hazard than an equivalent mass of the same material in larger form. In addition, the number of *particles* per unit mass is far greater than the number of particles in bulk material per unit mass, resulting in significantly different inhalational hazards between the two forms. Because of their size, nanoparticles can penetrate deep into the lungs, and with a large number of particles in a small volume, can overwhelm the organ and disrupt normal clearance processes. The greater surface reactivity also plays a role in this disruption. Once inside the lungs, nanoparticles may translocate to other organs via pathways not demonstrated in studies with larger particles. In addition, at the interface of the nanoparticle and human cell surface, bioactivity may occur. For example, nanometal particles have been demonstrated to produce reactive oxygen species, implicating the presence of free radicals, and causing the biological effects of inflammation and fibrosis.

The nanoparticulate forms of some materials show unusually high reactivity, especially for fire, explosion, and catalytic reactions. Engineered nanoparticles and nanostructured porous materials have been used effectively for many years as catalysts for increasing the rate of reactions or decreasing the temperature needed for reactions in liquids and gases. Depending on their composition and structure, some nanomaterials initiate catalytic reactions that would not otherwise be anticipated from their chemical composition. Note also that nanomaterials may be attached to the surface of larger particles. In those cases, the larger material may take on the higher reactivity features of the engineered nanoscale material, even though it is not in the form of a particle in the 1- to 100-nm size range.

As noted above, because material properties can change at the nanoscale, nanomaterials should not be assumed to present only those hazards known to be associated with bulk forms of material having the same composition. Instead, they must be handled as though toxic and reactive until credible evidence eliminates uncertainty. Hazard information is available on a limited number of nano-size materials. For example, NIOSH has proposed special exposure limits for nano-size titanium dioxide that are significantly more restrictive than for larger particles of titanium

---

[3]Note that this definition is slightly different from the definition of the International Organization for Standardization, where "*nano-object* is defined as material with one, two, or three external dimensions in the size range of approximately 1–100 nm. Subcategories of nano-object are (1) *nanoplate*, a nano-object with one external dimension at the nanoscale; (2) *nanofiber*, a nano-object with two external dimensions at the nanoscale with a nanotube defined as a hollow nanofiber and a nanorod as a solid nanofiber; and (3) *nanoparticle*, a nano-object with all three external dimensions at the nanoscale. Nano-objects are commonly incorporated in a larger matrix or substrate referred to as a *nanomaterial*" (HHS/CDC/NIOSH, 2009a).

dioxide. Determination of EHS issues is an ongoing effort. The CHO assisting with protection from the EHS hazards will need special education and training to adequately assist in risk assessment and control of nanomaterial risks. Specialized monitoring equipment is required to evaluate potential exposures or release of nanomaterials.

Although there is limited specific guidance on evaluation and control of risks posed by nanomaterials, preliminary research suggests that a well-designed ventilation system with high-efficiency filtration is effective at capturing nanoparticles. However, recent studies (Ellenbecker and Tsai, 2008) have demonstrated that conventional laboratory chemical hoods may create turbulence that can push the materials back into the laboratory space. Lower flow hoods with less turbulence may be more appropriate. (For more information about engineering controls for handling of nanoparticles, see Chapter 9, section 9.E.5. For further information on transportation, see Chapter 5, section 5.F.2 and Chapter 6, section 6.J for information about working with nanoparticles.)

4.G BIOHAZARDS

Biohazards are a concern in laboratories in which microorganisms, or material contaminated with them, are handled. Anyone who is likely to come in contact with blood or potentially infectious materials at work is covered under OSHA's Bloodborne Pathogen Standard, 29 CFR § 1910.1030. These hazards are usually present in clinical and infectious disease research laboratories but may also be present in any laboratory in which bodily fluids, tissues, or primary or immortalized cell lines of human or animal origin are handled. Biohazards are also present in any laboratory that uses microorganisms, including replication-deficient viral vectors, for protein expression or other *in vitro* applications. Occasionally, biohazards are present in testing and quality control laboratories, particularly those associated with water and sewage treatment plants and facilities involved in the production of biological products and disinfectants. Teaching laboratories may introduce low-risk infectious agents as part of a course of study in microbiology.

Synthetic biology makes it possible to synthesize microorganisms from basic chemical building blocks, and these microorganisms may have different hazards from their naturally occurring relatives. If a microorganism identical or very similar to one found in nature is synthesized, the risks are assumed to be similar to those of the naturally occurring microorganism. If a novel microorganism is synthesized, however, extra caution must be used until the characteristics of the agent are well understood.

Risk assessment for biohazardous materials can be complicated because of the number of factors that must be considered. The things that must be accounted for are the organism being manipulated, any alterations made to the organism, and the activities that will be performed with the organism. Risk assessment for biological toxins is similar to that for chemical agents and is based primarily on the potency of the toxin, the amount used, and the procedures in which the toxin is used. An example of a risk assessment for a material with unknown biological risks can be found in Backus et al. (2001). See Box 4.3 for a quick guide to assessing risks from biohazards in the laboratory.

Certain biological toxins and agents are classified as select agents under 42 CFR Part 73 and have additional regulatory and security requirements that must be considered when receiving and working with these agents. For detailed information on risk assessment of biohazards, consult the fifth and most recent edition of *Biosafety in Microbiological and Biomedical Laboratories* (BMBL; HHS/CDC/NIH, 2007a) and the *NIH Guidelines for Research Involving Recombinant DNA Molecules* (NIH, 2009). BMBL is considered the consensus code of practice for identifying and controlling biohazards and was first produced by the CDC and the National Institutes of Health in 1984. (Also see Chapter 6, section 6.E, and Chapter 11.)

4.H HAZARDS FROM RADIOACTIVITY

This section provides a brief primer on the potential hazards arising from the use of radioactivity in a laboratory setting. A comprehensive treatment of this topic is given in *Radiation Protection: A Guide for Scientists, Regulators, and Physicians* (Shapiro, 2002). For an introduction to health physics, see Cember and Johnson (2008). Note that the receipt, possession, use, transfer, and disposal of most radioactive materials is strictly regulated by the U.S. Nuclear Regulatory Commission (USNRC; see 10 CFR Part 20, Standards for Protection Against Radiation) and/or by state agencies who have "agreements" with the USNRC to regulate the users within their own states. Radioactive materials may be used only for purposes specifically described in licenses issued by this agency to licensees. Individuals working with radioactive materials should thus be aware of the restrictions and requirements of these licenses. Consult your radiation safety officer or other designated EHS professional for training, policies, and procedures specific to uses at your institution.

Unstable atomic nuclei eventually achieve a more stable form by emission of some type of radiation. These nuclei or isotopes are termed radioactive. The emitted radiation may be characterized as particulate (α, β, proton, or neutron) or electromagnetic (γ rays

---

BOX 4.3
Quick Guide to Risk Assessment for Biological Hazards in the Laboratory

The following steps are provided to assist trained laboratory personnel in performing a risk assessment of activities involving biohazardous materials. This is not intended as a flowchart; rather, all questions should be considered before arriving at the final risk assessment.

1. **Identify the risk group of the parent organism** (the American Biological Safety Association has a database of risk groups from various sources online at http://absa.org).

2. **This Risk Group assignment is only a starting point**; the actual biosafety level at which the work is performed may be higher or lower depending on the remainder of the risk assessment.

3. **Refer to the Agent Summary Statement in Biosafety in Microbiological and Biomedical Laboratories (BMBL; HHS/CDC/NIH, 2007a) if available.** This will provide a recommended biosafety level as well as personal protective and containment equipment to use while handling the organism. It will also summarize the frequency and route of laboratory-acquired infections.

4. **Identify the natural route of transmission for the parent organism.** This will indicate the most likely route of laboratory-acquired infections. Be aware that at the volumes and concentrations used in the laboratory, however, organisms can often be transmitted in ways other than their normal route of exposure (e.g., aerosol transmission of agents normally only transmitted via mosquito bite).

5. **Consider any modifications made to the organism.** Has the host range been modified? Have virulence factors been inserted or removed? Has the organism been rendered replication-defective? If so, is there any possibility of recombination events with wild-type organisms to restore replication competency?

6. **Consider the transgenes expressed by the organism.** Think about the effect of aberrant expression of that protein if personnel were accidentally exposed to the organism. Is the organism now expressing oncogenes or toxins, or knocking down expression of a tumor suppressor?

7. **What volume and concentration of agent are handled at any one time?** In general, higher volumes or concentrations call for stricter safety precautions. Recombinant organisms in volumes greater than 10 L in a single vessel have additional regulatory requirements as well.

8. **Will research animals be used in any procedures?** If so, will they be restrained or anesthetized? What is the immune status of the research animals? Is there a possibility that research animals will excrete potentially infectious or otherwise harmful substances?

9. **Are sharps utilized in any procedures?** Use of sharps should be minimized whenever possible. If sharps must be used with potentially infectious materials, use caution.

10. **Will any manipulations generate aerosols (e.g., vortexing, centrifuging)?** If so, perform these operations in a biosafety cabinet or other appropriate containment equipment whenever possible.

11. **Are vaccinations or treatments available for the agents in question?** Consult with an occupational health provider to determine the necessity for vaccination or postexposure management.

12. **Are the organisms used particularly hazardous for certain groups of people (e.g., pregnant women, immunocompromised individuals)?** Notify any personnel who will work with or around the organism(s) of these special concerns.

NOTE: For a quick guide for assessing the toxicity hazards associated with laboratory chemicals, see Box 4.1. For a quick guide for assessing physical flammable, explosive, and reactive hazards in the laboratory, see Box 4.2.

---

or X rays). Particulate radiations have both mass and electromagnetic radiations, which are sometimes referred to as photons. Radiation that has enough energy to ionize atoms and create ion pairs is referred to as ionizing radiation. Ionizing radiation not only comes from unstable nuclei, but can also be produced by machines such as particle accelerators, cyclotrons, and X-ray machines.

Alpha particles are charged particles containing two protons and two neutrons and are emitted from certain heavy atoms such as uranium and thorium. These particles are relatively large, slow, heavy, and easily stopped by a sheet of paper, a glove, a layer of clothing or even a dead layer of skin cells, and thus present virtually no external exposure hazard to people. However, because of the very large number of ionizations that α particles produce in short distances, α emitters can present a serious hazard when they come in contact with internal living cells and tissues. Special precautions are thus taken to ensure that α emitters are not inhaled, ingested, or injected. Care must be taken with unsealed α-emitting sources to control contamination and minimize the potential for internal uptakes.

A β particle (see Table 4.10) is an electron emitted from the nucleus of a radioactive atom. Positively charged counterparts of β particles are called positrons. Beta particles are much less massive and less charged than α particles and interact less intensely with atoms in the materials through which they pass, which gives them a longer range than α particles. Examples of β emitters commonly used in biological research are hydrogen-3 (tritium) (^{3}H), carbon-14 (^{14}C), phosphorus-32 (^{32}C), phosphorus-33 (^{33}P), and sulfur-35 (^{35}P). Although low-energy β particles are usually stopped by the dead layer of skin, higher energy β particles can penetrate more deeply and cause high exposures to the skin and eyes. The energy level of the β particle thus determines if shielding and exposure monitoring is required when working with these materials, as well as how contamination surveys are performed. Table 4.10 provides typical examples of high-energy, low-energy, and extremely low-energy β–particle handling precautions. When shielding is used to reduce external exposures from β emitters, a low-density shielding material such as Plexiglas, Lucite, or acrylic works best.

Gamma rays, x rays, and photon radiations have no mass or charge. Gamma rays are generally emitted from the nucleus during nuclear decay, and x rays are emitted from the electron shells. Extremely dense material such as lead typically makes the best shields for these electromagnetic forms of radiation. Iodine-125 (^{125}I), indium-111 (^{111}In), and chromium-51 (^{51}Cr) are a few examples of radionuclides sometimes used in research laboratories.

Neutrons are emitted from the nucleus during decay, have no electrical charge, and are one-fourth the mass of an α particle. Exposure to neutrons can be hazardous because the interaction of neutrons with molecules in the body can cause disruption to molecules and atoms. Because of its lack of charge, the neutron is difficult to shield, can penetrate deeply into tissues, and can travel hundreds of yards in air depending on the kinetic energy of the neutron. A neutron is slowed when it collides with the nucleus of other atoms. This transfers kinetic energy of the neutron to the nucleus of the atom. As the mass of the nucleus approaches the mass of the neutron, this reaction becomes more effective in slowing the neutron. Therefore water and other hydrogen-rich materials, such as paraffin or concrete, are often used as shielding material.

Radioactive decay rates are reported in curies (1 curie [Ci] = 3.7×10^{10} disintegrations per second [dps]) or in the International System of Units (SI) in becquerels (1 Bq = 1 dps). The decay rate provides a characterization of a given source but is not an absolute guide to the hazard of the material. The hazard depends on the nature, as well as the rate of production, of the ionizing radiation. In characterizing human exposure to ionizing radiation, it is assumed that the damage is proportional to the energy absorbed. The radiation absorbed dose (rad) is defined in terms of energy absorbed per unit mass: 1 rad = 100 ergs/g (SI: 1 Gy = 1 J/kg = 100 rads). For electromagnetic energy, the roentgen (R) produces 1.61×10^{12} ion pairs per gram of air (SI: 1 C/kg = 3.876 R).

Acceptable limits for occupational exposure to ionizing radiation are set by the USNRC based on the potential amount of tissue damage that can be caused by the exposure. This damage is expressed as a dose equivalent; the common unit for dose equivalent is the roentgen equivalent man (rem). The dose equivalent is determined by the rad multiplied by a weighting factor, called a quality factor, to account for the differences in the nature of the ionizing radiation from different types of radiation. Table 4.11 shows the quality factors for different types of radiation. For γ rays and X rays, rad and rem are virtually equivalent.

Damage may occur directly as a result of the radiation interacting with a part of the cell or indirectly by the formation of toxic substances within the cell. The extent of damage incurred depends on many factors, including the dose rate, the size of the dose, and the site of exposure. Effects may be short term or long term. Acute short-term effects associated with large

TABLE 4.10 Examples of β Emitters

	High-Energy Emitters	Low-Energy Emitters	Extremely Low Energy Emitters
Examples	Cl-35 P-32 Sr-90	C-14 S-35 P-33	H-3
Shielding	Shielding required (Plexiglas)	Not required	Not required
Contamination survey type	Survey meter	Survey meter	Wipe sample
Exposure dosimetry	Recommended	Not required	Not required

TABLE 4.11 Radiation Quality Factors

Type of Radiation	Quality Factor (Q)
x, γ, or β radiation	1
α particles	20
Neutrons of unknown energy	10

doses and high dose rates—for example, 100,000 mrad (100 rad) in less than 1 week—may include nausea, diarrhea, fatigue, hair loss, sterility, and easy bruising. In appropriately managed workplaces, such exposures are impossible unless various barriers, alarms, and other safety systems are deliberately destroyed or bypassed. Single-dose exposures higher than 500 rem are probably fatal. A single dose of ~100 rem may cause a person to experience nausea or skin reddening, although recovery is likely. However, if these doses are cumulative over a period of time rather than a single dose, the effects are less severe. Long-term effects, which develop years after a high-dose exposure, are primarily cancer. Exposure of the fetus in utero to radiation is of concern, and the risk of damage to the fetus increases significantly when doses exceed 15,000 mrem. The USNRC has set limits for whole-body occupational exposure at 5,000 mrem/year, with minors and declared pregnant workers allowed only 500 mrem/year (or 9-month gestation period), and members of the public allowed only 100 mrem/year (see Table 4.12). Exposure limits are lower in facilities operated by the U.S. Department of Energy and other agencies. Note that properly managed work with radioactive materials in the vast majority of laboratory research settings can be performed without any increase in a worker's exposure to radiation.

As with all laboratory work, protection of laboratory personnel against the hazard consists of good facility design, operation, and monitoring, as well as good work practices. The ALARA (as low as reasonably achievable) exposure philosophy is central to both levels of protection. The amount of radiation or radioactive material used should be minimized. Exposures should be minimized by shielding radiation sources, laboratory personnel, and visitors and by use of emergency alarm and evacuation procedures. The amount of time spent working with radioactive mate-

TABLE 4.12 U.S. Nuclear Regulatory Commission Dose Limits

Area of Dose	Occupational Dose Limits (mrem/year)	Public Dose Limits (mrem/year)
Total effective dose equivalent (or whole body: external + internal)	5,000	100*
Committed dose equivalent (or any organ dose)	50,000	NA
Eye dose equivalent (or lens of the eye)	15,000	NA
Shallow dose equivalent (or skin dose)	50,000	NA
Extremity dose (or shallow dose to any extremity)	50,000	NA
Minor (less than 18 years of age)	10% of occupational limits for adults	NA
Embryo/fetus of declared pregnant woman (limit taken over time of pregnancy)	500	NA

Personnel dosimetry is required if occupational dose is likely to exceed 10% of the limit (for embryo/fetus, it is required if worker's dose is likely to exceed 100 mrem during gestation period)

*NOTE: For 10 CFR § 35.75 patient release, limit is 500 mrem.

rials should be minimized. Physical distance between personnel and radiation sources should be maximized, and whenever possible, robotic or other remote operations should be used to reduce exposure of personnel. (Also see Chapter 6, section 6.E.)

5 Management of Chemicals

5.A	INTRODUCTION	84
5.B	GREEN CHEMISTRY FOR EVERY LABORATORY	84
	5.B.1 Prevent Waste	84
	5.B.2 Microscale Work and Wet Chemistry Elimination	84
	5.B.2.1 Design Less Hazardous Laboratory Processes and Reaction Conditions	85
	5.B.3 Use Safer Solvents and Other Materials	85
	5.B.4 Design Experimental Products for Degradation After Use	86
	5.B.5 Include Real-Time Controls to Prevent Pollution	86
	5.B.6 Minimize the Potential for Accidents	87
	5.B.7 Green Chemistry Principles Avoid Multihazardous Waste Generation	87
	5.B.8 Mercury Replacements in the Laboratory	87
	5.B.8.1 Thermometers	88
	5.B.8.2 Digital Thermometers	88
	5.B.8.3 Differential Manometers	88
5.C	ACQUISITION OF CHEMICALS	88
	5.C.1 Ordering Chemicals	88
	5.C.2 Receiving Chemicals	89
5.D	INVENTORY AND TRACKING OF CHEMICALS	90
	5.D.1 General Considerations	90
	5.D.2 Exchange of Chemicals Between Laboratories and Stockrooms	92
	5.D.3 Recycling of Chemicals and Laboratory Materials	93
	5.D.3.1 General Considerations	93
	5.D.3.2 Solvent Recycling	93
	5.D.3.3 Recycling Containers, Packaging, and Labware	93
	5.D.4 Labeling Commercially Packaged Chemicals	94
	5.D.5 Labeling Other Chemical Containers	94
	5.D.6 Labeling Experimental Materials	94
	5.D.7 Use of Inventory and Tracking Systems in Emergency Planning	94
5.E	STORAGE OF CHEMICALS IN STOCKROOMS AND LABORATORIES	94
	5.E.1 General Considerations	95
	5.E.2 Storage According to Compatibility	96
	5.E.3 Containers and Equipment	97
	5.E.4 Cold Storage	98
	5.E.5 Storing Flammable and Combustible Liquids	98
	5.E.6 Storing Gas Cylinders	100
	5.E.7 Storing Highly Reactive Substances	100
	5.E.8 Storing Highly Toxic Substances	101
5.F	TRANSFER, TRANSPORT, AND SHIPMENT OF CHEMICALS	101
	5.F.1 Materials of Trade Exemption	102
	5.F.2 Transfer, Transport, and Shipment of Nanomaterials	102
	5.F.2.1 Off-Site Transport and Shipments of Nanomaterials	103
	5.F.2.2 On-Site Transfer and Transport of Nanomaterials	104

5.A INTRODUCTION

This chapter organizes the discussion of managing laboratory chemicals into six main topics: reducing and eliminating the use and generation of hazardous substances (green chemistry); acquisition; inventory and tracking; storage in stockrooms and laboratories; recycling of chemicals and laboratory materials; and transfer, transport, and shipment of chemicals. As Chapter 1 makes clear, prudence in these areas requires knowledge of the hazards posed by laboratory chemicals and the formulation of reasonable measures to control and minimize the risks associated with their handling and disposal. Not all risk can be eliminated, but through informed risk assessment and careful risk management, laboratory safety is greatly enhanced.

Trained laboratory personnel, laboratory supervisors, and individuals who handle chemicals will find essential information in this chapter. Each person has an important role to play in a chemical's life cycle at an institution, and each one of them should be aware that the wise management of that life cycle not only minimizes risks to humans and to the environment but also decreases costs. Acknowledging this role and giving it due consideration is one element of the culture of safety within a laboratory.

5.B GREEN CHEMISTRY FOR EVERY LABORATORY

Green chemistry is the philosophy of designing products and processes that reduce or eliminate the use and generation of hazardous substances, which fits well with the overall goals of a culture of safety. The 12 principles of green chemistry (Anastas and Warner, 1998) can be applied in the laboratory as guidelines for prudent experimental design and execution. Some of the principles are explained in more detail below, with examples of their broader application. A wealth of green chemistry resources exists online in the form of reports, databases, and other Web applications and tools. These resources assist the development of green synthetic methods by providing information about the redesign of processes at the molecular level, the reduction or elimination of the use of hazardous materials, and the modification of chemical substances to make them safer.

5.B.1 Prevent Waste

Prudent laboratory chemical management begins with adopting the first green chemistry principle of waste prevention, which is considered before the ordering of the chemicals. When experiments have been carefully planned, trained laboratory personnel can be confident that they have chosen procedures that minimize the quantities of chemicals to be used and minimize the disposal of hazardous materials.

Experiment planning in the culture of laboratory safety includes minimization of the material used at each step of an experiment. Consider two simple examples: (1) Transferring a liquid reaction mixture or other solution from one flask to another container usually requires the use of a solvent to rinse out the flask. During this procedure, laboratory personnel should use the smallest amount of solvent possible that enables a complete transfer. (2) Celite is often used during filtrations to keep the pores of filter papers or filter frits from becoming clogged. When positioning the Celite, carefully determine the minimum amount needed to be effective. Other examples of such strategies include

- considering how a reaction product will be used and making only the amount needed for that use;
- appreciating the cost of making and storing unneeded material;
- thinking about minimization of material used in each step of an experiment;
- searching for ways to reduce the number of steps in an experiment;
- improving yields;
- recycling and reusing materials when possible;
- coordinating work with co-workers who may be using some of the same chemicals (section 5.D.2);
- considering the amount of reagents, solvents, and hazardous materials used by automated laboratory equipment when purchasing a new system;
- isolating nonhazardous waste from hazardous waste; and
- using a column purification system for recycling of used solvent (section 5.D.3)

These steps are increasingly important because of the changing requirements and economics of laboratory management.

5.B.2 Microscale Work and Wet Chemistry Elimination

One successful method of reducing hazards is to carry out chemical reactions and other laboratory procedures on a smaller scale (i.e., microscale) when feasible. In microscale chemistry the amounts of materials used are reduced to 25 to 100 mg for solids and 100 to 200 μL for liquids, compared with the usual 10 to 50 g for solids or 100 to 500 mL for liquids. Smaller scale synthetic methods save money because they require less reagent and result in less waste. Of course, not all laboratory procedures can be scaled down. Multigram laboratory preparation is often required to provide sufficient material for further work. Whether large or small scale, exercise precaution appropriate to the scale, as well as the inherent hazards, of the procedure.

Similarly, in many cases instrumental analyses—which require little reagent and generate very little waste in themselves—can be substituted for wet chemistry. Consider the waste reduction inherent in spectroscopic organic analysis versus chemical derivatization. And, hazardous waste reduction also reduces both compliance and disposal costs. When purchasing equipment to automate laboratory processes, choose equipment that is efficacious for the job at hand, but uses the least amount of reagents or solvents, or uses materials that are least hazardous. (See Vignette 5.1.)

5.B.2.1 Design Less Hazardous Laboratory Processes and Reaction Conditions

The third principle of green chemistry suggests that, where possible, syntheses should be designed using less toxic reagents. Although the use of a toxic reagent does not necessarily imply generation of a toxic waste, in line with the first principle, chemists should evaluate potential sources of hazardous waste expected from the proposed synthesis and incorporate strategies to minimize them.

VIGNETTE 5.1
Pollution prevention
reduces solvent waste

A pollution prevention assessment of one organic chemistry research laboratory at a university revealed that each of the 25 researchers in the group used 1 L of solvent, usually acetone, every week to clean and/or rinse glassware, spatulas, and other items used in their procedures. For example, a researcher might rinse a spatula with acetone at the end of a procedure or use a solvent to speed the drying process after cleaning with soap and water. The excuses for using the solvent ranged from not having enough glassware available (thus the need to expedite drying) to lack of good brushes for cleaning residue to simply taking a shortcut to the cleaning process.

The lab purchased more glassware, better brushes, and an ultrasonicator that uses a mild detergent. The savings in solvent purchase and disposal paid back the price of the new purchases within 3 months. Later, the lab installed under-the-bench lab dishwashers, which resulted in even further reductions in solvent use for cleaning.

5.B.3 Use Safer Solvents and Other Materials

Traditionally, chemists have chosen reagents and materials to meet scientific criteria without always giving careful consideration to waste minimization or environmental objectives. In synthetic procedures, overall yield and purity of the desired product are important factors, because better yield implies lower cost. On the other hand, material substitution can be an important consideration in manufacturing process design because of the large quantity, and potential cost, of chemicals involved. The following questions should be considered when choosing a material to be used as a reagent or solvent in an experimental procedure:

- Can this material be replaced by one that will expose the experimenter, and others who handle it, to less potential hazard?
- Can this material be replaced by one that will reduce or eliminate the hazardous waste and the resulting cost of waste disposal?
- Can these steps be taken in conjunction with yield maximization and minimization of overall waste and cost?

All things being equal, laboratories are safer when they substitute nonhazardous, or less hazardous, chemicals where possible by considering alternative synthetic routes and alternative procedures for working up reaction mixtures. The following additional examples illustrate the application of this principle to common laboratory procedures:

- To reduce the amount of copper released to the sewer, use iron complexes rather than copper when studying spectrophotometry in general chemistry.
- In liquid scintillation counting of low-level radioactive samples, where possible, use nonflammable, lower toxicity, water-miscible solvents rather than xylene, toluene, or dioxane, so as to eliminate fire hazard and waste that must be incinerated.
- Substitute solid or liquid reagents for hazardous gases that must be used at elevated pressure. As an example, phosgene is a highly toxic gas occasionally used as a reagent in organic transformations. Its use requires proper precautions to contain the gas and handle and dispose of cylinders. Commercially available products such as diphosgene (trichloromethyl) chloroformate, a liquid, or triphosgene bis(trichloromethyl) carbonate, a low-melting solid, are often substituted for phosgene by appropriate adjustment of experimental conditions or are used to generate

phosgene only on demand. Both chemicals are highly toxic themselves, and their use in any event should be considered carefully, but solids avoid the problems associated with handling a toxic gas.

- Consider carefully the use of reagents containing toxic heavy metals. For example, proprietary detergents for glassware (used, if necessary, with ultrasonic baths) are a safer substitute for chromic acid cleaning solutions. Various chromium(VI) and other metal oxidants have been important in synthetic organic chemistry, but other oxidants are possible substitutes. When planning a reaction, consider the cost of disposal of heavy metal waste in addition to its utility. Search the literature for other oxidation reagents tailored to the specific needs of a given transformation. (For information about reducing the use of mercury in laboratory equipment, see section 5.B.8.)

- F-TEDA-BF4, or 1-chloromethyl-4-fluoro-1,4-diazoniabicyclo[2.2.2]octane bis(tetrafluoroborate), substitutes for more hazardous reagents in many fluorination procedures. To reduce the reactivity and toxicity risks associated with perchloryl fluoride, fluorine, and other fluorinating reagents, search the literature for appropriate substitutes.

- Avoid solvents listed as select carcinogens (for a definition of select carcinogens, see Chapter 4, section 4.C.3.4), reproductive toxins, or hazardous air pollutants. Choose solvents with relatively high American Conference of Governmental Industrial Hygienists threshold limit values. Recognizing that not all hazards can always be reduced simultaneously, the best substitute solvent meets needed experimental constraints but has physiochemical properties, such as boiling point, flash point, and dielectric constant that are similar to the original solvent. Although cost can be a factor, consider the benefits of safety, health, and the environment as well. For example, heptane is more costly than hexane, but is very similar physiochemically and is not listed by the U.S. Environmental Protection Agency (EPA) as a hazardous air pollutant. Toluene usually can substitute for the carcinogen benzene. Chemical suppliers now highlight solvents with lower hazards including reduced flammability and potential for peroxide formation.

- Supercritical fluids present an interesting case in conflicting green chemistry principles. Supercritical CO_2 as a solvent involves a chemically relatively benign material, carbon dioxide. Reaction workup requires only ambient heat, and there is no hazardous waste. On the other hand, it requires elevated pressure. Supercritical solvents for chromatography and synthesis require special-

ized equipment for handling, but because of the ubiquity of chromatography methods operating at elevated pressure and the common nature of the pumps and vessels necessary, much of the hazard has been mitigated. The technology for using supercritical fluids has developed rapidly in recent years. Consider use of these materials, but with appropriate precaution and dedicated permanent equipment.

5.B.4 Design Experimental Products for Degradation After Use

Green chemistry practitioners plan synthesis and other processes so that, as part of the experiment, the products and byproducts are rendered safe or less hazardous. For example, they include in the experimental plan reaction workup steps that deactivate hazardous materials or reduce their toxicity.

5.B.5 Include Real-Time Controls to Prevent Pollution

To cut costs, firms are increasingly asking for just-in-time delivery of raw materials and using other real-time controls. Green chemistry laboratories can borrow this strategy. A quantity of hazardous chemical not ordered is one to which trained laboratory personnel are not exposed, for which appropriate storage need not be found, which need not be tracked in an inventory control system, and which will not end up requiring costly disposal when it becomes a waste.

Part of acquiring a chemical is a life-cycle analysis. All costs associated with the presence of each chemical at an institution must be considered. The purchase cost is only the beginning; the handling costs, human as well as financial, and the disposal costs must be taken into account. Without close attention to these aspects of managing chemicals in a laboratory, orders are not likely to be minimized, and unused chemicals become a significant fraction of the laboratory's hazardous waste.

The American Chemical Society's booklet *Less Is Better: Laboratory Chemical Management for Waste Reduction* (Task Force on Laboratory Waste Management, 1993) gives several reasons for ordering chemicals in smaller containers, even if that means using several containers of a material for a single experiment:

- Consequence of breakage is substantially reduced for small package sizes.
- Risk of accident and exposure to hazardous material is less when handling smaller containers.
- Storeroom space needs are reduced when only a single size is inventoried.

- Containers are emptied faster, resulting in less chance for decomposition of reactive compounds.
- Use of the so-called "economy size" often dictates a need for other equipment, such as transfer containers, funnels, pumps, and labels. Added labor to subdivide the larger quantities into smaller containers, as well as additional personal protective equipment for the hazards involved, also may be needed. In most cases, it is safer, and may be less costly, to allow commercial providers to break bulk rather than "doing it yourself."
- If unused hazardous material must be disposed of, the disposal cost per container is less for smaller containers.

An institution should also minimize the amount of chemical accepted as a gift or as part of a research contract. More than one laboratory has been burdened with the cost of disposing of a donated chemical that was not needed.

Donated material can easily become a liability. A chemical engineering researcher accepted a 55-gallon drum of an experimental diisocyanate as part of a research contract. The ensuing research project used less than 1 gallon of the material, and the grantor would not take the material back for disposal. No commercial incinerator would handle the material in its bulk form. The remaining material had to be transferred to 1-liter containers and sent as lab packs for disposal, at significant cost.

In section 5.D.2, the exchange or transfer of chemicals to other trained laboratory personnel is discussed. Smaller containers increase the chance that chemicals to be transferred are in sealed containers, which increases the receiver's confidence that the chemicals are pure.

5.B.6   Minimize the Potential for Accidents

Green chemistry also means designing to reduce accidents, injuries, and exposures to laboratory, storeroom, and receiving personnel. Chapters 4 and 6 explain planning and risk assessment for laboratory personnel. Be sure that hazardous properties are understood before a material is purchased, synthesized, or otherwise acquired. Search references and the literature to be cognizant of the properties of explosivity, water and air reactivity, instability, age-related degradation, and pressurization when contained. Searches of historical laboratory accident data reveal risks associated with experimental setups, procedures, equipment, facilities, inadequate training, and noncompliance with safety rules. Trained laboratory personnel with this knowledge should communicate it to co-workers and material handling personnel. New laboratory personnel deserve a special orientation.

5.B.7   Green Chemistry Principles Avoid Multihazardous Waste Generation

Because the management of multihazardous waste is often difficult, prudent green chemistry principles minimize its generation. Chapter 8, sections 8.C.2 and 8.C.3, provides information on eliminating or minimizing the components of waste that are biological or radioactive hazards, respectively. For chemical–biological waste, the primary strategy for minimizing the multihazardous waste is to maintain segregation of chemical and biological waste streams as much as possible. For reduction of radioactive hazards, the strategies discussed include substituting nonradioactive materials for radioactive materials, substituting radioisotopes having shorter decay times (e.g., when radioactive iodine is specified, using iodine-131, with a half-life of 8 days, instead of iodine-125, with a half-life of 60 days), and carrying out procedures with smaller amounts of materials.

5.B.8   Mercury Replacements in the Laboratory

Chronic exposure to mercury (Chemical Abstracts Service [CAS] No. 7439-97-6) through any route can produce central nervous system damage (Mallinkrodt Baker, Inc., 2008). Common exposure routes include inhalation, ingestion, and skin or eye contact. Thermometers and manometers are the most common laboratory uses of elementary mercury, and in many cases, there are suitable nonmercury alternatives available. Broken thermometers and manometers create a health hazard in the laboratory and, where possible, should be replaced with mercury-free substitutes.

The consequences of broken mercury-filled equipment (thermometers, manometers, diffusion pumps, bubblers, etc.) can include personnel exposure, laboratory and environmental contamination, mercury spill cleanup, and disposal of mercury and mercury-contaminated debris. Mercury spills are challenging to clean up completely and require training and special spill control materials (see Chapter 6, section 6.C.10.8, for more information about mercury spill cleanup). Elemental mercury is very heavy and can be expensive to dispose as waste (Foster, 2005a). Replacing mercury-filled equipment in the laboratory ensures compliance with 2 of the 12 principles of green chemistry: No. 1, "Prevent Waste: Design chemical syntheses to prevent waste, leaving no waste to treat or clean up"; and No. 12, "Minimize the potential for accidents: Design chemicals and their forms (solid, liquid, or gas) to minimize the potential for chemical accidents including explosions, fires, and releases to the environment" (Anastas and Warner, 1998).

5.B.8.1 Thermometers

Design a mercury thermometer replacement program to provide safe, suitable substitutes for use in laboratories. Factors that should be considered during the mercury replacement process are various applications in the laboratories, required temperature range, thermometer length, immersion depth, scale divisions, cost, accuracy in relation to application, and durability upon exposure to corrosive solutions. In some cases, these alternative thermometers have a more limited temperature range than a mercury thermometer. Perform tests for accuracy in the laboratory prior to a total replacement program to ensure that the mercury substitutes will be suitable for the methods that will be employed in that particular laboratory. To ensure accuracy, thermometers must be calibrated using approved methods such as ASTM E 77 (ASTM International, 2007a) and must be traceable to the National Institute of Standards and Technology (NIST).

There is a wide selection of mercury-free liquid-filled thermometers available, including spirit thermometers (filled with biodegradable petroleum-based mineral spirits and dyes) and alcohol-based thermometers. When broken, these thermometers present no hazardous material disposal problems. Some spirit thermometers had a history of the thread breaking more easily than a mercury thermometer, but many of the newer formulations have overcome this problem. In the event that the thread breaks, the simplest and safest method to reunite the liquid is to use a centrifuge. Carefully insert the thermometer, bulb down, in the centrifuge. Use cotton wadding at the bottom of the cup to prevent any damage to the bulb. Turn on the centrifuge and in just a few seconds all the liquid will be forced past the separation. Note that if the cup is not deep enough, and all the centrifuge force is not below the column, the column will split, forcing half the liquid in the bulb and half the liquid in the expansion chamber (Izzo, 2002).

For liquid-filled thermometers used to measure the temperature of liquids, accuracy will also depend on choosing the correct immersion depth. This is less of an issue for mercury thermometers because mercury generally has better thermoconductivity. A **total immersion** thermometer is designed to indicate temperatures correctly when the bulb and all but 12 mm of the liquid column are immersed in the bath medium. The top 12 mm of the liquid column should be above the bath medium so that the thermometer can be read and the material does not distill at high temperatures. Thermometers that have been graduated for total immersion usually have no markings on the back pertaining to immersion. A **partial immersion** thermometer is designed to indicate temperatures correctly when the bulb and a specified portion of the stem are exposed to

the temperature being measured and the remainder of the stem is exposed to the ambient temperature. Partial immersion thermometers are clearly marked with a permanently placed line on the stem to indicate the proper immersion depth (ASTM International, 2007a).

In addition to thermometers that are filled with mercury-alternative liquids, long-stem digital thermometers are available with probes that are resistant to most laboratory chemicals, including acids, bases, and solvents. The bright displays, usually ¼ in. high, are easy to read and display the temperature in both degrees Fahrenheit and degrees Celsius, with ranges from −58 to 302 °F and −50 to 150 °C. Long-stem thermometers are constructed of plastic and stainless steel and do not contain glass or mercury, which make them ideal thermometers for use in academic laboratories. The stems are generally 8 inches long with an overall thermometer length of 11 in.

5.B.8.2 Digital Thermometers

Where a mercury thermometer is the only option, armor cases, which protect against breakage without affecting accuracy, or Teflon-coated mercury thermometers are recommended. These are particularly useful in high-temperature ovens, oil baths, and autoclaves, where cleaning up a mercury spill can be challenging and the spill creates a serious health hazard.

5.B.8.3 Differential Manometers

Depending on the measurement range, labs can substitute water or calibrated oils for mercury. Pressure transducers or electronic pressure gauges may also be an alternative to a conventional manometer.

5.C   ACQUISITION OF CHEMICALS

5.C.1   Ordering Chemicals

Authority to place orders for chemicals may be centralized in one purchasing office or may be dispersed to varying degrees throughout the institution. The advent of highly computerized purchasing systems, and even online ordering, has made it feasible to allow ordering at the departmental or research group level. However, the ability to control ordering of certain types of materials through a central purchasing system (e.g., prohibiting flammables in containers over a certain size or ensuring appropriate licensing of radioactive material users) is almost completely lost when the purchasing function is decentralized. In these cases, other creative ways of exercising control need to be found.

Before purchasing a chemical, prudent laboratory personnel ask several questions:

- Is the material already available from another laboratory within the institution or from a surplus-chemical stockroom? If so, waste is reduced, and the purchase price is saved. The tendency to use only new chemicals because of their purity should be scrutinized, and that tendency should be carefully justified to ensure that materials already on hand are used whenever possible.

- What is the minimum quantity that will suffice for current use? Chemical purchases should not be determined by the cheaper unit price basis of large quantities but rather by the amount needed for the experiment. The cost of disposing of the excess is likely to exceed any potential savings gained in a bulk purchase (i.e., the cost of getting rid of a chemical may exceed its acquisition cost). If a quantity smaller than the minimum offered by a supplier is needed, the supplier should be contacted and repackaging requested. Compressed gas cylinders, including lecture bottles, should normally be purchased from suppliers who accept return of empty cylinders. If paying demurrage charges, the laboratory may want to return partially filled cylinders that will not be used in the near future.

- What is the maximum size container allowed in the areas where the material will be used and stored? Fire codes and institutional policies regulate quantities of certain chemicals, most notably flammables and combustibles. For these materials, a maximum allowable quantity for laboratory storage has been established (see also sections 5.E.5 and 5.E.6).

- Can the chemical be managed safely when it arrives? Does it require special storage, such as in a drybox, refrigerator, or freezer? Do receiving personnel need to be notified of the order and given special instructions for receipt? Will any special equipment necessary to use the chemical be ready when it arrives? An effort should be made to order chemicals for just-in-time delivery by purchasing all unstable or extremely reactive materials from the same supplier with a request for one delivery at the best time for performing an experiment.

- Does the chemical present any unique security risks? Is it a controlled substance? Is there a risk of potential intentional misuse of the chemical? Will the quantity ordered affect compliance with the U.S. Department of Homeland Security (DHS) Chemical Facility Anti-Terrorism Standard (CFATS)? (See Chapter 10 for a discussion of laboratory security.)

- Is the chemical unstable? Inherently unstable materials may have very short storage times and should be purchased just before use to avoid los-

ing a reagent and creating an unnecessary waste of material and time. Some materials may require express or overnight delivery and will not tolerate being held in transit over a weekend or holiday.

- Can the waste be managed satisfactorily? A chemical that produces a new category of waste may cause problems for the waste management program. An appropriate waste characterization and method for proper disposal should be identified before the chemical is ordered.

Within an institution or organization, one of the advantages of computerization of ordering is that information about deliveries of chemicals can be retrieved from the chemical supplier, which provides a clear picture of the purchasing history and distribution of chemicals across buildings. Some institutions include in their annual contracts with suppliers a requirement to report on a monthly, a quarterly, or an annual basis the quantity of each type of chemical purchased and the location to which it was delivered. This information can be helpful in preparing the various annual reports on chemical use that may be required by federal, state, or local agencies. For example, centralized ordering may assist the institution in complying with the Controlled Substances Act and with CFATS. In addition, such a system is also useful for tracking the use of flammables, locations of Food and Drug Administration drug precursors, and DHS chemicals of interest. [See *Handbook of Chemical Health and Safety* (Alaimo, 2001); *Code of Federal Regulations*, 1998.]

A purchase order for a chemical should include a request for a material safety data sheet (MSDS). However, many of the larger laboratory chemical suppliers send each MSDS only when an organization first orders the chemical. Subsequent orders of the same chemical are not accompanied by the MSDS. Therefore, a central network of accessible MSDSs should be established. This collection of MSDSs can be electronic if computer access is available to all employees at all times.

5.C.2   Receiving Chemicals

Chemicals arrive at institutions in a variety of ways, including U.S. mail, commercial package delivery, express mail services, and direct delivery from chemical warehouses. Deliveries of chemicals should be confined to areas that are equipped to handle them, usually a loading dock, receiving room, or laboratory. Proper equipment for receipt of chemicals includes chains for temporarily holding cylinders and carts designed to safely move various types of chemical containers. Shelves, tables, or caged areas should be designated for packages to avoid damage by receiving room vehicles. Chemical deliveries should not be made to departmental offices because, in general, offices are unlikely to be equipped to receive these packages. However, if delivery to such an office is the only option, a separate undisturbed location, such as a table or shelf, should be identified for chemical deliveries, and the person ordering the material should be notified immediately on its arrival.

Receiving room, loading dock, and clerical personnel should to be trained adequately to recognize hazards that may be associated with chemicals coming into the facility. They need to know what to do if a package is leaking or if there is a spill in the receiving facility, and they need to know who to call for assistance when a problem develops. They should also be trained to identify activity that could suggest a security risk, such as unauthorized personnel near the loading dock or unwarranted interest in their activities. The Department of Transportation (DOT) requires training for anyone involved in the movement of hazardous materials, including individuals who have been designated to receive hazardous materials on behalf of the organization (see Chapter 11, sections 11.E.1.5, and 11.F.1).

Your firm or institution should decide if stockroom or laboratory personnel are responsible for unpacking incoming chemicals. Incoming packages should be promptly opened and inspected to ensure that containers are sealed in good condition and to confirm what was ordered. The unpacked chemicals should be stored safely. In particular, reactive chemicals shipped in metal containers (e.g., lithium aluminum hydride, sodium peroxide, phosphorus)—which are often sealed—must be promptly unpacked and stored to prevent degradation and corrosion and to be available for periodic inspection.

Transportation of chemicals within the facility, whether by internal staff or outside delivery personnel, must be done safely. Single boxes of chemicals in their original packaging can be hand carried to their destination if they are light enough to manage easily. Groups of packages or heavy packages should be transported on a cart that is stable, has straps or sides to contain packages securely, and has wheels large enough to negotiate uneven surfaces easily. Suitable carriers (e.g., secondary containment) should be used when transporting individual containers of liquids.

Cylinders of compressed gases should always be secured on specially designed carts and never be dragged or rolled. The cap should always be securely in place. Whenever possible, chemicals and gas cylinders should be moved on freight elevators that are not used for public occupancy, especially when moving toxic, cryogenic, or asphyxiating gases.

If outside delivery personnel do not handle materials according to the receiving facility's standards, immediate correction should be sought, or other carriers

or suppliers should be used. Original purchase order should specify delivery criteria. Some examples of delivery criteria would be that the gas cylinder must have a cap and the cap must not be stuck, and damaged containers may not be accepted without the inspection and approval of a technically qualified individual on-site.

When packages are opened in the laboratory, laboratory personnel should verify that the container is intact and is labeled, at a minimum, with an accurate name on a well-adhered label. For unstable materials, and preferably for all materials, the date of receipt should be on the label. Labels placed by the manufacturer should remain intact. New chemicals should be entered into the laboratory's inventory promptly and moved to the appropriate storage area.

5.D INVENTORY AND TRACKING OF CHEMICALS

5.D.1 *General Considerations*

Prudent management of chemicals in any laboratory is greatly facilitated by keeping an accurate inventory of the chemicals stored. An inventory is a record (usually a database) that lists the chemicals in the laboratory, along with information essential for their proper management. Chemical inventories are also a vital tool, and in some cases are required, for maintaining regulatory compliance. An organization cannot adequately manage safety, security, emergency planning, waste disposal, and the like without knowing what chemicals are on-site and where they are stored. Without an up-to-date inventory of chemicals, many important questions pertinent to prudent management of chemicals can be answered only by visually scanning container labels. A well-managed inventory system promotes economical use of chemicals by making it possible to determine immediately what chemicals are on hand. An inventory is not limited to materials obtained from commercial sources but includes chemicals synthesized in a laboratory. If a chemical is on hand, the time and expense of procuring new material are avoided. Information on chemicals that present particular storage or disposal problems facilitates appropriate planning for their handling. Although a detailed list of hundreds or thousands of chemicals stored in a particular location may not be directly useful to emergency responders, it can be used to prepare a summary of the types of chemicals stored and the hazards that might be encountered. In larger organizations where chemicals are stored in multiple locations, the inventory system should include the storage location for each container of each chemical. An inventory system is also of use when considering laboratory security concerns. It can assist in ensuring compliance with regulations, such as CFATS (see Chapter 10), tracking of materials to ensure that they are not intercepted en route, and in identification of unusual orders within the department or organization.

If procedures for the facile updating of information on storage locations are developed, the system becomes a tracking system. Such a system promotes the sharing of chemicals originally purchased by different research groups or laboratories. The more laboratories in an organization agree to share chemicals, the greater the likelihood that items unneeded in one location will be used elsewhere. Tracking systems are more complex to establish than simple inventories and require more effort to maintain, but their favorable impact on the economics and efficiency of chemical use in a large organization often justify their use.

Each record in a chemical inventory database generally corresponds to a single container of a chemical rather than merely to the chemical itself. This approach allows for a more logical correspondence between the records in the database and the chemicals stored in the laboratory. The following data fields for each item are recommended for any system:

- name as printed on the container;
- molecular formula, for further identification and to provide a simple means of searching;
- CAS registry number, for unambiguous identification of chemicals despite the use of different naming conventions;
- source; and
- size of container or original quantity of chemical.

In addition, the following information may be useful:

- hazard classification, as a guide to safe storage, handling, and disposal;
- date of acquisition, to ensure that unstable chemicals are not stored beyond their useful life;
- storage location, in laboratories where multiple locations exist; and
- on-site owner or staff member responsible for the sample.

In a chemical tracking system, how the consumption of chemicals is tracked must be considered. The effort involved in maintaining data on the precise contents of each container must be weighed against the potential benefit such a system would provide. Many tracking systems omit this information and record only the container size.

A simple inventory system records the above information for each container on index cards, which are then kept in an accessible location in some logical order, such as by molecular formula. The ease of searching

such a card file is limited by its size and the order in which it is sorted. This type of system has obvious advantages in terms of simplicity and low cost, but it suffers several limitations. Listings of chemicals must be prepared manually, and the integrity of the database depends on how well the card file is maintained.

For an inventory of more than a few hundred chemicals, a computer-based system offers advantages. Many spreadsheet and database programs maintain an effective chemical inventory system, cross-referenced by different scientific or common names. The integrity of the inventory system is enhanced by the ease of making backup copies of the database. Searches for desired chemicals are carried out in a number of ways, depending on the software. The ability to search and sort the database, for example, by hazard classification, acquisition date, owner, or other parameters, and to prepare lists of the results of such a sort contribute to efficiency in a variety of chemical management tasks. Section 5.C.1 notes the prudence of establishing a central network of MSDSs. Including MSDSs and laboratory chemical safety summaries (LCSSs) (see Chapter 4 and accompanying CD) in the inventory's database is highly desirable. Alternatively, the inventory could be linked to other databases containing safety and environmental information about the chemicals. The quality of MSDSs varies significantly from one manufacturer to another. LCSSs, which are targeted to the needs of typical trained laboratory personnel, are a useful supplement to the information provided by MSDSs.

Having a fully capable chemical tracking system depends on careful selection of database software. Such a package should permit access from multiple terminals or networked computers and, most importantly, have a foolproof efficient method for rapidly recording the physical transfer of a chemical from one location to another. Bar-code labeling of chemical containers as they are received provides a means of rapid error-free entry of information for a chemical tracking system. If reagent chemical suppliers were to adopt a system in which chemical containers were labeled with bar codes providing essential information on their products, the maintenance of chemical tracking systems would be greatly facilitated. Proprietary software packages for tracking chemicals are available. Organizations operating under good laboratory practice regulations may even want to track the quantity of material in each container. The investment in hardware, software, and personnel to set up and maintain a chemical inventory tracking system is considerable but pays significant dividends in terms of economical and prudent management of chemicals.

As with any database, the usefulness of an inventory or chemical tracking system depends on the integrity of the information it contains. If an inventory system is used to locate chemicals for use or sharing in the laboratory, even a moderate degree of inaccuracy erodes confidence in the system and discourages use. The need for high fidelity of data is greater for a tracking system, because trained laboratory personnel will rely on it to save time locating chemicals rather than physically searching. For these reasons, appropriate measures should be taken periodically to purge any inventory or tracking system of inaccurate data. A physical inventory of chemicals stored, verification of the data on each item, and reconciliation of differences are performed annually. This procedure coincides with an effort to identify unneeded, outdated, or deteriorated chemicals and to arrange for their disposal. The following guidelines for culling inventory may be helpful:

- Consider disposing of materials not expected to be used within a reasonable period, for example, 2 years. For stable, relatively nonhazardous substances with indefinite shelf lives, a decision to retain them in storage should take into account their economic value, scarcity, availability, and storage costs.
- Make sure that deteriorating containers or containers in which evidence of a chemical change in the contents is apparent are inspected and handled by someone experienced in the possible hazards inherent in such situations.
- Dispose of or recycle chemicals before the expiration date on the container.
- Replace deteriorating labels before information is obscured or lost to ensure traceability and appropriate storage and disposal of the chemicals.
- Because many odoriferous substances make their presence known despite all efforts to contain them, aggressively purge such items from storage and inventory.
- Aggressively cull the inventory of chemicals that require storage at reduced temperature in environmental rooms or refrigerators. Because these chemicals may include air- and moisture-sensitive materials, they are especially prone to problems that are exacerbated by the effects of condensation.
- Dispose of all hazardous chemicals at the completion of the laboratory professional's tenure or transfer to another laboratory. The institution's cleanup policy for departing laboratory researchers and students should be enforced strictly to avoid abandoned unknowns that pose unknown hazards to remaining personnel and have high disposal costs.
- Develop and enforce procedures for transfer or disposal of chemicals and other materials when decommissioning laboratories because of reno-

vation or relocation. Try to avoid receiving entire chemical inventories from decommissioned laboratories and do not donate entire chemical inventories to schools or small businesses.

Chemical inventory challenges have not changed since the first use of index card files. The initial challenge is ensuring that every laboratory chemical gets entered into the inventory. This task often requires the concerted effort of many staff members. The second challenge is keeping the inventory current. Meeting this challenge usually requires designating one or more responsible individuals to enter new materials into the system; these individuals are the only personnel who should have write/edit access to the inventory. Facility procedures must make sure that notice of all new materials is presented to these designated individuals for entry into the inventory. Assuming that every staff member will faithfully enter new chemicals into the system results in an obsolete inventory. A third challenge is making sure that consumed chemicals, that is, empty containers, are removed from the active inventory.

Inventories are valuable to laboratory operations if everyone supports and contributes to the inventory. Managers with budgetary responsibilities appreciate the value of an established inventory system in reducing procurement and operating costs. Laboratory waste coordinators favor more efficient use of in-house materials resulting in reduced quantities of waste.

More information about chemical management systems can be found in Chapter 2, section 2.D.4.

5.D.2 Exchange of Chemicals Between Laboratories and Stockrooms

The exchange or transfer of chemicals between laboratories at an institution depends on the kind of inventory system and central stockroom facilities in place. Some institutions encourage laboratory personnel to return materials to the central stockroom for redistribution to others. The containers are sealed or open with a portion of the material used. Containers that have been opened are often of sufficient purity to be used as is in many procedures. If the purity is in doubt, the person who returned the material should be consulted. The stockroom personnel can update the central inventory periodically to indicate what is available for exchange or transfer. For an exchange program to be effective, all contributors to and users of the facility must reach a consensus on the standards to be followed concerning the labeling and purity of stored chemicals.

A word of caution is offered in regard to surplus-chemical stockrooms; they must be managed with the same degree of control as a new-chemical storage area.

The surplus-chemical stockroom is not a depository for any chemical that will not be wanted in the laboratory within a reasonable period (e.g., 2 to 3 years); such materials are to be disposed of properly. Rooms that are used as general depositories of unwanted chemicals become mini-Superfund sites because of lack of control.

Academic institutions could recycle common organic solvents from one research laboratory to another, or from research laboratories to teaching laboratories. For example, chromatography effluents such as toluene could be collected from research laboratories, distilled, and checked for purity before reuse. Commercial distillation systems are available for such purposes, but laboratory personnel performing the distillations or working in the immediate vicinity need appropriate training. (See Chapter 7 for hazards associated with distillation.)

Laboratory-to-laboratory exchange can be an effective alternative to a central surplus-chemical stockroom in organizations unwilling or unable to manage a central storeroom properly. In such a system, trained laboratory personnel retain responsibility for the storage of unwanted chemicals but notify colleagues periodically of available materials. A chemical tracking system as described above facilitates an exchange system greatly. If colleagues within the same laboratory are using the same hazardous material, particularly one that is susceptible to decomposition on contact with air or water, they should try to coordinate the timing of their experiments.

5.D.3 Recycling of Chemicals and Laboratory Materials

5.D.3.1 General Considerations

Chemical recycling takes many forms. In each case a material that is not quite clean enough to be used as is must be brought to a higher level of purity or changed to a different physical state.

Recycling occurs on-site or off-site. On-site recycling occurs at the laboratory or at a central location that collects recyclables from several laboratories. Because on-site recycling can be very time and energy intensive, it may not be economically justifiable. In some cases, although the amount of waste may be quite small, it can require very expensive disposal if a commercial vendor must be used. Before a decision on recycling is made, the cost of avoided waste disposal should be calculated. Because of the difficulty of maintaining the needed level of cleanliness and safety, on-site recycling of mercury and other toxic metals is no longer recommended. Another significant issue is whether recycling activities require a waste treatment permit under the Resource Conservation and Recovery Act (RCRA).

More information about this regulation can be found in Chapters 8 and 11. State and local regulations must also be considered.

Off-site commercial firms recycle, reclaim, purify, and stabilize vacuum pump oil, solvents, mercury, rare materials, and metals. Off-site recycling is preferable to disposal, and sometimes is less expensive. Another off-site option is to work with suppliers of laboratory chemicals who accept return of unopened chemicals, including highly reactive chemicals. Gas suppliers sometimes accept returns of partially used cylinders.

A general comment applicable to all recycling is that a recyclable waste stream needs to be kept as clean as possible. If a laboratory produces a large quantity of waste xylene, small quantities of other organic solvents should be collected in a separate container, because the distillation process gives a better product with fewer materials to separate. Steps should also be taken to avoid getting mercury into oils used in vacuum systems, and oil baths. Similarly, certain ions in a solution of waste metal salts have a serious negative impact on the recrystallization process. Identify users for a recycled product before time and energy are wasted on producing a product that must still be disposed of as a waste. Recycling some of the chemicals used in large undergraduate courses is especially cost-effective because the users are known well in advance.

Many recycling processes result in some residue that is not reusable and will probably have to be handled as a hazardous waste.

5.D.3.2 Solvent Recycling

Because the choice of a distillation unit for solvent recycling is controlled largely by the level of purity desired in the solvent, know the intended use of the redistilled solvent before equipment is purchased. A simple flask, column, and condenser setup may be adequate for a solvent that will be used for crude separations or for initial glassware cleaning. For a much higher level of purity, a spinning band column is probably required. Stills with automatic controls that shut down the system under conditions such as loss of cooling or overheating of the still pot are highly recommended, because they enhance the safety of the distillation operation greatly. Overall, distillation is likely to be most effective when fairly large quantities (roughly 5 L) of relatively clean single-solvent waste are accumulated before the distillation process is begun.

5.D.3.3 Recycling Containers, Packaging, and Labware

Laboratory materials other than chemicals, such as containers or packaging materials and parts of laboratory instruments, can also be recycled. Examples include certain clean glass and plastic containers, drums and pails, plastic and film scrap, cardboard, office paper, lightbulbs, circuit boards, other electronics, and metals such as steel and aluminum. Note that an empty container may still be subject to management requirements. See the following regulations: 40 CFR § 261.7 (EPA "empty"); 49 CFR § 173.29 (DOT "empty"); 49 CFR §§ 173.12(c) and 173.28 (DOT "reuse").

5.D.4 Labeling Commercially Packaged Chemicals

Warning: Do not remove or deface any existing labels on incoming containers of chemicals and other materials.

Commercially packaged (by U.S. manufacturers) chemical containers received from 1986 onward generally meet current labeling requirements. The label usually includes the name of the chemical and any necessary handling and hazard information. Inadequate labels on older containers should be updated to meet current standards. To avoid ambiguity about chemical names, many labels carry the CAS registry number as an unambiguous identifier and this information should be added to any label that does not include it. On receipt of a chemical, the manufacturer's label is supplemented by the date received and possibly the name and location of the individual responsible for purchasing the chemical. If chemicals from commercial sources are repackaged into transfer vessels, the new containers should be labeled with all essential information on the original container.

5.D.5 Labeling Other Chemical Containers

The overriding goal of prudent practice in the identification of laboratory chemicals is to avoid abandoned containers of unknown materials that may be expensive or dangerous to dispose of. The contents of all chemical containers and transfer vessels, including, but not limited to, beakers, flasks, reaction vessels, and process equipment, should be properly identified. The labels should be understandable to trained laboratory personnel and members of well-trained emergency response teams. Labels or tags should be resistant to fading from age, chemical exposure, temperature, humidity, and sunlight.

Chemical identification and hazard warning labels on containers used for storing chemicals should include the following information:

- identity of the owner,
- chemical identification and identity of hazard component(s), and
- appropriate hazard warnings.

Materials transferred from primary (labeled) bulk containers to transfer vessels (e.g., safety cans and squeeze bottles) should be labeled with chemical identification and synonyms, precautions, and first-aid information.

Label containers in immediate use, such as beakers and flasks, with the chemical contents. All reactants should be labeled with enough information to avoid confusion between them.

5.D.6 Labeling Experimental Materials

Labeling all containers of experimental chemical materials is prudent. Because the properties of an experimental material are generally not completely known, do not expect its label to provide all necessary information to ensure safe handling.

The most important information on the label of an experimental material is the name of the researcher responsible, as well as any other information, such as a laboratory notebook reference, that can readily lead to what is known about the material. For items that are to be stored and retained within a laboratory where the properties of materials are likely to be well understood, only the sample identification and name are needed.

(For information about labeling samples for transport and shipping, see section 5.F.)

5.D.7 Use of Inventory and Tracking Systems in Emergency Planning

The most important information to have in an emergency is how to access a researcher who is knowledgeable about the chemical(s) involved. In addition, an organization's emergency preparedness plan should include what to do in the event of a hazardous material release. The inventory and tracking systems and the ability to access and make use of them are essential to proper functioning of the plan in an emergency. The care taken in labeling chemicals is also extremely important. (See Chapter 6, section 6.C.10, for a detailed discussion of what to do in laboratory emergencies.)

5.E STORAGE OF CHEMICALS IN STOCKROOMS AND LABORATORIES

The storage requirements and limitations for stockrooms and laboratories vary widely depending on

- level of expertise of the employees,
- level of safety features designed into the facility,
- level of security designed into the facility,
- location of the facility and neighboring homes or buildings,
- nature of the chemical operations,

- accessibility of the stockroom,
- local and state regulations,
- insurance requirements, and
- building and fire codes.

Many local, state, and federal regulations have specific requirements that affect the handling and storage of chemicals in laboratories and stockrooms. For example, radioactive materials, consumable alcohol, explosives, dual-use materials, and hazardous waste have requirements ranging from locked storage cabinets and controlled access to specified waste containers and regulated areas. Stringent requirements may also be placed on an institution by its insurance carriers.

Controlled substances (e.g., narcotics and other controlled prescription drugs) used in research or with research animals have special requirements. The laboratory director must first register with the U.S. Drug Enforcement Agency (DEA) and with the relevant state agency to purchase, possess, or use a Schedule 1–5 controlled substance. Schedule 1 and 2 drugs (e.g., morphine, pentobarbital) must be stored in a safe that is bolted to the floor or wall. Schedule 3–5 drugs (e.g., chloral hydrate, phenobarbital) must be stored in a locked drawer or cabinet. Access should be limited to the laboratory director and, if necessary, no more than the one or two laboratory members who will be using the substance. Detailed inventory records must be kept up-to-date, including amounts purchased, used, left on hand, and disposed of. Contact your local DEA office for disposal instructions. In some cases a DEA agent must witness disposal or packaging for shipment to a disposal facility.

5.E.1 General Considerations

In general, store materials and equipment in cabinets and on shelving designated for such storage:

- Avoid storing materials and equipment on top of cabinets. With all stored items, maintain a clearance of *at least 18 inches from the sprinkler heads* to allow proper functioning of the sprinkler system [see National Fire Protection Association Standard 13 (NFPA, 2010)].
- To make chemicals readily accessible and to reduce accidents caused by overreaching, do not store materials on shelves higher than 5 ft (~1.5 m). If retrieving materials stored above head level, use a step stool.
- Store heavy materials on lower shelves. While recommended for all laboratories, this is particularly important in areas where seismic activity is possible because items may fall during an earthquake.
- Keep exits, passageways, areas under tables or

benches, and emergency equipment areas free of stored equipment and materials to allow for ease of egress and access in case of emergency.

Storing chemicals in stockrooms and laboratories requires consideration of a number of health and safety factors. In addition to the inventory control and storage area considerations discussed above, proper use of containers and equipment is crucial (see section 5.E.3).

In addition to the basic storage area guidelines above, follow these general guidelines when storing chemicals:

- Label all chemical containers appropriately to ensure that chemicals will be stored safely.
- Place the user's name and the date received on all purchased materials to facilitate inventory control.
- To assist in maintaining a clean work environment and to ensure that segregation of incompatible chemicals is maintained, provide a definite storage place for each chemical and return the chemical to that location after each use.
- To avoid clutter, avoid storing chemicals on benchtops, except for those chemicals being used currently.
- To avoid clutter and to maintain adequate airflow, avoid storing chemicals in chemical hoods, except for those chemicals in current use.
- Store volatile toxic or odoriferous chemicals in a ventilated cabinet. Check with the institution's environmental health and safety officer.
- Provide ventilated storage near laboratory chemical hoods.
- If a chemical does not require a ventilated cabinet, store it inside a closable cabinet or on a shelf that has a lip to prevent containers from sliding off in the event of a fire, serious accident, or earthquake.
- Do not expose stored chemicals to heat or direct sunlight.
- Observe all precautions regarding the storage of incompatible chemicals.
- Separate chemicals into compatible groups and store alphabetically within compatible groups. See Table 5.1 and Figure 5.1 for one suggested method for arranging chemicals. Because chemicals in storage are contained, their separation by compatibility groups can be simplified. The color-coded system described here allows for ease of storage. As explained in Chapter 6, compatibility precautions for mixing chemicals are far more complex.
- Store flammable liquids in approved flammable-liquid storage cabinets.
- Consider the security needs for the materials.

TABLE 5.1   Examples of Compatible Storage Groups

A: Compatible Organic Bases

Diethylamine
Piperidine
Triethanolamine
Benzylamine
Benzyltrimethylammonium hydroxide

B: Compatible Pyrophoric & Water-Reactive Materials

Sodium borohydride
Benzoyl chloride
Zinc dust
Alkyl lithium solutions such as methyl lithium in tetrahydrofuran
Methanesulfonyl chloride
Lithium aluminum hydride

C: Compatible Inorganic Bases

Sodium hydroxide
Ammonium hydroxide
Lithium hydroxide
Cesium hydroxide

D: Compatible Organic Acids

Acetic acid
Citric acid
Maleic acid
Propionic acid
Benzoic acid

E: Compatible Oxidizers Including Peroxides

Nitric acid
Perchloric acid
Sodium hypochlorite
Hydrogen peroxide
3-Chloroperoxybenzoic acid

F: Compatible Inorganic Acids not Including Oxidizers or Combustibles

Hydrochloric acid
Sulfuric acid
Phosphoric acid
Hydrogen fluoride solution

J: Poison Compressed Gases

Sulfur dioxide
Hexafluoropropylene

K: Compatible Explosives or Other Highly Unstable Materials

Picric acid dry(<10% H_2O)
Nitroguanidine
Tetrazole
Urea nitrate

L: Nonreactive Flammables and Combustibles, Including Solvents

Benzene
Methanol
Toluene
Tetrahydrofuran

X: Incompatible with ALL Other Storage Groups

Picric acid moist (10-40% H_2O)
Phosphorus
Benzyl azide
Sodium hydrogen sulfide

NOTE: A larger list of examples can be found on the CD that accompanies this book.
SOURCE: Adapted from Stanford University's Chem Tracker Storage System. Used with permission from Lawrence M. Gibbs, Stanford University.

Some chemicals are regulated by federal agencies and require locked cabinets or storage in secure areas.

5.E.2   Storage According to Compatibility

It is prudent to store containers of incompatible chemicals separately. Separation of incompatibles will reduce the risk of mixing in case of accidental breakage, fire, earthquake, or response to a laboratory emergency. Even when containers are tightly closed, fugitive vapors can cause deleterious incompatibility reactions that degrade labels, shelves, cabinets, and containers themselves. As discussed in Chapter 4, a far more detailed review of incompatibilities needs to be done when chemicals are deliberately mixed,

Figure 5.1 (also available on the CD accompanying this book) and Table 5.1 show an example of a detailed classification system for the storage of groups of chemicals by compatibility. The system classifies chemicals into 11 storage groups. Each group should be separated by secondary containment (e.g., plastic trays) or, ideally, stored in its own storage cabinet. According to this system, it is most important to separate storage groups B (compatible pyrophoric and water-reactive chemicals) and X (incompatible with all other storage groups). These two groups merit their own storage cabinets. The accompanying compact disc includes a spreadsheet of hundreds of chemicals listed according to these storage groups.

There are other good classification systems for storing chemicals according to compatibility. At a minimum, always store fuels away from oxidizers. In other systems, the following chemical groups are kept separate by using secondary containment, cabinets, or distance:

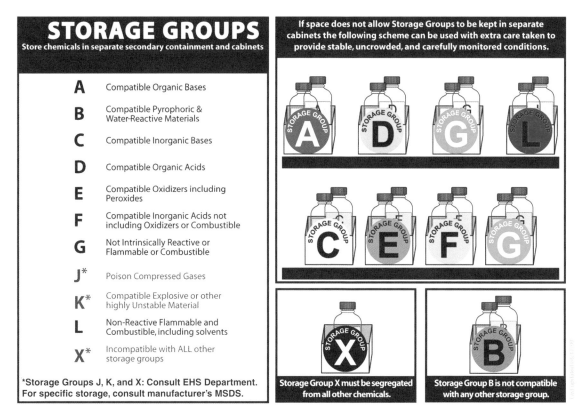

FIGURE 5.1 Compatible storage group classification system. This system should be used in conjunction with specific storage conditions taken from the manufacturer's label and material safety data sheet. SOURCE: Adapted from Stanford University's ChemTracker Storage System. Used with permission from Lawrence M. Gibbs, Stanford University. NOTE: Also available on the CD accompanying this book.

- oxidizers, including peroxides;
- corrosives—inorganic bases;
- corrosives—inorganic acids, not including oxidizers or combustibles;
- flammable materials;
- reproductive toxins;
- select carcinogens; and
- substances with a high degree of acute toxicity.

Depending on the chemicals, their amounts, and the activities of your laboratory, it may make sense to separate these alternative storage groups. Also be sure to follow any storage information on the container's label or on the chemical's MSDS.

In seismically active regions, storage of chemicals requires additional stabilization of shelving and containers. Shelving and other storage units should be secured and contain a front-edge lip to prevent containers from falling. Ideally, containers of liquids are placed on a metal or plastic tray that could hold the liquid if the container broke while on the shelf. All laboratories, not only those in seismically active regions, benefit from these additional storage precautions.

5.E.3 Containers and Equipment

Specific guidelines regarding containers and equipment to use in storing chemicals are as follows:

- Use of corrosion-resistant storage trays as secondary containment for spills, leaks, drips, or weeping is a good idea. Polypropylene trays are suitable for most purposes.
- Use secondary containment (i.e., an overpack) to retain materials if the primary container breaks or leaks.
- Provide vented cabinets beneath chemical hoods for storing hazardous materials. (This encourages the use of the hoods for transferring such materials.)
- Seal containers to minimize escape of corrosive, flammable, or toxic vapors.

5.E.4 Cold Storage

Safe storage of chemicals, biologicals, and radioactive materials in refrigerators, cold rooms, or freezers requires good labels, organization, and active management. The laboratory director assigns responsibility for keeping these units safe, clean, and organized and monitors their proper operation. Extra care is required because frost and condensation not only obscure labels but also make containers hard to hold and easy to drop. Too often, research materials are stored haphazardly in cold storage areas. To ensure safety:

- Use chemical storage refrigerators *only* for storing chemicals.
- Use waterproof tape and markers to label laboratory refrigerators and freezers with the following:

NO FOOD—LAB CHEMICAL STORAGE ONLY

- Do not store flammable liquids in a refrigerator unless it is approved for such storage. Such refrigerators are designed not to spark inside the refrigerator. If refrigerated storage is needed inside a flammable-storage room, it is advisable to choose an explosion-proof refrigerator. Do not store oxidizers or highly reactive materials in the same unit as flammables.
- All containers must be closed and stable to reduce the risk of a spill. Round-bottom flasks need secondary containment.
- Label all materials in the refrigerator with contents, owner, date of acquisition or preparation, and nature of any potential hazard.
- Organize contents by owner but keep incompatibles separate. Organize by labeling shelves and posting the organization scheme on the outside of the unit.
- Secondary containment, such as plastic trays, is highly recommended for all containers. Secondary containment captures spills and leaks and facilitates organization and labeling.
- Every year, review the entire contents of each cold storage unit. Dispose of all unlabeled, unknown, or unwanted materials.
- When any trained laboratory personnel leaves, review the contents of each cold storage unit to identify that person's material, so that it can be disposed of or reassigned.

5.E.5 *Storing Flammable and Combustible Liquids*

NFPA Standard 45 (NFPA, 2004) limits the quantity of flammable and combustible liquids in laboratories. (International, state, and local building codes and regulations should also be consulted.) The quantity allowed depends on a number of factors, including

- construction of the laboratory,
- number of fire control zones in the building,
- floor level where the laboratory is located,
- fire protection systems built into the laboratory,
- storage of flammable liquids in flammable-liquid storage cabinets or safety cans, and
- type of laboratory (i.e., instructional or research and development).

Many laboratories have a business (B) classification with sprinkler systems and a flammable and combustible liquid storage limitation, as shown in Table 5.2. Note that laboratory unit fire hazard classes are based on the quantities of flammable and combustible liquids in the space. This classification significantly affects the fire separation requirements for the laboratory. Most research laboratories fall under Class B, C, or D.

Note that some laboratories may be in jurisdictions that refer to the International Code Agency rather than NFPA, and state and local regulations may be more stringent than those cited here. Laboratory personnel and organization should be sure to check the requirements specific to their area.

The container size for storing flammable and combustible liquids is limited both by NFPA Standards 30 and 45 and by the Occupational Safety and Health Administration (OSHA). Limitations are based on the type of container and the flammability of the liquid, as shown in Table 5.3.

Label all chemical containers with the identity of the contents and hazard warning information. All chemical waste containers must have appropriate waste labels. Flammable liquids that are not stored in safety cans should be placed in storage cabinets rated for flammable storage. When space allows, store combustible liquids in flammable-storage cabinets. Otherwise, store combustible liquids in their original containers. Store 55-gal drums of flammable and combustible liquids in special storage rooms for flammable liquids. Keep flammable and combustible liquids away from strong oxidizing agents, such as nitric or chromic acid, permanganates, chlorates, perchlorates, and peroxides. Keep flammable and combustible liquids away from any ignition sources. Remember that many flammable vapors are heavier than air and can travel to ignition sources. Take the following additional precautions when storing flammable liquids:

- When possible, store quantities of flammable liquids greater than 1 L (approximately 1 qt, or 32 oz) in safety cans. Refer to Table 5.3.
- Store combustible liquids either in their original (or other NFPA- and DOT-approved) containers or in safety cans. Refer to Table 5.3.

TABLE 5.2 Storage Limits for Flammable and Combustible Liquids for Laboratories with Sprinkler System (maximum 100 ft² laboratory space)

Laboratory Unit Fire Hazard Class	Class of Liquid	Excluding Quantities in Rated Storage Cabinets/Safety Cans (max per 100 ft²)		Including Quantities in Rated Storage Cabinets/Safety Cans (max per 100 ft²)	
		gal	L	gal	L
A (high fire hazard)	Class I flammable (flash point <100 °F)	10	38	20	76
	Combined Class I, II, IIIA (flash point <200 °F)	20	76	40	150
B (moderate fire hazard)	Class I flammable (flash point <100 °F)	5	20	10	38
	Combined Class I, II, IIIA (flash point <200 °F)	10	38	20	76
C (low fire hazard)	Class I flammable (flash point <100 °F)	2	7.5	4	15
	Combined Class I, II, IIIA (flash point <200 °F)	4	15	8	30
D (minimal fire hazard)	Class I flammable (flash point <100 °F)	1	4	2	7.5
	Combined Class I, II, IIIA (flash point <200 °F)	1	4	2	7.5

NOTE: Limits for laboratories in health care occupancies and in K-12 educational facilities may be significantly lower.
SOURCE: Reproduced with permission from NFPA 45, *Fire Protection for Laboratories Using Chemicals*, Copyright 2004, National Fire Protection Association. This reprinted material is not the complete and official position of NFPA on the referenced subject, which is represented only by the standard in its entirety.

TABLE 5.3 Container Size for Storage of Flammable and Combustible Liquids

Container	Flammable Liquids[a]						Combustible Liquids[b]			
	Class IA		Class IB		Class IC		Class II		Class IIIA	
	L	gal	L	gal	L	gal	L	gal	L	gal
Glass[c,d]	0.5	0.12	1	0.25	4	1	4	1	20	5
Metal/approved plastic[d]	4	1	20	5	20	5	20	5	20	5
Safety cans[d]	10	2.6	20	5	20	5	20	5	20	5

NOTE: Label safety cans with contents and hazard warning information. Safety cans containing flammable or combustible liquid waste must have appropriate waste labels. Place 20-L (5-gal) and smaller containers of flammable liquids that are not in safety cans in storage cabinets for flammable liquids. Do not vent these cabinets unless they also contain volatile toxics or odoriferous chemicals. Aerosol cans that contain 21% (by volume), or greater, alcohol or petroleum-based liquids are considered Class IA flammables. When space allows, store combustible liquids in storage cabinets for flammable liquids. Otherwise, store combustible liquids in their original (or other Department of Transportation–approved) containers according. Store 55-gal drums of flammable and combustible liquids in special storage rooms for flammable liquids. Keep flammable and combustible liquids away from strong oxidizing agents, such as nitric or chromic acid, permanganates, chlorates, perchlorates, and peroxides. Keep flammable and combustible liquids away from an ignition source. Remember that most flammable vapors are heavier than air and can travel to ignition sources.
[a]Class IA includes those flammable liquids having flash points <73 °F and having a boiling point <100 °F, Class IB includes those having flash points <73 °F and having a boiling point ≥100 °F, and Class IC includes those having flash points ≥73 °F and <100 °F. Aerosol cans that contain 21% (by volume), or greater, alcohol or petroleum-based liquids are considered Class IA flammables.
[b]Class II includes those combustible liquids having flash points at ≥100 °F and <140 °F, Class IIIA includes those having flash points ≥140 °F and <200 °F, and Class IIIB includes those having flash points ≥200 °F.
[c]Glass containers as large as 1 gal can be used if needed and if the required purity would be adversely affected by storage in a metal or approved plastic container, or if the liquid would cause excessive corrosion or degradation of a metal or approved plastic container.
[d]In educational and institutional laboratory work areas, containers for Class I or Class II liquids should not exceed 8 L (32.1 gal) for safety cans or 4 L (1 gal) for other containers.
SOURCE: Reproduced with permission from NFPA 45, *Fire Protection for Laboratories Using Chemicals*, Copyright© 2004, National Fire Protection Association. This reprinted material is not the complete and official position of NFPA on the referenced subject, which is represented only by the standard in its entirety.

5.E.6 Storing Gas Cylinders

Check applicable international, regional, or local building and fire codes to determine the maximum amount of gas to be stored in a laboratory. These limits vary by storage conditions and type of chemical.

With toxic and reactive gases, or large quantities of asphyxiating gases, a special gas cabinet may be required. Gas cabinets are designed for leak detection, safe change-outs, ventilation, and emergency release.

The following general precautions should be taken when storing compressed gas cylinders or lecture bottles:

- Always label cylinders with their contents; do not depend on the manufacturer's color code. They may vary across companies.
- Securely strap or chain gas cylinders to a wall or benchtop. In seismically active areas, use more than one strap or chain.
- When cylinders are no longer in use, shut the valves, relieve the pressure in the gas regulators, remove the regulators, and cap the cylinders.
- Segregate gas cylinder storage from the storage of other chemicals.
- Do not store corrosives near gas cylinders or lecture bottles. Corrosive vapors from mineral acids can deface markings and damage valves.
- Keep incompatible classes of gases stored separately. Keep flammables away from reactives, which include oxidizers and corrosives. (For more information on storage of flammable gases, see Chapter 7, section 7.D.3.3.)
- Segregate empty cylinders from full cylinders.
- Keep in mind the physical state—compressed, cryogenic, or liquefied—of the gases.
- Do not abandon cylinders in the dock storage areas.
- Return cylinders to the supplier when you are finished with them.

For commonly used laboratory gases, consider the installation of in-house gas systems. Such systems remove the need for transport and in-laboratory handling of compressed gas cylinders. Chapter 6, section 6.H, provides additional information on working with compressed gases in the laboratory.

5.E.7 Storing Highly Reactive Substances

Check applicable international, regional, or local building and fire codes to determine the maximum amount of highly reactive chemicals that can be stored in a laboratory. These limits vary by storage conditions and type of chemical. Follow these additional guidelines when storing highly reactive substances:

- Consider the storage requirements of each highly reactive chemical prior to bringing it into the laboratory.
- Consult the MSDSs or other literature in making decisions about storage of highly reactive chemicals.
- Bring into the laboratory only the quantities of material needed for immediate purposes (<3- to 6-month supply, depending on the nature and sensitivity of the materials).
- Label, date, and inventory all highly reactive materials as soon as received. Make sure the label states

DANGER! HIGHLY REACTIVE MATERIAL!

- Do not open a container of highly reactive material that is past its expiration date. Call your institution's hazardous waste coordinator for special instructions.
- Do not open a liquid organic peroxide or peroxide former if crystals or a precipitate are present. Call your institution's hazardous waste coordinator for special instructions.
- For each highly reactive chemical, determine a review date to reevaluate its need and condition and to dispose of (or recycle) material that degrades over time.
- Segregate the following materials:
  - oxidizing agents from reducing agents and combustibles,
  - powerful reducing agents from readily reducible substrates,
  - pyrophoric compounds from flammables, and
  - perchloric acid from reducing agents.
- Store highly reactive liquids in trays large enough to hold the contents of the bottles.
- Store perchloric acid bottles in glass or ceramic trays.
- Store peroxidizable materials away from heat and light.
- Store materials that react vigorously with water away from possible contact with water.
- Store thermally unstable materials in a refrigerator. Use a refrigerator with these safety features:
  - all spark-producing controls on the outside,
  - a magnetically locked door,
  - an alarm to warn when the temperature is too high, and
  - a backup power supply.
- Store liquid organic peroxides at the lowest possible temperature consistent with the solubility or freezing point. Liquid peroxides are particularly sensitive during phase changes. Follow the manufacturer's guidelines for storage of these

highly hazardous materials. (See Chapter 4, section 4.D.)

- Inspect and test peroxide-forming chemicals periodically (these should be labeled with an acquisition or expiration date), and dispose of chemicals that have exceeded their safe storage lifetime.
- Store particularly sensitive materials or larger amounts of explosive materials in explosion relief boxes.
- Restrict access to the storage facility.
- Assign responsibility for the storage facility and the above responsibilities to one primary person and a backup person. Review this responsibility at least yearly.

5.E.8 *Storing Highly Toxic Substances*

Take the following precautions when storing carcinogens, reproductive toxins, and chemicals with a high degree of acute toxicity:

- Store chemicals known to be highly toxic in ventilated storage in unbreakable, chemically resistant secondary containment.
- Keep quantities at a minimum working level.
- Label storage areas with appropriate warning signs, such as

CAUTION! REPRODUCTIVE TOXIN STORAGE
or
CAUTION! CANCER-SUSPECT AGENT STORAGE

- Limit access to storage areas.
- Maintain an inventory of all highly toxic chemicals. Keep records of acquisition, use, possession, and disposal. Some localities require that inventories be maintained of all hazardous chemicals in laboratories.

Note: Facilities covered by the OSHA Laboratory Standard must use and store carcinogens, reproductive toxins, and chemicals with a high degree of acute toxicity in designated areas.

5.F TRANSFER, TRANSPORT, AND SHIPMENT OF CHEMICALS

U.S. and international regulations apply to the movement of chemicals, samples, and other research materials on public roads, by airplane, or by mail or other carrier. When moving these materials on-site, anyone personally transporting regulated materials between adjacent or neighboring buildings within an

institution should walk. (Secondary containment, such as a rubber bucket, should always be used for carrying bottled chemicals.) Organizations located in a larger campus setting should have guidelines indicating if special courier or designated vehicles are to be used to transport regulated materials according to applicable regulations.

Samples of experimental material to be transferred outside the laboratory, or that may be handled by individuals not generally familiar with the type of material involved, should be labeled as completely as possible. In addition, hazardous samples sent to individuals at another institution must be accompanied by appropriate labeling and an MSDS, according to OSHA's Hazard Communication Standard amendments and OSHA's Laboratory Standard hazard identification provision, including the name, address, and contact information of the sender and recipient for samples in transit. When available, the following information should accompany experimental materials:

- **Originator:** List the name of the owner or individual who first obtained the material. If sending the material to another facility, add contact information for the person who can provide safe handling information.
- **Identification:** Include, at least, the laboratory notebook reference.
- **Hazardous components:** List primary components that are known to be hazardous.
- **Potential hazards:** Indicate all known or potential hazards.
- **Date:** Note the date that the material was placed in the container and labeled.
- **Ship to:** Indicate the name, location, and telephone number of the person to whom the material is being transferred.

When transporting or shipping most chemicals, biological agents, and radioactive materials, even small amounts or samples preserved in solvents or alcohol, domestically or internationally, please note that the DOT or the International Air Transport Association (IATA) regulations may apply. Before preparing any packages for shipment, personnel must have documented evidence that they have complete DOT and IATA training. DOT controls shipment of chemicals by a specific set of hazardous materials regulations, 49 CFR Parts 100-199 (updated 2006). These regulations contain detailed instructions on how to identify, package, mark, label, document, and placard hazardous materials. Shipments not in compliance with the applicable regulations may not be offered or accepted for transportation. The regulations on training for safe transportation of hazardous materials are located in

49 CFR § 172.700–172.704 (updated Oct. 1, 2006). All individuals who are preparing hazardous materials for shipment must communicate with their institution's transportation coordinators. Shipment of experimental materials is also discussed in Chapter 11, sections 11.F.1 and 11.F.2.

The use of personal vehicles, company or institutional vehicles (including airplanes), and customer vehicles for transporting regulated materials, which may be hazardous, is a major concern. In many cases, handling regulated materials in this manner is prohibited by DOT or will require shipping papers, placarding, and/or other conditions. *Most businesses and academic institutions forbid the use of privately owned personal vehicles, because of the serious insurance consequences if an accident occurs. Most individuals will find that their personal vehicle insurance does not cover them when they are transporting hazardous materials.*

Shipping chemicals by air is regulated by IATA. An individual who holds IATA certification must inspect the packaging, review the paperwork, and sign the shipping papers. For domestic shipping by ground or rail, DOT regulations apply and may require a bill of lading or manifest, placarding, special packaging, and other conditions.

Be aware that international transfer of chemicals and research materials is regulated by EPA, the Department of Commerce, and the U.S. Customs Bureau as imports and exports. Federal and international laws strictly regulate domestic and international transport of samples, specimens, drugs, and genetic elements, as well as research equipment, technologies, and supplies—even if the material is not hazardous, valuable, or uncommon. Mail, shipments, and luggage are being screened for these materials. Packages to or from research institutions receive additional scrutiny, as well as any package that appears to contain bottles or liquids.

Chapter 8 describes the requirements for shipping hazardous waste.

5.F.1 Materials of Trade Exemption

DOT has an exception to many requirements for transportation of hazardous materials, referred to as the "materials of trade" (MOT) exemption, which applies to the transportation of small quantities of hazardous materials that are part of your business. Examples include the following:

- facilities maintenance services (i.e., paints and paint thinners for painters and gasoline for groundskeepers),
- researchers (i.e., preservatives for field samples), and

- educational demonstrations (i.e., chemicals for public school outreach education programs).

Under this exemption, it is permissible to transport your own hazardous materials as long as certain conditions are met. These include proper packaging according to DOT requirements. The packaging must be the manufacturer's original packaging or a package of equal or greater strength and integrity. The packaging must be marked with a common name or a proper shipping name from the Hazardous Materials Table. Other requirements are

- Packagings must be leaktight for liquids and gases, and siftproof for solids.
- Packages must be securely closed, secured against movement, and protected against damage.
- Outer packagings are not required for receptacles (such as cans or bottles) that are secured against movement in cages, bins, boxes, or compartments.
- Cylinders and pressure vessels must conform to DOT's hazardous materials regulations (49 CFR Parts 171–180) except that outer packagings are not required. These cylinders must be marked with the proper shipping name and identification number and have a hazard class warning label.
- If the package contains a reportable quantity of a hazardous substance, it must be marked "RQ." Reportable quantities are found in Appendix A of 49 CFR § 172.101.

5.F.2 Transfer, Transport, and Shipment of Nanomaterials

This guidance applies to the movement of material from a laboratory to and from off-site locations. Personnel who package and prepare nanomaterials for shipment off-site must be current on hazardous material employee training required by 49 CFR Part 172, Subpart H. Consult your institution's shipping department for assistance and routing of your materials. Although the guidelines provided here are for nanomaterials, the procedures are worth considering for shipping any material.

Any nanomaterial that meets the definition of a hazardous material according to 49 CFR § 171.8 and is classified as a hazardous material in accordance with 49 CFR §§ 173.115–173.141 and 173.403–173.436 must be packaged and marked, and labeled shipping papers must be prepared. The package must be shipped in accordance with 49 CFR Parts 100–185 and all applicable regulations.

Any nanomaterial shipped by air that meets the definition of a dangerous good according to the International Civil Aviation Organization (ICAO) must be packaged, marked, labeled, and shipped with an accompanying properly prepared dangerous goods declaration in accordance with the ICAO technical instructions.

Nanomaterials that are suspected to be hazardous (e.g., toxic, reactive, flammable) should be classified, labeled, marked, and manifested as though that hazard exists. These materials should be classified and shipped as samples according to 49 CFR § 172.101(c)(11) unless the material is specifically prohibited by sections 173.21, 173.54, 173.56(d), 173.56(e), 173.224(c), or 173.225(b). These suspect materials should be packaged in accordance with sections 5.F.2.1 and 5.F.2.2, below.

Nanomaterials that do not meet DOT's criteria listed above may pose health and safety risks to personnel handling the materials if the materials are released during transport. Therefore, all shipments of nanomaterials, regardless of whether they meet the definition for hazardous materials, should be consistently packaged using the equivalent of a DOT-certified packing group I (PG I) container and labeled as described in section 5.F.2.1, below.

5.F.2.1 Off-Site Transport and Shipments of Nanomaterials

This section applies to nanomaterials that are sent to a laboratory and from a laboratory to off-site locations and that do not otherwise meet the DOT definition of hazardous material.

The outer and inner package should meet the definition of PG I–type package. The innermost container should be tightly sealed to prevent leakage of nanomaterials. It should have a secondary seal, such as a tape seal, or a wire tie to prevent a removable closure from inadvertently opening during transport.

The outer package should be filled with shock-absorbing material that can

- protect the inner sample container(s) from damage and
- absorb liquids that might leak from the inner container(s) during normal events in transport.

As depicted in Figure 5.2, the inner package should be labeled (not to be confused with DOT hazard labeling).

If the nanomaterial is in the form of dry dispersible particles, add the following line of text:

Nanomaterials can exhibit unusual reactivity and toxicity. Avoid breathing dust, ingestion, and skin contact.

CAUTION

Nanomaterials Sample

Consisting of [Technical Description Here].

In Case of Container Breakage

Contact: [Point of Contact]

at [Contact's telephone number].

FIGURE 5.2 Recommended inner packaging label for on-site transfer of nanomaterials.

Documentation and notifications for off-site transfer of nanomaterials should include the following:

- a signed and complete dangerous goods declaration or shipping papers prepared in accordance with ICAO and DOT regulations by certified/qualified hazardous material employees who are authorized to release materials from the site;
- available descriptions of the material (e.g., MSDSs) (researchers should prepare a document for the samples that describes known properties and other properties that are reasonably likely to be exhibited by samples); and
- a notification to the receiving facility of the incoming shipment.

All materials should be transported by a qualified carrier.

- Shipments of nanomaterials classified as other materials (neither recognized HazMat or suspected DOT HazMat) may be transported using the most expeditious method provided they are packaged according to the requirements in section 5.F.1.
- The driver must possess basic hazard information on the commodity being transported, that is, material name, quantity, form, and MSDS if available.

5.F.2.2 On-Site Transfer and Transport of Nanomaterials

The on-site transfer of nanomaterials should follow the site-specific transportation safety document or other institutional document (i.e., Chemical Hygiene Plan); in lieu of such a document, the transfer should fully comply with DOT requirements. The site's transportation authority (e.g., transportation safety officer or equivalent) should be the authority having jurisdiction over the requirements for packaging, marking, and documenting necessary for on-site transfers. For nanomaterials, the following is suggested:

- Assess and record the hazards posed by the material(s) following a graded approach that takes into account the form of the material(s) (e.g., free particle versus fixed on substrate).
- Use packaging consistent with the recommendations for off-site shipment or that affords an equivalent level of safety.

- Mark the transfer containers in accordance with the recommendations for off-site shipments.
- Include the following documents in the package:
  o results of the safety assessment and
  o an MSDS, if available, or a similar form detailing possible hazards associated with the material.
- Notify the receiving facility of the incoming shipment.

6 Working with Chemicals

6.A   INTRODUCTION   107

6.B   PRUDENT PLANNING   107

6.C   GENERAL PROCEDURES FOR WORKING WITH HAZARDOUS
      CHEMICALS   108
      6.C.1   Personal Behavior   108
      6.C.2   Minimizing Exposure to Hazardous Chemicals   108
               6.C.2.1   Engineering Controls   108
               6.C.2.2   Avoiding Eye Injury   108
               6.C.2.3   Avoiding Ingestion of Hazardous Chemicals   109
               6.C.2.4   Avoiding Inhalation of Hazardous Chemicals   110
               6.C.2.5   Avoiding Injection of Hazardous Chemicals   111
               6.C.2.6   Minimizing Skin Contact   111
      6.C.3   Housekeeping   113
      6.C.4   Transport of Chemicals   114
      6.C.5   Storage of Chemicals   114
      6.C.6   Use and Maintenance of Equipment and Glassware   114
      6.C.7   Working with Scaled-Up Reactions   115
      6.C.8   Responsibility for Unattended Experiments and Working
               Alone   116
      6.C.9   Chemistry Demonstrations and Magic Shows   116
      6.C.10   Responding to Accidents and Emergencies   117
               6.C.10.1   General Preparation for Emergencies   117
               6.C.10.2   Handling the Accidental Release of Hazardous
                          Substances   117
               6.C.10.3   Notification of Personnel in the Area   117
               6.C.10.4   Treatment of Injured and Contaminated Personnel   117
               6.C.10.5   Spill Containment   120
               6.C.10.6   Spill Cleanup   120
               6.C.10.7   Handling Leaking Gas Cylinders   120
               6.C.10.8   Handling Spills of Elemental Mercury   121
               6.C.10.9   Responding to Fires   121

6.D   WORKING WITH SUBSTANCES OF HIGH TOXICITY   122
      6.D.1   Planning   122
      6.D.2   Experiment Protocols Involving Highly Toxic Chemicals   123
      6.D.3   Designated Areas   123
      6.D.4   Access Control   123
      6.D.5   Special Precautions for Minimizing Exposure to Highly
               Toxic Chemicals   124
      6.D.6   Preparing for Accidents with and Spills of Substances of
               High Toxicity   125
      6.D.7   Storage and Waste Disposal   125
      6.D.8   Multihazardous Materials   125

6.E   WORKING WITH BIOHAZARDOUS AND RADIOACTIVE
        MATERIALS   126
        6.E.1   Biohazardous Materials   126
        6.E.2   Radioactive Materials   127

6.F   WORKING WITH FLAMMABLE CHEMICALS   127
        6.F.1   Flammable Materials   129
        6.F.2   Flammable Liquids   129
        6.F.3   Flammable Gases   129
        6.F.4   Catalyst Ignition of Flammable Materials   130

6.G   WORKING WITH HIGHLY REACTIVE OR EXPLOSIVE
        CHEMICALS   130
        6.G.1   Overview   130
        6.G.2   Reactive or Explosive Compounds   131
                6.G.2.1   Protective Devices   131
                6.G.2.2   Personal Protective Apparel   132
                6.G.2.3   Evaluating Potentially Reactive Materials   132
                6.G.2.4   Determining Reaction Quantities   132
                6.G.2.5   Conducting Reaction Operations   133
        6.G.3   Organic Peroxides   133
                6.G.3.1   Peroxidizable Compounds   134
                6.G.3.2   Peroxide Detection Tests   134
                6.G.3.3   Disposal of Peroxides   134
        6.G.4   Explosive Gases and Liquefied Gases   135
        6.G.5   Hydrogenation Reactions   135
        6.G.6   Materials Requiring Special Attention Because of Toxicity,
                Reactivity, Explosivity, or Chemical Incompatibility   135
        6.G.7   Chemical Hazards of Incompatible Chemicals   140

6.H   WORKING WITH COMPRESSED GASES   140
        6.H.1   Chemical Hazards of Compressed Gases   140
        6.H.2   Specific Chemical Hazards of Select Gases   140

6.I   WORKING WITH MICROWAVE OVENS   141

6.J   WORKING WITH NANOPARTICLES   141
        6.J.1   Controls for Research and Development Laboratory
                Operations That Utilize or Synthesize Nanomaterials   141
                6.J.1.1   Nanomaterial Work Planning and Hazard
                          Assessment   142
                6.J.1.2   A Graded Approach to Determining Appropriate
                          Nanomaterial Controls   142
                6.J.1.3   Engineering Controls for Nanomaterials Research 143

6.A   INTRODUCTION

Prudent execution of experiments requires not only sound judgment and an accurate assessment of the risks involved in the laboratory, but also the selection of appropriate work practices to reduce risk and protect the health and safety of trained laboratory personnel as well as the public and the environment. Chapter 4 provides specific guidelines for evaluating the hazards and assessing the risks associated with laboratory chemicals, equipment, and operations. Chapter 5 demonstrates how to control those risks when managing the inventory of chemicals in the laboratory. The use of the protocols outlined in Chapter 4 in carefully planned experiments is the subject of this chapter.

This chapter presents general guidelines for laboratory work with hazardous chemicals rather than specific standard operating procedures for individual substances. Hundreds of thousands of chemicals are encountered in the research conducted in laboratories, and the specific health hazards associated with most of these compounds are generally not known. Also, laboratory work frequently generates new substances that have unknown properties and unknown toxicity. Consequently, the only prudent course is for laboratory personnel to conduct their work under conditions that minimize the risks from both known and unknown hazardous substances. The general work practices outlined in this chapter are designed to achieve this purpose.

Specifically, section 6.C provides guidelines that are the standard operating procedures where hazardous chemicals are stored or are in use. In section 6.D, additional special procedures for work with highly toxic substances are presented. How to determine when these additional procedures are necessary is discussed in detail in Chapter 4, section 4.C. Section 6.E gives detailed special procedures for work with substances that pose risks due to biohazards and radioactivity, section 6.F addresses flammability, and section 6.G, reactivity and explosivity. Special considerations for work with compressed gases are the subject of section 6.H. Section 6.I covers microwave ovens, and section 6.J describes working with nanoparticles.

Chapter 7 provides precautionary methods for handling laboratory equipment commonly used in conjunction with hazardous chemicals. Chapters 4, 6, and 7 should all be consulted before working with hazardous chemicals.

Four fundamental principles underlie all the work practices discussed in this chapter. Consideration of each should be encouraged before beginning work as part of the culture of safety within the laboratory.

- **Plan ahead.** Determine the potential hazards associated with an experiment before beginning.

- **Minimize exposure to chemicals.** Do not allow laboratory chemicals to come in contact with skin. Use laboratory chemical hoods and other ventilation devices to prevent exposure to airborne substances whenever possible.
- **Do not underestimate hazards or risks.** Assume that any mixture of chemicals will be more toxic than its most toxic component. Treat all new compounds and substances of unknown toxicity as toxic substances. Consider how the chemicals will be processed and whether changing states or forms (e.g., fine particles vs. bulk material) will change the nature of the hazard.
- **Be prepared for accidents.** Before beginning an experiment, know what specific action to take in the event of accidental release of any hazardous substance. Post telephone numbers to call in an emergency or accident in a prominent location. Know the location of all safety equipment and the nearest fire alarm and telephone, and know who to notify in the event of an emergency. Be prepared to provide basic emergency treatment. Keep your co-workers informed of your activities so they can respond appropriately.

Virtually every laboratory experiment generates some waste, which may include such items as used disposable labware, filter media and similar materials, aqueous solutions, and hazardous chemicals. (For more information about disposal of chemical waste, see Chapter 8.)

6.B   PRUDENT PLANNING

Before beginning any laboratory work, determine the hazards and risks associated with the experiment or activity and implement the necessary safety precautions. Ask yourself a hypothetical question before starting work: "What would happen if . . . ?" Consider the possible contingencies and make preparations to take appropriate emergency actions. For example, what would be the consequences of a loss of electrical power or water pressure? Within each laboratory, all personnel should know the location of emergency equipment and how to use it, be familiar with emergency procedures, and know how to obtain help in an emergency. Laboratories should have a standing operational plan that describes how reactions, chemicals, and other laboratory processes will be handled in the case of a natural disaster or in the event that the individual responsible for laboratory activities is unavailable indefinitely (i.e., in the case of illness or death). Included in the plan should be emergency procedures and actions to be taken in the event that laboratory personnel experience a sudden medical emergency while performing an experiment.

Pay attention to the potential safety implications of subtle changes to experimental procedures. Slight changes to commonly performed operations often present unrecognized hazards. Changing solvents, suppliers, reagent concentration, reaction scale, and materials of construction may bring unintended consequences.

Determine the physical and health hazards associated with chemicals before working with them. This determination may involve consulting literature references, laboratory chemical safety summaries (LCSSs), material safety data sheets (MSDSs), or other reference materials (see also Chapter 4, section 4.B) and may require discussions with the laboratory supervisor, safety personnel, and industrial hygienists. Check every step of the waste minimization and removal processes against federal, state, and local regulations. Before producing mixed chemical-radioactive-biological waste (see Chapter 8, section 8.C.1.3) consult your institution's or firm's environmental health and safety (EHS) personnel.

Many of the general practices applicable to working with hazardous chemicals are given elsewhere in this volume (see Chapter 2). (See Chapter 5, section 5.F for detailed instructions on the transport of chemicals and section 5.E on storage; Chapter 7 for information on use and maintenance of equipment and glassware; and Chapter 8 for information on disposal of chemicals.)

6.C GENERAL PROCEDURES FOR WORKING WITH HAZARDOUS CHEMICALS

6.C.1 Personal Behavior

Demonstrating prudent behavior within the laboratory is a critical part of a culture of safety. This includes following basic safety rules and policies (see Chapter 2, section 2.C.1), being cognizant of the hazards within the laboratory (see Chapter 4), and exhibiting professionalism with co-workers. Maintaining an awareness of the work being performed in nearby hoods and on neighboring benches and any risks posed by that work is also important.

6.C.2 Minimizing Exposure to Hazardous Chemicals

Take precautions to avoid exposure by the principal routes, that is, contact with skin and eyes, inhalation, and ingestion (see Chapter 4, section 4.C, for a detailed discussion).

The preferred methods for reducing chemical exposure are, in order of preference,

1. substitution of less hazardous materials or processes (see Chapter 5, section 5.B, *Green Chemistry for Every Laboratory*),
2. engineering controls (Chapter 9),
3. administrative controls (Chapter 2), and
4. personal protective equipment (PPE)

See also the Occupational Safety and Health Administration's (OSHA) Safety and Health Management eTool, Hazard Prevention and Control module available at www.osha.gov. Before beginning work, review all proposed laboratory procedures thoroughly to determine potential health and safety hazards. Refer to the MSDS for guidance on exposure limits, health hazards and routes of entry into the body, and chemical storage, handling, and disposal. Avoid underestimating risk when handling hazardous materials.

6.C.2.1 Engineering Controls

Engineering controls are measures that eliminate, isolate, or reduce exposure to chemical or physical hazards through the use of various devices. Examples include laboratory chemical hoods and other ventilation systems, shields, barricades, and interlocks. Engineering controls must always be considered as the first and primary line of defense to protect personnel and property. When possible, PPE is not to be used as a first line of protection. For instance, a personal respirator should not be used to prevent inhalation of vapors when a laboratory chemical hood (formerly called fume hoods) is available. (See Box 6.1 and Chapter 9 for more information about laboratory design and ventilation.)

6.C.2.2 Avoiding Eye Injury

Eye protection is required for all personnel and visitors in all locations where laboratory chemicals are stored or used, whether or not one is actually performing a chemical operation. Visitor eye protection should be made available at the entrances to all laboratories.

Researchers should assess the risks associated with an experiment and use the appropriate level of eye protection:

- Safety glasses with side shields provide the minimum protection acceptable for regular use. They must meet the American National Standards Institute (ANSI) Z87.1-2003 Standard for Occupational and Educational Eye and Face Protection, which specifies minimum lens thickness and impact resistance requirements.
- Chemical splash goggles are more appropriate

**BOX 6.1
A Simple Qualitative Method
to Verify Adequate Laboratory
Chemical Hood Ventilation**

Materials
  200 g (approximately 250 mL) of dry ice pellets (5-
    to 10-mm diam)
  Shallow bowl, approximately 3-L volume
  1 L water at 43 °C (mix hot and cold water as
    needed to obtain the target temperature)
  Thermometer

Procedure
  1. Open the chemical fume hood sash to simulate
     actual operation. Position laboratory equipment
     as close as possible to where it will be used.
  2. Place the shallow bowl approximately 15 cm into
     the chemical fume hood and in the center of the
     sash opening.
  3. Add 1 L of the warm water to the bowl.
  4. Add the dry ice pellets to the water.
  5. After approximately 5 s, observe the vapor flowing
     from the bowl.
  6. Repeat the observation while a colleague walks
     past or moves around the chemical fume hood to
     simulate actual operation.
  7. If vapors are observed escaping the chemical fume
     hood face, the result is a fail; none escaping is a
     pass.

In the event of a failure or if there is any concern about
proper operation, contact appropriate personnel and
take corrective action. Adjustment of the sash opening and the baffles and relocation of equipment in the
chemical fume hood should be considered.

NOTE: In addition, airflow should be measured on an
annual basis.

---

goggles when conducting particularly hazardous laboratory operations (e.g., working with
glassware under vacuum or handling potentially
explosive compounds). In addition, glassblowing
and the use of laser or ultraviolet light sources
require special glasses or goggles.

• Operations at risk of explosion or that present the
possibility of projectiles must have engineering
controls as a first line of protection. For instance,
in addition to chemical splash goggles or full-
face shields, these operations must be conducted
behind blast shields, in rubber-coated or taped
glassware.

Ordinary prescription glasses do not provide adequate protection against injury because they lack side
shields and are not resistant to impact, but prescription safety glasses and chemical splash goggles are
available.

Similarly, contact lenses offer no protection against
eye injury and do not substitute for safety glasses and
chemical splash goggles. They should not be worn
where chemical vapors are present or a chemical splash
or chemical dust is possible because contact lenses can
be damaged under these conditions. If, however, an
individual chooses to wear contact lenses in the laboratory, chemical splash goggles must be worn. Note
that there has been a change in recommended guidance regarding the wearing of contact lenses since the
previous edition. Many organizations, including the
National Institute for Occupational Safety and Health
(NIOSH) (HHS/CDC/NIOSH, 2005) and the American
Chemical Society (Ramsey and Breazeale, 1998) have
removed most restrictions on wearing contact lenses
in the laboratory.

6.C.2.3 Avoiding Ingestion of Hazardous Chemicals

Eating, drinking, smoking, gum chewing, applying
cosmetics, and taking medicine in laboratories where
hazardous chemicals are used or stored should be
strictly prohibited. Food, beverages, cups, and other
drinking and eating utensils should not be stored in areas where hazardous chemicals are handled or stored.
Glassware used for laboratory operations should never
be used to prepare or consume food or beverages.
Laboratory refrigerators, ice chests, cold rooms, and
ovens should not be used for food storage or preparation. Laboratory water sources and deionized laboratory water should not be used as drinking water. Never
wear gloves or laboratory coats outside the laboratory
or into areas where food is stored and consumed, and
always wash laboratory apparel separately from personal clothing.

---

than regular safety glasses to protect against hazards such as projectiles, as well as when working
with glassware under reduced or elevated pressures (e.g., sealed tube reactions), when handling
potentially explosive compounds (particularly
during distillations), and when using glassware
in high-temperature operations.

• Chemical splash goggles or face shields should be
worn when there is a risk of splashing hazardous
materials or flying particles.

• Because chemical splash goggles offer little protection to the face and neck, full-face shields
should be worn in addition to safety glasses or

Laboratory chemicals should never be tasted. A pipet bulb, aspirator, or mechanical device must be used to pipet chemicals or to start a siphon. To avoid accidental ingestion of hazardous chemicals, pipetting should never be done by mouth. Hands should be washed with soap and water immediately after working with any laboratory chemicals, even if gloves have been worn.

6.C.2.4 Avoiding Inhalation of Hazardous Chemicals

Only in certain controlled situations should any laboratory chemical be sniffed.[1] In general, the practice is not encouraged. Toxic chemicals or compounds of unknown toxicity should never be deliberately sniffed. Conduct all procedures involving volatile toxic substances and operations involving solid or liquid toxic substances that may result in the generation of aerosols in a laboratory chemical hood. Air-purifying respirators are required for use with some chemicals if engineering controls cannot control exposure. Significant training, along with a medical evaluation and respirator fit, are necessary for the use of respirators. For further guidance on the use of respirators with specific chemicals refer to Chapter 7, section 7.F.2.4 of this book, the OSHA Respiratory Protection Standard (29 CFR § 1910.134), and ANSI Standard Z88.2-1992.

Laboratory chemical hoods should not be used for disposal of hazardous volatile materials by evaporation. Such materials should be treated as chemical waste and disposed of in appropriate containers according to institutional procedures and government regulations. (See Chapter 8 for information on waste handling.)

6.C.2.4.1 General Rules for Laboratory Chemical Hoods

Detailed information regarding laboratory ventilation can be found in Chapter 9. The information here is intended to provide a brief overview. These general rules should be followed when using laboratory chemical hoods:

- Before using a laboratory chemical hood, learn how it operates. They vary in design and operation.
- For work involving hazardous substances, use only hoods that have been evaluated for adequate

face velocity and proper operation. They should be inspected regularly and the inspection certification displayed in a visible location.
- Review the MSDS and the manufacturer's label before using a chemical in the laboratory or hood. Observe the permissible exposure limit, threshold limit value, the primary routes of exposure, and any special handling procedures described within the document. Confirm that the experimental methods and available engineering controls are capable of controlling personnel exposure to the hazardous chemicals being used.
- Keep reactions and hazardous chemicals at least 6 in. (15 cm) behind the plane of the sash, farther if possible.
- Never put your head inside an operating hood to check an experiment. The plane of the sash is the barrier between contaminated and uncontaminated air.
- On hoods where sashes open vertically, work with the sash in the *lowest possible position*. Where sashes open horizontally, position one of the doors to act as a shield in the event of an accident. When the hood is not in use, the sash should be kept at the recommended position to maintain laboratory airflow.
- Keep laboratory chemical hoods clean and clear; do not clutter with bottles or equipment. If there is a grill along the bottom slot or a baffle in the back, clean it regularly so it does not become clogged with papers and dirt. Allow only materials actively in use to remain in the hood. Following this rule provides optimal containment and reduces the risk of extraneous chemicals being involved in any fire or explosion. Support any equipment in hoods on racks or feet to provide airflow under the equipment.
- Do not remove the airfoil, alter the position of inner baffles, block exterior grills, or make any other modifications without the approval of the appropriate staff.
- Report suspected laboratory chemical hood malfunctions promptly to the appropriate office, and confirm that the problems are corrected.
- If working in a glovebox, check the seals and pressures on the box before use.

Post the name of the individual responsible for the hood in a visible location. Clean hoods before maintenance personnel work on them.

(See Chapter 9, section 9.C, for more information on laboratory chemical hoods.)

---

[1]In a controlled instructional setting, students may be told to sniff the contents of a container. In such cases, the chemical being sniffed should be screened ahead of time to ensure that it is safe to do so. If instructed to sniff a chemical, gently waft the vapors toward your nose using a folded sheet of paper. Do not directly inhale the vapors.

6.C.2.5 Avoiding Injection of Hazardous Chemicals

Solutions of chemicals are often transferred in syringes, which for many uses are fitted with sharp needles. The risk of inadvertent injection is significant, and vigilance is required to avoid an injury. Use special care when handling solutions of chemicals in syringes with needles. When accompanied by a cap, syringe needles should be placed onto syringes with the cap in place and remain capped until use. Do not recap needles, especially when they have been in contact with chemicals. Remove the needle and discard it immediately after use in the appropriate sharps containers. Blunt-tip needles, including low-cost disposable types, are available from a number of commercial sources and should be used unless a sharp needle is specifically required to puncture rubber septa or for subcutaneous injection.

6.C.2.6 Minimizing Skin Contact

6.C.2.6.1 Gloves

The OSHA Personal Protective Equipment (PPE) Standard (29 CFR §§ 1910.132–1910.138) requires completion of a hazards assessment for each work area, including an evaluation of the hazards involved and selection of appropriate hand protection. Wear gloves whenever handling hazardous chemicals, sharp-edged objects, very hot or very cold materials, toxic chemicals, and substances of unknown toxicity. **No single glove material provides effective protection for all uses.** Before starting, carefully evaluate the type of protection required in order to select the appropriate glove. The discussion presented here is geared toward gloves that protect against chemical exposure. (For information about gloves that protect against other types of hazards, see Chapter 7, section 7.F.1.4.)

Select gloves carefully to ensure that they are impervious to the chemicals being used and are of correct thickness to allow reasonable dexterity while also ensuring adequate barrier protection. Choosing an improper glove can itself be a serious hazard in handling hazardous chemicals. If chemicals do penetrate glove material, they could be held in prolonged contact with the hand and cause more serious damage than in the absence of a proper glove. The degradation and permeation characteristics of the selected glove material must be appropriate for protection from the hazardous chemicals that are handled. Double gloves provide a multiple line of defense and are appropriate for many situations. Find a glove or combination of gloves that addresses all the hazards present. For example, operations involving a chemical hazard and sharp objects may require the combined use of a chemical-resistant (butyl, viton, or neoprene) glove and a cut-resistant (e.g., leather, Kevlar®) glove. Reusable gloves should be washed and inspected before and after each use. Be sure to wash your hands after wearing gloves and handling laboratory chemicals, to remove any skin contamination that might have occurred.

Gloves that might be contaminated with toxic materials should not be removed from the immediate area (usually a laboratory chemical hood) in which the chemicals are located. To prevent contamination of common surfaces that others might touch bare-handed, never wear gloves when handling common items such as doorknobs, handles, or switches on shared equipment, or outside the laboratory. Along the same lines, consider, before touching a surface while wearing gloves, whether it would be common for people to touch the surface with or without gloves and use appropriate precautions. For example, controls for hood nitrogen or water may be located outside the hood itself but may well be contaminated.

When working with chemicals in the laboratory, wear gloves of a material known to be resistant to permeation by the substances in use. Glove selection guides for a wide array of chemicals are available from most glove manufacturers and vendors. In general, nitrile gloves are suitable for incidental contact with chemicals. Both nitrile and latex gloves provide minimum protection from chlorinated solvents and should not be used with oxidizing or corrosive acids. Latex gloves protect against biological hazards but offer poor protection against acids, bases, and most organic solvents. In addition, latex is considered a sensitizer and triggers allergic reactions in some individuals. (For more information, see section 6.C.2.6.1.1.) Neoprene and rubber gloves with increased thickness are suggested for use with most caustic and acidic materials. Barrier creams and lotions can provide some skin protection but are never a substitute for gloves, protective clothing, or other protective equipment. Use these creams only to supplement the protection offered by PPE.

According to the National Ag Safety Database (www.nasdonline.org), a program supported by NIOSH and the Centers for Disease Control and Prevention, materials that are used in the manufacture of gloves designed to provide chemical resistance include the following:

- **Butyl** is a synthetic rubber with good resistance to weathering and a wide variety of chemicals.
- **Natural rubber latex** is a highly flexible and conforming material made from a liquid tapped from rubber plants. It is a known allergen. (See section 6.C.2.6.1.1 for more information.)

- **Neoprene** is a synthetic rubber having chemical and wear-resistance properties superior to those of natural rubber.
- **Nitrile** is a copolymer available in a wide range of acrylonitrile content; chemical resistance and stiffness increase with higher acrylonitrile content.
- **Polyethylene** is a fairly chemical-resistant material used as a freestanding film or a fabric coating.
- **Poly(vinyl alcohol)** is a water-soluble polymer that exhibits exceptional resistance to many organic solvents that rapidly permeate most rubbers.
- **Poly(vinyl chloride)** is a stiff polymer that is made softer and more suitable for protective clothing applications by the addition of plasticizers.
- **Polyurethane** is an abrasion-resistant rubber that is either coated into fabrics or formed into gloves or boots.
- **4H® or Silvershield®** is a registered trademark of North Hand Protection; it is highly chemical-resistant to many different class of chemicals.
- **Viton®**, a registered trademark of DuPont, is a highly chemical-resistant but expensive synthetic elastomer.

When choosing an appropriate glove, consider the required thickness and length of the gloves as well as the material. Consult the glove manufacturer for chemical-specific glove recommendations and information about degradation and permeation times. Certain disposable gloves should not be reused. (For more information, see OSHA PPE Standard, 29 CFR § 1910.138, regarding hand protection.)

The following general guidelines apply to the selection and use of protective gloves:

- Do not use a glove beyond its expiration date. Gloves degrade over time, even in an unopened box.
- When not in use, store gloves in the laboratory but not close to volatile materials. To prevent chemical contamination of nonlaboratory areas by people coming to retrieve them, gloves must not be stored in offices or in break rooms or lunchrooms.
- Inspect gloves for small holes, tears, and signs of degradation before use.
- Replace gloves periodically because they degrade with use, depending on the frequency of use and their permeation and degradation characteristics relative to the substances handled.
- Replace gloves immediately if they become contaminated or torn.
- Replace gloves periodically, depending on the frequency of use. Regular inspection of their serviceability is important. If they cannot be cleaned,

dispose of contaminated gloves according to institutional procedures.
- Decontaminate or wash gloves appropriately before removing them. [Note: Some gloves, e.g., leather and poly(vinyl alcohol), are water permeable. Unless coated with a protective layer, poly(vinyl alcohol) gloves will degrade in the presence of water.]
- Do not wear gloves outside the laboratory, to avoid contamination of surfaces used by unprotected individuals.
- Gloves on a glovebox should be inspected with the same care as any other gloves used in the laboratory. Disposable gloves appropriate for the materials being handled within the glovebox should be used in addition to the gloves attached to the box. Protect glovebox gloves by removing all jewelry prior to use.

6.C.2.6.1.1   Latex Gloves

Although natural rubber latex gloves can be used as protective equipment to prevent transmission of infectious diseases and for skin protection against contact with some chemicals, they can also cause allergic reactions. In addition to causing skin contact allergic reactions to individuals wearing the gloves, they can also cause allergic reactions through inhalation of latex proteins that may be released into the air when the powders used to lubricate the interior of the glove are dispersed as gloves are removed. Thus the risk of exposure via inhalation presents a risk both to the wearer of latex gloves and to sensitized individuals who may be working nearby.

Latex exposure symptoms include skin rash, respiratory irritation, asthma, and, in rare cases, anaphylactic shock. The amount of exposure needed to sensitize an individual to natural rubber latex is not known, but when exposures are reduced, sensitization decreases. Individuals with known latex allergies should never wear latex gloves and may not be able to work in areas where latex gloves are used. Persons with known latex allergies should follow their organization's procedures to ensure that they are not exposed.

To help minimize the risk of exposure to latex allergens, NIOSH issued an alert, *Preventing Allergic Reactions to Latex in the Workplace* (HHS/CDC/NIOSH, 1997). NIOSH recommends the following to reduce exposure to latex:

- Whenever possible, substitute another glove material.
- If latex gloves are the best choice, use reduced-protein, powder-free gloves.
- Wash hands with mild soap and water after removing latex gloves.

6.C.2.6.2 Clothing and Protective Apparel

Protective clothing should be used when there is significant potential for skin-contact exposure to chemicals. Protective clothing does not offer complete protection to the wearer and should not be used as a substitute for engineering controls. The protective characteristics of any protective clothing must be matched to the hazard. As with gloves, no single material that provides protection to all hazards is available. When multiple hazards are present, multiple layers of protective clothing may be required. Some types of PPE, such as aprons of reduced permeability and disposable laboratory coats, offer additional safeguards when working with toxic materials. (See also Chapter 7, section 7.F.1.1.)

Commercial lab coats are fabricated from a variety of materials, such as cotton, polyester, cotton-polyester blends, polyolefin, and polyaramid. Selection of the proper material to deal with the particular hazards present is critical. For example, although cotton is a good material for laboratory coats, it reacts rapidly with acids. Plastic or rubber aprons can provide good protection from corrosive liquids but can be inappropriate in the event of a fire. Because plastic aprons can also accumulate static electricity, they should not be used around flammable solvents, explosives sensitive to electrostatic discharge, or materials that can be ignited by static discharge. Because many synthetic fabrics are flammable and can adhere to the skin, they increase the severity of a burn and should not be worn if working with flammable materials or an open flame. When working with flammable materials or pyrophorics, use laboratory coats made from flame-resistant, nonpermeable materials (polyaramids). Disposable garments may be a good option if handling carcinogenic or other highly hazardous materials. However, these provide only limited protection from vapor or gas penetration. Take care to remove disposable garments without exposing any individual to toxic materials and dispose of as hazardous waste.

To prevent chemical exposure from spilled materials in the laboratory, wear shoes that cover the entire foot. Perforated shoes, open-toe and open-heel shoes, sandals, or clogs should not be permitted. Shoes should have stable soles that provide traction in slippery or wet environments to reduce the chance of falling. Socks should cover the ankles so as to protect against chemical splashes. High heels should not be worn in the laboratory.

Once they have been used, laboratory coats and other protective apparel may become contaminated. Therefore, they must be stored in the laboratory and not in offices or common areas. Institutions should provide a commercial laundry service for laboratory coats and uniforms; they should not be laundered at home.

6.C.3 Housekeeping

A definite correlation exists between orderliness and the level of safety in the laboratory. In addition, a disorderly laboratory can hinder or endanger emergency response personnel. The following housekeeping rules should be adhered to:

- Never obstruct access to exits and emergency equipment such as fire extinguishers and safety showers. Comply with local fire codes for emergency exits, electrical panels, and minimum aisle width.
- Store coats, bags, and other personal items in the proper area, not on the benchtops or in the aisles.
- Do not use floors, stairways, and hallways as storage areas. Items stored in these areas can become hazards in the event of an emergency.
- Keep drawers and cabinets closed when not in use, to avoid accidents.
- Properly label (see Chapter 4, section 4.B.5) in permanent marker and store (see Chapter 5, section 5.E) all chemicals appropriately by compatibility.
- Label transfer vessels[2] with the full chemical name, manufacturer's name, hazard class, and any other special warnings.
- Store chemical containers in order and neatly. Face labels outward for easy viewing. Containers themselves should be clean and free of dust. Containers and labels that have begun to degrade should be replaced, repackaged, or disposed of in the proper location. Do not store materials or chemicals on the floor because these may present trip and spill hazards.
- Keep chemical containers closed when not in use.
- Secure all compressed gas cylinders to walls or benches in accordance with the guidance provided in Chapter 5, section 5.E.6.
- Secure all water, gas, air, and electrical connections in a safe manner.
- Return all equipment and laboratory chemicals to their designated storage location at the end of the day.
- To reduce the chance of accidentally knocking containers to the floor, keep bottles, beakers, flasks, and the like at least 2 in. from the edge of benchtops.
- Keep work areas clean (including floors) and uncluttered. Wipe up all liquid and ice on the floor promptly. Accumulated dust, chromatography adsorbents, and other chemicals pose respira-

---

[2]Transfer vessels may also be known as "secondary containers." The term "transfer vessel" is used here to avoid confusion with secondary containment, which is a tray, bucket, or other container used to control spills from a primary container in the event of breakage.

tory hazards. To avoid formation of aerosols, dry sweeping should not be used in the laboratory. Remove broken glass, spilled chemicals, and paper litter from benchtops and laboratory chemical hoods.

- To avoid flooding, do not block the sink drains. Place rubber matting in the bottom of the sinks to prevent breakage of glassware and to avoid injuries.
- Do not pile up dirty glassware in the laboratory. Wash glassware carefully. Remember that dirty water can mask glassware fragments. Handle and store laboratory glassware with care. Discard cracked or chipped glassware promptly.
- Dispose of all waste chemicals properly and in accordance with organizational policies.
- Dispose of broken glass and in a specially labeled container for broken glass. Treat broken glassware contaminated with a hazardous substance as a hazardous substance.
- Dispose of sharps (e.g., needles and razor blades) in a specially labeled container for sharps. Treat sharps contaminated with a hazardous substance as hazardous substances.

Formal housekeeping and laboratory inspections should be conducted on a regular basis by the Chemical Hygiene Officer or a designee.

6.C.4 Transport of Chemicals

For more detailed information about transfer and transport of chemicals, see Chapter 5, section 5.F.

When transporting chemicals outside the laboratory or between stockrooms and laboratories, use only break-resistant secondary containment. Commercially available secondary containment is made of rubber, metal, or plastic, with carrying handle(s), and is large enough to hold the contents of the chemical containers in the event of breakage. Resealable plastic bags serve as adequate secondary containment for small samples.

When transporting cylinders of compressed gases, the cylinder must always be strapped in a cylinder cart and the valve protected with a cover cap. When cylinders must be transported between floors, passengers should not be in the elevator.

6.C.5 Storage of Chemicals

Avoid the accumulation of excess chemicals by acquiring the minimum quantities necessary for each procedure or research project. Properly label all chemical containers. Indicate any special hazards on the label. For certain classes of compounds (e.g., ethers

as peroxide formers), write the date the container was opened on the label. For peroxide formers, write the test history and date of discard on the label as well.

Keep only small quantities (<1 L) of flammable liquids at workbenches. Larger quantities should be stored in approved storage cabinets. Store large containers (>1 L) below eye level on low shelves. Unless additional protection and secondary containment are provided, never store hazardous chemicals and waste on the floor. Be aware that fire codes dictate the total volume of flammable liquids, liquefied gases, and flammable compressed gases in a given work area. Ask your institution's EHS expert for the fire code's maximum flammable liquid and gas load for your laboratory, and ensure that your laboratory is in compliance with this code.

Refrigerators used for storage of significant quantities of flammable chemicals must be explosion-proof laboratory-safe units. Explosion-proof refrigerators are sold for this purpose and are labeled and hard-wire installed. Such a refrigerator is mandatory for a renovated or new laboratory where flammable materials need refrigeration. Because of the expense of an explosion-proof refrigerator, a modified sparkproof refrigerator is sometimes found in older laboratories and laboratories using very small amounts of flammable materials. However, a modified sparkproof refrigerator cannot meet the standards of an explosion-proof refrigerator. Where they exist, a plan to phase out the sparkproof refrigerator is recommended.

Materials placed in refrigerators should be clearly labeled with water-resistant labels. Storage trays or secondary containment should be used to minimize the distribution of material in the event a container should leak or break. Retaining the shipping can for secondary containment is good practice. Regularly inspect storage trays, shipping cans, and secondary containment for primary container leaks and degradation. Laboratory refrigerators should have permanent labels warning against the storage of food and beverages for human consumption.

All chemicals should be stored with attention to incompatibilities so that if containers break in an accident, reactive materials do not mix and react violently. (See Chapter 5, section 5.E, and Chapter 8, section 8.C.1.2, for more information.)

6.C.6 Use and Maintenance of Equipment and Glassware

Good equipment maintenance is essential for safe and efficient operations. Laboratory equipment should be regularly inspected, maintained, and serviced on schedules that are based on the manufacturer's recommendations, as well as the likelihood and hazards of equipment failure. Maintenance plans should ensure that any lockout procedures cannot be violated.

Carefully handle and store glassware to avoid damage. Discard or repair chipped or cracked items. Handle vacuum-jacketed glassware with extreme care to prevent implosions. Evacuated equipment such as Dewar flasks or vacuum desiccators should be taped, shielded, or coated. Only glassware designed for vacuum work should be used for that purpose.

Use tongs, a tweezer, or puncture-proof hand protection when picking up broken glass. Small pieces should be swept up with a brush into a dustpan. Glassblowing operations should not be attempted unless an area has been made safe for both fabrication and annealing. Protect your hands and body when performing forceful operations involving glassware. For instance, leather or Kevlar® gloves should be used when placing rubber tubing on glass hose connections. Cuts from forcing glass tubing into stoppers or plastic tubing are a common laboratory accident and are often serious. (See Vignette 6.1.) Constructing adaptors from glass tubing and rubber or cork stoppers is obsolete; instead, use fabricated, commercial adaptors made from plastic, metal, or other materials.

(See Chapter 7 for more discussion.)

6.C.7 Working with Scaled-Up Reactions

Special care and planning is necessary to ensure safe scaled-up work. Scale-up of reactions from those producing a few milligrams or grams to those producing more than 100 g of a product may magnify risks by several orders. Although the procedures and controls for large-scale laboratory reactions may be the same as those for smaller-scale procedures, significant differences may exist in heat transfer, stirring effects, times

---

VIGNETTE 6.2
Runaway reaction during scale-up

A researcher scaled up the cycloaddition reaction of maleic anhydride with quadricyclane, a strained high-energy hydrocarbon. This reaction is reported in the literature and was also previously performed in the researcher's laboratory without incident, albeit at small scale (<10 g). No solvent is used in the procedure. The researcher combined the reagents (approximately 250 g total, a 20-fold scale-up) and began heating to the 60–70 °C target temperature. On reaching 50–60 °C the internal temperature rose very rapidly to more than 220 °C. The subsequent rapid boiling of the reagents dislodged the reflux condenser and expelled some liquid and solid into the chemical fume hood. There was no fire. The materials were fully contained within the chemical fume hood, with no injuries, personnel exposure, or equipment damage.

The likelihood of runaway exothermic reactions must be considered whenever conducting a reaction on a larger scale than previous experience. In the present example this possibility was increased by the use of ultrapure reagents and the lack of solvent. When using high-energy reagents, it is preferable to run them as dilute as possible in a solvent. This practice significantly lowers the energy density and significantly adds to the thermal mass, which help to decrease the chance of a runaway reaction. Slow addition of one reagent also limits the effects of an exothermic reaction.

---

for dissolution, and the effects of concentration—all of which need to be considered. (See Vignette 6.2.) When planning large-scale work, practice requires consulting with experienced workers and considering all possible risks.

Although one cannot always predict whether a scaled-up reaction has increased risk, hazards should be evaluated if the following conditions exist:

- The starting material and intermediates contain functional groups that have a history of being explosive (e.g., N—N, N—O, N—halogen, O—O, and O—halogen bonds) or that could explode to give a large increase in pressure.
- A reactant or product is unstable near the reaction or workup temperature. A preliminary test to determine the temperature and mode of de-

---

VIGNETTE 6.1
Finger laceration from broken tubing connector

A technician planned to replace the rubber vacuum tubing leading from a vacuum pump to a glass cold trap. While attempting to remove the old rubber tubing from the trap, the glass nipple broke and the broken glass cut the employee's thumb. The technician did not don protective gloves or attempt to precut the rubber tubing to ease removal. The employee received three sutures.

composition consists of heating a small sample in a melting-point tube.

- A reactant is capable of self-polymerization.
- A reaction is delayed; that is, an induction period is required.
- Gaseous byproducts are formed.
- A reaction is exothermic. What can be done to provide, or regain, control of the reaction if it begins to run away?
- A reaction requires a long reflux period. What will happen if solvent is lost owing to poor condenser cooling?
- A reaction requires temperatures less than 0 °C. What will happen if the reaction warms to room temperature?
- A reaction involves stirring a mixture of solid and liquid reagents. Will magnetic stirring be sufficient at large scale or will overhead mechanical stirring be required? What will happen if stirring efficiency is not maintained at large scale?

In addition, thermal phenomena that produce significant effects on a larger scale may not have been detected in smaller-scale reactions and therefore could be less obvious than toxic or environmental hazards. Thermal analytical techniques should be used to determine whether any process modifications are necessary.

Consider scaling up the process in multiple small steps, evaluating the above issues at each step. Be sure to review the literature and other sources to fully understand the reactive properties of the reactants and solvents, which may not have been evident at a smaller scale.

(See sections 6.D.1 and 6.G.1 and Chapter 5, section 5.B, for more information.)

6.C.8 Responsibility for Unattended Experiments and Working Alone

It is prudent practice to avoid working alone at the bench in a laboratory building. Individuals working in separate laboratories outside normal working hours should make arrangements to check on each other periodically, or ask security guards to check on them. Experiments known to be hazardous should not be undertaken by a person who is alone in a laboratory. Under unusually hazardous conditions, special rules, precautions, and alert systems may be necessary. (See also Chapter 2, section 2.C.2.)

Laboratory operations involving hazardous substances are sometimes carried out continuously or overnight with no one present. Although unattended operations should be avoided when possible, personnel are responsible for designing experiments to prevent the release of hazardous substances if utility services such as electricity, cooling water, and flow of inert gas are interrupted.

For unattended operations, laboratory lights should be left on, and signs should be posted identifying the nature of the experiment and the hazardous substances in use. If appropriate, arrangements should be made for other workers to periodically inspect the operation. Information should be posted indicating how to contact the responsible individual in the event of an emergency.

(See also Chapter 4, section 4.A.)

6.C.9 Chemistry Demonstrations and Magic Shows

All planned demonstrations and chemistry magic shows that will be performed by chemistry personnel that are not a part of normal laboratory activities should be preapproved and authorized by the organization and should follow all institutional policies. Activity organizers should obtain safety advice from experts as necessary. Experienced chemists who are interested in participating in such activities and want to use the organization's chemicals and apparatus should submit an activity plan in advance of the event. This plan should include

- location of the demonstration,
- date of the event,
- age of the intended audience,
- number of persons who will attend the event,
- degree of audience participation,
- demonstrations that will be performed,
- list of chemicals that will be transported to the demonstration site, and
- PPE that will be worn and by whom.

All chemicals must be transported in accordance with U.S. Department of Transportation regulations, if applicable, and must be handled in a prudent manner, packaged appropriately, labeled properly, and transported back to the institution for disposal via the institution's chemical waste disposal system. Under no circumstances should any chemicals be left at the demonstration site or disposed of there. Prior to the planned event, organizers should ensure that, if an accident involving chemicals occurs in their personal vehicles, they will be covered under their private insurance policies.

[For more information about safety when performing chemistry demonstrations, see the American Chemical Society's NCW and Community Activity SAFETY GUIDELINES (available at http://portal.acs.org/).]

6.C.10 Responding to Accidents and Emergencies

6.C.10.1 General Preparation for Emergencies

Every laboratory should have a written emergency response plan that addresses injuries, spills, fires, accidents, and other possible emergencies and includes procedures for communication and response. All laboratory personnel should know what to do in an emergency. Laboratory work should not be undertaken without knowledge of the following points:

- how to report a fire, injury, chemical spill, or other emergency and how to summon emergency response;
- the location of emergency equipment such as safety showers and eyewash units;
- the location of fire extinguishers and spill control equipment;
- the locations of all available exits for evacuation from the laboratory; and
- how police, fire, and other emergency personnel respond to laboratory emergencies, and the role of laboratory personnel in emergency response.

The above information should be available in descriptions of laboratory emergency procedures and in the institution's Chemical Hygiene Plan. Laboratory supervisors should ensure that all trained laboratory personnel are familiar with this information.

Trained laboratory personnel should know their level of expertise with respect to using fire extinguishers and emergency equipment, dealing with chemical spills, and handling injuries. They should not take actions outside the limits of their expertise but instead should rely on trained emergency personnel. A U.S. Environmental Protection Agency (EPA) regulation, Hazardous Waste Operations and Emergency Response (HAZWOPER), 29 CFR § 1910.120, specifies the training required for various response actions.

Names and contact information for individuals responsible for laboratory operations should be posted on the laboratory door.

6.C.10.2 Handling the Accidental Release of Hazardous Substances

Experiments should always be designed to minimize the possibility of an accidental release of hazardous substances. Laboratory personnel should use the minimum amount of hazardous material possible and perform the experiment so that, as much as possible, any spill is contained.

In the event of an incidental, laboratory-scale spill, follow these general guidelines, in order:

1. Tend to any injured or contaminated personnel and, if necessary, request help (see section 6.C.10.4).
2. If necessary, evacuate the area (see section 6.C.10.3).
3. Notify other laboratory personnel of the accident.
4. Take steps to confine and limit the spill if this can be done without risk of injury or contamination (see section 6.C.10.5).
5. Clean up the spill using appropriate procedures, if this can be done without risk of injury and is allowed by institutional policy. (see section 6.C.10.6).
6. Dispose of contaminated materials properly, according to the procedures described in Chapter 8, section 8.B.6.

(See Chapter 7, section 7.G for more information on emergency procedures.)

6.C.10.3 Notification of Personnel in the Area

Other nearby laboratory personnel should be alerted to the accident and the nature of the chemicals involved. If a highly toxic gas or volatile material is released, the laboratory should be evacuated and personnel posted at entrances to prevent others from inadvertently entering the contaminated area. In some cases (e.g., incidents involving the release of highly toxic substances and spills occurring in nonlaboratory areas), it may be appropriate to activate a fire alarm to alert personnel to evacuate the entire building. The proper emergency responders should be called. Follow your institution's policies for such situations.

6.C.10.4 Treatment of Injured and Contaminated Personnel

If an individual is injured or contaminated with a hazardous substance, tending to him or her generally takes priority over implementing the spill control measures outlined in section 6.C.10.5. Obtain medical attention for the individual as soon as possible by calling emergency personnel. Provide a copy of the appropriate MSDS to the emergency responders or attending physician, as needed. **If you cannot assess the conditions of the environment well enough to be sure of your own safety, do not enter the area. Call emergency personnel and describe the situation as best you can.**

Every laboratory should develop specific procedures

for the highest-risk materials used in their laboratory. To identify these materials, consider past accidents, chemicals used in large volumes, and particularly hazardous chemicals. For example, laboratories in which hydrofluoric acid (HF) is used should establish special procedures for accidental exposures, and laboratory personnel should be trained in these emergency procedures. When specific procedures have not been established, the following steps provide general guidance.

For spills covering small areas of skin:

1. Immediately flush with flowing water for no less than 15 minutes; remove any jewelry or clothing as necessary to facilitate clearing of any residual materials.
2. If there is no visible burn, wash with warm water and soap.
3. Check the MSDS to determine if special procedures are needed or if any delayed effects should be expected.
4. Seek medical attention for even minor chemical burns.
5. Do not use creams, lotions, or salves, unless specifically called for.

For spills on clothes:

1. The emergency responder should wear appropriate PPE during emergency treatment to avoid exposure.
2. Do not attempt to wipe the clothes.
3. To avoid contamination of the victim's eyes, do not remove the victim's eye protection before emergency treatment.
4. Quickly remove all contaminated clothing, shoes, and jewelry while using the safety shower. Seconds count; do not waste time or limit the showered body areas because of modesty. Take care not to spread the chemical on the skin or, especially, in the eyes.
5. Cut off garments such as pullover shirts or sweaters to prevent spreading the contamination, especially to the eyes.
6. Immediately flood the affected body area with water for at least 15 minutes. Resume if pain returns.
7. Get medical attention as soon as possible. The affected person should be escorted and should not travel alone. Send a copy of the MSDS with the victim. If the institution's MSDS is digital, hardcopies of the relevant information should be provided to responders. If the MSDS is not immediately available, it is vitally important that the person in charge convey the name of

the chemical involved to the responders. The responders can then arrange for an MSDS to be available at the hospital, if necessary.
8. Discard contaminated clothes or have them laundered separately from other clothing.

For splashes into the eye:

1. Immediately flush with tepid potable water from a gently flowing source for at least 15 minutes. Use an eyewash unit if one is available. If not, place the injured person on his or her back and pour water gently into the eyes for at least 15 minutes.
2. Hold the individual's eyelids away from the eyeball, and instruct him or her to move the eye up and down and sideways to wash thoroughly behind the eyelids.
3. Follow first aid by prompt treatment by medical personnel or an ophthalmologist who is acquainted with chemical injuries.
4. Send a copy of the MSDS with the victim. If the institution's MSDS is digital, hardcopies of the relevant information should be provided to responders. If the MSDS is not immediately available, it is vitally important that the person in charge convey the name of the chemical involved to the responders. The responders can then arrange for an MSDS to be available at the hospital, if necessary.

For cuts:

1. WARNING: Always wear gloves as a precaution when there is risk of contact with blood or other potentially infectious fluids to prevent the transmission of bloodborne pathogens. (See OSHA 29 CFR § 1910.1030 for more information.)
2. If the injured person has experienced a minor cut, flush the wound with tepid running water to remove any possible chemical contaminants. If there is a cut on a gloved hand, remove the glove after thoroughly washing the affected area to avoid contamination of the cut with chemicals.
3. Apply a bandage and advise the victim to report any signs of infection to a physician. If there is a possibility that the wound is contaminated by broken glass or chemicals, the victim should seek immediate medical attention.
4. If the injured person has experienced a serious injury (if sutures will be necessary), call emergency personnel (911) and apply sterile gauze pads to the wound. If necessary, apply direct pressure to the wound to stop the bleeding.

5. Apply additional pads if blood soaks through the first sterile pad. If bleeding continues, encourage the victim to lie down and elevate the wound area to a position above the heart. If you are unable to stop the bleeding, remain calm and carefully explain the situation to the emergency dispatcher (911). The dispatcher will advise you on further action.
6. Send a copy of the MSDS with the victim. If the institution's MSDS is digital, hardcopies of the relevant information should be provided to responders. If the MSDS is not immediately available, it is vitally important that the person in charge convey the name of the chemical involved to the responders. The responders can then arrange for an MSDS to be available at the hospital, if necessary.

For ingestion:

1. Call emergency personnel (911).
2. Do not encourage vomiting except under the advice of a physician. Call the Poison Control Center **(800-222-1222)** immediately and consult the MSDS for the appropriate action.
3. Save all chemical containers and a small amount of vomitus, if possible, for analysis.
4. Stay with the victim until emergency medical assistance arrives.
5. Send a copy of the MSDS with the victim. If the institution's MSDS is digital, hardcopies of the relevant information should be provided to responders. If the MSDS is not immediately available, it is vitally important that the person in charge convey the name of the chemical involved to the responders. The responders can then arrange for an MSDS to be available at the hospital, if necessary.

If the victim is unconscious:

1. Call emergency personnel (911).
2. **If it is safe for you to enter the area**, place the victim on his or her back and cover with a blanket. Do not attempt to remove the victim from the area unless there is immediate danger.
3. Clear the area of any chemical spill or broken glassware.
4. If the victim begins to vomit, turn the head so that the stomach contents are not aspirated into the lungs.
5. Stay with the victim until emergency medical assistance arrives.
6. If the incident involves a chemical exposure, send a copy of the MSDS with the victim. If the institu-

tion's MSDS is digital, hardcopies of the relevant information should be provided to responders. If the MSDS is not immediately available, it is vitally important that the person in charge convey the name of the chemical involved to the responders. The responders can then arrange for an MSDS to be available at the hospital, if necessary.

For convulsions:

1. Call emergency personnel (911).
2. **If it is safe for you to enter the area**, remove anything that might cause harm to the victim. Clear the area of any chemical spills or broken glassware.
3. If the victim begins to vomit, turn the head so that the stomach contents are not aspirated into the lungs.
4. Try to protect the victim from further danger with as little interference as possible. Do not attempt to restrain the victim.
5. Stay with the victim until emergency medical assistance arrives.
6. If the incident involves a chemical exposure, send a copy of the MSDS with the victim.

For burns from heat:

1. Call emergency personnel (911).
2. For first-degree burns, flush with copious amounts of tepid running water. Apply a moist dressing and bandage loosely.
3. For second-degree (with open blisters) and third-degree burns, do not flush with water. Apply a dry dressing and bandage loosely. Immediately seek medical attention.
4. Do not apply ointments or ice to the wound.

For cold burns:

1. Call emergency personnel (911).
2. **Do not apply heat.**
3. If it is not in the area involved, loosen any clothing that may restrict circulation.
4. Cryogenic liquids produce tissue damage similar to that associated with thermal burns and cause severe deep freezing with extensive destruction of tissue.
5. Flush affected areas with large volumes of tepid water (41–46 °C [105–115 °F]) to reduce freezing.
6. Cover the affected area with a sterile protective dressing or with clean sheets if the area is large, and protect the area from further injury.
7. Seek medical attention.

6.C.10.5 Spill Containment

All personnel who work in a laboratory in which hazardous substances are used should be familiar with their institution's policy regarding spill control. For non-emergency[3] spills, spill control kits may be available that are tailored to the potential risk associated with the materials being used in the laboratory. These kits are used to confine and limit the spill if such actions can be taken without risk of injury or contamination. An individual should be assigned to maintain the kit. Store spill kits near areas where spills may occur. Typical spill control kits include these items:

- spill control pillows, which are commercially available and generally can be used for absorbing solvents, acids, and caustic alkalis, but not HF;
- inert absorbents such as vermiculite, clay, sand, kitty litter, and Oil Dri, but not paper because it is not an inert material and should not be used to clean up oxidizing agents such as nitric acid;
- neutralizing agents for acid spills such as sodium carbonate and sodium bicarbonate;
- neutralizing agents for alkali spills such as sodium bisulfate and citric acid;
- large plastic scoops and other equipment such as brooms, pails, bags, and dustpans; and
- appropriate PPE, warnings, barricade tapes, and protection against slips or falls on the wet floor during and after cleanup.

In an emergency,[4] follow institutional guidelines regarding spill containment.

6.C.10.6 Spill Cleanup

Specific procedures for cleaning up spills vary depending on the location of the accident, the amount and physical properties of the spilled material, the degree and type of toxicity, and the training of the personnel involved. Any cleanup should be performed while wearing appropriate PPE and in line with institutional guidance. General guidelines for handling several common incidental, non-emergency spills follow:

- **Materials of low flammability that are not volatile or that have low toxicity.** This category of hazardous substances includes inorganic acids (e.g., sulfuric and nitric acid) and caustic bases (e.g., sodium and potassium hydroxide). For cleanup, appropriate PPE, including gloves, chemical splash goggles, and (if necessary) shoe coverings, should be worn. Absorption of the spilled material with an inert absorbent and appropriate disposal are recommended. The spilled chemicals can be neutralized with materials such as sodium bisulfate (for alkalis) and sodium carbonate or bicarbonate (for acids), absorbed on Floor-Dri or vermiculite, scooped up, and disposed of according to the procedures detailed in Chapter 8, section 8.B.6.

- **Flammable solvents.** Fast action is crucial when a flammable solvent of relatively low toxicity is spilled. This category includes acetone, petroleum ether, pentane, hexane, diethyl ether, dimethoxyethane, and tetrahydrofuran. Other personnel in the laboratory should be alerted, all flames extinguished, and any spark-producing equipment turned off. In some cases the power to the laboratory should be shut off with the circuit breaker, but the ventilation system should be kept running. The spilled solvent should be soaked up with spill absorbent or spill pillows as quickly as possible. If this cannot be done quickly, evacuation should occur, and emergency personnel (911) should be called. Used absorbent and pillows should be sealed in containers and disposed of properly. Nonsparking tools should be used in cleanup.

- **Highly toxic substances.** The cleanup of highly toxic substances should not be attempted alone. Emergency responders should be notified, and the appropriate EHS expert should be contacted to obtain assistance in evaluating the hazards involved. These professionals will know how to clean up the material and may perform the operation.

- **Debris management.** Debris from the cleanup should be handled as hazardous waste if the spilled material falls into that category.

6.C.10.7 Handling Leaking Gas Cylinders

Leaking gas cylinders constitute serious hazards that may require an immediate evacuation of the area and a call to emergency responders. If a leak occurs, do not apply extreme tension to close a stuck valve. Wear appropriate PPE, which usually includes a self-contained breathing apparatus or an air-line respirator, when entering the area with the leak. (See also section 6.D.6.) The following guidelines cover leaks of various types of gases:

- **Flammable, inert, or oxidizing gases.** If safe to do so, move the cylinder to an isolated area, away from combustible material if the gas is flammable or an oxidizing agent. Post signs that describe

---

[3]A non-emergency response is appropriate in the case of an incidental release of hazardous substances where the substance can be absorbed, neutralized, or otherwise controlled at the time of release by personnel in the immediate area or by maintenance personnel.

[4]An emergency is a situation that poses an immediate threat to personal safety and health, the environment, or property that cannot be controlled and corrected safely and easily by individuals at the scene.

the hazards and state warnings. Take care when moving leaking cylinders of flammable gases so that accidental ignition does not occur. If feasible, move the leaking cylinder into a laboratory chemical hood until it is exhausted.

- **Corrosive gases.** Corrosive gases may increase the size of the leak as they are released, and some corrosives are also oxidants, flammable, or toxic. Move the cylinder to an isolated well-ventilated area, and, if possible, use suitable means to direct the gas into an appropriate chemical neutralizer. If there is apt to be a reaction with the neutralizer that could lead to a suck-back into the valve (e.g., aqueous acid into an ammonia tank), place a trap in the line before starting neutralization. Post signs that describe the hazards and state warnings.
- **Toxic gases.** The same procedure should be followed for toxic gases as for corrosive gases. Be sure to warn others of exposure risks. Move the cylinder to an isolated well-ventilated area. Direct the gas into an appropriate chemical neutralizer. Post signs that describe the hazards and state warnings.

Contact the supplier for specific information and guidance.

6.C.10.8 Handling Spills of Elemental Mercury

When spilled in a laboratory, mercury can become trapped beneath floor tiles, under cabinets, and even between walls. Even at very low levels, chronic mercury exposure can be a serious risk, especially in older laboratory facilities, where multiple historic spills may have occurred. Government and standard-setting organizations have established cleanup standards for laboratory spills. These stringent standards ensure the safety of trained laboratory personnel, students, and future occupants of the space.

A portable atomic absorption spectrophotometer with a sensitivity of at least 2 ng/m^3 or other suitable instruments are used to find mercury residues and reservoirs that result from laboratory spills, and for the final clearance survey. Follow institutional procedures in cleaning up spills. General guidelines for handling incidental, non-emergency elemental mercury spills are as follows:

- First, isolate the spill area. Keep people from walking through and spreading the contamination.
- Wear protective gloves, booties, and a Tyvek® suit when necessary, while performing cleanup activities.
- Collect the droplets on wet toweling, which con-

solidates the small droplets to larger pieces, or with a piece of adhesive tape. Do not use sulfur; the practice is ineffective and the resulting waste creates a disposal problem.

- Consolidate large droplets by using a scraper or a piece of cardboard.
- Use commercial mercury spill cleanup sponges and spill control kits.
- Use specially designed mercury vacuum cleaners that have special collection traps and filters to prevent the emission of mercury vapors. A standard vacuum cleaner should never be used to pick up mercury.
- Waste mercury should be treated as a hazardous waste. Place it in a thick-walled high-density polyethylene bottle and transfer it to a central depository for reclamation.
- Decontaminate the exposed work surfaces and floors by using an appropriate decontamination kit.
- Verify decontamination to the current standards by using a portable atomic absorption spectrophotometer or other suitable instrument as described above.

Prevent mercury spills by using supplies and equipment that do not contain mercury. (For information about reducing the use of mercury in laboratories, see Chapter 5, section 5.B.8.)

6.C.10.9 Responding to Fires

Fires are one of the most common types of laboratory accidents. All personnel should be familiar with the general guidelines below to prevent and minimize injury and damage from fires. Hands-on experience with common types of extinguishers and the proper choice of extinguisher should be part of basic laboratory training. (See also Chapter 7, section 7.F.2.)

Be prepared to respond to a fire:

- Preparation is essential! Make sure all laboratory personnel know the locations of all fire extinguishers in the laboratory, what types of fires they can be used for, and how to operate them correctly. Also ensure that they know the location of the nearest fire-alarm pull station, telephone, emergency contact list, safety showers, and emergency blankets.
- In case of fire, immediately notify emergency response personnel by activating the nearest fire alarm. After initial containment, it is also important to report all fires to appropriate personnel for possible follow-up action.
- Even though a small fire that has just started can

sometimes be extinguished with a laboratory fire extinguisher, attempt to put out such fires only if you are trained to use that type of extinguisher, confident that you can do it successfully and quickly, and from a position in which you are always between the fire and an exit to avoid being trapped. Do not attempt to extinguish fires of any size if the institution's policy prohibits this. A fire can spread and surround you in seconds. Toxic gases and smoke present additional hazards. When in doubt, evacuate immediately instead of attempting to extinguish the fire. Only attempt to extinguish fires of any size if the institution's policy allows.

- Put out fires in small vessels by covering the vessel loosely. Never pick up a flask or container of burning material.
- Extinguish small fires involving reactive metals and organometallic compounds (e.g., magnesium, sodium, potassium, and metal hydrides) with Met-L-X or Met-L-Kyl extinguishers or by covering with dry sand. Apply additional fire suppression techniques if solvents or combustibles become involved. Because these fires are very difficult to extinguish, sound the fire alarms before you attempt to put out the fire.
- In the event of a more serious fire, evacuate the laboratory and activate the nearest fire alarm. When the fire department and emergency response team arrive, tell them what hazardous substances are in the laboratory.
- If a person's clothing catches fire, douse him or her immediately in a safety shower. The drop-and-roll technique is also effective. Use fire blankets only as a last resort because they tend to hold in heat and to increase the severity of burns by creating a chimney-like effect. Remove contaminated clothing quickly. Wrap the injured person in a blanket to avoid shock, and get medical attention promptly.

6.D WORKING WITH SUBSTANCES OF HIGH TOXICITY

Individuals who work with highly toxic chemicals, as identified in Chapter 4 (see section 4.C, Tables 4.1, 4.2, and 4.3), should be thoroughly familiar with the general guidelines for the safe handling of chemicals in laboratories (see section 6.C). They should also have acquired through training and experience the knowledge, skill, and discipline to carry out safe laboratory practices consistently. However, these guidelines alone are not sufficient when handling substances that are known to be highly toxic and chemicals that, when combined in an experimental reaction, may generate

highly toxic substances or produce new substances with the potential for high toxicity. Additional precautions are needed to set up *multiple lines of defense to minimize the risks* posed by these substances. As discussed in section 6.B, preparations for handling highly toxic substances must include sound and thorough planning of the experiment, an understanding of the intrinsic hazards of the substances and the risks of exposure inherent in the planned processes, selection of additional precautions that may be necessary to minimize or eliminate these risks, and review of all emergency procedures to ensure appropriate response to unexpected spills and accidents. Each experiment must be evaluated individually because assessment of the level of risk depends on how the substance will be used. Therefore, a prudent planner does not rely solely on a list of highly toxic chemicals to determine the level of the risk; under certain conditions, chemicals not on these lists may react to form highly toxic substances.

In general, the guidelines in section 6.C reflect the minimum standards for handling hazardous substances and should become standard practice when handling highly toxic substances. For example, although working alone in laboratories should be avoided, it is essential that more than one person be present when highly toxic materials are handled. All people working in the area must be familiar with the hazards of the experiments being conducted and with the appropriate emergency response procedures.

Use engineering controls to minimize the possibility of exposure (see section 6.D.5). The use of appropriate PPE to safeguard the hands, forearms, and face from exposure to chemicals is essential in handling highly toxic materials. Cleanliness, order, and general good housekeeping practices create an intrinsically safer workplace. Compliance with safety rules should be maintained scrupulously in areas where highly toxic substances are handled. Source reduction is always a prudent practice, but in the case of highly toxic chemicals it may mean the difference between working with toxicologically dangerous amounts of materials and working with quantities that can be handled safely with routine practice. Emergency response planning and training are very important when working with highly toxic compounds. Additional hazards from these materials (e.g., flammability and high vapor pressures) can complicate the situation, making operational safety all the more important.

6.D.1 Planning

Careful planning should precede any experiment involving a highly toxic substance whenever the substance is to be used for the first time or whenever an experienced user carries out a new protocol that increases the risk of exposure substantially. Planning should include consultations with colleagues who have experience in handling the substance safely and in protocols of use. Experts in the institution's EHS program are a valuable source of information on the hazardous properties of chemicals and safe practice. They also need to be consulted for guidance regarding those chemicals that are regulated by federal, state, and local agencies or by institutional policy. Thoroughly review the wealth of information available in the MSDS, the literature, and toxicological and safety references.

When planning, always consider substituting less toxic substances for highly toxic ones. Also, be sure to use the smallest amount of material that is practicable for the conduct of the experiment. Other important factors to be considered in determining the need for additional safeguards are the likelihood of exposure inherent in the proposed experimental process, the toxicological and physical properties of the chemical substances being used, the concentrations and amounts involved, the duration of exposure, and known toxicological effects. Plan for careful management of the substances throughout their life cycle—from acquisition and storage through destruction or safe disposal. Document these plans, and review them with personnel doing the work, as well as others in the laboratory. Finally, include a method for receiving feedback that can be incorporated into policy revisions, allowing for continuous improvement of the procedures.

6.D.2 Experiment Protocols Involving Highly Toxic Chemicals

Before the experiment begins, prepare an experiment plan that describes the additional safeguards that will be used for all phases of the experiment from acquisition of the chemical to its final safe disposal. The amounts of materials used and the names of the people involved should be included in the written summary and recorded in the laboratory notebook.

The planning process may demonstrate that monitoring is necessary to ensure the safety of the experimenters. Such a determination is made when there is reason to believe that exposure levels for the substances planned to be used could exceed OSHA-established regulatory action levels, similar guidelines established by other authoritative organizations, or when the exposure level is uncertain.

People who conduct the work should know the signs and symptoms of acute and chronic exposure, including delayed effects. Arrange ready access to an occupational health physician, and consult with the physician to determine if health screening or medical surveillance is appropriate.

6.D.3 Designated Areas

Experimental procedures involving highly toxic chemicals, including their transfer from storage containers to reaction vessels, should be confined to a designated work area in the laboratory. This area, which may be a laboratory chemical hood or glovebox, a portion of a laboratory, or the entire laboratory module, should be recognized by everyone in the laboratory or institution as a place where special training, precautions, laboratory skill, and safety discipline are required.

Post signs conspicuously to indicate the designated areas. It may also be prudent to post any relevant LCSS outside the laboratory door. The designated area may be used for other purposes, as long as all laboratory personnel comply with training, safety, and security requirements, and they are familiar with the emergency response protocols of the institution.

In consultation with the institution's EHS experts, the laboratory supervisor should determine which procedures and highly toxic chemicals need to be confined to designated areas. The general guidelines (section 6.C) for handling hazardous chemicals in laboratories may be sufficient for procedures involving very low concentrations and small amounts of highly toxic chemicals, depending on the experiment, the reagents, and their toxicological and physical properties.

6.D.4 Access Control

Restrict access to laboratories where highly toxic chemicals are in use to personnel who are authorized for this laboratory work and trained in the special precautions that apply. Administrative procedures or even physical barriers may be required to prevent unauthorized personnel from entering these laboratories.

Keep laboratory doors closed and locked to limit access to unattended areas where highly toxic materials are stored or routinely handled. However, security measures must not prevent emergency exits from the laboratory. Be sure to make special arrangements for emergency response, including after normal work hours. Use locks to secure refrigerators, freezers, and other storage areas. Keep track of authorized personnel, and be sure to retrieve keys and change locks and access when these people no longer work in the area.

Keep a detailed inventory of highly toxic chemicals. The date, amount, location, and responsible individual should be recorded for all acquisitions, syntheses, access, use, transport, distribution to others, and disposal. Perform a physical inventory every year to verify active inventory records. A procedure should be in place to report security breaches, inventory discrepancies, losses, diversions, or suspected thefts.

When long-term experiments involving highly toxic compounds require unattended operations, securing the laboratory from access by untrained personnel is essential. These operations should also include fail-safe backup options such as shutoff devices in case a reaction overheats or pressure builds up. Additionally, equipment should include interlocks that shut down experiments by turning off devices such as heating baths or reagent pumps, or that close solenoid valves if cooling water stops flowing through an apparatus or if airflow through a laboratory chemical hood becomes restricted or stops. An interlock should be constructed in such a way that if a problem develops, it places the experiment in a safer mode and will not reset even if the hazardous condition is reversed. Protective devices should include alarms that indicate their activation. Security guards and untrained personnel should never be asked or allowed to check on the status of unattended experiments involving highly toxic materials. Warning signs on locked doors should list the trained laboratory personnel to be contacted in case an alarm sounds within the laboratory.

6.D.5 Special Precautions for Minimizing Exposure to Highly Toxic Chemicals

The practices listed below help establish the necessary precautions to enable laboratory work with highly toxic chemicals to be conducted safely:

1. Conduct procedures involving highly toxic chemicals that can generate dust, vapors, or aerosols in a laboratory chemical hood, glovebox, or other suitable containment device. Check hoods for acceptable operation prior to conducting experiments with toxic chemicals. If experiments are to be ongoing over a significant period of time, the hood should be rechecked at least quarterly for proper operation and be equipped with flow-sensing devices that show at a glance or by an audible signal whether they are performing adequately. When toxic chemicals are used in a glovebox, it should be operated under negative pressure, and the gloves should be checked for integrity and appropriate composition before use. Consider if reactive or toxic effluents may be generated by the procedure. If so, scrubbing may be necessary. If dusts or aerosols are generated, consider using high-efficiency particulate air (HEPA) filters prior to discharge to the atmosphere. Hoods should not be used as waste disposal devices, particularly when toxic substances are involved. To offer maximum protection, they should be operated with sashes at their proper level whenever possible. Monitoring equipment might include both active and passive devices to sample laboratory working environments. The experimenter must ensure that the hood exhaust will not present a hazard to anyone outside the immediate laboratory environment. For instance, rooftop access may need to be eliminated during certain operations or, when rooftop access is required, work with highly toxic materials must not be allowed. (See Chapter 9, section 9.C, for detailed discussion on laboratory chemical hoods and environmental control.) When available, alarmed detection devices are another engineering control that should be used for highly toxic materials. Air dispersion modeling may be necessary to determine if exhaust ventilation will affect nearby air intakes or other sensitive receptors.

2. Gloves must be worn when working with toxic liquids or solids to protect the hands and forearms. Select gloves carefully to ensure that they are impervious to the chemicals being used and are of correct thickness to allow reasonable dexterity while also ensuring adequate barrier protection. (See section 6.C.2.6.1 for more information on gloves.)

3. Face and eye protection is necessary to prevent ingestion, inhalation, and skin absorption of toxic chemicals. Safety glasses with side shields are a minimum standard for all laboratory work. When using toxic substances that could generate vapors, aerosols, or dusts, additional levels of protection, including full-face shields and respirators, are appropriate, depending on the degree of hazard represented. Transparent explosion shields in hoods offer additional protection from splashes. Medical certification, training, and fit-testing are required if respirators are worn.

4. Equipment used for the handling of highly toxic chemicals should be isolated from the general laboratory environment. Consider venting laboratory vacuum pumps used with these substances via high-efficiency scrubbers or an exhaust hood. Motor-driven vacuum pumps are recommended because they are easy to decontaminate (decontamination should be conducted in a designated hood).

5. Always practice good laboratory hygiene where highly toxic chemicals are handled. After using toxic materials, trained laboratory personnel should wash their face, hands, neck, and arms. Equipment (including PPE such as gloves) that might be contaminated must never be removed from the environment reserved for handling toxic materials without complete decontamination. Choose laboratory equipment and glassware that are easy to clean and decontaminate. Work

surfaces should also be easy to decontaminate or covered with appropriate protective material, which can be properly disposed of when the procedure is complete. Mixtures that contain toxic chemicals or substances of unknown toxicity must never be smelled or tasted.

6. Carefully plan the transportation of very toxic chemicals. Handling these materials outside the specially designated laboratory area should be minimized. When these materials are transported, the transporter should wear the full complement of PPE appropriate to the chemicals and the type of shipping containers being transported. Samples should be carried in unbreakable secondary containment. (See Chapter 5 for more information about transporting laboratory chemicals.)

6.D.6 Preparing for Accidents with and Spills of Substances of High Toxicity

Be sure that emergency response procedures cover highly toxic substances. Spill control and appropriate emergency response kits should be nearby, and laboratory personnel should be trained in their proper use. These kits should be marked, contained, and sealed to avoid contamination and to be accessible in an emergency. Essential contents include spill control absorbents, impermeable surface covers (to prevent the spread of contamination while conducting emergency response), warning signs, emergency barriers, first-aid supplies, and antidotes. Before starting experiments, the kit contents should be validated. Safety showers, eyewash units, and fire extinguishers should be readily available nearby. Self-contained impermeable suits, a self-contained breathing apparatus, and cartridge respirators may also be appropriate for spill response preparedness, depending on the physical properties and toxicity of the materials being used (see section 6.C.2.4).

Experiments conducted with highly toxic chemicals should be carried out in work areas designed to contain accidental releases (see also section 6.D.3). Trays and other types of secondary containment should be used to contain inadvertent spills. Careful technique must be observed to minimize the potential for spills and releases.

Prior to work, all toxicity and emergency response information should be posted outside the immediate area to ensure accessibility in emergencies. All laboratory personnel who could potentially be exposed must be properly trained on the appropriate response in the event of an emergency. Conducting occasional emergency response drills is always a good idea. Such dry runs may involve medical personnel as well as emergency cleanup crews.

(See also sections 6.C.10.5 and 6.C.10.6.)

6.D.7 Storage and Waste Disposal

Use unbreakable secondary containment for the storage of highly toxic chemicals. If the materials are volatile (or could react with moisture or air to form volatile toxic compounds), containers should be in a ventilated storage area. All containers of highly toxic chemicals should be clearly labeled with chemical composition, known hazards, and warnings for handling. Chemicals that can combine to make highly toxic materials (e.g., acids and inorganic cyanides, which can generate hydrogen cyanide) should not be stored in the same secondary containment. A list of highly toxic compounds, their locations, and contingency plans for dealing with spills should be displayed prominently at any storage facility. Highly toxic chemicals that have a limited shelf life need to be tracked and monitored for deterioration in the storage facility. Those that require refrigeration should be stored in a ventilated refrigeration facility.

Procedures for disposal of highly toxic materials should be established before experiments begin, preferably before the chemicals are ordered. The procedures should address methods for decontamination of all laboratory equipment that comes into contact with highly toxic chemicals. Waste should be accumulated in clearly labeled impervious containers that are stored in unbreakable secondary containment. Volatile or reactive waste must always be covered to minimize release.

Follow procedures established by the institution's EHS experts for commercial waste disposal. Alternatively, consider the possibility of pretreatment of waste either before or during accumulation. In-laboratory destruction may be the safest and most effective way of dealing with waste, but regulatory requirements may affect this decision.

(For further information about disposal of hazardous waste, see Chapter 8. For information about regulatory requirements, see Chapter 11.)

6.D.8 Multihazardous Materials

Some highly toxic materials present additional hazards because of their flammability (see Chapter 4, section 4.D.1, and Chapter 6, section 6.F), volatility (see sections 6.F and 6.G.6), explosivity (see Chapter 4, section 4.D.3; see also section 6.G.4), or reactivity (see Chapter 4, section 4.D.2; see also section 6.G.2). These materials warrant special attention to ensure that risks are minimized and that plans to deal effectively with all potential hazards and emergency response are implemented. (Table 5.1 provides information regarding incompatible chemicals and substances requiring extreme caution.)

6.E WORKING WITH BIOHAZARDOUS AND RADIOACTIVE MATERIALS

6.E.1 Biohazardous Materials

For even the most experienced laboratory personnel, careful review of the 2009 publication *Biosafety in Microbiological and Biomedical Laboratories* (BMBL; HHS/CDC/NIH, 2009) should be a prerequisite for beginning any laboratory activity involving a microorganism, whether naturally or synthetically derived. It defines four levels of control that are appropriate for safe laboratory work with microorganisms that present occupational and public health risks, ranging from no risk of disease for normal healthy adults to high risk of life-threatening disease. The BMBL provides guidance for handling specific agents and a tiered approach to control and containment for each biosafety level.

The four levels of control, referred to as biosafety levels (BSLs) 1 through 4, describe microbiological practices, safety equipment, and features of laboratory facilities for the corresponding level of risk associated with handling a particular agent. The designation of a BSL is influenced by several characteristics of the infectious agent, the most important of which are the severity of the disease, the mode and efficiency of transmission of the infectious agent, the availability of protective immunization or effective therapy, and the relative risk of exposure created by manipulations used in handling the agent. Novel synthetic agents should be handled at a higher BSL until the characteristics of the agent are better understood. Biological toxins are generally safely handled using BSL 2 practices and procedures with strict attention to sharps safety, PPE, and appropriate use of containment equipment. Certain agents and toxins designated as select agents under 42 CFR Part 73 have security requirements that must be met in addition to the biosafety requirements addressed through the application of BSL.

BSL 1 is the basic level of protection appropriate only for agents that are not known to cause disease in normal healthy adult humans.

BSL 2 is appropriate for handling a broad spectrum of moderate-risk agents that cause human disease by ingestion or through percutaneous or mucous membrane exposure. Hepatitis B virus, salmonellae, *Toxoplasma* spp., and human blood and body fluids are representative BSL 2 agents. Extreme precaution with needles or sharp instruments is emphasized at this level.

BSL 3 is appropriate for agents with potential for respiratory transmission, agents that may cause serious and potentially lethal infections, and agents that have a moderate risk to the outside community as well as the individual. Emphasis is placed on the control of aerosols by performing all manipulations within a biological safety cabinet or other containment equipment. At this level, the facility is at least two doors from general building traffic, has a dedicated exhaust fan designed to operate the facility under a negative pressure gradient, and usually is equipped with HEPA filters to purify the air before it is exhausted to the outside. Air from these laboratories cannot be recirculated to other areas of the building. These requirements are designed to control access to the laboratory and to minimize the release of infectious aerosols from the laboratory. The bacterium *Mycobacterium tuberculosis* is an example of an agent for which this higher level of control is appropriate.

Exotic agents that pose a high individual risk of life-threatening disease by the aerosol route and for which no treatment is available are restricted to high-containment laboratories that meet BSL 4 standards. These agents represent a higher risk to the community because of their higher morbidity and mortality rates. Protection for personnel in these laboratories includes physically sealed gloveboxes or fully enclosed barrier suits that supply breathing air.

Several authoritative reference works are available that provide excellent guidance for the safe handling of infectious microorganisms in the laboratory, including BMBL (HHS/CDC/NIH, 2009), the *NIH Guidelines for Research Involving Recombinant DNA Molecules* (NIH, 2011), and *Biosafety in the Laboratory—Prudent Practices for the Handling and Disposal of Infectious Materials* (NRC, 1989). Standard microbiological practices described in these references are consistent with the prudent practices used for the safe handling of chemicals. *Biosafety in the Laboratory* lists seven foundational work practices in biosafety:

1. Do not eat, drink, or smoke in the laboratory. Do not store food in the laboratory. Keep your hands away from your face; avoid touching your eyes, nose, or mouth with gloved hands.
2. Do not pipette liquids by mouth; use mechanical pipetting devices.
3. Wear personal protective clothing in the laboratory (e.g., eye protection, laboratory coats, gloves, and face protection).
4. Eliminate or work very carefully with sharp objects (such as needles, scalpels, Pasteur pipettes, and capillary tubes).
5. Work carefully to minimize the potential for aerosol formation. Confine aerosols as close as possible to their source of generation (i.e., use a biosafety cabinet).
6. Disinfect work surfaces and equipment after use.
7. Wash your hands after removing protective clothing, after contact with contaminated materials, and before leaving the laboratory.

Other practices that are most helpful for preventing laboratory-acquired infections or intoxications are as follows:

- Keep laboratory doors closed when experiments are in progress.
- Use leakproof secondary containment to move or transfer cultures.
- Deactivate, disinfect, or sterilize infectious waste before disposal.

6.E.2 Radioactive Materials

The receipt, possession, use, transfer, and disposal of most radioactive materials is strictly regulated by the U.S. Nuclear Regulatory Commission (USNRC) (see 10 CFR Part 20, Standards for Protection Against Radiation) and by state agencies who have agreements with the USNRC to regulate the users within their own states. Radioactive materials may only be used for purposes specifically described in licenses issued by these agencies. Individuals working with radioactive materials need to be aware of the restrictions and requirements of these licenses. Consult your institution's radiation safety officer or other designated EHS expert for training, policies, and procedures specific to uses at your institution. Prudent practices for working with radioactive materials are similar to those needed to reduce the risk of exposure to toxic chemicals (section 6.C has similar information) and to biohazards:

- Know the characteristics of the radioisotopes that are being used, including half-life, type and energy of emitted radiation, potential for and routes of exposure, and annual exposure limit. Know how to detect contamination.
- Protect against exposure to airborne and ingestible radioactive materials.
- Never eat, drink, smoke, handle contact lenses, apply cosmetics, or take or apply medicine in the laboratory. Keep food, drinks, cosmetics, and tobacco products out of the laboratory entirely to avoid contamination.
- Do not pipet by mouth.
- Provide for safe disposal of waste radionuclides and their solutions.
- Use PPE (e.g., eye protection, gloves, protective clothing, respirators) to minimize exposures.
- Use shielding and gloveboxes to minimize exposure.
- If possible, use equipment that can be operated remotely.
- Plan experiments to minimize exposure by reducing the time, using shielding, increasing your distance from the radiation, and paying attention to monitoring and decontamination.
- Keep an accurate inventory of radioisotopes.
- Record all receipts, transfers, and disposals of radioisotopes.
- Record surveys.
- Check personnel and the work area each day that radioisotopes are used.
- Plan procedures to use the smallest amount of radioisotope possible.
- Minimize radioactive waste.
- Check waste materials for contamination before discarding.
- Place only materials with known or suspected radioactive contamination in appropriate radioactive waste containers.
- Do not generate multihazardous waste (combinations of radioactive, biological, and chemical waste) without first consulting with the designated radiation and chemical safety officers.

(See Chapter 8 for more information on waste and disposal.)

6.F WORKING WITH FLAMMABLE CHEMICALS

Flammable and combustible materials are a common laboratory hazard. Always consider the risk of fire when planning laboratory operations.

To prepare for fire, become familiar with your institution's response and evacuation procedures. Some institutions have a policy that forbids attempts to control or extinguish a fire. Laboratory personnel need to be trained in the necessary steps to take in case of a fire, including knowing the locations of fire alarms, pull stations, fire extinguishers, safety showers, and other emergency equipment (see section 6.C.10, above). Exit routes should be reviewed. Fire extinguishers in the immediate vicinity of an experiment should be appropriate to the particular fire hazards. Fires can be exacerbated by use of an inappropriate extinguisher. Post telephone numbers to call in an emergency or accident in a prominent location.

(Refer to section 6.F and Chapter 7, section 7.F.2, for further information.)

To minimize the risk of fire, all laboratory personnel should know the properties of chemicals they are handling as well as have a basic understanding of how these properties might be affected by the variety of conditions found in the laboratory. As stated in section 6.B, MSDSs or other sources of information should be consulted for information such as vapor pressure, flash point, and explosive limit in air. The use of flammable substances is common, and their properties are also discussed in Chapter 4, section 4.D.

For a fire to start, an ignition source, a fuel, and an oxidizer must be present. Eliminating ignition sources is difficult and takes careful planning, but avoiding the combined presence of fuel and an oxidizer is possible. Keep fuel sources in closed vessels. Control, contain, and minimize the amount of fuels and oxidizers. Containers with large openings (e.g., beakers, baths, vats) should not be used with highly flammable liquids or with liquids above their flash point. Although all flammable substances should be handled prudently, the extreme flammability of some materials requires additional precautions. Consider using inert gases to blanket or purge vessels containing flammable liquids.

Plan to both prevent and respond to a flammable liquid spill, especially one that occurs during a laboratory operation. Place distillation apparatuses and heated reaction flasks in secondary containment to prevent the spread of flammable liquid in the event of breakage. For example, secondary containment can be a large tray in which the apparatus stands. This precaution is particularly important for distillations and similar operations where the breakage of the still pot would result in the release of large quantities of flammable liquid, which may be at its boiling point.

Eliminate ignition sources from areas where flammable substances are handled. Open flames, such as Bunsen burners and matches, are obvious ignition sources. Gas burners should not be used as a source of heat in any laboratory where flammable substances are used. Less obvious ignition sources include gas-fired space heating or water-heating equipment and electrical equipment, such as stirring devices, motors, relays, and switches, which can all produce sparks that will ignite flammable vapors. Nonsparking, explosion-proof devices should be used. Alternatively, actions can be taken to minimize the potential contact of flammable vapors with ignition sources and air by ensuring that vessels containing flammables are closed and maintained under a blanket of inert gas. In situations where large volumes of flammable liquids are stored or in use, fire codes may legally mandate the use of nonsparking, explosion-proof equipment and electrical fixtures.

Even low-level sources of ignition, such as hot plates, static discharge from clothing, steam lines or other hot surfaces, provide a sufficiently energetic ignition source for the most flammable substances in general laboratory use, such as diethyl ether and carbon disulfide (see Chapter 4, section 4.D.1.3 and Vignette 6.3). Flammable substances that require low-temperature storage should be stored only in refrigerators designed for that purpose. Ordinary refrigerators are a hazard because of the presence of potential ignition sources, such as switches, relays, and, possibly, sparking fan motors, and should never be used for storing flammable chemicals. When transferring flammable liquids in metal containers, sparks from accumulated static charge must be avoided by grounding.

Laboratory fires are also caused by hot plates, oil baths, and heating mantles that can melt and combust plastic materials (e.g., vials, containers, tubing). Dry and concentrated residues can ignite when overheated in stills, ovens, dryers, and other heating devices. Do not operate this equipment unattended. When purchasing these devices, choose those models with automatic high-temperature shutoffs. (See section 7.C.5 for more information about the hazards posed by heating devices.)

Fire hazards posed by water-reactive substances such as alkali metals and metal hydrides, by pyrophoric substances such as metal alkyls, by strong

---

VIGNETTE 6.3
Solvent fire

A researcher placed a 1,000-mL beaker containing 60 g of a reagent on a magnetic stirrer located in the chemical fume hood. Carbon disulfide, 500 mL, was added from a second 1,000-mL beaker with stirring. During the solvent addition, the contents of both beakers ignited. The fire was contained in the beakers. The researcher stopped the addition, set the beakers down in the chemical fume hood and unplugged the power cord to the stirrer. The employee immediately retrieved the nearest fire extinguisher (Halon 1121) from the hall and reentered the laboratory to extinguish the fire. The first discharge of approximately 5 seconds caused the beakers to spill but appeared to stop the fire. After a second or two, the solvent in both beakers and on the chemical fume hood bench reignited. A second discharge was applied and the fire from the beakers and spilled material extinguished and did not reignite.

An investigation failed to determine the ignition source but was most likely due to the magnetic stirrer. Static discharge from clothing or the pouring liquid (the incident occurred on a very low humidity day in January) could not be ruled out. The researcher did not appreciate that carbon disulfide is extremely flammable or that its vapors are heavier than air and may travel considerable distances to an ignition source and flash back. Large open beakers should not have been used to contain this solvent. Finally, a dry-powder fire extinguisher should have been used to minimize the chance of reignition.

oxidizers such as perchloric acid, and by flammable gases such as acetylene require procedures beyond the standard prudent practices for handling chemicals described here (see sections 6.C and 6.D) and should be researched in LCSSs or other references before work begins. In addition, emergency response plans must address these substances and their special hazards.

6.F.1 Flammable Materials

The basic precautions for safe handling of flammable materials include the following:

- As much as possible minimize or eliminate the combined presence of flammable material and oxidizer (air). Cap bottles and vessels not in use. Use inert gas blankets when possible.
- Handle flammable substances only in areas free of ignition sources. In addition to open flames, ignition sources include electrical equipment (especially motors), static electricity, and, for some materials (e.g., carbon disulfide), hot surfaces. Check the work area for flames or ignition sources before using a flammable substance. Before igniting a flame, check for the presence of other flammable substances.
- Never heat flammable substances with an open flame. Preferred heat sources include steam baths, water baths, oil and wax baths, salt and sand baths, heating mantles, and hot-air or nitrogen baths.
- Provide ventilation until the vapors are dilute enough to no longer be flammable. This is one of the most effective ways to prevent the formation of flammable gaseous mixtures. Use appropriate and safe exhaust whenever appreciable quantities of flammable substances are transferred from one container to another, allowed to stand in open containers, heated in open containers, or handled in any other way. In using dilution techniques, make certain that equipment (e.g., fans) is explosion-proof and that sparking items are located outside the airstream.
- Use only refrigeration equipment certified for storage of flammable materials.
- Use the smallest quantities of flammable substances compatible with the need, and, especially when the flammable liquid must be stored in glass, purchase the smallest useful size bottle.

6.F.2 Flammable Liquids

Flammable liquids burn only when their vapor is mixed with air in the appropriate concentration. Therefore, such liquids should always be handled so as to minimize the creation of flammable vapor concentrations. Dilution of flammable vapors by ventilation is an important means of avoiding flammable concentrations. Containers of liquids should be kept closed except during transfer of contents. Transfers should be carried out only in laboratory chemical hoods or in other areas where ventilation is sufficient to avoid a buildup of flammable vapor concentrations. Spillage or breakage of vessels or containers of flammable liquids or sudden eruptions from nucleation of heated liquid can result in a sudden release of flammable vapor.

Metal lines and vessels discharging flammable liquids should be properly grounded to disperse static electricity. For instance, when transferring flammable liquids in metal equipment, static-generated sparks can be eliminated by proper grounding with ground straps. Development of static electricity is related closely to the level of humidity and may become a problem on cold dry winter days. When nonmetallic containers (especially plastic) are used, contact with the grounding device should be made directly to the liquid rather than to the container. In the rare circumstance that static electricity cannot be avoided, all processes should be carried out as slowly as possible to give the accumulated charge time to disperse or should be handled in an inert atmosphere.

Note that vapors of many flammable liquids are heavier than air and capable of traveling considerable distances along the floor. This possibility should be recognized, and special note should be taken of ignition sources at a lower level than that at which the substance is being used. Close attention should be given to nearby potential sources of ignition.

6.F.3 Flammable Gases

Leakage or escape of flammable gases can produce an explosive atmosphere in the laboratory. Acetylene, hydrogen, ammonia, hydrogen sulfide, propane, and carbon monoxide are especially hazardous. Acetylene, methane, and hydrogen have a wide range of concentrations at which they are flammable (flammability limits),[5] which adds greatly to their potential fire and explosion hazard. Installation of flash arresters on hydrogen cylinders is recommended. Prior to introduction of a flammable gas into a reaction vessel, the equipment should be purged by evacuation or with an inert gas. The flush cycle should be repeated three times to reduce residual oxygen to approximately 1%.

(See section 6.H for specific precautions on the use of compressed gases.)

---

[5]Acetylene, lower flammability limit (LFL) = 2.5%, upper flammability limit (UFL) = 82%; methane, LFL = 5%, UFL = 15%; hydrogen, LFL = 4%, UFL = 75% (NFPA, 2004).

6.F.4 Catalyst Ignition of Flammable Materials

Palladium or platinum on carbon, platinum oxide, Raney nickel, and other hydrogenation catalysts should be filtered carefully from hydrogenation reaction mixtures. The recovered catalyst is usually saturated with hydrogen, is highly reactive, and, thus, inflames spontaneously on exposure to air. Especially for large-scale reactions, the filter cake should not be allowed to become dry. The funnel containing the still-moist catalyst filter cake should be put into a water bath immediately after completion of the filtration. Use of a purge gas (nitrogen or argon) is strongly recommended for hydrogenation procedures so that the catalyst can be filtered and handled under an inert atmosphere.

6.G WORKING WITH HIGHLY REACTIVE OR EXPLOSIVE CHEMICALS

An explosion occurs when a material undergoes a rapid reaction that results in a violent release of energy. Such reactions can happen spontaneously or be initiated and can produce pressures, gases, and fumes that are hazardous. Highly reactive and explosive materials used in the laboratory require appropriate procedures. In this section, techniques for identifying and handling potentially explosive materials are discussed.

6.G.1 Overview

Light, mechanical shock, heat, and certain catalysts can be initiators of explosive reactions. Hydrogen and chlorine react explosively in the presence of light. Examples of shock-sensitive materials include many acetylides, azides, organic nitrates, nitro compounds, azo compounds, perchlorates, and peroxides. Acids, bases, and other substances may catalyze explosive polymerizations. The catalytic effect of metallic contamination leads to explosive situations. Many metal ions catalyze the violent decomposition of hydrogen peroxide.

Whenever possible, use a safer alternative. For example, perchlorate used as a counteranion to crystallize salts can often be replaced with safer alternatives such as fluorophosphate or fluoroborate.

Many highly reactive chemicals polymerize vigorously, decompose, condense, or become self-reactive. The improper handling of these materials may result in a runaway reaction that could become violent. Careful planning is essential to avoid serious accidents. When highly reactive materials are in use, emergency equipment should be at hand. The apparatus should be assembled in such a way that if the reaction begins to run away, immediate removal of any heat source, cooling of the reaction vessel, cessation of reagent addition,

and closing of laboratory chemical hood sashes are possible. Restrict access to the area until the reaction is under control, and consider remote operating controls. A heavy transparent plastic explosion shield should be in place to provide extra protection in addition to the laboratory chemical hood window.

Highly reactive chemicals lead to reactions with rates that increase rapidly as the temperature increases. If the heat evolved is not dissipated, the reaction rate increases until an explosion results. Such an event must be prevented, particularly when scaling up experiments. Sufficient cooling and surface for heat exchange should be provided to allow control of the reaction. It is also important to consider the impact of solution concentration, especially when a reaction is being attempted or scaled up the first time. Use of too highly concentrated reagents has led to runaway conditions and to explosions. Particular care must also be given to the rate of reagent addition versus its rate of consumption, especially if the reaction is subject to an induction period. A chemical reaction with an induction period has a reaction rate that is slow initially but accelerates over time.

Large-scale reactions with organometallic reagents and reactions that produce flammables as products or are carried out in flammable solvents require special attention. Active metals, such as sodium, lithium, potassium, calcium, and finely divided magnesium are serious fire and explosion risks because of their reactivity with water, alcohols, and other compounds or solutions containing acidic hydrogens. These materials require special storage, handling, and disposal procedures. Where active metals are present, Class D fire extinguishers that use special extinguishing materials such as a plasticized graphite–based powder or a sodium chloride–based powder (Met-L-X) are required.

Some chemicals decompose when heated. Slow decomposition may not be noticeable on a small scale, but on a large scale with inadequate heat transfer, or if the evolved heat and gases are confined, an explosive situation may develop. The heat-initiated decomposition of some substances, such as certain peroxides, is almost instantaneous. In particular, reactions that are subject to an induction period can be dangerous because there is no initial indication of a risk, but after induction, a violent process can result.

Oxidizing agents may react violently when they come in contact with reducing materials, trace metals, and sometimes ordinary combustibles. These compounds include the halogens, oxyhalogens, peroxyhalogens, permanganates, nitrates, chromates, and persulfates, as well as peroxides (see also section 6.G.3). Even though inorganic peroxides are generally considered to be stable, they may generate organic peroxides and hydroperoxides in contact with organic compounds, react violently with water (alkali metal peroxides), or form superoxides and ozonides (alkali metal peroxides). Perchloric acid and nitric acid are powerful oxidizing agents with organic compounds and other reducing agents. Perchlorate salts can be explosive and should be treated as potentially hazardous compounds. Dusts—suspensions of oxidizable particles (e.g., magnesium powder, zinc dust, carbon powder, or flowers of sulfur) in the air—constitute a powerful explosive mixture.

Scale-up reactions create difficulties in dissipation of heat that are not evident on a smaller scale. Evaluation of observed or suspected exothermicity can be achieved by differential thermal analysis to identify exothermicity in open reaction systems; differential scanning calorimetry, using a specially designed sealable metal crucible, to identify exothermicity in closed reaction systems; or syringe injection calorimetry and reactive systems screening tool calorimetry to determine heats of reaction on a microscale and small scale. [For an expanded discussion of identifying process hazards using thermal analytical techniques, see Tuma (1991).] When it becomes apparent that exothermicity exists at a low temperature or a large exothermicity occurs that might present a hazard, large-scale calorimetry determination of exothermic onset temperatures and drop weight testing are advisable. In situations where formal operational hazard evaluation or reliable data from any other source suggest a hazard, review or modification of the scale-up conditions by an experienced group is recommended to avoid the possibility that an individual might overlook a hazard or the most appropriate procedural changes. Finally, to verify the proper use of equipment and safeguards, it is recommended that researchers calculate the adiabatic temperature and pressure change for the reaction at a fixed volume to estimate the heat and pressure that may be generated.

Any sample of a highly reactive material may be dangerous. The greatest risk is due to the remarkably high rate of a detonation reaction rather than the total energy released. A high-order explosion of even milligram quantities can drive small fragments of glass or other matter deep into the body. It is important to use minimum amounts of hazardous materials with adequate shielding and personal protection.

Not all explosions result from chemical reactions. A dangerous physically caused explosion occurs if a hot liquid is brought into sudden contact with a lower-boiling-point one. The instantaneous vaporization of the lower-boiling-point substance can be hazardous to personnel and destructive to equipment. The presence or inadvertent addition of water to the hot fluid of a heating bath is an example of such a hazard. Explosions can also occur when warming a cryogenic material in a closed container or overpressurizing glassware with nitrogen (N_2) or argon when the regulator is incorrectly

set. Violent physical explosions have also occurred when a collection of very hot particles is suddenly dumped into water. For this reason, dry sand should be used to catch particles during laboratory thermite reaction demonstrations.

6.G.2 Reactive or Explosive Compounds

Occasionally, it is necessary to handle materials that are known to be explosive or that may contain explosive impurities such as peroxides. Because mechanical shock, elevated temperature, or chemical action might result in explosion with forces that release large volumes of gases, heat, and often toxic vapors, they must be treated with special care.

The proper handling of highly energetic substances without injury demands attention to the most minute details. The unusual nature of work involving such substances requires special safety measures and handling techniques that must be understood thoroughly and followed by all persons involved. The practices listed in this section are a guide for use in any laboratory operation that might involve explosive materials.

Work with explosive (or potentially explosive) materials generally requires the use of special protective apparel (e.g., face shields, gloves, and laboratory coats) and protective devices such as explosion shields, barriers, or even enclosed barricades or an isolated room with a blowout roof or window (see Chapter 7, sections 7.F.1 and 7.F.2).

Before work with a potentially explosive material is begun, the experiment should be discussed with a supervisor or an experienced co-worker, and the relevant literature consulted (see Chapter 4, sections 4.B.2, 4.B.5, and 4.B.6). A risk assessment should be carried out.

Various state and federal regulations cover the transportation, storage, and use of explosives. Along with EHS and transportation experts, these regulations should be consulted before explosives (and related dangerous materials) are used or generated in the laboratory. Explosive materials should be brought into the laboratory only as required and in the smallest quantities adequate for the experiment (see Chapter 5, section 5.B). Insofar as possible, direct handling should be minimized. Explosives should be segregated from other materials that could create a serious risk to life or property should an accident occur.

6.G.2.1 Protective Devices

Barriers such as shields, barricades, and guards should be used to protect personnel and equipment from injury or damage from a possible explosion or fire. The barrier should completely surround the hazardous area. On benches and laboratory chemical hoods, a 0.25-in.-thick acrylic sliding shield, which is screwed together in addition to being glued, can effectively protect trained laboratory personnel from glass fragments resulting from a laboratory-scale explosion. The shield should be in place whenever hazardous reactions are in progress or whenever hazardous materials are being stored temporarily. However, such shielding is not effective against metal shrapnel. The hood sash provides a safety shield only against chemical splashes or sprays, fires, and minor explosions. If more than one hazardous reaction is carried out, the reactions should be shielded from each other and separated as far as possible.

Dryboxes should be fitted with safety glass windows overlaid with 0.25-in.-thick acrylic when potentially explosive materials capable of explosion in an inert atmosphere are to be handled. This protection is adequate against most internal 5-g explosions. Protective gloves should be worn over the rubber drybox gloves to provide additional protection. Other safety devices that allow remote manipulation should be used with the gloves. Explosions of high-energy chemicals from static sparks can be a considerable problem in dryboxes, and so adequate grounding is essential, and an antistatic gun or antistatic ionizer is recommended.

Armored laboratory chemical hoods or barricades made with thick (1.0 in.) poly(vinyl butyral) resin shielding and heavy metal walls give complete protection against explosions less than the acceptable 20-g limit. They are designed to contain a 100-g explosion, but an arbitrary 20-g limit is usually set because of the noise level in the event of an explosion. Such chemical hoods should be equipped with mechanical hands that enable the operator to manipulate equipment and handle adduct containers remotely. A sign should be posted, for example,

CAUTION: NO ONE MAY ENTER AN ARMORED LABORATORY CHEMICAL HOOD FOR ANY REASON DURING THE COURSE OF A HAZARDOUS OPERATION.

Miscellaneous protective devices such as both long- and short-handled tongs for holding or manipulating hazardous items at a safe distance and remote control equipment (e.g., mechanical arms, stopcock turners, labjack turners, remote cable controllers, and closed-circuit television monitors) should be available as required to prevent exposure of any part of the body to injury.

6.G.2.2 Personal Protective Apparel

When explosive materials are handled, the following items of personal protective apparel are needed:

- Safety glasses that have solid side shields or chemical splash goggles must be worn by all personnel, including visitors, in the laboratory.
- Full-length shields that fully protect the face and throat must be worn whenever trained laboratory personnel are in a hazardous or exposed position. Special care is required when operating or manipulating synthesis systems that may contain explosives (e.g., diazomethane), when bench shields are moved aside, and when handling or transporting such systems. In view of the special hazard to life that results from severing the jugular vein, extra shielding around the throat is recommended.
- Heavy leather gloves must be worn if it is necessary to reach behind a shielded area while a hazardous experiment is in progress or when handling reactive compounds or gaseous reactants. Proper planning of experiments should minimize the need for such activities.
- Laboratory coats should be worn at all times. The coat should be made of flame-resistant material and should be quickly removable. A coat can help reduce minor injuries from flying glass as well as the possibility of injury from an explosive flash.

6.G.2.3 Evaluating Potentially Reactive Materials

Potentially reactive materials must be evaluated for their possible explosive characteristics by consulting the literature and considering their molecular structures. The presence of functional groups or compounds listed in Chapter 4, sections 4.D.2 and 4.D.3 and section 6.G.6 indicates a possible explosion hazard. Urben (2007) notes three methods for determining the sensitivity of very explosive compounds. The first is a commonly used drop test that measures sound levels, which as the editor notes, is "not entirely satisfactory." The second is an electrostatic test, and the third uses friction generated by grinding two porcelain surfaces together under load. See Dou et al. (1999) for more information. Highly reactive chemicals should be segregated from materials that might interact with them to create a risk of explosion. They should not be used past their expiration date.

6.G.2.4 Determining Reaction Quantities

When a possibly hazardous reaction is attempted, small quantities of reactants should be used. When handling highly reactive chemicals, use the smallest quantities needed for the experiment. In conventional explosives laboratories, no more than 0.1 g of product should be prepared in a single run. During the actual reaction period, no more than 0.5 g of reactants should be present in the reaction vessel: The diluent, the substrate, and the energetic reactant must all be considered when determining the total explosive power of the reaction mixture. Special formal risk assessments should be established to examine operational and safety problems involved in scaling up a reaction in which an explosive substance is used or could be generated.

6.G.2.5 Conducting Reaction Operations

The most common heating devices are heating tapes and mantles and sand, water, steam, wax, silicone oil, and air (or nitrogen) baths. They should be used in such a way that if an explosion were to occur the heating medium would be contained. Heating baths should consist of nonflammable materials. All controls for heating and stirring equipment should be operable from outside the shielded area. (See Chapter 7, section 7.C.5, for further information.)

Vacuum pumps should carry tags indicating the date of the most recent oil change. Oil should be changed once a month, or sooner if it is known that the oil has been unintentionally exposed to reactive gases. All pumps should be either vented into a hood or trapped. Vent lines may be Tygon, rubber, or copper. If Tygon or rubber lines are used, they should be supported so that they do not sag and cause a trap for condensed liquids. (See Chapter 7, section 7.C.2, for details.)

When potentially explosive materials are being handled, the area should be posted with a sign such as

WARNING: VACATE THE AREA AT THE FIRST INDICATION OF [the indicator for the specific case] **AND STAY OUT.**
CALL [responsible person] **AT** [phone number].

When condensing explosive gases, the temperature of the bath and the effect on the reactant gas of the condensing material selected must be determined experimentally (see Chapter 7, section 7.D). Very small quantities should be used because explosions may occur. A taped and shielded Dewar flask should always be used when condensing reactants. Maximum quantity limits should be observed. A dry-ice solvent bath is not recommended for reactive gases; liquid nitrogen is recommended. (See also Chapter 4, section 4.D.3.1.)

6.G.3 Organic Peroxides

Organic peroxides are a special class of compounds with unusually low stability that makes them among the most hazardous substances commonly handled in laboratories, especially as initiators for free-radical reactions. Although they are low-power explosives, they are hazardous because of their extreme sensitivity to shock, sparks, and other forms of accidental detonation. Many peroxides that are used routinely in laboratories are far more sensitive to shock than most primary explosives (e.g., TNT), although many have been stabilized by the addition of compounds that inhibit reaction. Nevertheless, even low rates of decomposition may automatically accelerate and cause a violent explosion, especially in bulk quantities of peroxides (e.g., benzoyl peroxide). These compounds are sensitive to heat, friction, impact, and light, as well as to strong oxidizing and reducing agents. All organic peroxides are highly flammable, and fires involving bulk quantities of peroxides should be approached with extreme caution.

Precautions for handling peroxides include the following:

- Limit the quantity of peroxide to the minimum amount required. Do not return unused peroxide to the container.
- Clean up all spills immediately. Solutions of peroxides can be absorbed on vermiculite or other absorbing material and disposed of harmlessly according to institutional procedures.
- Reduce the sensitivity of most peroxides to shock and heat by dilution with inert solvents, such as aliphatic hydrocarbons. However, do not use aromatics (such as toluene), which are known to induce the decomposition of diacyl peroxides.
- Do not use solutions of peroxides in volatile solvents under conditions in which the solvent might vaporize because this will increase the peroxide concentration in the solution.
- Do not use metal spatulas to handle peroxides because contamination by metals can lead to explosive decomposition. Magnetic stirring bars can unintentionally introduce iron, which can initiate an explosive reaction of peroxides. Ceramic, Teflon, or wooden spatulas and stirring blades may be used if it is known that the material is not shock sensitive.
- Do not permit open flames and other sources of heat near peroxides. It is important to label areas that contain peroxides so that this hazard is evident.
- Avoid friction, grinding, and all forms of impact near peroxides, especially solid peroxides. Glass containers that have screw-cap lids or glass stoppers should not be used. Polyethylene bottles that have screw-cap lids may be used.
- To minimize the rate of decomposition, store peroxides at the lowest possible temperature consistent with their solubility or freezing point. Do not store liquid peroxides or solutions at or lower than the temperature at which the peroxide freezes or

precipitates because peroxides in these forms are extremely sensitive to shock and heat.

6.G.3.1 Peroxidizable Compounds

Certain common laboratory chemicals form peroxides on exposure to oxygen in air. Over time, some chemicals continue to build peroxides to potentially dangerous levels, whereas others accumulate a relatively low equilibrium concentration of peroxide, which becomes dangerous only after being concentrated by evaporation or distillation. The peroxide becomes concentrated because it is less volatile than the parent chemical.

Excluding oxygen by storing potential peroxide formers under an inert atmosphere (N_2 or argon) greatly increases their safe storage lifetime. Purchasing the chemical stored under nitrogen in septum-capped bottles is also possible. In some cases, stabilizers or inhibitors (free-radical scavengers that terminate the chain reaction) are added to the liquid to extend its storage lifetime. Because distillation of the stabilized liquid removes the stabilizer, the distillate must be stored with care and monitored for peroxide formation. Furthermore, high-performance liquid chromatography–grade solvents generally contain no stabilizer, and the same considerations apply to their handling.

- If a container of Class B and C peroxidizables (see Chapter 4, Table 4.8 and section 4.D.3.2) is past its expiration date, and there is a risk that peroxides may be present, open it with caution and dispose of it according to institutional procedures (see section 6.G.3.3). **If a container of a Class A peroxidizable is past its expiration date, or if the presence of peroxides is suspected or proven, do not attempt to open the container.** Because of their explosivity, these compounds can be deadly when peroxidized, and the act of unscrewing a cap or dropping a bottle can be enough to trigger an explosion. Such containers should only be handled by experts. Contact your organization's safety personnel for assistance.
- Test for the presence of peroxides if there is a reasonable likelihood of their presence and the expiration date has not passed (see section 6.G.3.2).

6.G.3.2 Peroxide Detection Tests

Warning: Do not test Class A peroxidizables suspected of or known to contain peroxides. Contact your safety coordinator.

The following tests detect most (but not all) peroxy compounds, including all hydroperoxides:

- Peroxide test strips, which turn to an indicative color in the presence of peroxides, are available commercially. Note that these strips must be air dried until the solvent evaporates and exposed to moisture for proper indication and quantification.
- Add 1 to 3 mL of the liquid to be tested to an equal volume of acetic acid, add a few drops of 5% aqueous potassium iodide solution, and shake. The appearance of a yellow to brown color indicates the presence of peroxides. Alternatively, addition of 1 mL of a freshly prepared 10% solution of potassium iodide to 10 mL of an organic liquid in a 25-mL glass cylinder produces a yellow color if peroxides are present.
- Add 0.5 mL of the liquid to be tested to a mixture of 1 mL of 10% aqueous potassium iodide solution and 0.5 mL of dilute hydrochloric acid to which has been added a few drops of starch solution just prior to the test. The appearance of a blue or blue-black color within 1 minute indicates the presence of peroxides.

None of these tests should be applied to materials (such as metallic potassium) that may be contaminated with inorganic peroxides.

6.G.3.3 Disposal of Peroxides

Check with state and federal environmental agencies before attempting to treat any chemical for the purpose of disposal without a permit. Pure peroxides should never be disposed of directly but must be diluted before disposal. Small quantities (≤25 g) of peroxides are generally disposed of by dilution with water to a concentration of 2% or less, after which the solution is transferred to a polyethylene bottle containing an aqueous solution of a reducing agent, such as ferrous sulfate or sodium bisulfite. The material can then be handled as a waste chemical; however, it must not be mixed with other chemicals for disposal. Spilled peroxides should be absorbed on vermiculite or other absorbent as quickly as possible. The vermiculite–peroxide mixture can be burned directly or may be stirred with a suitable solvent to form a slurry that can be handled according to institutional procedures. Organic peroxides should never be flushed down the drain.

Large quantities (>25 g) of peroxides require special handling and should only be disposed of by an expert or a bomb squad. Each case should be considered separately, and handling, storage, and disposal procedures should be determined by the physical and chemical properties of the particular peroxide [see also Hamstead (1964)].

Peroxidized solvents such as tetrahydrofuran (THF), diethyl ether, and 1,4-dioxane may be disposed of in the same manner as the nonautoxidized solvent. Care should be taken to ensure that the peroxidized solvent is not allowed to evaporate and thus concentrate the peroxide during handling and transport.

(Also see Chapter 4, section 4.D.3.2.)

6.G.4 Explosive Gases and Liquefied Gases

Rapid evaporation of liquefied gases may present a serious hazard. Liquid oxygen, in particular, may introduce extreme risk due to the combined hazard of rapid overpressurization or volume expansion and the high concentration of a potent oxidizer. Liquefied air is almost as dangerous as liquid oxygen because the nitrogen boils away, leaving an increasing concentration of oxygen. Other cryogenic liquids, such as nitrogen and helium, if they have been open to air, may have absorbed and condensed enough atmospheric oxygen to be very hazardous. When a liquefied gas is used in a closed system, pressure may build up, so that adequate venting is required. Relief devices are required to prevent this dangerous buildup of pressure. If the liquid or vapor is flammable (e.g., hydrogen), explosive concentrations in air may develop. Because flammability, toxicity, and pressure buildup may become serious when gases are exposed to heat, gases should be stored only in specifically designed and designated areas (see Chapter 9, section 9.C.3.7).

6.G.5 Hydrogenation Reactions

Hydrogenation reactions pose additional risks because they are often carried out under pressure with a reactive catalyst. Hydrogenations conducted at atmospheric pressure at some point require the use of a pressurized cylinder unless the hydrogen is generated chemically (e.g., $NaBH_4$) as needed. Take all precautions for the gas cylinders and flammable gases, plus the additional precautions for reactions at pressures greater than 1 atm. The following precautions are applicable:

- Choose a pressure vessel appropriate for the experiment, such as an autoclave or pressure bottle. For example, most preparative hydrogenations of substances such as alkenes are carried out safely in a commercial hydrogenation apparatus using a heterogeneous catalyst (e.g., platinum and palladium) under moderate (<80 psi H_2) pressure.
- Review the operating procedures for the apparatus, and inspect the container before each experiment. Glass reaction vessels with scratches or chips are at risk to break under pressure; impaired vessels should not be used.
- Never fill the vessel to capacity with the solution; filling it half full (or less) is much safer.
- Remove as much oxygen from the solution as pos-

sible before adding hydrogen. This is one of the most important precautions to be taken with any reaction involving hydrogen. Failure to do this could result in an explosive oxygen–hydrogen (O_2–H_2) mixture. Normally, the oxygen in the vessel is removed by pressurizing the vessel with inert gas (N_2 or argon), followed by venting the gas. If available, a vacuum can be applied to the solution. Repeat this procedure of filling with inert gas and venting several times before hydrogen or other high-pressure gas is introduced.

- Stay well below the rated safe-pressure limit of the bottle or autoclave; a margin of safety is needed if heat or gas is generated. A limit of 75% of the rating in a high-pressure autoclave is advisable. If this limit is exceeded accidentally, replace the rupture disk on completion of the experiment.
- Monitor the pressure of the high-pressure device periodically as the heating proceeds, to avoid excessive pressure.
- Purge the system of hydrogen by repeated rinsing with inert gas at the end of an experiment to avoid producing hydrogen–oxygen mixtures in the presence of the catalyst during workup. Handle catalyst that has been used in a reaction with special care because it can be a source of spontaneous ignition on contact with air.

(Also see section 6.C.)

6.G.6 Materials Requiring Special Attention Because of Toxicity, Reactivity, Explosivity, or Chemical Incompatibility

The following list is not intended to be all-inclusive. Further guidance on reactive and explosive materials should be sought from pertinent sections of this book (see Chapter 4, sections 4.D.2 and 4.D.3) and other sources of information (note sources included in Chapter 4, section 4.B.6).

Acetylenic compounds, both organic and inorganic (especially heavy metal salts), can be explosive and shock sensitive. At pressures of 2 atm or greater and moderate temperature, acetylene (C_2H_2) has been reported to decompose explosively, even in the absence of air. Because of these dangers, **acetylene must be handled in acetone solution and never stored alone in a cylinder.**

Alkyllithium compounds are highly reactive and pyrophoric. Violent reactions may occur on exposure to water, carbon dioxide, and other materials. Alkyllithium compounds are highly corrosive to the skin and eyes. *tert*-Butyllithium solutions are the most pyrophoric and may ignite spontaneously on exposure to air. Contact with water or moist materials can lead to fires and explosions. These compounds should be stored and handled under an inert atmosphere in areas that are free from ignition sources. Detailed information about handling of organolithium compounds is provided by Schwindeman et al. (2002).

Aluminum chloride ($AlCl_3$) should be considered a potentially dangerous material. If moisture is present, sufficient decomposition may form hydrogen chloride (HCl) and build up considerable pressure. If a bottle is to be opened after long storage, it should first be completely enclosed in a heavy towel.

Ammonia and amines. Ammonia (NH_3) reacts with iodine to give nitrogen triiodide, which explodes on touch. Ammonia reacts with hypochlorites (bleach) to give chlorine. Mixtures of ammonia and organic halides sometimes react violently when heated under pressure. Ammonia is combustible. Inhalation of concentrated fumes can be fatal. Ammonia and amines can react with heavy metal salts to produce explosive fulminates.

Azides, both organic and inorganic, and some azo compounds can be heat and shock sensitive. Azides such as sodium azide can displace halide from chlorinated hydrocarbons such as dichloromethane to form highly explosive organic polyazides; this substitution reaction is facilitated in solvents such as dimethyl sulfoxide.

Boron halides are powerful Lewis acids and hydrolyze to strong protonic acids.

Carbon disulfide (CS_2) is both very toxic and very flammable; mixed with air, its vapors can be ignited by a steam bath or pipe, a hot plate, or a lightbulb.

Chlorine (Cl_2) is highly toxic and may react violently with hydrogen (H_2) or with hydrocarbons when exposed to sunlight.

Diazomethane (CH_2N_2) and related diazo compounds should be treated with extreme caution. They are very toxic, and the pure gases and liquids explode readily even from contact with sharp edges of glass. Solutions in ether are safer from this standpoint. An ether solution of diazomethane is rendered harmless by dropwise addition of acetic acid.

Diethyl and other ethers, including tetrahydrofuran and 1,4-dioxane and particularly the branched-chain type of ethers, may contain peroxides that have developed from air autoxidation. Concentration of these peroxides during distillation may lead to explosion. Ferrous salts or sodium bisulfite can be used to decompose these peroxides, and passage over basic active alumina can remove most of the peroxidic material. In general, however, dispose of old samples of ethers if they test positive test for peroxide.

Diisopropyl ether is a notoriously dangerous, Class A (see Chapter 4, Table 4.8 and section 4.D.3.2) peroxide former. The peroxide is not completely soluble in the mother liquor. Peroxide concentrations from autoxidation may form saturated solutions that then crystallize the peroxide as it is being formed. There are numerous reports of old bottles of diisopropyl ether being found with large masses of crystals settled at the bottom of the bottle. These crystals are extremely shock sensitive, even while wetted with the diisopropyl ether supernatant. Mild shock (e.g., bottle breakage, removing the bottle cap) is sufficient to result in explosion. This ether should not be stored in the laboratory. Only the amount required for a particular experiment or process should be purchased; any leftover material should be disposed of immediately.

Dimethyl sulfoxide (DMSO), $(CH_3)_2SO$, decomposes violently on contact with a wide variety of active halogen compounds, such as acyl chlorides. Explosions from contact with active metal hydrides have been reported. DMSO does penetrate and carry dissolved substances through the skin membrane.

Dry benzoyl peroxide $(C_6H_5CO_2)_2$ is easily ignited and sensitive to shock. It decomposes spontaneously at temperatures greater than 50 °C. It is reported to be desensitized by addition of 20% water.

Dry ice should not be kept in a container that is not designed to withstand pressure. Containers of other substances stored over dry ice for extended periods generally absorb carbon dioxide (CO_2) unless they have been carefully sealed. When such containers are removed from storage and allowed to come rapidly to room temperature, the CO_2 may develop sufficient pressure to burst the container with explosive violence. On removal of such containers from storage, the stopper should be loosened or the container itself should be wrapped in towels and kept behind a shield. Dry ice can produce serious burns, as is also true for all types of dry-ice cooling baths.

Drying agents, such as Ascarite® (sodium hydroxide–coated silica), should not be mixed with phosphorus pentoxide (P_2O_5) because the mixture may explode if it is warmed with a trace of water. Because the cobalt salts used as moisture indicators in some drying agents may be extracted by some organic solvents, the use of these drying agents should be restricted to drying gases.

Dusts that are suspensions of oxidizable particles (e.g., magnesium powder, zinc dust, carbon powder, and flowers of sulfur) in the air can constitute powerful explosive mixtures. These materials should be used with adequate ventilation and should not be exposed to ignition sources. When finely divided, some solids, including zirconium, titanium, Raney nickel, lead (such as prepared by pyrolysis of lead tartrate), and catalysts (such as activated carbon containing active metals and hydrogen), can combust spontaneously if allowed to dry while exposed to air and should be handled wet.

Ethylene oxide (C_2H_4O) has been known to explode when heated in a closed vessel. Experiments using ethylene oxide under pressure should be carried out behind suitable barricades.

Fluorine (F_2) is an extremely toxic reactive oxidizing gas with extremely low permissible exposure levels. Only trained personnel should be authorized to work with fluorine. (See Vignette 6.4.) **Anyone planning to work with fluorine must be knowledgeable of proper first-aid treatment and have the necessary supplies on hand before beginning.**

Halogenated compounds, such as chloroform ($CHCl_3$), carbon tetrachloride (CCl_4), and other halogenated solvents, should not be dried with sodium, potassium, or other active metals; violent explosions usually result. Many halogenated compounds are toxic. Oxidized halogen compounds—chlorates, chlorites, bromates, and iodates—and the corresponding peroxy compounds may be explosive at high temperatures.

Hydrogen fluoride and hydrogen fluoride generators. Anhydrous HF or hydrogen fluoride is a colorless liquid that boils at 19.5 °C. It has a pungent irritating odor, and a time-weighted average exposure of 3 ppm for routine work. Aqueous HF is a colorless very corrosive liquid that fumes at concentrations greater than 48%. It attacks glass, concrete, and some metals, especially cast iron and alloys containing silica as well as

VIGNETTE 6.4
Fluorine inhalation

A graduate student was filling the chamber of an excimer laser with fluorine gas. The gas was connected to the laser with copper tubing. Over the course of 1 hour, the student noticed the chamber was not filling even though the gas continued to flow. There was an odor in the room but the student was concerned that the chamber was not filling as expected and remained in the room to try and determine what the problem was. That evening the student experienced chest pain and difficulty breathing and went to the emergency room. She was diagnosed with pulmonary edema due to the prolonged exposure to fluorine gas.

Fluorine is exceedingly toxic, with allowable exposure levels of 1 ppm or less. The fluorine cylinder, laser, and piping should have been contained in a ventilated enclosure. An alarmed fluorine gas detector should have been used in the work area. The student was not adequately trained to recognize the signs or hazards of fluorine exposure.

organic materials such as leather, natural rubber, wood, and human tissue. Although HF is nonflammable, its corrosive action on metals can result in the formation of hydrogen in containers and piping, creating a fire and explosion hazard. HF should be stored in tightly closed polyethylene containers. HF attacks glass and therefore should **never** be stored in a glass container. Containers of HF may be hazardous when empty because they retain product residues. HF and related materials (e.g., NaF, SF_4, acyl fluorides) capable of generating HF upon exposure to acids, water, or moisture are of major concern because of their potential for causing serious burns.

HF causes severe injury via skin and eye contact, inhalation, and ingestion. It is very aggressive physiologically because the fluoride ion readily penetrates the skin and may cause decalcification of the bones and systemic toxicity, including pulmonary edema, cardiac arrhythmia and death. Burns from HF may not be painful or visible for several hours and even moderate exposure to concentrated HF can result in fatality. Unlike other acids which are rapidly neutralized, this process may continue for days if left untreated.

Strong HF acid concentrations (over 50%), particularly anhydrous HF, cause immediate, severe, burning pain and a whitish discoloration of the skin that usually proceeds to blister formation. In contrast to the immediate effects of concentrated HF, the onset of effects of contact with more dilute solutions or their vapors may be delayed. Skin contact with acid concentrations in the 20% to 50% range may not produce clinical signs or symptoms for 1 to 8 hours. With concentrations less than 20%, the latent period may be up to 24 hours. The usual initial signs of a dilute solution HF burn are redness, swelling, and blistering, accompanied by severe throbbing pain. **Burns larger than 25 in.^{2} (160 cm^2) may result in *serious systemic toxicity*.**

When exposed to air, concentrated solutions and anhydrous HF produce pungent fumes which are especially dangerous. Acute symptoms of inhalation of HF include coughing, choking, chest tightness, chills, fever, and cyanosis (blue lips and skin). All individuals suspected of having inhaled HF should seek medical attention and observation for pulmonary effects. This includes any individuals with HF exposure to the head, chest, or neck areas. **HF exposures require immediate and specialized first aid and medical treatment.**

For skin exposure:

1. Immediately start rinsing under safety shower or other water source and flush affected area thoroughly with large amounts of water, removing contaminated clothing while rinsing. Speed and thoroughness in washing off the acid is of primary importance.

2. Call for emergency response.
3. While wearing neoprene or butyl rubber gloves to avoid a secondary HF burn, massage 2.5% (w/w) calcium gluconate gel onto the affected area after 5 minutes of flushing with water. If calcium gluconate gel is unavailable, continue flushing the exposed areas with water until medical assistance arrives.
4. Send a copy of the MSDS with the victim.

For eye exposure:

1. Immediately flush the eyes, holding eyelids open, for at least 15 minutes with large amounts of gently flowing water, preferably using an eye-wash station.
2. Do not apply calcium gluconate gel directly onto the eye.
3. Seek medical attention.
4. Send a copy of the MSDS with the victim.

For inhalation:

1. Immediately move to fresh air.
2. Call emergency responders.
3. Send a copy of the MSDS with the victim.

For ingestion:

1. Ingestion of HF is a life-threatening emergency. Seek immediate medical attention.
2. Drink large amounts of water or milk as quickly as possible to dilute the acid.
3. Do not induce vomiting. Do not ingest emetics or baking soda. **Never give anything by mouth to an unconscious person.**
4. If medical attention must be delayed and the materials are available, drink several ounces of milk of magnesia or other antacids.
5. Send a copy of the MSDS with the victim.

Laboratory personnel should be trained in first-aid procedures for HF exposure before beginning work. Calcium gluconate gel (2.5% w/w) must be readily accessible in work areas where any potential HF exposure exists. Check the expiration date of your supply of commercially obtained calcium gluconate gel and reorder as needed to ensure a supply of fresh stock. Note that homemade calcium gluconate gel has a shelf life of approximately 4 months.

There are a number of ways to prevent HF exposure:

- Only use HF when necessary. Consider substitution of a less hazardous substance whenever possible.
- Establish written standard operating procedures for work with HF.
- Ensure that all workers in a lab where HF is used are informed about the hazards and first-aid procedures involved.
- Only use HF in a chemical hood.
- Depending on the concentration used, workers should wear butyl rubber, neoprene, 4H® or Silvershield® gloves. Protective lab coats or aprons are also recommended.
- At a minimum, workers should wear chemical splash goggles when working with HF. A face shield is also recommended when there is a significant splash hazard.

Hydrogen peroxide (H_2O_2) stronger than 3% can be dangerous; in contact with skin, it causes severe burns. Thirty percent H_2O_2 may decompose violently if contaminated with iron, copper, chromium, or other metals or their salts. Stirring bars may inadvertently bring metal into a reaction and should be used with caution.

Liquid nitrogen–cooled traps open to the atmosphere condense liquid air rapidly. When the coolant is removed, an explosive pressure buildup occurs, usually with enough force to shatter glass equipment if the system has been closed. Hence, only sealed or evacuated equipment should be so cooled. Vacuum traps must not be left under static vacuum; liquid nitrogen in Dewar flasks must be removed from these traps when the vacuum pumps are turned off.

Lithium aluminum hydride ($LiAlH_4$) should not be used as a drying agent for solvents that are hygroscopic and may contain high concentrations of water, such as methyl ethers and tetrahydrofuran; fires from reaction with damp ethers are often observed. Predrying these solvents with a less efficient drying agent, followed by $LiAlH_4$ treatment is recommended. The reaction of $LiAlH_4$ with carbon dioxide has reportedly generated explosive products. Carbon dioxide or bicarbonate extinguishers should not be used for $LiAlH_4$ fires; instead, such fires should be smothered with sand or some other inert substance.

Nitric acid is a strong acid, very corrosive, and decomposes to produce nitrogen oxides. The fumes are very irritating, and inhalation may cause pulmonary edema. Nitric acid is also a powerful oxidant and reacts violently, sometimes explosively reducing agents (e.g., organic compounds) with liberation of toxic nitrogen oxides. Contact with organic matter must be avoided. Extreme caution must be taken when cleaning glassware contaminated with organic solvents or material with nitric acid. Toxic fumes of NO_x are generated and explosion may occur.

Nitrate, nitro, and nitroso compounds may be explosive, especially if more than one of these groups is present in the molecule. Alcohols and polyols may form highly explosive nitrate esters (e.g., nitroglycerine) from reaction with nitric acid.

Organometallics may be hazardous because some organometallic compounds burn vigorously on contact with air or moisture. For example, solutions of *tert*-butyllithium ignite some organic solvents on exposure to air. The pertinent information should be obtained for a specific compound.

Oxygen tanks should be handled with care because serious explosions have resulted from contact between oil and high-pressure oxygen. Oil or grease should not be used on connections to an O_2 cylinder or gas line carrying O_2.

Ozone (O_3) is a highly reactive toxic gas. It is formed by the action of ultraviolet light on oxygen (air), and therefore certain ultraviolet sources may require venting to the exhaust hood. Ozonides can be explosive.

Palladium (Pd) or platinum (Pt) on carbon, platinum oxide, Raney nickel, and other catalysts presents the danger of explosion if additional catalyst is added to a flask in which an air-flammable vapor mixture or hydrogen is present. The use of flammable filter paper should be avoided.

Perchlorates should be avoided whenever possible. Perchlorate salts of organic, organometallic, and inorganic cations are potentially explosive and may detonate by heat or shock. Whenever possible, perchlorate should be replaced with safer anions such as fluoroborate, fluorophosphates, and trifluoromethanesulfonate (triflate).

Perchlorates should not be used as drying agents if there is a possibility of contact with organic compounds or of proximity to a dehydrating acid strong enough to concentrate the perchloric acid ($HClO_4$) (e.g., in a drying train that has a bubble counter containing sulfuric acid). Safer drying agents should be used.

Seventy percent $HClO_4$ boils safely at approximately 200 °C, but contact of the boiling undiluted acid or the hot vapor with organic matter, or even easily oxidized inorganic matter, leads to serious explosions. Oxidizable substances must never be allowed to contact $HClO_4$. This includes wooden benchtops or laboratory chemical hood enclosures, which may become highly flammable after absorbing $HClO_4$ liquid or vapors. Beaker tongs, rather than rubber gloves, should be used when handling fuming $HClO_4$. **Perchloric acid** evaporations should be carried out in a chemical hood that has a good draft.

The hood and ventilator ducts should be washed with water frequently (weekly; but see also Chapter 9, section 9.C.2.10.5) to avoid danger of spontaneous combustion or explosion if this acid is in common use. Special $HClO_4$ hoods are available from many manufacturers. Disassembly of such chemical hoods must be preceded by washing the ventilation system to remove deposited perchlorates.

Permanganates are explosive when treated with sulfuric acid. If both compounds are used in an absorption train, an empty trap should be placed between them and monitored for entrapment.

Peroxides (inorganic) should be handled carefully. When mixed with combustible materials, barium, sodium, and potassium peroxides form explosives that ignite easily.

Phenol is a corrosive and moderately toxic substance that affects the central nervous system and can cause damage to the liver and kidneys. Phenol-formaldehyde reactions are used in creation of phenolic resins, and can be highly exothermic. These reactions have been implicated in a number of plant-scale accidents when runaway reactions caused a sudden rise in pressure and rupturing of pressure disks or vessels (Urben and Bretherick, 1999). Care should be taken if performing such reactions in the laboratory.

Phenol is readily absorbed through the skin and can cause severe burns to the skin and eyes. Phenol is irritating to the skin, but has a local anesthetic effect, so that no pain may be felt on initial contact. A whitening of the area of contact generally occurs and severe burns may develop hours after exposure. Exposure to phenol vapor can cause severe irritation of the eyes, nose, throat, and respiratory tract. In the event of skin exposure to phenol, do not immediately rinse the site with water. Instead, treat the site with low-molecular-weight poly(ethylene glycol) (PEG) such as PEG 300 or PEG 400. This will safely deactivate phenol. Irrigate the site with PEG for at least 15 minutes or until there is no detectable odor of phenol.

Phosphorus (P) (red and white) forms explosive mixtures with oxidizing agents. White phosphorus should be stored underwater because it ignites spontaneously in air. The reaction of phosphorus with aqueous hydroxides gives phosphine, which is toxic and also may either ignite spontaneously or explode in air.

Phosphorus trichloride (PCl_3) reacts with water to form phosphorous acid with HCl evolution; the phosphorous acid decomposes on heating to form phosphine, which may either ignite spontaneously or explode. Care should be taken in opening containers of PCl_3, and samples that have been exposed to moisture should not be heated without adequate shielding to protect the operator.

Piranha solution is a mixture of concentrated sulfuric acid and 30% hydrogen peroxide. It is a powerful oxidant and strong acid used to remove organic residues from various surfaces. Many instances of explosions have been reported with this solution upon contact with reducing agents, especially organics. The solution slowly evolves oxygen, and therefore containers must be vented at all times.

Potassium (K) is much more reactive than sodium; it ignites quickly on exposure to humid air, and therefore should be handled under the surface of a hydrocarbon solvent such as mineral oil or toluene (see *Sodium*, below). Potassium can form a crust of the superoxide (KO_2) or the hydrated hydroxide ($KOH \cdot H_2O$) on contact with air. If this happens, the act of cutting a surface crust off the metal or of melting the encrusted metal can cause a severe explosion due to oxidation of the organic oil or solvent by superoxide, or from reaction of the potassium with water liberated from the hydrated hydroxide (Yarnell, 2002).

Residues from vacuum distillations have been known to explode when the still was vented suddenly to the air before the residue was cool. To avoid such explosions, vent the still pot with nitrogen, cool it before venting, or restore pressure slowly. Sudden venting may produce a shock wave that explodes sensitive materials.

Sodium (Na) should be stored in a closed container under kerosene, toluene, or mineral oil. Scraps of sodium or potassium should be destroyed by reaction with *n*-butyl alcohol. Contact with water should be avoided because sodium reacts violently with water to form hydrogen (H_2) with evolution of sufficient heat to cause ignition. Carbon dioxide, bicarbonate, and carbon tetrachloride fire extinguishers should not be used on alkali metal fires. Metals such as sodium become more reactive as the surface area of the particles increases. Prudence dictates using the largest particle size consistent with the task at hand. For example, use of sodium balls or cubes is preferable to use of sodium sand for drying solvents.

Sodium amide ($NaNH_2$) can undergo oxidation on exposure to air to give sodium nitrite in a mixture that is unstable and may explode.

Sulfuric acid (H_2SO_4) should be avoided, if possible, as a drying agent in desiccators. If it must be used, glass beads should be placed in it to help prevent splashing when the desiccator is moved. To dilute H_2SO_4, the acid should be added slowly to cold water. Addition of water to the denser H_2SO_4 can cause localized surface boiling and spattering on the operator.

tert-**Butyllithium.** See *Alkyllithium compounds*, above.

Trichloroethylene (Cl_2CCHCl) reacts under a variety of conditions with potassium or sodium hydroxide to form dichloroacetylene, which ignites spontaneously in air and explodes readily even at dry-ice temperatures. The compound itself is highly toxic, and suitable precautions should be taken when it is used.

6.G.7 Chemical Hazards of Incompatible Chemicals

For each chemical, follow specific storage recommendations in MSDSs and other references with respect to containment and compatibility. Keep incompatibles separate during transport, storage, use, and disposal (see Chapter 4, section 4.D; Chapter 5; and section 6.C). Contact could result in a serious explosion or the formation of substances that are highly toxic or flammable. Store oxidizers, reducing agents, and fuels separately to prevent contact in the event of an accident. Some reagents pose a risk on contact with the atmosphere.

6.H WORKING WITH COMPRESSED GASES

6.H.1 Chemical Hazards of Compressed Gases

Compressed gases expose laboratory personnel to both chemical and physical hazards. If the gas is flammable, flash points lower than room temperature, compounded by rapid diffusion throughout the laboratory, present the danger of fire or explosion. Additional hazards arise from the reactivity and toxicity of the gas. Asphyxiation can be caused by high concentrations of even inert gases such as nitrogen. An additional risk of simple asphyxiants is head injury from falls due to rapid loss of oxygen to the brain. Death can also occur if oxygen levels remain too low to sustain life. Finally, the large amount of potential energy resulting from the compression of the gas makes a highly compressed gas cylinder a potential rocket or fragmentation bomb.

Monitoring for leaks and proper labeling are essential for the prudent use of compressed gases. If relatively small amounts are needed, consider on-site chemical gas generation as an alternative to compressed gas. Reduce risks by monitoring compressed gas inventories and disposing of or returning gases for which there is no immediate need. The equipment required for the safe use of compressed gases is discussed in Chapter 7, section 7.D.

6.H.2 Specific Chemical Hazards of Select Gases

Workers are advised to consult LCSSs and MSDSs for specific gases. Certain hazardous substances that may be supplied as compressed gases are listed below:

Boron trifluoride and boron trichloride (BF_3 and BCl_3, respectively) react with water to give HF and HCl, respectively. Their fumes are corrosive, toxic, and irritating to the eyes and mucous membranes.

Chlorine trifluoride (ClF_3) in liquid form is corrosive and very toxic. It is a potential source of explosion and causes deep penetrating burns on contact with the body. The effect may be delayed and progressive, as in the case of burns caused by hydrogen fluoride.

Chlorine trifluoride reacts vigorously with water and most oxidizable substances at room temperature, frequently with immediate ignition. It reacts with most metals and metal oxides at elevated temperatures. In addition, it reacts with silicon-containing compounds and thus can support the continued combustion of glass, asbestos, and other such materials. Chlorine trifluoride forms explosive mixtures with water vapor, ammonia, hydrogen, and most organic vapors. The substance resembles elemental fluorine in many of its chemical properties and handling procedures, which include precautionary steps to prevent accidents.

Hydrogen selenide (H_2Se) is a colorless gas with an offensive odor. It is a dangerous fire and explosion risk and reacts violently with oxidizing materials. Hydrogen selenide is an irritant to eyes, mucous membranes, and the pulmonary system. Acute exposures can cause symptoms such as pulmonary edema, severe bronchitis, and bronchial pneumonia. Symptoms also include gastrointestinal distress, dizziness, increased fatigue, and a metallic taste in the mouth.

Hydrogen sulfide (H_2S) is a highly toxic and flammable gas. Although it has a characteristic odor of rotten eggs, it fatigues the sense of smell. This could result in failure to notice the seriousness of the situation before health becomes at risk and is problematic for rescuers who think danger has passed when the odor disappears.

Methyl chloride (CH_3Cl) has a slight, not unpleasant, odor that is not irritating and may pass unnoticed unless a warning agent has been added. Exposure to excessive concentrations is indicated by symptoms similar to those of alcohol intoxication, that is, drowsiness, mental confusion, nausea, and possibly vomiting. Methyl chloride may, under certain conditions, react with aluminum or magnesium to form materials that ignite or fume spontaneously with air, and contact with these metals should be avoided.

Phosphine (PH_3) is a spontaneously flammable and explosive poisonous colorless gas with the foul odor of decaying fish. The liquid can cause frostbite. Phosphine is a dangerous fire hazard and ignites in the presence of air and oxidizers. It reacts with water, acids, and halogens. If heated, it forms hydrogen phosphides, which are explosive and toxic. There may be a delay between exposure and the appearance of symptoms.

Silane (SiH_4) is a pyrophoric colorless gas that ignites spontaneously in air. It is incompatible with water, bases, oxidizers, and halogens. The gas has a choking repulsive odor.

Silyl halides are toxic colorless gases with a pungent odor. They are corrosive irritants to the skin, eyes, and mucous membranes. When silyl halides are heated, toxic fumes can be emitted.

6.I WORKING WITH MICROWAVE OVENS

Do not use domestic microwave ovens for laboratory work. Metal-based or volatile, flammable, and explosive compounds pose a significant hazard when used in a domestic microwave oven. Domestic microwave ovens do not provide mechanisms for monitoring temperature and pressure and contain no safeguards against explosion. Instead, use an industrial grade instrument (equipped with explosion-proof chambers, exhaust lines, and temperature and pressure monitors) suitable for such experiments.

Although industrial ovens may reduce the risk of such hazards, significant caution is required in their use. In general, the use of closed vessels should be avoided. Any reactions conducted in a microwave oven should be regarded with the same caution as those conducted with highly reactive and explosive chemicals. Reactions should use the smallest scale possible to determine the potential for explosions and fires (refer to sections 6.F and 6.G). Precautions should be taken for proper ventilation and potential explosion. (See Chapter 7, section 7.C.5.7 for more information about the use of microwave ovens in laboratories.)

6.J WORKING WITH NANOPARTICLES

6.J.1 Controls for Research and Development Laboratory Operations That Utilize or Synthesize Nanomaterials

Nanoparticles are dispersible particles between 1 and 100 nm in size that may or may not exhibit a size-related intensive property. The U.S. Department of Energy (DOE, 2008, 2009) states that *engineered nanomaterials* are intentionally created (in contrast with natural or incidentally formed) and engineered to be between 1 and 100 nm. This definition excludes biomolecules (proteins, nucleic acids, and carbohydrates).

Nanoparticles and nanomaterials have different reactivities and interactions with biological systems than bulk materials, and understanding and exploiting these differences is an active area of research. However, these differences also mean that the risks and hazards associated with exposure to engineered nanomaterials are not well known. At the time this book was written, NIOSH had only defined occupational exposure limits for one nanomaterial, titanium dioxide. Until material-specific guidance can be issued, NIOSH, DOE, the British Standards Institute, and others have issued general guidelines for management of engineered nanomaterials. The procedures outlined here are based on those guidelines, which were developed from accepted chemical hygiene protocols for handling compounds of unknown toxicity.

Because this is an area of ongoing research, consult trusted sources to ensure that the methods described here are not obsolete, and check for any applicable material-specific guidance. Sources include

- *Approaches to Safe Nanotechnology: Managing the Health and Safety Concerns Associated with Engineered Nanomaterials* (HHS/CDC/NIOSH, 2009a), and the NIOSH nanotechnology topic Web page, www.cdc.gov/niosh;
- *Nanoscale Science Research Centers: Approach to Nanomaterial ES&H* (DOE, 2008);
- *ASTM E 2535-07: Standard Guide for Handling Unbound Engineered Nanoscale Particles in Occupational Settings* (ASTM International, 2007b);
- the National Nanotechnology Initiative, www.nano.gov; and
- the United Kingdom's Health and Safety Executive Web site, at www.hse.gov.uk.

6.J.1.1 Nanomaterial Work Planning and Hazard Assessment

Before beginning any work with or intended to produce nanomaterials, perform a safety assessment for the laboratory. Involve the organization's EHS personnel in the process, and consult subject-matter experts as needed. The assessment should

- include a well-defined description of the work;
- identify the state of the nanomaterials at each stage of the work (i.e., dry and dispersible, in a slurry or solution, affixed to a matrix, or embedded in a solid);
- identify recognized and suspected hazards and uncertainties, both biological and physical, at each stage;
- specify hazard controls including
  - ○ engineering controls,
  - ○ identification of appropriate PPE,
  - ○ training plans for laboratory personnel,
  - ○ emergency procedures, including spill or release response,
  - ○ experimental design elements to minimize risk of exposure, and
  - ○ any other administrative controls;
- evaluate the potential for generating new nanomaterial-bearing waste streams and define waste management protocols for these streams; and
- consider the potential for reactions involving

nanomaterials or other incompatible materials already captured in exhaust air filters.

When developing controls for the nanomaterials, consider, but do not unquestioningly rely on, chemical hazard information for bulk or raw materials and any new information specific to the material at the scale being used. When evaluating the hazards and uncertainties for the materials, consider the recognized and foreseeable hazards of the precursor materials and intermediates as well as those of the resulting nanomaterials. Note that the higher reactivity of many nanoscale materials suggests that they should be treated as potential sources of ignition, accelerants, and fuel that could result in fire or explosion.

The risk of exposure may continue after laboratory work has been completed if, for example, a laboratory chemical hood was used to house a reaction. Before removing, remodeling, servicing, maintaining, or repairing laboratory equipment and exhaust systems, evaluate the potential for trained laboratory personnel's (including laboratory and maintenance workers) exposure to nanomaterials and escape of the materials into the environment.

6.J.1.2 A Graded Approach to Determining Appropriate Nanomaterial Controls

When performing the assessment described above, follow a graded approach in specifying controls. For example, from the perspective of managing the health of laboratory personnel, easily dispersed dry nanomaterials pose the greatest health hazard because of the risk of inhalation, and operations involving these nanomaterials deserve more attention and more stringent controls than those where the nanomaterials are embedded in solid or suspended in liquid matrixes. The list below and Figure 6.1 describe the graded risk posed by the state of the material. Preference should be given to handling materials in the lower risk forms (top of the list).

1. solid materials with embedded nanostructures,
2. solid nanomaterials with nanostructures fixed to the material's surface,
3. nanoparticles suspended in liquids, and
4. dry dispersible (engineered) nanoparticles, nanoparticle agglomerates, or nanoparticle aggregates.

Be sure to consider all routes of possible exposure to nanomaterials including inhalation, ingestion, injection, and dermal contact (including eye and mucous membranes). Avoid handling nanomaterials in the open air in a free-particle state. Whenever possible,

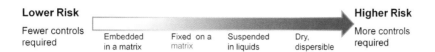

FIGURE 6.1 U.S. Department of Energy graded exposure risk for nanomaterials. This figure assumes no disruptive force (e.g., sonication, grinding, burning) is applied to the matrix.
SOURCE: Adapted from Karn (2008).

handle and store dispersible nanomaterials, whether suspended in liquids or in a dry particle form, in closed (tightly sealed) containers. Unless cutting or grinding occurs, nanomaterials that are not in a free form (encapsulated in a solid or a nanocomposite) typically will not require engineering controls. If a synthesis is being performed to create nanomaterials, it is not enough to only consider the final material in the risk assessment. Consider the hazardous properties of the precursor materials as well as those of the resulting nanomaterial product.

6.J.1.3 Engineering Controls for Nanomaterials Research

6.J.1.3.1 Work Area Design

When evaluating the work area, consider the need for additional engineering or procedural controls to ensure trained laboratory personnel are protected in areas where engineered nanoparticles will be handled. Additional controls to ensure that engineered nanoparticles are not brought out of the work area on clothing or other surfaces may be advisable. Examples of possible additional controls include installing step-off pads to trap dust, creating a buffer area around the work zone, and ensuring the availability of decontamination facilities (possibly for daily use) for laboratory personnel.

6.J.1.3.2 Ventilation Preferences

To minimize laboratory personnel exposure, conduct any work that could generate engineered nanoparticles in an enclosure that operates at a negative pressure differential compared to the laboratory personnel breathing zone. Examples of such enclosures include gloveboxes, glovebags, and laboratory benchtop or floor-mounted chemical hoods. Do not use horizontal laminar-flow hoods (clean benches) that direct a flow of HEPA-filtered air into the user's face for operations involving engineered nanomaterials. If the air reactivity of precursor materials may make it unsafe to perform a synthesis in a negative-pressure glovebox, a positive-pressure box may be used if it has passed a helium leak test. If a process (or subset of a process) cannot be enclosed, use other engineering systems to control

fugitive emissions of nanomaterials or hazardous precursors that might be released. For example, use a local exhaust system such as a snorkel hood. Laboratory ventilation and exhaust systems should be chosen on the basis of what is known about nanoparticle motion in air. (For more information about ventilation options, see Chapter 9, section 9.E.5.)

Do not exhaust unfiltered effluent (air) that has been demonstrated or strongly suspected to contain engineered nanoparticles to the laboratory. Whenever practical, filter it or otherwise clean (scrub) it before releasing it to the outdoors. Although HEPA filtration appears to effectively remove nanoparticles from air, the filters must be held in well-designed housings. A poorly seated filter can allow nanomaterials to escape through the gaps. If it is not practicable to contain the nanoparticles with such a system, conduct and document the results of a hazard analysis before using alternative hazard controls.

Exhaust the effluent from the ventilated enclosure outside the building whenever feasible. Filters, scrubbers, or bubblers used to treat unreacted precursors appropriately may also be effective in reducing nanomaterial emissions. If using portable benchtop HEPA-filtered units, exhaust them through ventilation systems that carry the effluent outside the building whenever possible.

If it is not feasible to duct HEPA-filtered treated exhaust air outside the building, follow the guidance in ANSI Z9.7-2007, *American National Standard for Recirculation of Air from Industrial Process Exhaust Systems*, and conduct a hazard assessment to identify appropriate engineering controls. Examples of such controls include periodic air monitoring and an accurate warning or signal capable of initiating corrective action or process shutdown before nanoparticles are exhausted or reenter the work area. If using a Type II biological safety cabinet for work with nanomaterials, consider exhausting directly to the exterior (hard ducted) or through a thimble connection over the cabinet's exhaust.

All exhaust systems should be maintained and tested as specified by the manufacturer. Before beginning any maintenance, however, evaluate equipment for contamination and chemical incompatibilities.

6.J.1.3.3 Clothing and PPE

Minimal data exist regarding the efficacy of PPE against exposure to nanoparticles. However, until further information is available, it is prudent to follow standard chemical hygiene practices. Conduct a hazard evaluation to determine PPE appropriate for the level of hazard according to the requirements set forth in 29 CFR § 1910.132. Protective clothing that would typically be required for a wet-chemistry laboratory would be appropriate and could include but is not limited to

- closed-toed shoes made of a low-permeability material (disposable over-the-shoe booties may be necessary to prevent tracking nanomaterials from the laboratory);
- long pants without cuffs;
- long-sleeved shirt;
- gauntlet-type gloves or nitrile gloves with extended sleeves; and
- laboratory coats.

Wear polymer (e.g., nitrile rubber) gloves when handling engineered nanomaterials and particulates in liquids. Choose gloves only after considering the resistance of the glove to chemical attack by both the nanomaterial and, if suspended in liquids, the liquid.

- Recognize that exposure to nanomaterials is not known to have good warning properties, and change gloves routinely to minimize potential exposure hazards. Alternatively, double glove.
- Keep contaminated gloves in a plastic bag or other sealed container until disposed of.
- Dispose of contaminated gloves in accordance with organizational requirements.
- Wash hands and forearms after wearing gloves.
- Follow any additional institutional rules regarding nanoparticles, such as proper waste disposal.

Wear eye protection, for example, spectacle-type safety glasses with side shields (meeting basic impact resistance of ANSI Z87.1-2003), face shields, chemical splash goggles, or other safety eyewear appropriate to the type and level of hazard. Do not consider face shields or safety glasses to provide sufficient protection against unbound dry materials that could become airborne.

Contact the organization's EHS professionals for an evaluation of airborne exposures to engineered nanomaterials. If respirators are to be used for protection against engineered nanoparticles, NIOSH-certified respirators should provide the expected levels of protection if properly selected and fit-tested as part of a complete respiratory protection program. Refer to the NIOSH publications *Approaches to Safe Nanotechnology*

(HHS/CDC/NIOSH, 2009a) and *Respirator Selection Logic* (Bollinger, 2004) for guidance on choosing appropriate air-purifying particulate respirators. These documents are available online at www.cdc.gov/niosh. If employees are required to wear respirators, consideration must be given to the OSHA regulation 29 CFR § 1910.134.

Keep potentially contaminated clothing and PPE in the laboratory or change-out area to prevent engineered nanoparticles from being transported into common areas.

Clean and dispose of all potentially contaminated clothing and PPE in accordance with the laboratory procedures.

6.J.1.3.4 Monitoring and Characterization

The NIOSH publication *Approaches to Safe Nanotechnology* (HHS/CDC/NIOSH, 2009a) describes an emission assessment technique that can be used for identification of sources and releases of engineered nanomaterials. The technique includes determining the particle number concentration using direct-reading, handheld particle counters at potential emission sources and comparing those data to background particle number concentrations. If elevated concentrations of suspected nanoparticles are detected at potential emission sources, relative to the background particle number concentrations, then a pair of filter-based, source-specific air samples are collected with one sample analyzed by transmission electron microscopy or scanning electron microscopy-for particle identification and characterization, and the other used for determining the elemental mass concentration.

If resources allow, a more comprehensive and quantitative approach using additional aerosol sampling equipment (such as impactors or diffusion charges) may be performed.

6.J.1.3.5 Housekeeping

Practice good housekeeping in laboratories where nanomaterials are handled. Follow a graded approach paying attention where dispersible nanomaterials are handled. Insofar as practicable, maintain all working surfaces (i.e., benches, glassware, apparatus, laboratory chemical hoods, support equipment) free of engineered nanoparticle contamination and otherwise limit laboratory personnel exposure to engineered nanoparticles and associated hazards. In areas where engineered nanoparticles might settle, perform precautionary cleaning, for example, by wiping horizontal surfaces with a moistened disposable wipe, no less frequently than at the end of each shift or day.

Before selecting a cleaning method, consider the potential for complications due to the physical and chemical properties of the engineered nanoparticles, particularly in the case of larger spills. Complications could include reactions with cleaning materials and other materials in the locations where the waste will be held. Such locations include vacuum cleaner filters and canisters.

Clean up dry engineered nanomaterials using

- Wet wiping.
- A dedicated approved HEPA vacuum with verified filtration effectiveness. (Note: Consider possible pyrophoric hazards associated with vacuuming up nanoparticles.) If using the vacuum for multiple types of nanomaterials, keep a log of the materials captured and check for chemical incompatibilities prior to use.
- Other facility-approved methods that do not involve an energetic cleaning method. Avoid dry sweeping or the use of compressed air, to prevent suspension of particles into the air.

Note that vacuum brushes may generate electrostatic charges that could make cleaning of charged particles difficult. Consider using vacuum cleaners with electrostatic-charge-neutralization features (such as those used for cleaning copier and printer toners.) Again, be sure that the vacuum is exhausted through a properly fitted and maintained HEPA filter.

Clean up spills of liquids containing nanomaterials using absorbent materials. If the size of the spill is large, place absorbent pads at all points of egress from the room to reduce tracking the spill into the other parts of the building. Use plastic sheeting to reduce ventilation in the area of a liquid spill to reduce the chance that it will dry prior to cleanup. As noted above, dry nanomaterials pose a greater hazard than those suspended in liquid.

Dispose of used cleaning materials and wastes in accordance with the laboratory's hazardous waste procedures.

6.J.1.3.6   Work Practices

Evaluate hazards and implement work practices to control potential contamination and exposure hazards, if engineered nanoparticle powders must be handled without the use of exhaust ventilation (i.e., laboratory chemical hood, local exhaust) or enclosures (i.e., glovebox). Take reasonable precautions to minimize the likelihood of skin contact with engineered nanoparticles or nanoparticle-containing materials likely to release nanoparticles (nanostructures). Transfer engineered nanomaterial samples between workstations such as laboratory chemical hoods, gloveboxes, furnaces in closed labeled containers (e.g., marked zip-lock bags). Handle nanomaterial-

bearing waste according to the laboratory's hazardous chemical waste guidelines.

6.J.1.3.7   Marking, Labeling, and Signage

Post signs indicating hazards, PPE requirements, and administrative control requirements at entry points into designated areas where dispersible engineered nanoparticles are handled. A designated area may be an entire laboratory, an area of a laboratory, or a containment device such as a laboratory chemical hood or glovebox. **Clearly label storage containers to indicate that the contents are in engineered nanoparticulate form** (e.g., nanoscale zinc oxide particles, or other identifier instead of simply zinc oxide).

When engineered nanoparticles are being moved outside a laboratory, use leakproof double containment. For example, use compatible double zip-lock bags or "Tupperware-type" containers, or proper shipping containers. Include label text that indicates that the particulates might be unusually reactive and vary in toxic potential, quantitatively and qualitatively, from normal size forms of the same material. (See Chapter 5, section 5.F.2, for more information about transport and shipping of nanomaterials.)

6.J.1.3.8   Disposal of Nanomaterial-Bearing Waste Streams

Do not put material containing nanomaterials down the drain or in the regular trash. Contact the organization's EHS personnel to assist in determining the appropriate waste disposal method. Using the guidelines provided by EPA (see Chapter 8), identify whether the material should be considered hazardous or nonhazardous. Remember that nanomaterials often have different reactivities than the bulk material, and while bulk material properties can be used as a guide, do not rely upon them to determine the properties of the nanomaterials. If the sample is in liquid, be sure to consider the hazards of the liquid as well as the nanoparticles.

As general guidance, DOE recommends collecting items that come in contact with nanomaterials, such as PPE, wipes, and the like in a sealable plastic bag or other sealable container under appropriate ventilation controls. When it is full, place the bag in a second sealable container before disposal. Label the waste container as containing nanomaterials, and note any particular hazards on the label. Notify the organization's hazardous waste handler that nanomaterials are in the waste stream.

6.J.1.3.9   Personnel Competency

Laboratory and support personnel who risk potential exposure to engineered nanoparticles should be given training on the risks of exposure and on safe handling procedures. Do not assume that laboratory personnel or visiting researchers are aware of the health and safety concerns posed by nanomaterials. At a minimum, provide personnel conducting hands-on work with an awareness-level orientation that will alert them to concerns (potential hazards) and to the laboratory's policies concerning prudent material handling.

Training should cover requirements and recommendations for

- employing engineered controls,
- using PPE,
- handling potentially contaminated laboratory garments and protective clothing,
- cleaning of potentially contaminated surfaces,
- disposal of spilled nanoparticles, and
- use of respirators, if applicable.

7 Working with Laboratory Equipment

7.A	INTRODUCTION	149
7.B	WORKING WITH WATER-COOLED EQUIPMENT	149
7.C	WORKING WITH ELECTRICALLY POWERED LABORATORY EQUIPMENT	149
	7.C.1 General Principles	149
	7.C.1.1 Outlet Receptacles	150
	7.C.1.2 Wiring	150
	7.C.1.3 General Precautions for Working with Electrical Equipment	151
	7.C.1.4 Personal Safety Techniques for Use with Electrical Equipment	152
	7.C.1.5 Additional Safety Techniques for Equipment Using High Current or High Voltage	152
	7.C.2 Vacuum Pumps	153
	7.C.3 Refrigerators and Freezers	153
	7.C.4 Stirring and Mixing Devices	154
	7.C.5 Heating Devices	154
	7.C.5.1 Ovens	156
	7.C.5.2 Hot Plates	157
	7.C.5.3 Heating Mantles	157
	7.C.5.4 Oil, Salt, or Sand Baths	158
	7.C.5.5 Hot Air Baths and Tube Furnaces	158
	7.C.5.6 Heat Guns	159
	7.C.5.7 Microwave Ovens	159
	7.C.6 Distillation	159
	7.C.6.1 Solvent Stills	159
	7.C.6.2 Column Purification Systems or "Push Stills"	160
	7.C.7 Ultrasonicators, Centrifuges, and Other Electrical Equipment	161
	7.C.7.1 Ultrasonicators	161
	7.C.7.2 Centrifuges	161
	7.C.7.3 Electrical Instruments	162
	7.C.8 Electromagnetic Radiation Hazards	162
	7.C.8.1 Visible, Ultraviolet, and Infrared Laser Light Sources	162
	7.C.8.2 Radio-Frequency and Microwave Sources	162
	7.C.8.3 X-Rays, Electron Beams, and Sealed Sources	162
	7.C.8.4 Miscellaneous Physical Hazards Presented by Electrically Powered Equipment	164
7.D	WORKING WITH COMPRESSED GASES	164
	7.D.1 Compressed Gas Cylinders	164
	7.D.1.1 Identification of Contents	165
	7.D.2 Equipment Used with Compressed Gases	165
	7.D.2.1 Records, Inspection, and Testing	165
	7.D.2.2 Assembly and Operation	165

	7.D.3	Handling and Use of Gas Cylinders		168
		7.D.3.1	Preventing and Controlling Leaks	169
		7.D.3.2	Pressure Regulators	169
		7.D.3.3	Flammable Gases	170
7.E	WORKING WITH HIGH OR LOW PRESSURES AND TEMPERATURES			170
	7.E.1	Pressure Vessels		170
		7.E.1.1	Records, Inspection, and Testing	170
		7.E.1.2	Pressure Reactions in Glass Equipment	171
	7.E.2	Liquefied Gases and Cryogenic Liquids		172
		7.E.2.1	Cold Traps and Cold Baths	173
		7.E.2.2	Selection of Low-Temperature Equipment	174
		7.E.2.3	Cryogenic Lines and Supercritical Fluids	174
	7.E.3	Vacuum Work and Apparatus		174
		7.E.3.1	Glass Vessels	174
		7.E.3.2	Dewar Flasks	174
		7.E.3.3	Desiccators	175
		7.E.3.4	Rotary Evaporators	175
		7.E.3.5	Assembly of Vacuum Apparatus	175
7.F	USING PERSONAL PROTECTIVE, SAFETY, AND EMERGENCY EQUIPMENT			175
	7.F.1	Personal Protective Equipment and Apparel		175
		7.F.1.1	Protective Clothing	175
		7.F.1.2	Foot Protection	175
		7.F.1.3	Eye and Face Protection	176
		7.F.1.4	Hand Protection	176
	7.F.2	Safety and Emergency Equipment		176
		7.F.2.1	Spill Control Kits and Cleanup	177
		7.F.2.2	Safety Shields	177
		7.F.2.3	Fire Safety Equipment	177
		7.F.2.4	Respiratory Protective Equipment	178
		7.F.2.5	Safety Showers and Eyewash Units	180
		7.F.2.6	Storage and Inspection of Emergency Equipment	180
7.G	EMERGENCY PROCEDURES			181

7.A   INTRODUCTION

Working safely with hazardous chemicals requires proper use of laboratory equipment. Maintenance and regular inspection of laboratory equipment are essential parts of this activity. Many of the accidents that occur in the laboratory can be attributed to improper use or maintenance of laboratory equipment. This chapter discusses prudent practices for handling equipment used frequently in laboratories.

The most common equipment-related hazards in laboratories come from devices powered by electricity, devices for work with compressed gases, and devices for high or low pressures and temperatures. Other physical hazards include electromagnetic radiation from lasers and radio-frequency generating devices. Seemingly ordinary hazards such as floods from water-cooled equipment, accidents with rotating equipment and machines or tools for cutting and drilling, noise extremes, slips, trips, falls, lifting, and poor ergonomics account for the greatest frequency of laboratory accidents and injuries. Understandably, injuries to the hands are very common in the laboratory. Care should be taken to use appropriate gloves when handling laboratory equipment to protect against electrical, thermal, and chemical burns, cuts, and punctures.

7.B   WORKING WITH WATER-COOLED EQUIPMENT

The use of water as a coolant in laboratory condensers and other equipment is common practice. Although tap water is often used for these purposes, this practice should be discouraged. In many localities conserving water is essential and makes tap water inappropriate. In addition, the potential for a flood is greatly increased. Refrigerated recirculators can be expensive, but are preferred for cooling laboratory equipment to conserve water and to minimize the impact of floods. To prevent freezing at the refrigeration coils, using a mixture of water and ethylene glycol as the coolant is prudent. Spills of this mixture are very slippery and must be cleaned thoroughly to prevent slips and falls.

Most flooding occurs when the tubing supplying the water to the condenser disconnects. Hoses can pop off when building water pressure fluctuates, causing irregular flows, or can break when the hose material has deteriorated from long-term or improper use. Floods also result when exit hoses jump out of the sink from a strong flow pulse or sink drains are blocked by an accumulation of extraneous material. Proper use of hose clamps and maintenance of the entire cooling system or alternative use of a portable cooling bath with suction feed can resolve such problems. Plastic locking disconnects can make it easy to unfasten water lines without having to unclamp and reclamp secured lines. Some quick disconnects also incorporate check valves, which do not allow flow into or out of either half of the connection when disconnected. This feature allows for disconnecting and reconnecting with minimal spillage of water. To reduce the possibility of overpressurization of fittings or glassware, consider installing a vented pressure relief device on the water supply. Interlocks are also available that shut off electrical power in the event of loss of coolant flow and are recommended for unattended operations.

7.C   WORKING WITH ELECTRICALLY POWERED LABORATORY EQUIPMENT

Electrically powered equipment is used routinely for laboratory operations requiring heating, cooling, agitation or mixing, and pumping. Electrically powered equipment found in the laboratory includes fluid and vacuum pumps, lasers, power supplies, both electrophoresis and electrochemical apparatus, x-ray equipment, stirrers, hot plates, heating mantles, microwave ovens, and ultrasonicators. Attention must be paid to both the mechanical and the electrical hazards inherent in using these devices. High-voltage and high-power requirements are increasingly prevalent; therefore prudent practices for handling these devices are increasingly necessary.

Electric shock is the major electrical hazard. Although relatively low current of 10 mA poses some danger, 80 to 100 mA can be fatal. In addition, if improperly used, electrical equipment can ignite flammable or explosive vapors. Most of the risks can be minimized by regular proper maintenance and a clear understanding of the correct use of the device. Before beginning any work, all personnel should be shown and trained in the use of all electrical power sources and the location of emergency shutoff switches. Information about emergency procedures can be found in section 7.G.

7.C.1   General Principles

Particular caution must be exercised during installation, modification, and repair, as well as during use of the equipment. To ensure safe operation, all electrical equipment must be installed and maintained in accordance with the provisions of the National Electrical Code (NEC) of the National Fire Protection Association (NFPA, 2008). Trained laboratory personnel should also consult state and local codes and regulations, which may contain special provisions and be more stringent than the NEC rules. All repair and calibration work on electrical equipment must be carried out by properly trained and qualified personnel. Before modification, installation, or even minor repairs of electrical equipment are carried out, the devices must be deenergized and all capacitors discharged safely. Furthermore, this deenergized and/or discharged condition must be verified before proceeding. Note that the Occupational Safety and Health Administration (OSHA) Control of Hazardous Energy Standard (29 CFR § 1910.147, Lock out/Tag out) applies.

All new electrical equipment should be inspected on receipt for a certification mark. If the device bears a certification mark from UL (Underwriters Laboratories Inc.), CSA (Canadian Standards Association), ETL (originally a mark of ETL Testing Laboratories, now a mark of Intertek Testing Services), or CE (Conformance European–Communaut Europenne or Conformit Europenne), detailed testing and inspection are not required. If the device does not bear one of these certification marks, the device should be inspected by an electrician before it is put into service.

Each person participating in any experiment involving the use of electrical equipment must be aware of all applicable equipment safety issues and be briefed on any potential problems. Trained laboratory personnel can significantly reduce hazards and dangerous behavior by following some basic principles and techniques: checking and rechecking outlet receptacles (section 7.C.1.1), making certain that wiring complies with national standards and recommendations (section 7.C.1.2), reviewing general precautions (section 7.C.1.3) and personal safety techniques (section 7.C.1.4), and ensuring familiarity with emergency procedures (section 7.G).

7.C.1.1   Outlet Receptacles

All 110-V outlet receptacles in laboratories should be of the standard design that accepts a three-prong plug and provides a ground connection. Replace two-prong receptacles as soon as feasible, and add a separate ground wire so that each receptacle is wired as shown in Figure 7.1.[1] The ground wire is preferably (but not required by code) on top to prevent anything falling onto a plug with exposed prongs, and will contact the ground before contacting the hot or the neutral line.

It is also possible to fit a receptacle with a ground-fault circuit interrupter (GFCI), which disconnects the current if a ground fault is detected. GFCI devices are required by local electrical codes for outdoor receptacles and for selected laboratory receptacles located less than 6 ft (1.83 m) from sinks if maintenance of a good ground connection is essential for safe operation. These devices differ in operation and purpose from fuses and circuit breakers, which are designed primarily to protect equipment and prevent electrical fires due to short circuits or other abnormally high current draw

---

[1]The outlet is always "female"; the plug is always "male."

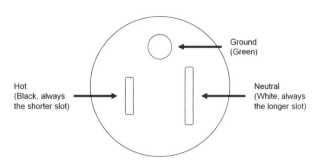

FIGURE 7.1   Representative design for a three-wire grounded outlet. The design shown is for 15-A, 125-V service. The specific design will vary with amperage and voltage.

situations. Certain types of GFCIs cause equipment shutdowns at unexpected and inappropriate times; hence, their selection and use need careful planning. Be aware that GFCIs are not fail-safe devices. They significantly reduce the possibility of fatal shock but do not entirely eliminate it.

Locate receptacles that provide electric power for operations in laboratory chemical hoods outside the hood. This location prevents the production of electrical sparks inside the chemical hood when a device is plugged in or disconnected, and it also allows trained laboratory personnel to disconnect electrical devices from outside the hood in case of an accident. Cords should not be routed in such a way that they can accidentally be pulled out of their receptacles or tripped over.

Simple inexpensive plastic retaining strips and ties can be used to route cords safely. For laboratory chemical hoods with airfoils, route the electrical cords under the bottom airfoil so that the sash can be closed completely. Most airfoils are easily removed and replaced with a screwdriver.

7.C.1.2   Wiring

Fit laboratory equipment plugged into a 110-V (or higher) receptacle with a standard three-conductor line cord that provides an independent ground connection to the chassis of the apparatus (see Figure 7.2). Ground all electrical equipment unless it is double-insulated. This type of equipment has a two-conductor line cord that meets national codes and standards. The use of two-pronged cheaters to connect equipment with three-prong grounded plugs to old-fashioned two-wire outlets is hazardous and should be prohibited.

Limit the use of extension cords to temporary (<1 day) setups, if they are permitted at all. Use a standard three-conductor extension cord of sufficient rating for the connected equipment with an independent ground connection. In addition, good practice uses only extension cords equipped with a GFCI. Install electrical

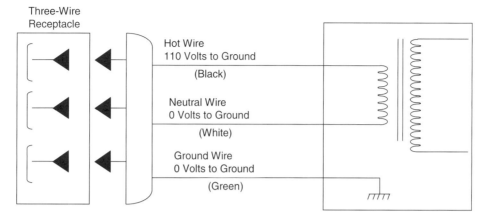

FIGURE 7.2 Standard wiring convention for 110-V electric power to equipment.

cables properly, even if only for temporary use, and keep them out of aisles and other traffic areas. Install overhead racks and floor channel covers if wires must pass over or under walking areas. Do not intermingle signal and power cables in cable trays or panels. Special care is needed when installing and placing water lines (used, for example, to cool equipment such as flash lamps for lasers) so that they do not leak or produce condensation, which can dampen power cables nearby.

Equipment plugged into an electrical receptacle should include a fuse or other overload protection device to disconnect the circuit if the apparatus fails or is overloaded. This overload protection is particularly useful for equipment likely to be left on and unattended for a long time, such as variable autotransformers (e.g., Variacs and powerstats),[2] vacuum pumps, drying ovens, stirring motors, and electronic instruments. If equipment does not contain its own built-in overload protection, modify it to provide such protection or replace it with equipment that does. Overload protection does not protect the trained laboratory personnel from electrocution but does reduce the risk of fire.

7.C.1.3 General Precautions for Working with Electrical Equipment

Laboratory personnel should be certain that all electrical equipment is well maintained, properly located, and safely used. To do this, review the following precautions and make the necessary adjustments prior to working in the laboratory:

- Insulate all electrical equipment properly. Visually inspect all electrical cords monthly, especially in any laboratory where flooding can occur. Keep in mind that rubber-covered cords can be eroded by organic solvents, ozone (produced by ultraviolet lamps), and long-term air oxidation.
- Properly replace all frayed or damaged cords before any further use of the equipment is permitted. Qualified personnel should conduct the replacement.
- Ensure the complete electrical isolation of electrical equipment and power supplies. Enclose all power supplies in a manner that makes accidental contact with power circuits impossible. In every experimental setup, including temporary ones, use suitable barriers or enclosures to protect against accidental contact with electrical circuits.
- Many laboratory locations are classified under fire and electrical codes with a mandate for nonsparking explosion-proof motors and electrical equipment. Areas where large amounts of flammable solvents are in use also require explosion-proof lighting and electrical fixtures. The owners of such facilities are responsible for ensuring that all electrical equipment and fixtures meet these codes and regulations.
- Equip motor-driven electrical equipment used in a laboratory where volatile flammable materials may be present with either nonsparking induction motors that meet Class 1, Division 2, Group C-D electrical standards (Earley, 2008; NFPA, 2008) or air motors instead of series-wound motors that use carbon brushes, such as those generally used in vacuum pumps, mechanical shakers, stirring motors, magnetic stirrers, and rotary evaporators. Do not use variable autotransformers to control the speed of an induction motor. The speed of an induction motor is determined by the AC frequency rather than the voltage. Thus, using a variable autotransformer that controls voltage and not frequency could cause the motor to overheat and presents a fire hazard.

---

[2]Commonly known as "variacs," variable autotransformers are devices that provide a voltage-adjustable output of AC electricity using a constant voltage input (e.g., the wall outlet).

- Because series-wound motors cannot be modified to make them spark-free, do not use appliances (e.g., kitchen refrigerators, mixers, and blenders) with such motors in laboratories where flammable materials may be present.
- When bringing ordinary electrical equipment such as vacuum cleaners and portable electric drills having series-wound motors into the laboratory for special purposes, take specific precautions to ensure that no flammable vapors are present before such equipment is used (see Chapter 6, section 6.G).
- Locate electrical equipment to minimize the possibility of spills onto the equipment or flammable vapors carried into it. If water or any chemical is spilled on electrical equipment, shut off the power immediately at a main switch or circuit breaker and unplug the apparatus using insulated rubber gloves.
- Minimize condensation that may enter electrical equipment if it is placed in a cold room or a large refrigerator. Cold rooms pose a particular risk in this respect because the atmosphere is frequently at a high relative humidity, and the potential for water condensation is significant.
- If electrical equipment must be placed in such areas, mount the equipment on a wall or vertical panel. This precaution reduces, but does not eliminate, the effects of condensation.
- Condensation can also cause electrical equipment to overheat, smoke, or catch fire. In such a case, shut off the power to the equipment immediately at a main switch or circuit breaker and unplug the apparatus using insulated rubber gloves.
- To minimize the possibility of electrical shock, carefully ground the equipment using a suitable flooring material, and install GFCIs.
- Always unplug equipment before undertaking any adjustments, modifications, or repairs (with the exception of certain instrument adjustments as indicated in section 7.C.7). When it is necessary to handle equipment that is plugged in, be certain hands are dry and, if feasible, wear nonconductive gloves and shoes with insulated soles.
- Ensure that all laboratory personnel know the location and operation of power shutoffs (i.e., main switches and circuit breaker boxes) for areas in which they work. Voltages in breaker boxes may present an arc or flash hazard. Only qualified personnel wearing proper personal protective equipment (PPE) are allowed to open these boxes to access the main switches and circuit breakers contained therein. Label high-voltage breaker boxes presenting an arc or flash hazard. Trained laboratory personnel should be familiar with,

and have in place, alternative power shutoffs (i.e., properly installed crash buttons, ready access to equipment power cord plugs).
- After making modifications to an electrical system or after a piece of equipment has failed, do not use it again until it has been cleaned and properly inspected.

All laboratories should have access to a qualified technician who can make routine repairs to existing equipment and modifications to new or existing equipment so that it will meet acceptable standards for electrical safety. The NFPA *National Electrical Code Handbook* (NFPA, 2008) provides guidelines.

7.C.1.4 Personal Safety Techniques for Use with Electrical Equipment

When operating or servicing electrical equipment, be sure to follow basic safety precautions as summarized below.

- Inform each individual working with electrical equipment of basic precautionary steps to take to ensure personal safety.
- Avoid contact with energized electrical circuits. Let only qualified individuals service electrical equipment.
- Before qualified individuals service electrical equipment in any way, disconnect the power source to avoid the danger of electric shock. Ensure that any capacitors are, in fact, discharged.
- Before reconnecting electrical equipment to its power source after servicing, check the equipment with a suitable tester, such as a multimeter, to ensure that it is properly grounded.
- Do not reenergize a circuit breaker until sure that the cause of the short circuit has been corrected.
- Install GCFIs as required by code to protect users from electric shock, particularly if an electrical device is handheld during a laboratory operation.
- If a person is in contact with a live electrical conductor, disconnect the power source before removing the person from the contact and administering first aid.

7.C.1.5 Additional Safety Techniques for Equipment Using High Current or High Voltage

Unless laboratory personnel are specially trained to install or repair high-current or high-voltage equipment, reserve such tasks for trained electrical workers. The following reminders are included for qualified personnel:

- Always assume that a voltage potential exists within a device while servicing it, even if it is deenergized and disconnected from its power source. A device may contain capacitors, for example, and could retain a potentially harmful electrical charge.
- Work with only one hand, if it is not awkward or otherwise unsafe to do so, while keeping the other hand at your side or in a pocket away from all conducting materials. This precaution reduces the likelihood of accidents that result in current passing through the chest cavity.
- Avoid becoming grounded by staying at least 6 in. away from walls, water, and all metal materials including pipes.
- Use voltmeters and test equipment with ratings and leads sufficient to measure the highest potential voltage to be found inside the equipment being serviced.

7.C.2 Vacuum Pumps

The use of water aspirators is discouraged. Their use in filtration or solvent-removal operations involving volatile organic solvents presents a hazard that volatile chemicals will contaminate the wastewater and the sewer, even if traps are in place. Water and sewer contamination may result in violation of local, state, or federal law. These devices also consume large volumes of water, present a flooding hazard, and can compromise local conservation measures.

Distillation or similar operations requiring a vacuum must use a trapping device to protect the vacuum source, personnel, and the environment. This requirement also applies to oil-free Teflon-lined diaphragm pumps. Normally the vacuum source is a cold trap cooled with dry ice or liquid nitrogen. Even with the use of a trap, the oil in a mechanical vacuum trap can become contaminated and the waste oil must be treated as a hazardous waste.

Vent the output of each pump to a proper air exhaust system. This procedure is essential when the pump is being used to evacuate a system containing a volatile toxic or corrosive substance. Failure to observe this precaution results in pumping the untrapped substances into the laboratory atmosphere. Scrubbing or absorbing the gases exiting the pump is also recommended. Even with these precautions, volatile toxic or corrosive substances may accumulate in the pump oil and thus be discharged into the laboratory atmosphere during future pump use. Avoid this hazard by draining and replacing the pump oil when it becomes contaminated. Follow procedures recommended by the institution's environmental health and safety office for the safe disposal of pump oil contaminated with toxic or cor-

rosive substances. General-purpose laboratory vacuum pumps should have a record of use to prevent cross-contamination or reactive chemical incompatibility problems.

Belt-driven mechanical pumps must have protective guards. Such guards are particularly important for pumps installed on portable carts or tops of benches where laboratory personnel might accidentally entangle clothing or fingers in the moving belt or wheels. Glassware under vacuum is at risk for implosion, which could result in flying glass. (For more information about working under vacuum, see Chapter 4, section 4.E.4.)

7.C.3 Refrigerators and Freezers

The potential hazards posed by laboratory refrigerators include release of vapors from the contents, the possible presence of incompatible chemicals, and spillage. As general precautions, laboratory refrigerators should be placed against fire-resistant walls, should have heavy-duty power cords, and preferably should be protected by their own circuit breaker. Enclose the contents of a laboratory refrigerator in unbreakable secondary containment. Because there is almost never a satisfactory way to continuously vent the interior atmosphere of a refrigerator, any vapors escaping from vessels placed in one will accumulate in the refrigerated space and gradually be absorbed into the surrounding insulation. Thus, the atmosphere in a refrigerator could contain an explosive mixture of air and the vapor of a flammable substance or a dangerously high concentration of the vapor of a toxic substance or both. The impact of exposure to toxic substances can be aggravated when a person inserts his or her head inside a refrigerator to search for a particular sample. Placing potentially explosive (see Chapter 6, sections 6.C and 6.G) or highly toxic substances (see Chapter 6, sections 6.D and 6.E) in a laboratory refrigerator is strongly discouraged. As noted in Chapter 6, section 6.C, laboratory refrigerators are never used to store food or beverages for human consumption. Add permanent labels warning against the storage of food and beverages to all laboratory refrigerators and freezers.

Potential ignition sources, (e.g., electrical sparks) must be eliminated from the inside of laboratory refrigerators used to store flammable chemicals. Use explosion-proof refrigerators for the storage of flammable materials; they are sold for this purpose and are labeled and hardwired. Only refrigerators that have been UL- or FM (Factory Mutual)-approved for flammable storage should be used for this purpose. A labeled hardwired explosion-proof refrigerator is mandatory for a renovated or new laboratory where flammable materials need refrigeration. Because of the expense of an explosion-proof refrigerator, a modified sparkproof refrigerator is sometimes found in older laboratories and laboratories using very small amounts of flammable materials. However, a modified sparkproof refrigerator cannot meet the standards of an explosion-proof refrigerator. Where they exist, a plan to phase them out is recommended.

Sparkproof refrigerators must have had the following modifications:

- Interior light and switch mounted on the door frame, if present, have been removed.
- Contacts of the thermostat controlling the fan and temperature have been moved outside the refrigerated compartment.

Permanently attach a prominent sign warning against the storage of flammable substances to the door of an unmodified refrigerator. Frost-free refrigerators are not suitable for laboratory use, owing to the problems associated with attempts to modify them. Many of these refrigerators have a drain tube or hole that carries water (and any flammable material present) to an area adjacent to the compressor and thus present a spark hazard. The electric heaters used to defrost the freezing coils are also a potential spark hazard (see section 7.C.5). To ensure its effective functioning, defrost a freezer manually when ice builds up.

Never place uncapped containers of chemicals in a refrigerator. Caps provide a vapor-tight seal to prevent a spill if the container is tipped over. Aluminum foil, corks, corks wrapped with aluminum foil, and glass stoppers do not meet this criterion, and their use is discouraged. The most satisfactory temporary seals are normally screw caps lined with either a conical polyethylene or a Teflon insert. The best containers for samples that are to be stored for longer periods of time are sealed nitrogen-filled glass ampoules. At a minimum, use catch pans for secondary containment.

Careful labeling of samples placed in refrigerators and freezers with both the contents and the owner's name is essential. Do not use water-soluble ink; labels should be waterproof or covered with transparent tape. Storing samples with due consideration of chemical compatibility is important in these often small crowded spaces.

7.C.4   Stirring and Mixing Devices

The stirring and mixing devices commonly found in laboratories include stirring motors, magnetic stirrers, shakers, small pumps for fluids, and rotary evaporators for solvent removal. These devices are often used in laboratory chemical hoods, and they must be operated such that they do not provide an ignition source for flammable vapors. Consider the use of air-driven stirrers and other spark-free devices. Furthermore, it is important that, in the event of an emergency, such devices can be turned on or off from outside the laboratory chemical hood. Heating baths associated with these devices (e.g., baths for rotary evaporators) should also be spark-free and controllable from outside the hood. (See sections 7.C.1 and 7.C.5.)

Use only spark-free induction motors in power stirring and mixing devices or any other rotating equipment used for laboratory operations. In some cases these devices may be required by fire and electrical codes. Although the motors in most of the currently marketed stirring and mixing devices meet this criterion, their on/off switches and rheostat-type speed controls can produce an electrical spark any time they are adjusted, because they have exposed contacts. Many of the magnetic stirrers and rotary evaporators currently on the market have this disadvantage. An effective solution is to remove any switch located on the device and insert a switch in the cord near the plug end; because the electrical receptacle for the plug should be outside the chemical hood, this modification ensures that the switch will also be outside. Do not control the speed of an induction motor operating under a load by a variable autotransformer.

Because stirring and mixing devices, especially stirring motors and magnetic stirrers, are often operated for fairly long periods without constant attention, consider the consequences of stirrer failure, electrical overload, or blockage of the motion of the stirring impeller. In good practice a stirring impeller is attached to the shaft of the stirring motor with lightweight rubber tubing. If the motion of the impeller is impeded, the rubber can twist away from the motor shaft, and the motor will not stall. Because this practice does not always prevent binding of the impeller, it is also desirable to fit unattended stirring motors with a suitable fuse or thermal protection device. (Also see section 7.C.1.) Take care when attaching an impeller shaft to an overhead motor. If the attachment fails, the impeller shaft could fall through the bottom of a glass vessel below, risking flying glass and a spill.

7.C.5   Heating Devices

Perhaps the most common types of electrical equipment found in a laboratory are the devices used to supply the heat needed to effect a reaction or separation. These include ovens, hot plates, heating mantles and tapes, oil baths, salt baths, sand baths, air baths, hot-tube furnaces, hot-air guns, and microwave ovens. The use of steam-heated devices rather than electrically heated devices is generally preferred whenever temperatures of 100 °C or less are required. Because they do not present shock or spark risks, they can be left unattended with assurance that their temperature will never exceed 100 °C. Use steam that is generated by units that are dedicated to laboratory use. Steam generated for general facility use may contain contaminants that could interfere with laboratory work.

Take a number of general precautions when working with heating devices in the laboratory. If using a variable autotransformer (variac), be sure to wire (or rewire) new or existing equipment, as illustrated in Figure 7.3, before use. However, temperature controllers with built-in safety interlock capability are available from commercial sources and are preferred to variable autotransformers. Enclose the actual heating element in any laboratory heating device in a glass, ceramic, or insulated metal case to prevent a metallic conductor or laboratory personnel from accidentally touching the wire carrying the electric current. This type of construction minimizes the risk of electric shock and of accidentally producing an electrical spark near a flammable liquid or vapor (see Chapter 6, section 6.G.1). It also diminishes the possibility that a flammable liquid or vapor will come into contact with wires at temperatures that might exceed its ignition temperature. Because many household appliances (e.g., hot plates and space heaters) do not meet this criterion, do not use them in a laboratory. Resistance devices used to heat oil baths should not contain bare wires. If any heating device becomes so worn or damaged that its heating element is exposed, either discard the device or repair it before it is used again.

Use laboratory heating devices with a variable autotransformer to control and limit the input voltage to some fraction of the total line voltage, typically 110 V. If a variable autotransformer is not wired in this manner, the switch on it may or may not disconnect both wires of the output from the 110-V line when it is switched to the off position. Also, if this wiring scheme has not been followed, and especially if the grounded three-prong plug is not used, even when the potential difference between the two output lines is only 10 V, each output line may be at a relatively high voltage (e.g., 110 V and 100 V) with respect to an electrical ground. *Because these potential hazards exist, whenever laboratory personnel use a variable autotransformer with an unknown wiring scheme, prudent practice assumes that either of the output lines carries a potential of 110 V and is capable of delivering a lethal electric shock.*

The external cases of all variable autotransformers have perforations for cooling and ventilation, and some sparking may occur whenever the voltage adjustment knob is turned. Therefore, locate these devices where water and other chemicals cannot be spilled onto them and where their movable contacts will not be exposed

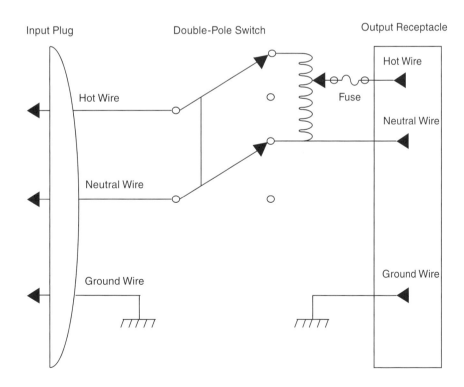

FIGURE 7.3 Schematic diagram of a properly wired variable autotransformer.

to flammable liquids or vapors. Mount variable auto-transformers on walls or vertical panels and outside laboratory chemical hoods; do not simply place them on laboratory benchtops.

Electrical input lines, including lines from variable transformers, to almost all laboratory heating devices have a potential of 110 V with respect to any electrical ground; always view these lines as potential shock and spark hazards. Connections from these lines to a heating device should be both mechanically and electrically secure and completely covered with insulating material. Do not use alligator clips to connect a line cord from a variable autotransformer to a heating device, especially to an oil bath or an air bath, because such connections pose a shock hazard. They also may slip off, creating an electrical spark and, perhaps, contacting other metal parts to create an additional hazard. Make all connections by using, preferably, a plug-and-receptacle combination, or wires with insulated terminals firmly secured to insulated binding posts.

Whenever an electrical heating device is used, either a temperature controller or a temperature-sensing device must be used that will turn off the electric power if the temperature of the heating device exceeds some preset limit. Similar control devices are available that will turn off the electric power if the flow of cooling water through a condenser is stopped owing to the loss of water pressure or loosening of the water supply hose to a condenser. Independent temperature sensors must be used for the temperature controller and shutoff devices. Fail-safe devices, which can be either purchased or fabricated, can prevent the more serious problems of fires or explosions that may arise if the temperature of a reaction increases significantly because of a change in line voltage, the accidental loss of reaction solvent, or loss of cooling. Use fail-safe devices for stills purifying reaction solvents, because such stills are often left unattended for significant periods of time. **Temperature-sensing devices absolutely must be securely clamped or firmly fixed in place, maintaining contact with the object or medium being heated at all times. If the temperature sensor for the controller is not properly located or has fallen out of place, the controller will continue to supply power until the sensor reaches the temperature setting, creating an extremely hazardous situation.** (See also Vignette 7.1.)

Hot plates, oil baths, and heating mantles that can melt and combust plastic materials (e.g., vials, containers, tubing) can cause laboratory fires, and the area around the equipment should be cleared of those hazards prior to use. Be aware that dry and concentrated residues can ignite when overheated in stills, ovens, dryers, and other heating devices.

(See section 7.C.1 for additional information.)

VIGNETTE 7.1
Oil bath fire as a result of a loose temperature sensor

A researcher walking past a laboratory noticed a flame burning behind the closed sashes of the chemical fume hood. He determined that the oil in an oil bath was burning. There was no other equipment in the oil bath and no other chemicals were in the vicinity. The researcher turned off electrical service to the chemical fume hood using the red Crash button on the front and deemed it safe to attempt to extinguish the fire with a B/C extinguisher. When the sash was opened slightly to extinguish the fire, the flames flared through the opening and singed the researcher's forehead and right forearm. The fire was extinguished immediately but continued to flare up because the oil was still above its autoignition temperature. A metal pan was placed over the oil bath to smother the fire.

An investigation determined that the thermocouple used by the oil bath temperature controller had fallen out of the oil bath. The controller, responding to the false temperature drop reading, continued to supply power to the bath, resulting in overheating and fire.

7.C.5.1 Ovens

Electrically heated ovens are commonly used in the laboratory to remove water or other solvents from chemical samples and to dry laboratory glassware. *Never use laboratory ovens to prepare food for human consumption.*

Purchase or construct laboratory ovens with their heating elements and their temperature controls physically separated from their interior atmospheres. Small household ovens and similar heating devices usually do not meet these requirements and, consequently, should not be used in laboratories. With the exception of vacuum drying ovens, laboratory ovens rarely prevent the discharge of the substances volatilized in them into the laboratory atmosphere. The volatilized substances may also be present in sufficient concentration to form explosive mixtures with the air inside the oven (see Chapter 6, section 6.G). This hazard can be reduced by connecting the oven vent directly to an exhaust system. (See Vignette 7.2.)

Do not use ovens to dry any chemical sample that has even moderate volatility and might pose a hazard because of acute or chronic toxicity unless special precautions have been taken to ensure continuous venting of the atmosphere inside the oven. (See Vignette 7.2.) Thus, do not dry most organic compounds in a conventional unvented laboratory oven.

To avoid explosion, do not dry glassware that has been rinsed with an organic solvent in an oven until it has been rinsed again with distilled water. Potentially explosive mixtures can be formed from volatile substances and the air inside an oven.

Bimetallic strip thermometers are preferred for monitoring oven temperatures. Do not mount mercury thermometers through holes in the tops of ovens with the bulb hanging into the oven. If a mercury thermometer is broken in an oven of any type, close the oven and turn it off immediately to avoid mercury exposure. Keep it closed until cool. Remove all mercury from the cold oven with the use of appropriate cleaning equipment and procedures (see Chapter 6, section 6.C.10.8). After removal of all visible mercury, monitor the heated oven in a laboratory chemical hood until the mercury vapor concentration drops below the threshold limit value. (For information about reducing the use of mercury in thermometers, see Chapter 5, section 5.B.8.)

VIGNETTE 7.2
Muffle furnace fire

A laboratory specializing in the analysis of paint samples was asked to analyze pigmented polypropylene. The first step of the analytical protocol called for ashing the sample in a muffle furnace. The technician loaded the furnace with four crucibles containing a total of approximately 110 g of polypropylene. The temperature was set to ramp up to 900 °C. At approximately 500 °C a fire erupted from the furnace, which was quickly extinguished.

Two major contributing factors to the fire were identified. First, the technician had no experience with the analysis of polypropylene-containing samples and did not recognize that polypropylene begins to decompose at approximately 500 °C to low-molecular-weight olefins. Second, the amount of organic matter placed in the furnace in the form of the polypropylene samples was significantly more than that in the usual paint samples.

7.C.5.2　Hot Plates

Laboratory hot plates are often used when solutions are to be heated to 100 °C or higher and the inherently safer steam baths cannot be used as the source of heat. As previously noted, use only hot plates that have completely enclosed heating elements in laboratories. Although almost all laboratory hot plates currently sold meet this criterion, many older ones pose an electrical spark hazard arising from either the on/off switch located on the hot plate, the bimetallic thermostat used to regulate the temperature, or both. Normally, these two spark sources are located in the lower part of the hot plate in a region where any heavier-than-air and possibly flammable vapors evolving from a boiling liquid on the hot plate would tend to accumulate. In principle, these spark hazards are alleviated by enclosing all mechanical contacts in a sealed container or by using solid-state circuitry for switching and temperature control. However, in practice, such modifications are difficult to incorporate into many of the hot plates now in use. Warn laboratory personnel of the spark hazard associated with these hot plates. Set up any newly purchased hot plates to avoid electrical sparks. In addition to the spark hazard, old and corroded bimetallic thermostats in these devices can eventually fuse shut and deliver full continuous current to a hot plate. This risk can be avoided by wiring a fusible coupling into the line inside the hot plate. If the device does overheat, the coupling will melt and interrupt the current (see section 7.C.1).

On many brands of combined stirrer/hot plates, the controls for the stirrer and temperature control are not easily differentiated. Care must be taken to distinguish their functions. A fire or explosion may occur if the temperature rather than the stirrer speed is increased inadvertently.

7.C.5.3　Heating Mantles

Heating mantles are commonly used to heat round-bottom flasks, reaction kettles, and related reaction vessels. These mantles enclose a heating element in layers of fiberglass cloth. As long as the fiberglass coating is not worn or broken and no water or other chemicals are spilled into the mantle (see section 7.C.1), heating mantles pose minimal shock hazard. They are normally fitted with a male plug that fits into a female receptacle on an output line from a variable autotransformer. This plug combination provides a mechanically and electrically secure connection.

Always use heating mantles with a variable autotransformer to control the input voltage. Never plug them directly into a 110-V line. Trained laboratory personnel should be careful not to exceed the input voltage recommended by the mantle manufacturer. Higher voltages will cause a mantle to overheat, melting the fiberglass insulation and exposing the bare heating element.

Some heating mantles are constructed by encasing the fiberglass mantle in an outer metal case that provides physical protection against damage to the fiberglass. If such metal-enclosed mantles are used, good practice is to ground the outer metal case either by using a grounded three-conductor cord from the variable autotransformer or by securely affixing one end of a heavy braided conductor to the mantle case and the other end to a known electrical ground. This practice protects the laboratory personnel against an electric shock if the heating element inside the mantle short-circuits against the metal case. Placing the heating mantle on a laboratory jack and holding the flask or container being heated by clamps attached to a separate ring stand or grid work is the recommended procedure. This allows for rapid removal of heat in the case of overheating or exothermicity.

7.C.5.4   Oil, Salt, or Sand Baths

When using oil, salt, or sand baths, take care not to spill water and other volatile substances into the baths. Such an accident can splatter hot material over a wide area and cause serious injuries.

Electrically heated oil baths are often used to heat small or irregularly shaped vessels or to maintain a constant temperature with a stable heat source. For temperatures below 200 °C, a saturated paraffin oil is often used; for temperatures up to 300 °C, a silicone oil should be used. Care must be taken with hot oil baths not to generate smoke or have the oil burst into flames from overheating. Always monitor an oil bath by using a thermometer or other thermal sensing device to ensure that its temperature does not exceed the flash point of the oil being used. For the same reason, fit oil baths left unattended with thermal-sensing devices that turn off the electric power if the bath overheats. Heat these baths by an enclosed heating element, such as a knife heater, a tubular immersion heater such as a calrod, or its equivalent. The input connection for this heating element is a male plug that fits a female receptacle from a variable autotransformer (e.g., Variac) output line. Alternatively, a temperature controller can be used to control the temperature of the bath precisely. Temperature controllers are available that provide a variety of heating and cooling options. **Thermocouples used by controlling devices must be clamped securely in place to maintain contact with the medium or object being heated at all times.**

Oil baths must be well mixed to ensure that there are no hot spots around the elements that take the surrounding oil to unacceptable temperatures. This problem can be minimized by placing the thermoregulator fairly close to the heater. Contain heated oil in either a metal pan or a heavy-walled porcelain dish; a Pyrex dish or beaker can break and spill hot oil if struck accidentally with a hard object. Mount the oil bath carefully on a stable horizontal support such as a laboratory jack that can be raised or lowered easily without danger of the bath tipping over. Always clamp equipment high enough above a hot plate or oil bath that if the reaction begins to overheat, the heater can be lowered immediately and replaced with a cooling bath without having to readjust the clamps holding the equipment setup. Never support a bath on an iron ring because of the greater likelihood of accidentally tipping the bath over. Provide secondary containment in the event of a spill of hot oil. Wear proper protective gloves when handling a hot bath.

Molten salt baths, like hot oil baths, offer the advantages of good heat transfer, commonly have a higher operating range (e.g., 200 to 425 °C), and may have a high thermal stability (e.g., 540 °C). The reaction container used in a molten salt bath must be able to withstand a very rapid heat rise to a temperature above the melting point of the salt. Care must be taken to keep salt baths dry, because they are hygroscopic, a property that can cause hazardous popping and splattering if the absorbed water vaporizes during heating.

7.C.5.5   Hot Air Baths and Tube Furnaces

Hot air baths can be useful heating devices. Nitrogen is preferred for reactions in which flammable materials are used. Electrically heated air baths are frequently used to heat small or irregularly shaped vessels. Because of their inherently low heat capacity, such baths normally must be heated considerably above the desired temperature (≥100 °C) of the vessel being heated. Purchase or construct these baths so that the heating element is completely enclosed and the connection to the air bath from the variable autotransformer is both mechanically and electrically secure. These baths can be constructed from metal, ceramic, or, less desirably, glass vessels. If a glass vessel is used, wrap it thoroughly with heat-resistant tape so that if the vessel breaks accidentally, the glass will be contained and the bare heating element will not be exposed. Fluidized sand baths are usually preferred over air baths.

Tube furnaces are often used for high-temperature reactions under reduced pressure. The proper choice of glassware or metal tubes and joints is required, and the procedures should conform to safe practice with electrical equipment and evacuated apparatus.

(See also section 7.C.1 and Chapter 6, section 6.G.2.5.)

7.C.5.6　Heat Guns

Laboratory heat guns are constructed with a motor-driven fan that blows air over an electrically heated filament. They are frequently used to dry glassware or to heat the upper parts of a distillation apparatus during distillation of high-boiling point materials. The heating element in a heat gun typically becomes red-hot during use and, necessarily, cannot be enclosed. Also, the on/off switches and fan motors are not usually spark-free. Furthermore, heat guns are designed to pull lab air into and across the red-hot heating elements, thereby increasing the ignition risk. For these reasons, heat guns almost always pose a serious spark hazard (see Chapter 6, section 6.G.1). Never use them near open containers of flammable liquids, in environments where appreciable concentrations of flammable vapors may be present, or in laboratory chemical hoods used to remove flammable vapors. Household hair dryers may be substituted for laboratory heat guns only if they have three-conductor line cords or are double-insulated. Any handheld heating device of this type that will be used in a laboratory should have GFCI protection to ensure against electric shock.

7.C.5.7　Microwave Ovens

Use microwave ovens specifically designed for laboratory use. Domestic microwave ovens are not appropriate.

Microwave heating presents several potential hazards not commonly encountered with other heating methods: extremely rapid temperature and pressure rise, liquid superheating, arcing, and microwave leakage. Microwave ovens designed for the laboratory have built-in safety features and operation procedures to mitigate or eliminate these hazards. Users of such equipment must be thoroughly knowledgeable of operation procedures and safety devices and protocols before beginning experiments, especially when there is a possibility of fire (flammable solvents), overpressurization, or arcing (Foster and Cournoyer, 2005).

To avoid exposure to microwaves, never operate ovens with the doors open. Do not place wires and other objects between the sealing surface and the door on the oven's front face. Keep the sealing surfaces absolutely clean. To avoid electrical hazards, the oven must be grounded. If use of an extension cord is necessary, use only a three-wire cord with a rating equal to or greater than that for the oven. To reduce the risk of fire in the oven, do not overheat samples. The oven must be closely watched when combustible materials are

in it. Do not use metal containers or metal-containing objects (e.g., stir bars) in the microwave, because they can cause arcing.

In general, do not heat sealed containers in a microwave oven, because of the danger of explosion. If sealed containers must be used, select their materials carefully and the containers properly designed. Commercially available microwave acid digestion bombs, for example, incorporate a Teflon sample cup, a self-sealing Teflon O-ring, and a compressible pressure-relief valve. Do not exceed the manufacturer's loading limits. For such applications, properly vent the microwave oven using an exhaust system. Placing a large item, such as a laboratory microwave or an oven, inside a chemical fume hood is not recommended.

Heating a container with a loosened cap or lid poses a significant risk. Microwave ovens can heat material (e.g., solidified agar) so quickly that, even though the container lid is loosened to accommodate expansion, the lid can seat upward against the threads and the container can explode. Screw caps must be removed from containers being microwaved. If the sterility of the contents must be preserved, screw caps may be replaced with cotton or foam plugs.

7.C.6　Distillation

Distillation of flammable and combustible solvents is dangerous due to the presence of heat and flammable vapors. Distillations should be maintained under inert atmosphere. At the completion of vacuum distillations, backfill the apparatus with inert gas. Perform such distillations in a chemical hood. Stills in use should be attended at all times and should have an automatic high-temperature shutoff. Distillation can sometimes be avoided by purchasing smaller quantities and high-purity solvents.

7.C.6.1　Solvent Stills

Solvent stills are used to produce dry, oxygen-free, high-purity solvents. Most high-purity solvents are commercially available in specialized kegs or may be obtained from column purification systems (see section 7.C.6.2); thus, thermal distillation processes should be a last resort. There have been numerous fires attributed to solvent stills, some resulting in serious injuries and extensive damage to the labs. [See, e.g., Yarnell (2002).]

The process involves reflux and distillation of organic solvents (many of which are flammable liquids) over drying materials, under nitrogen or argon gas. The most commonly used drying agents involve potentially pyrophoric metals: sodium metal/benzophenone and magnesium metal/iodine. The stills must be periodically quenched to prepare the still bottoms for disposal. This usually involves adding solvent to consume the scavenging agents. The process itself poses a risk of reactive metal adhering to the bottom of the flask, with the potential for exposure to air, potentially causing a spontaneous fire. Most thermal stills rely on electric heating mantles to heat the flammable solvents upward of 82 °C (180 °F), presenting a fire risk and potential ignition source.

Always set up stills in a chemical hood. Although many procedures suggest allowing the process to run overnight, it is prudent to ensure that it is not left completely unattended. Start the process at the beginning of the day and let it run as long as laboratory workers are present. Place Plexiglas shields around the still to protect workers in the event of a serious accident. Deactivate the stills under argon or nitrogen, never air. Do not add fresh solvent, drying agent, or indicator while the still is hot. Ensure that water cooling lines are in good condition. Do not allow material to accumulate at the bottom of the still; quench the still at the end of every procedure and clean thoroughly. Use caution when collecting the reactive materials as waste.

7.C.6.2 Column Purification Systems or "Push Stills"

Column purification systems offer a safer, more environmentally friendly process for providing dry, oxygen-free, high-purity solvents as compared with thermal distillation. The level of impurity (water, oxygen, peroxides) is comparable to thermal distillation.

The system is usually composed of refillable stainless steel "kegs" that hold high-purity solvent and act as a solvent reservoir. Inert gas (nitrogen, argon) is used to maintain an inert atmosphere as well as to force solvent through the packed columns that contain activated alumina (for water scavenging) and copper catalyst (for oxygen scavenging). For those solvents that are incompatible with copper (e.g., tetrahydrofuran, methylene chloride, acetonitrile), a second column of alumnia is used along with a dry nitrogen or argon purge to facilitate oxygen removal. The solvent product is dispensed from the columns into a variety of specialized containers for use in the laboratory (glass, stainless steel, etc.).

Column purification systems (Figure 7.4) present much less of a fire risk compared with thermal distillation, because they do not employ heating devices or reactive metals. Because glass containers are not needed, the potential for injury or spill related to breakage is also eliminated.

The column purification system significantly reduces utility usage compared with a thermal still. Thermal distillation uses an average of 70,000 gal of water per coolant line, per year; the column purification system uses no water. There is no need for heating mantles when solvent is present, and the intrinsically safe properties of the system allow it to be set up virtually anywhere in the laboratory, thus eliminating the need to place the apparatus in a chemical hood. As a result, there is a significant savings in electricity usage, although heating jackets may be required for installations where the water and oxygen scavengers are activated or regenerated.

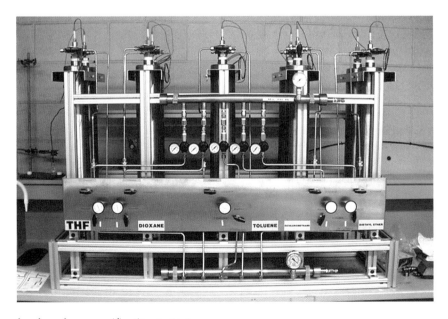

FIGURE 7.4 Example of a column purification system.

When using a column purification system, it is important not to draw down the column completely empty. Bubbling or splattering as the product is drawn from the column is an indication of breakthrough of argon. For the column to be functional again, a lengthy priming operation may be needed.

7.C.7 Ultrasonicators, Centrifuges, and Other Electrical Equipment

7.C.7.1 Ultrasonicators

The use of high-intensity ultrasound in the chemical laboratory has grown substantially during the past decade. Human exposure to ultrasound with frequencies of between 16 and 100 kHz can be divided into three distinct categories: airborne conduction, direct contact through a liquid-coupling medium, and direct contact with a vibrating solid.

Ultrasound through airborne conduction does not appear to pose a significant health hazard to humans. However, exposure to the associated high volumes of audible sound can produce a variety of effects, including fatigue, headaches, nausea, and tinnitus. When ultrasonic equipment is operated in the laboratory, the apparatus must be enclosed in a 2-cm-thick wooden box or in a box lined with acoustically absorbing foam or tiles to substantially reduce acoustic emissions (most of which are inaudible).

Avoid direct contact of the body with liquids or solids subjected to high-intensity ultrasound that promotes chemical reactions. Under some chemical conditions, cavitation is created in liquids that induces high-energy chemistry in liquids and tissues. Cell death from membrane disruption can occur even at relatively low acoustic intensities. Exposure to ultrasonically vibrating solids, such as an acoustic horn, can lead to rapid frictional heating and potentially severe burns.

7.C.7.2 Centrifuges

High-speed centrifuges and ultracentrifuges rely on rotors designed specifically for the particular make and model. These rotors are subject to high mechanical stresses from the forces of the rotation speed. Rotors are rated for a maximum speed and a load of specific weight. Improper loading and balancing can cause the rotors to dislodge while spinning. Failure of the rotors may present a number of hazards: violent movement of the unit itself may cause injury or damage to equipment, electrical lines, gas lines, etc.; flying shrapnel may cause personal injury or facility damage; and some units are susceptible to explosions due to the configuration and materials of construction. (See Vignette 7.3.)

The following precautions should be taken when operating and inspecting centrifuge rotors:

- Balance the load each time the centrifuge is used. The disconnect switch should automatically shut off the equipment when the top is opened.
- Do not overfill the centrifuge tubes. Ensure that they are hung properly.
- Ensure that the lid is closed before starting the centrifuge.
- Do not overload a rotor beyond the rotor's maximum mass without reducing the rated rotor speed.
- Follow the manufacturer's instructions for safe

VIGNETTE 7.3
Centrifuge explosion from use of improper rotor

Lab workers had left samples running unattended in an ultracentrifuge using a large aluminum rotor that previously had been used multiple times without incident. The rotor dislodged while spinning at 20,000 rpm. Friction generated heat and finely divided aluminum powder while at the same time, the refrigeration lines ruptured and released Freon. The mixture of aluminum powder, heat, and Freon confined in a large airtight area resulted in an explosion. The safety shielding within the unit did not contain all of the metal fragments. The flying shrapnel damaged a refrigerator and freezer and gouged holes in the walls and ceiling. The movement of the unit itself damaged cabinets and shelving that held more than 100 containers of chemicals. Fortunately, the cabinets had sliding doors that prevented the chemical containers from falling and breaking. The shock wave from the explosion shattered all four windows in the lab and caused structural damage to the walls. Fortunately, because the lab was unoccupied, no one was injured.

The cause of the incident was the use of a rotor that was not approved for the particular unit. There was a warning decal on the unit explaining which model rotors were acceptable. The unit was more than 25 years old and not designed to current safety standards, resulting in more physical damage than what would be expected. There was no use log or derating of the rotor, and the operator had not been fully trained. The manufacturer's instruction guide for the unit described similar incidents.

operating speeds. Do not run a rotor beyond its maximum rated speed.

- Check O-rings and grease the seals routinely with vacuum grease.
- Do not use harsh detergents to clean the rotors, especially aluminum rotors. Use a mild detergent and rinse with deionized water, if possible.
- Be sure to follow the manufacturer's guidelines for when to retire a rotor.
- For flammable and/or hazardous materials, keep the centrifuge under negative pressure to a suitable exhaust system.
- Keep a usage and maintenance log.
- Always use the rotor specified by the manufacturer.
- Inspect the components of the centrifuge each time it is used:
  - Look for signs of corrosion of the rotors. Metal fatigue will eventually cause any rotor to fail.
  - Ensure that the coating on the rotor is not damaged.
  - Check the cone area for cracks, because this area is highly stressed during rotation.
  - Look for corrosion or cracks in the tube cavity.

7.C.7.3 Electrical Instruments

Most modern electronic instruments have a cord that contains a separate ground wire for the chassis and are supplied with a suitable fuse or other overload protection. Modify any existing instrument that lacks these features to incorporate them. As is true for any electrical equipment, take special precautions to avoid possibility of water or other chemical spills into these instruments.

Under most circumstances, any repairs to, adjustments to, or alterations of electrical instruments should be made only by a qualified individual. Laboratory personnel should not undertake such adjustments unless they have received certification as well as specific training for the particular instrument to be serviced. If trained laboratory personnel do undertake repairs, always unplug the cord before any disassembly begins. However, certain adjustments require connection to a power source, and appropriate protective measures and due diligence are required when working on energized devices. Extra precautions are particularly important for instruments that incorporate high-voltage circuitry.

Many electrical instruments, such as lasers and X-ray, electron-beam, radioactive, photochemical, and electrophoresis equipment, emit potentially harmful radiation, and, therefore, special precautions must be taken when they are used. Only trained laboratory personnel should use and service this equipment. (See section 7.C.1 and Chapter 6, section 6.E.)

7.C.8 Electromagnetic Radiation Hazards

Laboratory equipment that can produce hazardous amounts of electromagnetic radiation include ultraviolet lamps, arc lamps, heat lamps, lasers, microwave and radio-frequency sources, and X-ray and electron-beam sources.

7.C.8.1 Visible, Ultraviolet, and Infrared Laser Light Sources

Seal or enclose direct or reflected ultraviolet light, arc lamps, and infrared sources to minimize overexposure whenever possible. Wear appropriately rated safety glasses, chemical splash goggles, and face shields for eye protection. Wear long-sleeved clothing and gloves to protect arms and hands from exposure. When lasers or deep UV light sources are in use, lights or highly visible signage should be posted outside the room.

Control measures for the safe use of lasers have been established by the American National Standards Institute and presented in *Safe Use of Lasers* (ANSI Z136.1-2007; ANSI, 2007), which describes the different types of laser hazards and the appropriate measures to control each type. Operate Class IIIB and IV lasers only in posted laser-controlled areas. No one but the authorized operator of a laser system should ever enter a posted laser-controlled laboratory when the laser is in use. (See Chapter 4, section 4.E.5.)

7.C.8.2 Radio-Frequency and Microwave Sources

Section 7.C.5.7 provides guidelines for the safe use of microwave ovens in the laboratory. Other devices in the laboratory can also emit harmful microwave or radio-frequency emissions. Train personnel working with these types of devices in their proper operation as well as in measures to prevent exposure to harmful emissions. Position shields and protective covers properly when the equipment is operating. Post warning signs on or near these devices to protect people wearing heart pacemakers.

7.C.8.3 X-Rays, Electron Beams, and Sealed Sources

X-rays and electron beams (E-beams) are used in a variety of laboratory applications but most often for analytical operations. The equipment is government regulated, and usually registration and licensing are required. Train personnel operating or working in the vicinity of these types of equipment appropriately to minimize the risk of exposing themselves and others in the laboratory to harmful ionizing radiation.

The beam from a low-energy X-ray diffraction

machine can cause cell destruction as well as genetic damage. The user must always be alert to the on/off status of the X-ray beam, keep aware of the location of the beam, and know how to work safely around the beam when aligning it in preparation for conducting an experiment. Machine warning lights indicate when the beam shutter is open. Users are required to wear a monitoring badge to measure any accumulated exposure.

7.C.8.4 Miscellaneous Physical Hazards Presented by Electrically Powered Equipment

7.C.8.4.1 Magnetic Fields

An object that moves into the attractive field of a strong magnet system, such as a nuclear magnetic resonance (NMR) system or any other instrument system requiring a superconducting magnet, can become a projectile that is pulled rapidly toward the magnet. For example, the large attractive force of an NMR requires that objects ranging from keys, scissors, knives, wrenches, other tools, oxygen cylinders, buffing machines, and wheelchairs, and other ferromagnetic objects are excluded from the immediate vicinity of the magnet to protect safety and data quality.

Magnetic fields of ~10 G can adversely affect credit cards, watches, and other magnetic objects (see Table 7.1). Computer and television screens in neighboring areas may be affected by shifts in small, peripheral magnetic fields as magnets are brought up

TABLE 7.1 Summary of Magnetic Field Effects

Effect	Field Strength at Which Effects Occur (G)
Effects on sensitive equipment such as electron microscopes, image intensifiers, and nuclear cameras	1
Disturbance of cathode ray tubes; possible detrimental effects on medical equipment, such as pacemakers, implants, surgical clips, or neurostimulators	5
Erasure of credit card and bank cards; disruption of small mechanical devices, such as analog watches and clocks; and disturbance of X-ray tubes	10
Destruction or corruption of magnetic storage material	20
Saturation of transformers and amplifiers	50

SOURCE: Adapted from *Site Planning Guide for Superconducting NMR Systems* (Bruker BioSpin GmbH, 2008b) and *General Safety Considerations for the Installation and Operation of Superconducting Magnet Systems* (Bruker BioSpin GmbH, 2008a).

to field or decommissioned. Prudent practices require posting warnings, cordoning off the area at the 5-G line, and limiting access to areas with more than 10 to 20 G to knowledgeable staff. Keep people wearing heart pacemakers and other electronic or electromagnetic prosthetic devices or other potentially magnetic surgical implants, such as aneurysm clips, away from strong magnetic sources. Repairs done in the vicinity of a strong magnet should be performed with nonferromagnetic tools.

Magnetic fields operate in three dimensions, and when considering the impact of an instrument, field strength should be checked on the floors above and below the floor where a superconducting magnet is installed. The 5-G line should be identified in all affected rooms, and appropriate warnings should be posted.

Because superconducting magnets use liquid nitrogen and liquid helium coolants, the precautions associated with the use of cryogenic liquids must be observed as well. (Also see section 7.E.2.) If the superconducting magnet loses superconductivity because of damage, physical shock, or for any other reason, the coil will heat the cryogenic liquid that surrounds it, the magnet will quench (lose field), and the helium will boil off rapidly into the surrounding space. Low-oxygen alarms are recommended in rooms where instruments with superconducting magnets are located. In the event of a quench, all personnel should leave the area and not return until oxygen levels return to normal. If emergency personnel must enter the area before the oxygen levels have been verified, they should wear a self-contained breathing apparatus (SCBA).

Rooms containing superconducting magnets should provide enough clearance for coolant fills to be performed safely.

If an object becomes stuck to a superconducting magnet, do not attempt to remove it, but call the vendor of the magnet for guidance. Attempting to remove the object could result in injury to personnel and damage to the magnet. It may also cause the magnet to quench, releasing dangerous quantities of gaseous helium into the area.

7.C.8.4.2 Rotating Equipment and Moving Parts

Injuries can result from bodily contact with rotating or moving objects, including mechanical equipment, parts, and devices. The risk of injury can be reduced through improved engineering, good housekeeping, and safe work practice and personal behavior. Trained laboratory personnel must know how to safely shut down equipment in the event of an emergency; must enclose or shield hazardous parts, such as belts, chains, gears, and pulleys, with appropriate guards; and must not wear loose-fitting clothing, jewelry, or unrestrained long hair around machinery with moving parts.

7.C.8.4.3   Cutting and Puncturing Tools

Hand injuries are the most frequently encountered injuries in laboratories. Many of these injuries can be prevented by keeping all sharp and puncturing devices fully protected, avoiding the use of razor blades as cutting tools, and using utility knives that have a spring-loaded guard that covers the blade. Appropriate cutting techniques and the use of the proper or specialized tools should also be considered. Dispose of razor blades, syringe needles, suture needles, and other sharp objects or instruments carefully in designated receptacles rather than throwing them into the trash bin unprotected. (See Chapter 4, Section 4.E.9.)

Minimize glass cuts by use of correct procedures (e.g., the procedure for inserting glass tubing into rubber stoppers and tubing, which is taught in introductory laboratories), through appropriate use of protective equipment, and by careful attention to manipulation. Protective equipment is not fail-safe and should not be relied on to prevent cutting injuries. A variety of adapters are available that render glass tubing and rubber stoppers largely obsolete. Technique is also important. In the case of a slip or a break, the resulting motion should not be in the direction of the person. For example, perform cutting operations with the cutting motion moving away from the body.

7.C.8.4.4   Noise Extremes

Any laboratory operation that exposes trained laboratory personnel to a significant noise source of 85 decibels or greater for an 8-hour average duration should have a hearing conservation program to protect from excessive exposure. Consult an audiologist or industrial hygienist to determine the need for such a program and to provide assistance in developing one.

7.C.8.4.5   Slips, Trips, and Falls

The risks of slips, trips, falls, and collisions between persons and objects are reduced by cleaning up liquid or solid spills immediately, keeping doors and drawers closed and passageways clear of obstructions, providing step stools, ladders, and lifts to reach high areas, and walking along corridors and on stairways at a deliberate pace. Floors that are likely to be wet, for example around ice, dry ice, or liquid nitrogen dispensers, should be slip resistant or have a slip-resistant floor covering. Make paper towel dispensers available for wiping up drops or small puddles as soon as they form. Avoid clutter in the laboratory to reduce the temptation to "make space" on the bench by storing items on the floor, which can create a trip hazard.

7.C.8.4.6   Ergonomics and Lifting

Both standing and sitting in a static posture and making repeated motions have been shown to cause a variety of musculoskeletal problems. Problems due to poor ergonomics include eyestrain, stiff and sore back, leg discomfort, and hand and arm injuries. Each situation needs to be evaluated individually. However, personnel who spend significant time working on video display terminals should use furniture appropriate for these tasks, proper posture, and perhaps special eyeglasses. Also, people who use the same tools and hand motions for extended periods of time should take breaks at appropriate intervals to help prevent injuries.

Lifting injuries are one of the more common types of injuries for trained laboratory personnel. The weight of the item to be lifted is a factor, but it is only one of several. The shape and size of an object as well as the lifting posture and the frequency of lifting are also key factors in determining the risks of lifting. The National Institute for Occupational Safety and Health (NIOSH) has developed a guide that should be consulted to help determine lifting safety (Waters et al., 1994). Personnel who are at risk for lifting injuries should receive periodic training.

7.D   WORKING WITH COMPRESSED GASES

7.D.1   Compressed Gas Cylinders

Precautions are necessary for handling the various types of compressed gases, the cylinders that contain them, the regulators used to control their delivery pressure, the piping used to confine them during flow, and the vessels in which they are ultimately used. *Regular inventories of cylinders and checks of their integrity with prompt disposal of those no longer in use are important.* (See Chapter 5, section 5.E.6 for information on storing gas cylinders, and Chapter 6, section 6.H, for discussion of the chemical hazards of gases.)

A compressed gas is defined as a material in a container with an absolute pressure greater than 276 kPa, or 40 psi at 21 °C or an absolute pressure greater than 717 kPa (104 psi) at 54 °C, or both, or any liquid flammable material having a Reid vapor pressure greater than 276 kPa (40 psi) at 38 °C. The U.S. Department of Transportation (DOT) has established codes that specify the materials to be used for the construction and the capacities, test procedures, and service pressures of the cylinders in which compressed gases are transported. However, regardless of the pressure rating of the cylinder, the physical state of the material within it determines the pressure of the gas. For example, liquefied gases such as propane and ammonia exert their own vapor pressure as long as liquid remains in the cylinder and the critical temperature is not exceeded.

Prudent procedures for the use of compressed gas cylinders in the laboratory include attention to appropriate purchase, especially selecting the smallest cylinder compatible with the need, as well as proper transportation and storage, identification of contents, handling and use, and marking and return of the empty cylinder to the company from which it was purchased. Empty compressed gas cylinders purchased for the laboratory should be returned to the company and should never be refilled by laboratory personnel.

Discourage the practice of purchasing unreturnable lecture bottles to avoid the accumulation of partially filled cylinders and cylinder disposal problems. Encourage trained laboratory personnel to lease the cylinders and, in essence, only purchase the contents.

7.D.1.1 Identification of Contents

Clearly label compressed gas cylinders so they are easily, quickly, and completely identified by trained laboratory personnel. Stencil or stamp identification on the cylinder itself, or provide *a durable label that cannot be removed from the cylinder*. Do not accept any compressed gas cylinder for use that does not identify its contents legibly by name. Color coding is not a reliable means of identification; cylinder colors vary from supplier to supplier, and labels on caps have no value because many caps are interchangeable. Care in the maintenance of cylinder labels is important because unidentified compressed gas cylinders may pose a high risk and present very high disposal costs. Good practice provides compressed gas cylinders with tags on which the names of users and dates of use can be entered. If the labeling on a cylinder becomes unclear or an attached tag is defaced and the contents cannot be identified, mark the cylinder as contents unknown and contact the manufacturer regarding appropriate procedures.

Clearly label all gas lines leading from a compressed gas supply to identify the gas, the laboratory served, and relevant emergency telephone numbers. The labels, in addition to being dated, should be color-coded to distinguish hazardous gases, such as flammable, toxic, or corrosive substances that are coded with a yellow background and black letters, and inert gases that are coded with a green background and black letters. Post signs conspicuously in areas in which flammable compressed gases are stored, identifying the substances and appropriate precautions, for example,

HYDROGEN—FLAMMABLE GAS NO SMOKING-NO OPEN FLAMES

7.D.2 Equipment Used with Compressed Gases

7.D.2.1 Records, Inspection, and Testing

Carry out high-pressure operations only with equipment specifically designed and built for this use and only by those personnel trained especially to use this equipment. Never carry out reactions in, or apply heat to, an apparatus that is a closed system unless it has been designed and tested to withstand pressure. To ensure that the equipment has been properly designed, each pressure vessel should have stamped on it, or on an attached plate, its maximum allowable working pressure, the allowable temperature at this pressure, and the material of construction. Similarly, the relief pressure—the pressure at which the safety system (e.g., rupture disk or safety vent) will be triggered—and setting data should be stamped on a metal tag attached to installed pressure-relief devices, and the setting mechanisms should be sealed. Relief devices used on pressure regulators do not require these seals or numbers.

Test or inspect all pressure equipment periodically. The frequency of tests and inspections varies, depending on the type of equipment, how often it is used, and the nature of its usage. Corrosive or otherwise hazardous service requires more frequent tests and inspections. Stamp inspection data on or attach it to the equipment. Testing the entire assembled apparatus with soap solution and air or nitrogen pressure to the maximum operating pressure of the weakest section of the assembled apparatus usually detects leaks at threaded joints, packings, and valves. Alternatively, the apparatus may be pressurized and monitored for pressure drop over time.

Before any pressure equipment is altered, repaired, stored, or shipped, vent it and completely remove all toxic, flammable, or other hazardous material so it can be handled safely. Especially hazardous materials may require special cleaning techniques, which should be solicited from the distributor.

(See section 7.E.1 for further information.)

7.D.2.2 Assembly and Operation

During the assembly of pressure equipment and piping, use only appropriate components, and take care to avoid strains and concealed fractures from the use of improper tools or excessive force. Do not support any significant weight with the tubing in place in a pressure apparatus.

Do not force threads that do not fit smoothly. (See Vignette 7.4.) Do not overtighten fittings. Thread connections must match; tapered pipe threads cannot be joined with parallel machine threads. Use Teflon tape or a suitable thread lubricant on appropriate fittings, (e.g., Teflon tape on pipe fittings only) when assembling the apparatus (see section 7.D.2.2.8). However, never use oil or lubricant on any equipment that will be used with oxygen. Reject parts having damaged or partly stripped threads (see also section 7.D.2.2.3).

In assembling copper-tubing installations, avoid sharp bends and allow considerable flexibility. Copper tubing hardens and cracks on repeated bending. Many metals can become brittle in hydrogen or corrosive gas service. In carbon monoxide atmospheres, some alloys containing nickel or iron can generate carbonyls [e.g., $Ni(CO)_4$] which are toxic when absorbed through the skin or inhaled. Inspect all tubing frequently and replace when necessary.

Stuffing boxes and gland joints are a likely source of trouble in pressure installations. Give particular attention to the proper installation and maintenance of these parts, including the proper choice of lubricant and packing material.

Shield all reactions under pressure and carry them out as remotely as possible, for example, with valve extensions and behind a heavy shield or with closed-circuit TV monitoring if needed.

Do not fill autoclaves and other pressure-reaction vessels more than half full to ensure that space remains for expansion of the liquid when it is heated. Do not make leak corrections or adjustments to the apparatus while it is pressurized; rather, depressurize the system before mechanical adjustments are made.

A regulator or step-down pressure valve should be used to pressurize low-pressure equipment from a high-pressure source. After pressurizing equipment with a high-pressure source, the equipment should either be disconnected or the connecting piping/tubing should be vented to atmospheric pressure. This will prevent the accidental buildup of excessive pressure in the low-pressure equipment due to leakage from the high-pressure source. For example, after completing the pressurization of an autoclave with a compressed gas cylinder, the cylinder valve should be closed, the delivery regulator backed off to 0 psig, and the lines between the cylinder and the autoclave vented.

Do not use vessels or equipment made partly or entirely of silver or copper or alloys containing more than 50% copper in contact with acetylene or ammonia. Do not let those vessels or equipment made of metals susceptible to amalgamation (e.g., copper, brass, zinc, tin, silver, lead, and gold) come into contact with mercury. This warning includes equipment that has soldered and brazed joints.

Place prominent warning signs in any area where a pressure reaction is in progress so that personnel entering the area will be aware of the potential risk.

7.D.2.2.1 Pressure-Relief Devices

Protect all pressure or vacuum systems and all vessels that may be subjected to pressure or vacuum by properly designed, installed, and tested pressure-relief devices. Experiments involving highly reactive materials that might explode or undergo rapid decomposition with gas evolution (tetrafluoroethylene and hydrogen cyanide are two examples) may also require the use of special pressure-relief devices and may need to be operated at a fraction of the permissible working pressure of the system.

Examples of pressure-relief devices include the rupture-disk type used with closed-system vessels and the spring-loaded safety valves used with vessels for transferring liquefied gases. The following precautions are advisable in the use of pressure-relief devices:

- In addition to the pressure setting, pressure-relief device and associated fittings (tubing, connectors, etc.) must be properly sized and configured to provide a sufficient rate of pressure relief while preventing overpressurization. The diameter of the relief device and fittings and the presence of bends and angles are important considerations that should be addressed by a qualified and trained person or persons.
- The materials of construction must be considered, taking into account the compatibility of the chemicals being handled with the relief components.
- The temperature rating of the relief device must be sufficient. Heat conduction via tubing and fittings can cause the relief device to reach high temperatures, depending on the apparatus design.
- Orient pressure-relief devices with the vent side of the device directed away from the operator or

other personnel. Also vent the relief device into an appropriate trap to catch flammable solvent, reaction solids, etc., avoiding spray into the workspace in the event of a release and minimizing the potential of a fire and aiding clean up. The relief device and trap must be supported so that they are not dislodged or thrown due the thrust resulting from sudden venting.

- The maximum setting of a pressure-relief device is the rated maximum allowable working pressure (MAWP) established for the vessel or for the weakest member of the pressure system at the operating temperature. The operating pressure should be less than the system MAWP. In the case of a system protected by a spring-loaded relief device, the maximum operating pressure should be from 5 to 25% lower than the rated working pressure, depending on the type of safety valve and the importance of leak-free operation. In a system protected by a rupture-disk device, the maximum operating pressure should be approximately two-thirds of the rated MAWP; the exact figure is governed by the fatigue life of the disk used, the temperature, and load pulsations.

- Vent pressure-relief devices that may discharge toxic, corrosive, flammable, or otherwise hazardous or noxious materials in a safe and environmentally acceptable manner such as scrubbing or diluting with nonflammable streams.

- Do not install valves or other shutoff devices between pressure-relief devices and the equipment they are to protect. Similarly, do not install shutoff valves downstream of the relief device and take care to ensure that the relief vent is not blocked or restricted. Tubing and piping downstream of such devices must be at least the same diameter as the fitting on the vent side of the relief device.

- Only qualified persons should perform maintenance work on pressure-relief devices.

- Inspect and replace pressure-relief devices periodically.

- Gas manifolds, compressors, and other sources of high-pressure gas used to supply an apparatus, and which can be isolated from the apparatus by valving, should also be protected by a properly designed pressure-relief device.

7.D.2.2.2 Pressure Gauges

The proper choice and use of a pressure gauge involve several factors, including the flammability, compressibility, corrosivity, toxicity, temperature, and pressure range of the fluid with which it is to be used. Generally, select a gauge with a range that is double the working pressure of the system.

A pressure gauge is normally a weak point in any pressure system because its measuring element must operate in the elastic zone of the metal involved. The resulting limited factor of safety makes careful gauge selection and use mandatory and often dictates the use of accessory protective equipment. The primary element of the most commonly used gauges is a Bourdon tube, which is usually made of brass or bronze and has soft-soldered connections. More expensive gauges are available that have Bourdon tubes made of steel, stainless steel, or other special metals and welded or silver-soldered connections. Accuracies vary from ±2% for less expensive pressure gauges to ±0.1% for higher quality gauges. Use a diaphragm gauge with corrosive gases or liquids or with viscous fluids that would destroy a steel or bronze Bourdon tube.

Consider alternative methods of pressure measurement that may provide greater safety than the direct use of pressure gauges. Such methods include the use of seals or other isolating devices in pressure tap lines, indirect observation devices, and remote measurement by strain-gauge transducers with digital readouts.

Mount pressure gauges so that they are easily read during operation.

Pressure gauges often have built-in pressure-relief devices. Care must be taken to ensure that, in the event of failure, this relief device is oriented away from personnel.

7.D.2.2.3 Piping, Tubing, and Fittings

The proper selection and assembly of components in a pressure system are critical safety factors. Considerations include the materials used in manufacturing the components, compatibility with the materials to be under pressure, the tools used for assembly, and the reliability of the finished connections. Use no oil or lubricant of any kind in a tubing system with oxygen because the combination produces an explosion hazard. Use all-brass and stainless steel fittings with copper or brass and steel or stainless steel tubings, respectively. Fitting of this type must be installed correctly. Do not mix different brands of tube fittings in the same apparatus assembly because construction parts are often not interchangeable.

7.D.2.2.4 Glass Equipment

Avoid glassware for work at high pressure whenever possible. Glass is a brittle material, subject to unexpected failures due to factors such as mechanical impact and assembly and tightening stresses. Poor annealing after glassblowing can leave severe strains. Glass equipment, such as rotameters and liquid-level gauges, incorporated in metallic pressure systems should be installed with shutoff valves at both ends to control the discharge of liquid or gaseous materials in the event of breakage. Mass flowmeters are available that can replace rotameters in desired applications.

7.D.2.2.5 Plastic Equipment

Except as noted below, avoid the use of plastic equipment for pressure or vacuum work unless no suitable substitute is available. These materials can fail under pressure or thermal stress. Only use materials that are appropriately rated or recommended for that particular service.

Tygon and similar plastic tubing have quite limited applications in pressure work. These materials can be used for hydrocarbons and most aqueous solutions at room temperature and moderate pressure. Reinforced plastic tubing that can withstand higher pressures is also available. However, loose tubing under pressure can cause physical damage by its own whipping action. Details of permissible operating conditions must be obtained from the manufacturer. Because of their very large coefficients of thermal expansion, some polymers have a tendency to expand greatly on heating and to contract on cooling. This behavior can create a hazard in equipment subjected to very low temperatures or to alternating low and high temperatures. Plastic tubing may also disrupt electrical grounding and thus present a static electricity hazard. The use of plastic tubing with flammable gases or liquids is not recommended if grounding is an issue.

7.D.2.2.6 Valves

Valves come in a wide range of materials of construction, pressure and temperature ratings, and type. The materials of construction (metal, elastomer, and plastic components) must be compatible with the gases and solvents being used. The valves must be rated for the intended pressure and temperature. Ball valves are preferred over needle valves because their status (on/off) can be determined by quick visual inspection. Use metering or needle valves only when careful flow control is important to the operation. Micrometers can sometimes be used with needle valves to allow quick determination of the status.

7.D.2.2.7 Gas Monitors

Electronic monitors and alarms are available to prevent hazards due to asphyxiant, flammable, and many toxic gases. Consider their use especially if large quantities or large cylinders of these gases are in use. Make sure the monitor is properly rated for the intended purpose as some detectors are subject to interference by other gases.

7.D.2.2.8 Teflon Tape Applications

Use teflon tape on tapered pipe thread where the seal is formed in the thread area. Tapered pipe thread is commonly found in applications where fittings are not routinely taken apart (e.g., general building piping applications).

Do not use Teflon tape on straight thread (e.g., Swagelok) where the seal is formed through gaskets or by other metal-to-metal contacts that are forced together when the fitting is tightened [e.g., Compressed Gas Association (CGA) gas cylinder fittings or compression fittings]. Metal-to-metal seals are machined to tolerances that seal without the need of Teflon tape or other gasketing materials. If used where not needed, as on CGA fittings, Teflon tape only spreads and weakens the threaded connections and can plug up lines that it enters accidentally.

7.D.3 Handling and Use of Gas Cylinders

Gas cylinders must be handled carefully to prevent accidents or damage to the cylinder. Leave the valve protection cap in place until the cylinder is secured and ready for use. Do not drag, roll, slide, or allow gas cylinders to strike each other forcefully. Always transport them on approved wheeled cylinder carts with retaining straps or chains. The plastic mesh sleeves sometimes installed by vendors are intended only to protect the paint on the cylinder and do not serve as a safety device.

Secure compressed gas cylinders firmly at all times. A clamp and belt or chain, holding the cylinder between waist and shoulder to a wall, are generally suitable for this purpose. In areas of seismic activity, secure gas cylinders both toward the top and toward the bottom. Individually secure cylinders; using a single restraint strap or chain around a number of cylinders is often not effective. Locate cylinders in well-ventilated areas. Although inert gases are not exposure hazards, they can produce conditions of oxygen depletion that could lead to asphyxiation. Vent pressure-relief devices protecting equipment that is attached to cylinders of flammable, toxic, or otherwise hazardous gases to a safe place. (See section 7.D.2.2.1 for details.)

Standard cylinder-valve outlet connections have been devised by CGA to prevent the mixing of incompatible gases due to an interchange of connections. Outlet threads used vary in diameter; some are male and some are female, some are right-handed and some are left-handed. In general, right-handed threads are used for nonfuel and water-pumped gases, and left-handed threads are used for fuel and oil-pumped gases. Information on the standard equipment assemblies for use with specific compressed gases is available from the supplier. To minimize undesirable connections that may result in a hazard, use only CGA standard combinations of valves and fittings in compressed gas installations. Avoid the assembly of miscellaneous parts (even of standard approved types). Do not use an adapter or cross-thread a valve fitting. Examine the threads on cylinder valves, regulators, and other fittings to ensure that they correspond to one another and are undamaged.

Place cylinders so that the rotary cylinder valve handle at the top is accessible at all times. Open cylinder valves slowly, and only when a proper regulator is firmly in place and the attachment has been shown to be leakproof by an appropriate test. Close the cylinder valve as soon as the necessary amount of gas has been released. Valves should be either completely open or completely closed. Install flow restrictors on gas cylinders to minimize the chance of excessive flows. Never leave the cylinder valve open when the equipment is not in use. This precaution is necessary not only for safety when the cylinder is under pressure but also to prevent the corrosion and contamination that would result from diffusion of air and moisture into the cylinder when it is emptied.

Most cylinders are equipped with hand-wheel valves. Those that are not should have a spindle key on the valve spindle or stem while the cylinder is in service. Use only wrenches or other tools provided by the cylinder supplier to remove a cylinder cap or to open a valve. Never use a screwdriver to pry off a stuck cap or pliers to open a cylinder valve. If valve fittings require washers or gaskets, check the materials of construction before the regulator is fitted.

If the valve on a cylinder containing an irritating or toxic gas is being opened outside, the worker should stand upwind of the cylinder with the valve pointed downwind, away from personnel, and warn those working nearby in case of a possible leak. If the work is being done inside, open the cylinder only in a laboratory chemical hood or specially designed cylinder cabinet. Install a differential pressure switch with an audible alarm in any chemical hood dedicated for use with toxic gases. In the event of chemical hood failure, the pressure switch should activate an audible alarm warning personnel.

7.D.3.1   Preventing and Controlling Leaks

Check cylinders, connections, and hoses regularly for leaks. Convenient ways to check for leaks include a flammable gas leak detector (for flammable gases only) or looking for bubbles after application of soapy water or a 50% glycerin–water solution. At or below freezing temperatures, use the glycerin solution instead of soapy water. Bubble-forming solutions designed for leak testing are commercially available. When the gas to be used in the procedure is a flammable, oxidizing, or highly toxic gas, check the system first for leaks with an inert gas (helium or nitrogen) before introducing the hazardous gas. Only leak-test solutions specifically designed for oxygen compatibility may be used to test for oxygen leaks; do not use soap solutions because they may contain oils that can react violently with the oxygen.

The general procedures discussed in Chapter 6, section 6.C, can be used for relatively minor leaks, when the indicated action can be taken without exposing personnel to highly toxic substances. The leaking cylinder can be moved through populated portions of the building, if necessary, by placing a plastic bag, rubber shroud, or similar device over the top and taping it (preferably with duct tape) to the cylinder to confine the leaking gas. If there is any risk of exposure, call the environmental health and safety office and evacuate the area before the tank is moved.

If a leak at the cylinder valve handle cannot be remedied by tightening a valve gland or a packing nut, take emergency action and notify the supplier. Never attempt to repair a leak at the junction of the cylinder valve and the cylinder or at the safety device; consult with the supplier for instructions.

When the nature of the leaking gas or the size of the leak constitutes a more serious hazard, an approved SCBA and protective apparel may be required, and personnel may need to be evacuated (see Chapter 6, section 6.C.2). If toxic gas is leaking from a cylinder, donning of protective equipment and evacuation of personnel are required. Cylinder coffins are also available to encapsulate leaking cylinders. (See Chapter 6, section 6.H for more information.)

7.D.3.2   Pressure Regulators

Pressure regulators are required to reduce a high-pressure supplied gas to a desirable lower pressure and to maintain a satisfactory delivery pressure and flow level for the required operating conditions. They are available to fit many operating conditions over a range of supply and delivery pressures, flow capacities, and construction materials. All regulators are typically of a diaphragm type and are spring-loaded or gas-loaded, depending on pressure requirements. They can be single-stage or two-stage. Under no circumstances should oil or grease be used on regulator valves or cylinder valves because these substances may react with some gases (e.g., oxygen).

Each regulator is supplied with a specific CGA standard inlet connection to fit the outlet connection on the cylinder valve for the particular gas. Never tamper with or adapt regulators for use with gases for which they are not designed. Likewise, never substitute the fittings that are on either the cylinder side or downstream (low-pressure) side of a vendor-supplied regulator. Instead, purchase a regulator designed for use with the specific cylinder, and use adapters only on the downstream side of the regulator. Unqualified persons must never attempt to repair or modify regulators.

Check regulators before use to verify they are free of foreign objects and to correct for the particular gas. Regulators for use with noncorrosive gases are usually made of brass. Special regulators made of corrosion-resistant materials are available for use with such gases as ammonia, boron trifluoride, chlorine, hydrogen chloride, hydrogen sulfide, and sulfur dioxide. Because of freeze-up and corrosion problems, regulators used with carbon dioxide gas must have special internal design features and be made of special materials. Regulators used with oxidizing agents must be cleaned specially to avoid the possibility of an explosion on contact of the gas with any reducing agent or oil left from the cleaning process.

All pressure regulators should be equipped with spring-loaded pressure-relief valves (see section 7.D.2.2.1 for further information on pressure-relief devices) to protect the low-pressure side. When used on cylinders of flammable, toxic, or otherwise hazardous gases, vent the relief valve to a laboratory chemical hood or other safe location. Avoid the use of internal-bleed-type regulators. When working with hazardous gases, installing flow-limiting devices after the regulator is recommended in order to add a level of control on the system. Remove regulators from corrosive gases immediately after use and flush with dry air or nitrogen. Bubblers of any type (e.g., mercury, oil) are not suitable for use as pressure regulators and should not be used. (For information about reducing the use of mercury in laboratories, see Chapter 5, section 5.B.8.)

7.D.3.3 Flammable Gases

Keep all sources of ignition away from cylinders of flammable gases and ensure that these cylinders will not leak. Always keep connections to piping, regulators, and other appliances tight to prevent leakage, and keep the tubing or hoses used in good condition. Perform leak checks periodically. Flash arrestors are recommended for flammable gases. Do not interchange regulators, hoses, and other appliances used with cylinders of flammable gases with similar equipment intended for use with other gases. Ground cylinders properly to prevent static electricity buildup, especially in very cold or dry environments. Separate cylinders containing flammable gases from cylinders of oxidizing gases by at least 20 ft or by a 5-ft-high fire-resistant partition with a minimum 30-minute fire rating. Store all cylinders containing flammable gases in a well-ventilated place. Never store reserve stocks of such cyl-

inders in the vicinity of cylinders containing oxidizing gases including oxygen, fluorine, and chlorine. Never store oxidizing gases near flammable liquids.

7.E WORKING WITH HIGH OR LOW PRESSURES AND TEMPERATURES

Work with hazardous chemicals at high or low pressures and high or low temperatures requires planning and special precautions. For many experiments, extremes of both pressure and temperature, such as reactions at elevated temperatures and pressures and work with cryogenic liquids and high vacuum, must be managed simultaneously. Carry out procedures at high or low pressures with protection against explosion or implosion by appropriate equipment selection and the use of safety shields. Provide appropriate temperature control and interlocks so that heating or cooling baths cannot exceed the desired limits even if the equipment fails. Take care to select and use glass apparatuses that can safely withstand thermal expansion or contraction at the designated pressure and temperature extremes.

7.E.1 Pressure Vessels

Perform high-pressure operations only in special chambers equipped for this purpose. Trained laboratory personnel should ensure that equipment and pressure vessels are appropriately selected, properly labeled and installed, and protected by pressure-relief and necessary control devices. Vessels must be strong enough to withstand the stresses encountered at the intended operating pressures and temperatures. The vessel material must not corrode when it is in contact with its contents. The material should not react with the process being studied, and the vessel must be of the proper size and configuration. Never carry out reactions in, or apply heat to, an apparatus that is a closed system unless it has been designed and tested to withstand the generated pressure.

Pressure-containing systems designed for use at elevated temperatures should have a positive-feedback temperature controller. Manual control using a simple variable autotransformer, such as a variac, is not good practice. The use of a backup temperature controller capable of both recording temperatures and shutting down an unattended system is strongly recommended. (See section 7.D.2, above.)

7.E.1.1 Records, Inspection, and Testing

In some localities, adherence to national codes such as the American Society of Mechanical Engineers

(ASME) Boiler and Pressure Vessel Code (ASME, 1992) is mandatory. Selection of containers, tubing, fittings, and other process equipment, along with the operational techniques and procedures, must conform to the constraints necessary for high-pressure service. The proper selection and assembly of components in a pressure system are critical safety factors. Compatibility of materials, tools used for assembly, and the reliability of connections are all key considerations.

Each pressure vessel in a laboratory should have a stamped number or fixed label plate that uniquely identifies it. Information such as the maximum allowable working pressure, allowable temperature at this pressure, material of construction, and burst diagram should be readily available. Information regarding the vessel's history should include temperature extremes it has experienced, any modifications and repairs made to the original vessel, and all inspections or test actions it has undergone. Similarly, the relieving pressure and setting data should be stamped on a metal tag attached to installed pressure-relief devices. Relief devices used on pressure regulators do not require these seals or numbers.

Test or inspect all pressure equipment periodically. The interval between tests or inspections is determined by the severity of the usage the equipment has received. Corrosive or otherwise hazardous service requires more frequent tests and inspections. Stamp inspection data on or attach it to the equipment. Pressure vessels may be subjected to nondestructive inspections such as visual inspection, penetrant inspection, acoustic emissions recording, and radiography. However, hydrostatic proof tests are necessary for final acceptance. They should be performed as infrequently as possible but before the vessel is placed into initial service, every 10 years thereafter, after a significant repair or modification, and if the vessel experiences overpressure or overtemperature.

Testing the entire apparatus with soap solution and air or nitrogen pressure to the maximum allowable working pressure of the weakest section of the assembled apparatus usually detects leaks at threaded joints, packings, and valves.

Pressure-test and leak-test final assemblies to ensure their integrity. Trained laboratory personnel are strongly advised to consult an expert on high-pressure work as they design, build, and operate a high-pressure process. Finally, exercise extreme care when disassembling pressure equipment for repair, modification, or decommissioning. (See Vignette 7.5.) Personnel should be familiar with the safe procedures for depressurizing the system, including the order in which to open valves or fittings. Wear protective equipment in

case a line or vessel that is opened contains material under pressure. Good practice is to cover the vessel or fitting being opened with a cloth or paper towel to contain any spray should the contents be unknowingly pressurized.

7.E.1.2   Pressure Reactions in Glass Equipment

Run reactions under pressure in metal equipment, not glass, if at all possible. For any reaction run on a large scale (>10 g total weight of reactants) or at a maximum pressure in excess of 690 kPa (100 psi), use only procedures involving a suitable high-pressure autoclave or shaker vessel. If glass is required because of material-of-construction concerns, use a metal reactor with a glass or Teflon liner instead of a glass vessel under pressure. Glass pressure reaction vessels are available from several vendors and are designed for use in the 0- to 200-psig range. However, it is sometimes convenient to run very small scale reactions at low pressures in a small sealed glass tube or in a thick-walled pressure bottle of the type used for catalytic hydrogenation. For any such reaction, laboratory personnel should be fully prepared for the significant possibility that the sealed vessel will burst. Gases must be vented properly and adequate precautions taken for ventilation. When using glass under pressure, assume that the glass will fail. Take every precaution to prevent injury from flying glass or from corrosive

---

VIGNETTE 7.5
Injury while working on equipment under pressure

A laboratory person connected a fresh helium cylinder to a gas manifold. When the cylinder valve was opened to pressurize the system, a slight hissing sound was heard from a fitting that connected a flexible metal hose to the manifold pressure regulator. An attempt was made to repair the leak while the system was still pressurized. On applying a wrench to the fitting, the flexible hose disconnected completely and whipped off the regulator, striking the individual on the head, cheek, and abdomen, causing bruises and lacerations.

This incident highlights the importance of deenergizing systems and processes prior to disassembly or maintenance.

or toxic reactants by using suitable shielding. Often a mesh is provided around the glassware to catch pieces should the vessel rupture. Seal centrifuge bottles with rubber stoppers clamped in place, wrapped with friction tape and shielded with a metal screen or wrapped with friction tape and surrounded by multiple layers of loose cloth toweling, and clamped behind a good safety shield. Some bottles are typically equipped with a head-containing inlet and exhaust gas valves, a pressure gauge, and a pressure-relief valve. If a pressure gauge is not used, estimate the maximum internal pressure by calculation prior to beginning the experiment to ensure that the maximum allowable pressure is not exceeded. When corrosive materials are used, use a Teflon pressure-relief valve. The preferred source of heat for such vessels is steam, because an explosion in the vicinity of an electrical heater could start a fire and an explosion in a liquid heating bath would scatter hot liquid around the area. Carry out any reaction of this type in a chemical hood, labeled with signs that indicate the contents of the reaction vessel and the explosion risk.

Fill glass tubes under pressure no more than three-quarters full. Appropriate precautions using the proper shielding must be taken for condensing materials and sealing tubes. Vacuum work can be carried out on a Schlenk line, an apparatus used for work with air-sensitive compounds, as long as proper technique is used. The sealed glass tubes can be placed either inside pieces of brass or iron pipe capped at one end with a pipe cap or in an autoclave containing some of the reaction solvent (to equalize the pressure inside and outside the glass tube). The tubes can be heated with steam or in a specially constructed, electrically heated sealed-tube furnace that is controlled thermostatically and located to direct the force of an explosion into a safe area. When the required heating has been completed, allow the sealed tube or bottle to cool to room temperature. Wrap sealed bottles and tubes of flammable materials with cloth toweling, place behind a safety shield, and cool slowly, first in an ice bath and then in dry ice. After cooling, the clamps and rubber stoppers can be removed from the bottles prior to opening. Use PPE and apparel, including shields, masks, coats, and gloves, during tube-opening operations. Note that NMR tubes are often thin-walled and should only be used for pressure reactions in a special high-pressure probe or in capillary devices.

Examine newly fabricated or repaired glass equipment for flaws and strains under polarized light. Never rely on corks, rubber stoppers, and rubber or plastic tubing as relief devices to protect glassware against excess pressure; use a liquid seal, Bunsen tube, or equivalent positive-relief device. With glass pipe, use only proper metal.

7.E.2 Liquefied Gases and Cryogenic Liquids

Cryogenic liquids are materials with boiling points of less than −73 °C (−100 °F). Liquid nitrogen, helium, argon, and slush mixtures of dry ice with isopropyl alcohol are the materials most commonly used in cold traps to condense volatile vapors from a gas or vapor stream. In addition, oxygen, hydrogen, and helium are often used in the liquid state.

The primary hazards of cryogenic liquids are frostbite, asphyxiation, fire or explosion, pressure buildup (either slowly or due to rapid conversion of the liquid to the gaseous state), and embrittlement of structural materials. The extreme cold of cryogenic liquids requires special care in their use. The vapor that boils off from a liquid can cause the same problems as the liquid itself.

The fire or explosion hazard is obvious when gases such as oxygen, hydrogen, methane, and acetylene are used. Air enriched with oxygen can greatly increase the flammability of ordinary combustible materials and may even cause some noncombustible materials to burn readily (see Chapter 6, sections 6.G.4 and 6.G.5). Oxygen-saturated wood and asphalt have been known to explode when subjected to shock. Because oxygen has a higher boiling point (−183 °C) than nitrogen (−195 °C), helium (−269 °C), or hydrogen (−252.7 °C), it can be condensed out of the atmosphere during the use of these lower boiling-point cryogenic liquids. With the use of liquid hydrogen particularly, explosive conditions may develop. (See Chapter 6, sections 6.F.3 and 6.G.2, for further discussion.)

Furnish all cylinders and equipment containing flammable or toxic liquefied gases (not vendor-owned) with a spring-loaded pressure-relief device (not a rupture disk) because of the magnitude of the potential risk that can result from activation of a nonresetting relief device. Commercial cylinders of liquefied gases are normally supplied only with a fusible-plug type of relief device, as permitted by DOT regulations. Protect pressurized containers that contain cryogenic material with multiple pressure-relief devices.

Cryogenic liquids must be stored, shipped, and handled in containers that are designed for the pressures and temperatures to which they may be subjected. Materials that are pliable under normal conditions can become brittle at low temperatures. Dewar flasks, which are used for relatively small amounts of cryogenic material, should have a dust cap over the outlet to prevent atmospheric moisture from condensing and plugging the neck of the tube. Special cylinders that are insulated and vacuum-jacketed with pressure-relief valves and rupture devices to protect the cylinder from pressure buildup are available in capacities of 100 to 200 L.

A special risk to personnel is skin or eye contact with the cryogenic liquid. Because these liquids are prone to splash owing to the large volume expansion ratio when the liquid warms up, wear eye protection, preferably chemical splash goggles and a face shield, when handling liquefied gases and other cryogenic fluids. Do not transfer liquefied gases from one container to another for the first time without the direct supervision and instruction of someone who is experienced in this operation. Transfer very slowly to minimize boiling and splashing.

Do not allow unprotected parts of the body to come in contact with uninsulated vessels or pipes that contain cryogenic liquids because extremely cold material may bond firmly to the skin and tear flesh if separation or withdrawal is attempted. Even very brief skin contact with a cryogenic liquid can cause tissue damage similar to that of frostbite or thermal burns, and prolonged contact may result in blood clots that have potentially very serious consequences. Gloves must be insulated, impervious to the fluid being handled, and loose enough to be tossed off easily in case the cryogenic liquid becomes trapped close to the skin. Never wear tight gloves when working with cryogenic liquids. Trained laboratory personnel are also encouraged to wear long sleeves when handling cryogenic fluids. Handle objects that are in contact with cryogenic liquids with tongs or potholders. Ventilate the work area well. Virtually all liquid gases present the threat of poisoning, explosion, or, at a minimum, asphyxiation in a confined space. Major harmful consequences of the use of cryogenic inert gases, including asphyxiation, are due to boiling off of the liquid and pressure buildup, which can lead to violent rupture of the container or piping.

Take special care when handling liquid hydrogen. In general, do not transfer liquid hydrogen in an air atmosphere because oxygen from the air can condense in the liquid hydrogen, presenting a possible explosion risk. Take all precautions to keep liquid oxygen from organic materials; spills on oxidizable surfaces can be hazardous.

Although nitrogen is inert, its liquefied form can be hazardous because of its cryogenic properties and because displacement of air oxygen in the vicinity can lead to asphyxiation followed by death with little warning. Fit rooms that contain appreciable quantities of liquid nitrogen (N_2) with oxygen meters and alarms. Do not store liquid nitrogen in a closed room because the oxygen content of the room can drop to unsafe levels.

Do not fill cylinders and other pressure vessels that are used for the storage and handling of liquefied gases to more than 80% capacity, to protect against possible thermal expansion of the contents and bursting of the vessel by hydrostatic pressure. If the possibility exists that the temperature outside of the cylinder may increase to greater than 30°C, a lower percentage (e.g., 60%) of capacity should be the limit.

7.E.2.1 Cold Traps and Cold Baths

Choose cold traps that are large enough and cold enough to collect the condensable vapors. Check cold traps frequently to make sure they do not become plugged with frozen material. After completion of an operation in which a cold trap has been used, isolate the trap from the source, remove from the coolant, and vent to atmospheric pressure in a safe and environmentally acceptable way. Otherwise, pressure could build up, creating a possible explosion or sucking pump oil into a vacuum system. Cold traps under continuous use, such as those used to protect inert atmosphere dryboxes, should be electrically cooled, and their temperature should be monitored with low-temperature probes.

Use appropriate gloves and a face shield to avoid contact with the skin when using cold baths. Wear dry gloves when handling dry ice. Do not lower the head into a dry ice chest because carbon dioxide is heavier than air and asphyxiation can result. The preferred liquids for dry-ice cooling baths are isopropyl alcohol or glycols; add dry ice slowly to the liquid portion of the cooling bath to avoid foaming. Avoid the common practice of using acetone–dry ice as a coolant; the alternatives are less flammable, less prone to foaming and splattering with dry ice, and less likely to damage some trap components (O-rings, plastic). Dry ice and liquefied gases used in refrigerant baths should always be open to the atmosphere. Never use them in closed systems, where they may develop uncontrolled and dangerously high pressures.

Exercise extreme caution in using liquid nitrogen as a coolant for a cold trap. If such a system is opened while the cooling bath is still in contact with the trap, *oxygen may condense from the atmosphere*. The oxygen could then combine with any organic material in the trap to create a highly explosive mixture. Therefore, do not open a system that is connected to a liquid nitrogen trap to the atmosphere until the liquid nitrogen Dewar flask or container has been removed. A liquid nitrogen–cooled trap must never be left under static vacuum. Also, if the system is closed after even a brief exposure to the atmosphere, some oxygen may have already condensed. Then, when the liquid nitrogen bath is removed or when it evaporates, the condensed gases will vaporize, producing a pressure buildup and the potential for explosion. The same explosion hazard can be created if liquid nitrogen is used to cool a flammable mixture that is exposed to air. Caution must be applied when using argon, for instance as an inert gas for Schlenk or vacuum lines, because it condenses as a colorless solid at liquid nitrogen temperature. A trap containing frozen argon is indistinguishable from one containing condensed solvent or other volatiles and presents an explosion hazard if allowed to warm without venting.

7.E.2.2 Selection of Low-Temperature Equipment

Select equipment used at low temperatures carefully because temperature can dramatically change characteristics of materials. For example, the impact strength of ordinary carbon steel is greatly reduced at low temperatures, and failure can occur at points of weakness, such as notches or abrupt changes in the material of construction. When combinations of materials are required, consider the temperature dependence of their volumes so that leaks, ruptures, and glass fractures are avoided. For example, O-rings that provide a good seal at room temperature may lose resilience and fail to function on chilled equipment.

Stainless steels containing 18% chromium and 8% nickel retain their impact resistance down to approximately −240 °C; the exact value depends heavily on special design considerations. The impact resistance of aluminum, copper, nickel, and many other nonferrous metals and alloys increases with decreasing temperatures. Use special alloy steels for liquids or gases containing hydrogen at temperatures greater than 200 °C or at pressures greater than 34.5 MPa (500 psi) because of the danger of weakening carbon steel equipment by hydrogen embrittlement.

7.E.2.3 Cryogenic Lines and Supercritical Fluids

Design liquid cryogen transfer lines so that liquid cannot be trapped in any nonvented part of the system. Experiments in supercritical fluids include high pressure and should be carried out with appropriate protective systems.

7.E.3 Vacuum Work and Apparatus

Vacuum work can result in an implosion and the possible hazards of flying glass, spattering chemicals, and fire. Set up and operate all vacuum operations with careful consideration of the potential risks. Although a vacuum distillation apparatus may appear to provide some of its own protection in the form of heating mantles and column insulation, this is not sufficient be-cause an implosion could scatter hot flammable liquid. Use an explosion shield and a full-face shield to protect laboratory personnel, and carry the procedure out in a laboratory chemical hood. Glassware under vacuum should be kept behind a shield or hood sash, taped, or resin (plastic) coated.

Equipment at reduced pressure is especially prone to rapid pressure changes, which can create large pressure differences within the apparatus. Such conditions can push liquids into unwanted locations, sometimes with undesirable consequences.

Do not allow water, solvents, and corrosive gases to be drawn into a building vacuum system. When the potential for such a problem exists, use a cold trap. Water aspirators are not recommended.

Protect mechanical vacuum pumps by cold traps, and vent their exhausts to an exhaust hood or to the outside of the building. If solvents or corrosive substances are inadvertently drawn into the pump, change the oil before any further use. (Oil contaminated with solvents, mercury, and corrosive substances must be handled as hazardous waste.) It may be desirable to maintain a log of pump usage as a guide to length of use and potential contaminants in the pump oil. Cover the belts and pulleys on vacuum pumps with guards.

(See section 7.C.2 for a discussion of vacuum pumps.)

7.E.3.1 Glass Vessels

Although glass vessels are frequently used in low-vacuum operations, evacuated glass vessels may collapse violently, either spontaneously from strain or from an accidental blow. Therefore, conduct pressure and vacuum operations in glass vessels behind adequate shielding. Check for flaws such as star cracks, scratches, and etching marks each time a vacuum apparatus is used. These flaws can often be noticed if the vessel is help up to a light. Use only round-bottom or thick-walled (e.g., Pyrex) evacuated reaction vessels specifically designed for operations at reduced pressure. Do not use glass vessels with angled or squared edges in vacuum applications unless specifically designed for the purpose (e.g., extra thick glass). Repaired glassware must be properly annealed and inspected with a cross-polarizer before vacuum or thermal stress is applied. Never evacuate thin-walled, Erlenmeyer, or round-bottom flasks larger than 1 L.

7.E.3.2 Dewar Flasks

Dewar flasks are under high vacuum and can collapse as a result of thermal shock or a very slight mechanical shock. Shield them, either by a layer of fiber-reinforced friction tape or by enclosure in a wooden or metal container, to reduce the risk of flying glass in case of collapse. Use metal Dewar flasks whenever there is a possibility of breakage.

Styrofoam buckets with lids can be a safer form of short-term storage and conveyance of cryogenic liquids than glass vacuum Dewar flasks. Although they do not insulate as well as Dewar flasks, they eliminate the danger of implosion.

7.E.3.3 Desiccators

If a glass vacuum desiccator is used, it should be made of Pyrex or similar glass, completely enclosed in a shield or wrapped with friction tape in a grid pattern that leaves the contents visible and at the same time guards against flying glass if the vessel implodes. Plastic (e.g., polycarbonate) desiccators reduce the risk of implosion and may be preferable but should also be shielded while evacuated. Solid desiccants are preferred. *Never carry or move an evacuated desiccator.* Take care opening the valve to avoid spraying the desiccator contents from the sudden inrush of gas.

7.E.3.4 Rotary Evaporators

Glass components of the rotary evaporator should be made of Pyrex or similar glass. Completely enclose in a shield to guard against flying glass should the components implode. Gradually increase rotation speed and application of vacuum to the flask whose solvent is to be evaporated.

7.E.3.5 Assembly of Vacuum Apparatus

Assemble vacuum apparatus to avoid strain. Joints must allow various sections of the apparatus to be moved if necessary without transmitting strain to the necks of the flasks. Support heavy apparatus from below as well as by the neck. Protect vacuum and Schlenk lines from overpressurization with a bubbler. Gas regulators and metal pressure-relief devices must not be relied on to protect vacuum and Schlenk lines from overpressurization. If a slight positive pressure of gas on these lines is desired, the recommended pressure range is not in excess of 1 to 2 psi. This pressure range is easily obtained by proper bubbler design (depth of the exit tubing in the bubbler liquid).

Place vacuum apparatus well back onto the bench or into the laboratory chemical hood where it will not be inadvertently hit. If the back of the vacuum setup faces the open laboratory, protect it with panels of suitably heavy transparent plastic to prevent injury to nearby personnel from flying glass in case of implosion.

7.F USING PERSONAL PROTECTIVE, SAFETY, AND EMERGENCY EQUIPMENT

As outlined in previous chapters, trained laboratory personnel must be proactive to ensure that the laboratory is a safe working environment. This attitude begins with wearing appropriate apparel and using proper eye, face, hand, and foot protection when working with hazardous materials. The institution is responsible for providing appropriate safety and emergency equipment for laboratory personnel and emergency personnel. (See also section 6.C.)

7.F.1 Personal Protective Equipment and Apparel

7.F.1.1 Protective Clothing

Clothing that leaves large areas of skin exposed is inappropriate in laboratories where hazardous chemicals are in use. Personal clothing should fully cover the body. Appropriate laboratory coats should be worn, buttoned, with the sleeves rolled down. Leave lab coats in the laboratory to minimize the possibility of spreading chemicals to public assembly, eating, or office areas, and clean them regularly. [For more information, see the OSHA Personal Protective Equipment Standard (29 CFR § 1910.132) and the OSHA Laboratory Standard (29 CFR § 1910.1450).]

Always wear protective apparel if there is a possibility that personal clothing could become contaminated or damaged with chemically hazardous material. Washable or disposable clothing worn for laboratory work with especially hazardous chemicals includes special laboratory coats and aprons, jumpsuits, special boots, shoe covers, and gauntlets, as well as splash suits. Protection from heat, moisture, cold, and radiation may be required in special situations. Among the factors to be considered in choosing protective apparel, in addition to the specific application, are resistance to physical hazards, flexibility and ease of movement, chemical and thermal resistance, and ease of cleaning or disposal.

(See also Chapter 6, section 6.C.2.6.2.)

7.F.1.2 Foot Protection

Not all types of footwear are appropriate in a laboratory where both chemical and mechanical hazards may exist. Wear substantial shoes in areas where hazardous chemicals are in use or mechanical work is being done. Clogs, perforated shoes, sandals, and cloth shoes do not provide protection against spilled chemicals. In many cases, safety shoes are advisable. Steel toes are recommended when working with heavy objects such as gas cylinders. Shoe covers may be required for work with especially hazardous materials. Shoes with conductive soles prevent buildup of static charge, and insulated soles can protect against electrical shock.

7.F.1.3 Eye and Face Protection

Appropriate eye protection is a requirement for working in a chemical laboratory. Requisite eye protection should be provided for laboratory personnel and visitors, and signs should be posted outside the laboratory indicating that eye protection is required where hazardous chemicals are in use. Ordinary prescription glasses with hardened lenses do not serve as eye protection in the laboratory. Appropriate laboratory eye and face protection includes impact goggles with splash protection (chemical splash goggles), full-face shields that also protect the throat, and specialized eye protection (i.e., protection against ultraviolet light or laser light). The following provides basic information regarding eye protection. (For more information, see Chapter 6, section 6.C.2.2.)

- Wear impact protection goggles if there is a danger of flying particles, and full-face shields with safety glasses and side shields for complete face and throat protection.
- Although safety glasses can provide satisfactory protection from flying particles, they do not fit tightly against the face and offer little protection against splashes or sprays of chemicals. Chemical splash goggles that conform to ANSI standard Z87.1-2003 are recommended when working in laboratories and, in particular, when working with hazardous chemicals that present a splash hazard, with vapors or particulates, and with corrosives. Chemical splash goggles have splash-proof sides to fully protect the eyes.
- When there is a possibility of liquid splashes, wear both a face shield and chemical splash goggles; this is especially important for work with highly corrosive liquids.
- Use full-face shields with throat protection and safety glasses with side shields when handling explosive or highly hazardous chemicals.
- Wear specialized eye protection if work in the laboratory could involve exposure to lasers, ultraviolet light, infrared light, or intense visible light.

7.F.1.4 Hand Protection

Use gloves that are appropriate to the degree and type of hazard. At all times pay special attention to the hands and any skin that is likely to be exposed to haz-

ardous chemicals. Wear proper protective gloves when handling hazardous chemicals, toxic materials, materials of unknown toxicity, corrosive materials, rough or sharp-edged objects, and very hot or very cold objects. (See Chapter 6, section 6.C.2.6.1, for more information about selecting and using gloves to prevent chemical exposure.) The following list highlights some basic information regarding protection of hands.

- Before using gloves, inspect them for integrity and check for discoloration, punctures, or tears.
- The thin latex surgical vinyl and nitrile gloves that are popular in many laboratories may not be appropriate for use with highly toxic chemicals or solvents because of their composition and thin construction.
- Cut-resistant gloves, such as Kevlar® or leather gloves, are appropriate for handling broken glassware, inserting tubing into stoppers, and handling sharp-edged objects if protection from chemicals is not needed.
- Wear insulated gloves when working with very hot or very cold materials. With cryogenic fluids the gloves must be impervious to fluid but loose enough to be tossed off easily. Absorbent gloves could freeze on the hand and intensify any exposure to liquefied gases.
- Wear insulating rubber gloves when working with electrical equipment.
- Wear a double set of gloves when a single glove material does not provide adequate protection for all the hazards encountered in a given operation. For instance, operations involving a chemical hazard and sharp objects may require the combined use of a chemical-resistant glove and a cut-resistant glove.
- Replace gloves immediately if they are contaminated or torn.
- Replace gloves periodically, depending on the frequency of use. Regular inspection of their serviceability is important. If they cannot be cleaned, dispose of contaminated gloves according to institutional procedures.
- Decontaminate or wash gloves appropriately before removing them; leave gloves in the work area, and do not touch any uncontaminated objects in the laboratory or any other area.

7.F.2 Safety and Emergency Equipment

Safety equipment, including spill control kits, safety shields, fire safety equipment, respirators, safety showers and eyewash units, and emergency equipment should be available in well-marked highly visible locations in all chemical laboratories. Fire-alarm pull stations and telephones with emergency contact numbers must be readily accessible. In addition to the standard items, other safety devices may also be needed. The laboratory supervisor is responsible for ensuring proper training and providing supplementary equipment as needed.

7.F.2.1   Spill Control Kits and Cleanup

All personnel who work in a laboratory in which hazardous substances are used should be familiar with their institution's policy regarding spill control. For non-emergency[3] spills, spill control kits may be available. Tailor them to deal with the potential risk associated with the materials being used in the laboratory. These kits are used to confine and limit the spill if such actions can be taken without risk of injury or contamination. If a spill exceeds the on-scene personnel's ability or challenges their safety, they should leave the spill site and call the emergency telephone number for help. Emergency response spill cleanup personnel should be provided with all available information about the spill.

Specific procedures for cleaning up spills vary depending on the location of the accident, the amount and physical properties of the spilled material, the degree and type of toxicity, and the training of the personnel involved. A typical cleanup kit may be a container on wheels that can be moved to the location of the spill and may include such items as instructions; absorbent pads; a spill absorbent mixture for liquid spills; a polyethylene scoop for dispensing spill absorbent, mixing it with the spill, and picking up the mixture; thick polyethylene bags for disposal of the mixture; and tags and ties for labeling the bags. Use any kit in conjunction with the appropriate PPE, and dispose of the material according to institutional requirements.
(Also see Chapter 6, section 6.C.10.5.)

7.F.2.2   Safety Shields

Use safety shields for protection against possible explosions or splash hazards. Shield laboratory equipment on all sides to avoid any line-of-sight exposure of personnel. The front sashes of laboratory chemical hoods provide shielding. Use a portable shield also when manipulations are performed, particularly with chemical hoods that have vertical-rising doors rather than horizontal-sliding sashes.

Use portable shields to protect against hazards of

---

[3]A non-emergency response is appropriate in the case of an incidental release of hazardous substances where the substance can be absorbed, neutralized, or otherwise controlled at the time of release by personnel in the immediate area or by maintenance personnel.

limited severity, such as small splashes, heat, and fires. A portable shield, however, provides no protection at the sides or back of the equipment, and if it is not sufficiently weighted for forward protection, the shield may topple toward personnel during a blast. A fixed shield that completely surrounds the experimental apparatus can afford protection against minor blast damage. Polymethyl methacrylate, polycarbonate, poly(vinyl chloride), and laminated safety plate glass are all satisfactory transparent shielding materials. Where combustion is possible, the shielding material should be nonflammable or slow burning; if it can withstand the working blast pressure, laminated safety plate glass may be the best material for such circumstances. When cost, transparency, high-tensile strength, resistance to bending loads, impact strength, shatter resistance, and burning rate are considered, poly(methyl methacrylate) offers an excellent overall combination of shielding characteristics.

Polycarbonate is much stronger and self-extinguishing after ignition but is readily attacked by organic solvents.

7.F.2.3   Fire Safety Equipment

7.F.2.3.1   Fire Extinguishers

All chemical laboratories should have carbon dioxide and dry chemical fire extinguishers. Other types of extinguishers should be available if required for the work that will be performed in the laboratory. The four types of most commonly used extinguishers are listed below, classified by the type of fire for which they are suitable. Note that multipurpose class A, B, and C extinguishers are available.

- Water extinguishers are effective against burning paper and trash (Class A fires). Do not use them for electrical, liquid, or metal fires.
- Carbon dioxide extinguishers are effective against burning liquids, such as hydrocarbons or paint, and electrical fires (Class B and C fires). They are recommended for fires involving computer equipment, delicate instruments, and optical systems because they do not damage such equipment. CO_2 extinguishers are less effective against paper and trash fires and *must not be used* against metal hydride or metal fires. Care must be taken in using these extinguishers, because the force of the compressed gas can spread burning combustibles such as papers and can tip over containers of flammable liquids.
- Dry powder extinguishers, which contain ammonium phosphate or sodium bicarbonate, are effective against burning liquids and electrical fires (Class B and C fires). They are less effective

against paper and trash or metal fires and are not recommended for fires involving delicate instruments or optical systems because of the cleanup problem. Computer equipment may need to be replaced if exposed to sufficient amounts of the dry powders. These extinguishers are generally used where large quantities of solvent may be present.

- Met-L-X extinguishers and others that have special granular formulations are effective against burning metal (Class D fires). Included in this category are fires involving magnesium, lithium, sodium, and potassium; alloys of reactive metals; and metal hydrides, metal alkyls, and other organometallics. These extinguishers are less effective against paper and trash, liquid, or electrical fires.

Every extinguisher should carry a label indicating what class or classes of fires it is effective against and the date it was last inspected. A number of other more specialized types of extinguishers are available for unusual fire hazard situations. All trained laboratory personnel are responsible for knowing the location, operation, and limitations of the fire extinguishers in the work area. The laboratory supervisor is responsible for ensuring that all personnel are aware of the locations of fire extinguishers and are trained in their use. After an extinguisher is used, designated personnel promptly recharge or replace it.

7.F.2.3.2   Heat Sensors and Smoke Detectors

Heat sensors and smoke detectors may be part of the building safety equipment. If designed into the fire alarm system, they may automatically sound an alarm and call the fire department, they may trigger an automatic extinguishing system, or they may only serve as a local alarm. Because laboratory operations may generate heat or vapors, the type and location of the detectors must be carefully evaluated to avoid frequent false alarms.

7.F.2.3.3   Fire Hoses

Fire hoses are intended for use by trained firefighters against fires too large to be handled by extinguishers and are included as safety equipment in some structures. Water has a cooling action and is effective against fires involving paper, wood, rags, and trash (Class A fires). Do not use water directly on fires that involve live electrical equipment (Class C fires) or chemicals such as alkali metals, metal hydrides, and metal alkyls that react vigorously with water (Class D fires).

Do not use streams of water against fires that involve oils or other water-insoluble flammable liquids (Class B fires). Water will not readily extinguish such

fires; instead, it can cause the fire to spread or float to adjacent areas. These possibilities are minimized by the use of a water fog. Water fogs are used extensively by the petroleum industry because of their fire-controlling and extinguishing properties. A fog can be used safely and effectively against fires that involve oil products, as well as those involving wood, rags, and rubbish.

Because of the potential risks involved in using water around chemicals, laboratory personnel should not use fire hoses except in extreme emergencies. Reserve them for trained firefighters. Extinguish clothing fires by immediately dropping to the floor and rolling; however, if a safety shower is nearby, use it to extinguish a clothing fire (as noted in section 7.F.2.5).

7.F.2.3.4   Automatic Fire-Extinguishing Systems

In areas where fire potential and the risk of injury or damage are high, automatic fire-extinguishing systems are often used. These may be of the water sprinkler, foam, carbon dioxide, halon, or dry chemical type. If an automatic fire-extinguishing system is in place, inform laboratory personnel of its presence and advise them of any safety precautions required in connection with its use (e.g., evacuation before a carbon dioxide total-flood system is activated, to avoid asphyxiation).

7.F.2.4   Respiratory Protective Equipment

The primary method for the protection of laboratory personnel from airborne contaminants is to minimize the amount of such materials entering the laboratory air. When effective engineering controls are not possible, use suitable respiratory protection after proper training. Respiratory protection may be needed in carrying out an experimental procedure, in dispensing or handling hazardous chemicals, in responding to a chemical spill or release in cleanup decontamination, or in hazardous waste handling.

Under OSHA regulations, only equipment listed and approved by the Mine Safety and Health Administration and NIOSH may be used for respiratory protection. Also under the regulations, each site on which respiratory protective equipment is used must implement a respirator program (including training and medical certification) in compliance with OSHA's Respiratory Protection Standard (29 CFR § 1910.134); see also ANSI standard Z88.2-1992, Practices for Respiratory Protection.

Respirators must fit snugly on the face to be effective. Conduct tests for a proper fit prior to selection of a respirator and verify before the user enters the area of contamination. Failure to achieve a good face-to-face piece seal (e.g., because of glasses or facial hair) can permit contaminated air to bypass the filter and create a dangerous situation for the user. For individuals with facial hair, do not use respirators requiring a face-to-face piece seal. In such cases, powered, air-purifying, or supplied-air respirators may be appropriate.

7.F.2.4.1 Types of Respirators

Several types of non-emergency respirators are available for protection in atmospheres that are not immediately dangerous to life or health but that could be detrimental after prolonged or repeated exposure. Other types of respirators are available for emergency or rescue work in hazardous atmospheres from which the wearer needs protection. Additional protection may be required if the airborne contaminant could be absorbed through or irritate the skin. For example, the possibility of eye or skin irritation may require the use of a full-body suit and a full-face mask rather than a half-face mask. For some chemicals the dose from skin absorption can exceed the dose from inhalation.

The choice of the appropriate respirator in a given situation depends on the type of contaminant and its estimated or measured concentration, known exposure limits, and hazardous properties. The degree of protection afforded by the respirator varies with the type. Six main types of respirators are currently available:

1. Chemical cartridge respirators are only for protection against particular individual (or classes of) vapors or gases as specified by the respirator manufacturer and cannot be used at concentrations of contaminants above that specified on the cartridge. Also, these respirators cannot be used if the oxygen content of the air is less than 19.5%, in atmospheres immediately dangerous to life, or for rescue or emergency work. These respirators function by trapping vapors and gases in a cartridge or canister that contains a sorbent material, with activated charcoal being the most common adsorbent. Because significant breakthrough can occur at a fraction of the canister capacity, knowledge of the potential workplace exposure and length of time the respirator will be worn is important. Replacing the cartridge after each use ensures the maximum available exposure time for each new use. Difficulty in breathing or the detection of odors indicates plugged or exhausted filters or cartridges or concentrations of contaminants higher than the absorbing capacity of the cartridge, and the user should immediately leave the area of contamination. Check and clean chemical cartridge respirators on a regular basis. Do not store new and used cartridges near chemicals because they are constantly filtering the air. Store them in sealed containers to prevent chemical contamination.

2. Organic vapor cartridges cannot be used for vapors that are not readily detectable by their odor or other irritating effects or for vapors that will generate substantial heat on reaction with the sorbent materials in the cartridge.

3. Dust, fumes, and mist respirators are used only for protection against particular, or certain classes of, dusts, fumes, and mists as specified by the manufacturer. The useful life of the filter depends on the concentration of contaminant encountered. Such particulate-removing respirators usually trap the particles in a filter composed of fibers; they are not 100% efficient. Respirators of this type are generally disposable. Examples are surgical masks and toxic-dust and nuisance-dust masks. Some masks are NIOSH-approved for more specific purposes such as protection against simple or benign dust and fibrogenic dusts and asbestos. Particulate-removing respirators afford no protection against gases or vapors and may give the user a false sense of security. They are also subject to the limitations of fit.

4. Supplied-air respirators deliver fresh air to the face piece of the respirator at a pressure high enough to cause a slight buildup relative to atmospheric pressure. As a result, the supplied air flows outward from the mask, and contaminated air from the work environment cannot readily enter the mask. This characteristic renders face-to-face piece fit less important than with other types of respirators. Fit testing is, however, required before selection and use.

5. Supplied-air respirators are effective protection against a wide range of air contaminants (gases, vapors, and particulates) and are used in oxygen-deficient atmospheres. Where concentrations of air contaminants could be immediately dangerous to life, such respirators can be used provided (a) the protection factor of the respirator is not exceeded and (b) the provisions of OSHA's Respiratory Protection Standard (which indicates the need for a safety harness and an escape system in case of compressor failure) are not violated. The air supply of this type of respirator must be kept free of contaminants (e.g., by use of oil filters and carbon monoxide absorbers). Most laboratory air is not suitable for use with these units because these units usually require the user to drag lengths of hose connected to the air supply and they have a limited range.

6. SCBA is the only type of respiratory protective equipment suitable for emergency or rescue work. Untrained personnel should not attempt to use one.

7.F.2.4.2 Procedures and Training

Each area where respirators are used should have written information available that shows the limitations, fitting methods, and inspection and cleaning procedures for each type of respirator available. Personnel who may have occasion to use respirators in their work must be thoroughly trained before initial use and annually thereafter in the fit testing, use, limitations, and care of such equipment. Training includes demonstrations and practice in wearing, adjusting, and properly fitting the equipment. OSHA regulations require that a worker be medically certified before beginning work in an area where a respirator must be worn [OSHA Respiratory Protection Standard, 29 CFR § 1910.134(b)(10)].

7.F.2.4.3 Inspections

Respirators for routine use should be inspected before each use by the user and periodically by the laboratory supervisor. Self-contained breathing apparatus should be inspected at least once a month and cleaned after each use.

7.F.2.5 Safety Showers and Eyewash Units

7.F.2.5.1 Safety Showers

Make safety showers available in areas where chemicals are handled; make sure they meet all installation and maintenance requirements (ANSI Z358.1 Emergency Eyewash and Shower Equipment; ANSI, 2004). Use them for immediate first-aid treatment of chemical splashes and for extinguishing clothing fires. All trained laboratory personnel should know where the safety showers are located in the work area and should learn how to use them. Test safety showers routinely to ensure that the valve is operable and to remove any debris in the system.

The shower should drench the subject immediately and be large enough to accommodate more than one person if necessary. It should have a quick-opening valve requiring manual closing; a downward-pull delta bar is satisfactory if long enough. Chain pulls are not advisable because they can hit the user and be difficult to grasp in an emergency. Install drains under safety showers to reduce the slip and fall risks and facility damage that is associated with flooding in a laboratory.

7.F.2.5.2 Eyewash Units

Eyewash units are required in research or instructional laboratories if substances used there present an eye hazard or if unknown hazards may be encountered. An eyewash unit provides a soft stream or spray of aerated water for an extended period (15 minutes). Locate these units close to the safety showers so that,

if necessary, the eyes can be washed while the body is showered.

7.F.2.5.3 Automatic External Defibrillators (AED)

AED owners should provide or arrange for training and refresher training. Staff that may be on-site during normal working hours and available to operate AED equipment should be selected for this training. The training should be an American Heart Association cardiopulmonary resuscitation (CPR)/AED course or a nationally acceptable equivalent. Competency is determined by the certified course instructor. Training records, including a description of the training program and refresher training schedule, should be documented. AED owners should be familiar with local laws concerning training and use of these devices.

7.F.2.6 Storage and Inspection of Emergency Equipment

Establish a central location for storage of emergency equipment. Include the following:

- SCBA (for use by trained personnel only),
- blankets for covering the injured,
- stretchers (generally best to wait for qualified medical help to move a seriously injured person),
- first-aid equipment (for unusual situations such as exposure to hydrofluoric acid or cyanide, where immediate first aid is required), and
- chemical spill cleanup kits and spill control equipment (e.g., spill pillows, booms, shoe covers, and a 55-gal drum in which to collect sorbed material). (Also consult Chapter 6, sections 6.C.10.5 and 6.C.10.6.)

Inspect safety equipment regularly (e.g., every 3 to 6 months) to ensure that it will function properly when needed. The laboratory supervisor or safety coordinator is responsible for establishing a routine inspection system and verifying that inspection records are appropriately maintained and archived as required by law.

Perform inspections of emergency equipment as follows:

- Inspect fire extinguishers for broken seals, damage, and low gauge pressure (depending on type of extinguisher). Check for proper mounting of the extinguisher and that it is readily accessible. Some types of extinguishers must be weighed annually, and periodic hydrostatic testing may be required.
- Check SCBA at least once a month and after each use to determine whether proper air pressure is

being maintained. Look for signs of deterioration or wear of rubber parts, harness, and hardware and make certain that the apparatus is clean and free of visible contamination. Periodically perform fit tests to ensure that the mask forms a good seal to an individual's face. Masks come in different sizes and cannot be considered universal or one-size-fits-all. Facial hair, especially beards, interferes with the mask seal and is not to permitted for SCBA users.

- Examine safety showers and eyewash units visually and test their mechanical function. Purge them as necessary to remove particulate matter from the water line.
- Inspect an AED periodically following the manufacturer's recommendations and procedures as well as after use and before returning to its storage location.

7.G   EMERGENCY PROCEDURES

The following general emergency procedures are recommended in the event of a fire, explosion, spill, or medical or other laboratory accident. These procedures are intended to limit injuries and minimize damage if an accident should occur. Post numbers to call in emergencies clearly at all telephones in hazard areas. Because emergency response (personnel, contact information, procedures) varies greatly from institution to institution, all laboratory personnel should be properly trained and informed of the protocols for their particular institution.

- Have someone call for emergency help, for instance, 911 or other number as designated by the institution. State clearly where the accident has occurred and its nature.
- Ascertain the safety of the situation. Do not enter or reenter an unsafe area.
- Without endangering yourself, render assistance to the personnel involved and remove them from exposure to further injury.
- Warn personnel in adjacent areas of any potential risks to their safety.

- Render immediate first aid; appropriate measures include washing under a safety shower, administration of CPR by trained personnel if heartbeat or breathing or both have stopped, and special first-aid measures.
- Put out small fires by using a portable extinguisher. Turn off nearby equipment and remove combustible materials from the area. For larger fires, contact the appropriate fire department promptly. Be aware that many organizations limit fire extinguisher use to designated trained personnel only.
- Provide emergency personnel with as much information as possible about the nature of the hazard, including a copy of the material safety data sheet (MSDS).
- In a medical emergency, laboratory personnel should remain calm and do only what is necessary to protect life.
- Summon medical help immediately.
- Do not move an injured person unless he or she is in danger of further harm.
- Keep the injured person warm. If feasible, designate one person to remain with the injured person. The injured person should be within sight, sound, or physical contact of that person at all times.
- If clothing is on fire and a safety shower is immediately available, douse the person with water; otherwise, roll the person on the floor to smother the flames.
- If harmful chemicals have been spilled on the body, remove the chemicals, usually by flooding the exposed area with the safety shower, and immediately remove any contaminated clothing.
- If a chemical has splashed into the eye, immediately wash the eyeball and the inner surface of the eyelid with water for 15 minutes. An eyewash unit should be used if available. Forcibly hold the eye open to wash thoroughly behind the eyelid.
- If possible, determine the identity of the chemical and inform the emergency medical personnel attending the injured person. Provide an MSDS for each chemical that is involved in the incident to the attending physician or emergency responders.

8 Management of Waste

8.A	INTRODUCTION	185
8.B	CHEMICAL HAZARDOUS WASTE	186
	8.B.1 In-Laboratory Hazard Reduction	186
	8.B.2 Characterization of Waste	186
	8.B.2.1 Characterization for Off-Site Management	186
	8.B.2.2 Identification Responsibilities of All Laboratory Personnel	187
	8.B.2.3 Characterization of Unknowns	187
	8.B.2.4 In-Laboratory Test Procedures for Unknowns	187
	8.B.3 Regulated Chemical Hazardous Waste	189
	8.B.3.1 Definition of Characteristic Waste	189
	8.B.3.2 Definition of Listed Waste	190
	8.B.3.3 Determining the Regulatory Status of a Waste	190
	8.B.3.4 Empty Containers	191
	8.B.4 Collection and Storage of Waste	191
	8.B.4.1 Accumulation of Waste at the Location of Generation	191
	8.B.4.2 Accumulation of Waste in a Central Area	192
	8.B.4.3 Special Regulations for Laboratories at Academic Institutions	194
	8.B.5 Disposal of Nonhazardous and Nonregulated Waste	194
	8.B.6 Treatment and Disposal Options	195
	8.B.6.1 Treatment and Recycling	195
	8.B.6.2 Disposal in the Sanitary Sewer	196
	8.B.6.3 Release to the Atmosphere	196
	8.B.6.4 Incineration	196
	8.B.7 Monitoring Waste Services, Transport, and Off-Site Treatment and Disposal	197
	8.B.7.1 Preparation for Off-Site Treatment or Disposal of Waste	198
	8.B.7.2 Choice of Transporter and Disposal Facility	198
	8.B.8 Liability Concerns	198
	8.B.9 Manifesting Hazardous Wastes	199
	8.B.10 Records and Record Keeping	199
8.C	MULTIHAZARDOUS WASTE	201
	8.C.1 Chemical–Radioactive (Mixed) Waste	202
	8.C.1.1 Minimization of Mixed Waste	203
	8.C.1.2 Safe Storage of Mixed Waste	203
	8.C.1.3 Hazard Reduction of Mixed Waste	204
	8.C.1.4 Commercial Disposal Services for Mixed Waste	204
	8.C.2 Chemical–Biological Waste	205
	8.C.2.1 Disposal of Chemically Contaminated Animal Tissue	206
	8.C.2.2 Sewer Disposal of Chemical–Biological Liquids	206

8.C.2.3   Disinfection and Autoclaving of Contaminated
          Labware   206
8.C.2.4   Disposal of Chemically Contaminated Medical
          Waste and Sharps   206
8.C.2.5   Minimization Methods for Chemical–Biological
          Waste   207
8.C.3 Radioactive–Biological Waste   207
8.C.3.1   Off-Site Management of Low-Level Radioactive
          Waste   207
8.C.3.2   Disposal of Radioactive Animal Carcasses and
          Tissue   207
8.C.3.3   Disposal of Radioactive–Biological Contaminated
          Labware   208
8.C.3.4   Sewer Disposal of Radioactive–Biological Liquids   208
8.C.4 Chemical–Radioactive–Biological Waste   208

8.D PROCEDURES FOR THE LABORATORY-SCALE TREATMENT OF
     SURPLUS AND WASTE CHEMICALS   209
8.D.1 Treatment of Acids and Bases   209
8.D.2 Treatment of Other Chemicals   209

8.A   INTRODUCTION

This chapter presents methods for the management and ultimate disposal of laboratory waste that may present chemical hazards, as well as those multihazardous wastes that contain some combination of chemical, radioactive, and biological hazards. The best strategy for managing laboratory waste aims to maximize safety and minimize environmental impact, and considers these objectives from the time of purchase. As suggested in previous chapters, there is a strategic hierarchy for managing chemicals and waste to accomplish these objectives.

The initial responsibility for implementing this hierarchy rests with trained laboratory personnel. These individuals are in the best position to know the chemical and physical properties of the materials they have used or synthesized. They are responsible for evaluating hazards, providing information necessary to make an accurate waste determination, and assisting in the evaluation of appropriate strategies for management, minimization, and disposal.

The overriding principle governing the prudent handling of laboratory waste is that *no activity should begin unless a plan for the disposal of nonhazardous and hazardous waste has been formulated*. Application of this simple principle ensures that the numerous state and federal regulatory requirements for waste handling are met and avoids unexpected difficulties, such as the generation of a form of waste (e.g., chemical, radioactive, biological) that the institution is not prepared to deal with.

There are four tiers to waste management to reduce its environmental impact: pollution prevention and source reduction; reuse or redistribution of unwanted, surplus materials; treatment, reclamation, and recycling of materials within the waste; and disposal through incineration, treatment, or land burial. The first tier of this strategic hierarchy incorporates the principles of green chemistry (see Chapter 5, section 5.B): pollution prevention and source reduction. Clearly, the best approach to laboratory waste is preventing its generation. Examples include reducing the scale of laboratory operations, reducing the formation of waste during laboratory operations, and substituting nonhazardous or less hazardous chemicals in chemical procedures.

The second strategic tier is to reuse unwanted material, redistribute surplus chemicals, and reduce hazards. Practices that implement this strategy include purchasing only what is needed, keeping chemical inventories to prevent the purchase of duplicates, and reusing excess materials. Sanitary sewer disposal of certain aqueous liquids is considered within this tier, although there are many restrictions (see section 8.B.6.2, below). At this tier it is important for laboratory personnel and environmental health and safety (EHS) staff to work cooperatively to determine the point at which the chemical becomes regulated as a waste and to ensure that requirements are met. In general terms, waste is defined as material that is discarded, is intended to be discarded, or is no longer useful for its intended purpose. This point may occur after the chemical has left the laboratory, however, if the organization has a way to reuse or redistribute the material or to use it in another procedure. Note that regulators may consider a material to be a waste if it is abandoned or is inherently wastelike (e.g., spilled materials). The determination of whether a waste is regulated as hazardous is usually made either by the institution's EHS staff or by employees of the waste disposal firm.

While the first two tiers are the preferred ways of managing chemical waste, the third strategic tier also provides safety and environmental benefits (see Section 8.B.6). If waste cannot be prevented or minimized, the organization should consider recycling chemicals that can be recovered safely from the waste and the potential for recovering energy from the waste (e.g., using solvent as a fuel). Although some laboratories distill waste solvents for reuse, these strategies are most commonly accomplished by sending the waste to a commercial recycling or reclamation facility or to a fuel blender. These strategies are described later in this chapter.

The fourth and final strategic tier for managing laboratory waste includes incineration, other treatment methods, and land disposal. Decisions within this tier consider the environmental fate of the waste and its constituents and process byproducts after it leaves the institution or firm. As with other tiers, the goal is to minimize risk to health and the environment. Land disposal is the least desirable disposal method. Although modern hazardous waste landfills can contain waste for many decades, there is always a future risk of leaking, contaminated runoff or other harmful releases to the environment. Laboratories that ship chemical waste off-site must address land disposal restrictions and treatability standards, which were put in place to discourage landfilling. Other reasons to consider environmental fate include exhibiting good environmental stewardship, teaching students and employees responsible waste management practices, and maintaining a good public image.

Of course, all laboratories wish to avoid fines and sanctions from federal, state, and local regulators. Because these potential penalties can be significant, this laboratory waste management guidance includes information on laws, regulations, rules, and ordinances that are likely to be most important to people who work in laboratories and support laboratory operations.

Please note, however, that this book is not a compliance manual, and as such, its compliance information is incomplete. In particular, this chapter focuses on federal rules that apply to laboratory waste but not the many different requirements particular to each state or locale. Chapter 11 contains additional information on the institutional regulation of laboratory waste (as well as other environmental requirements) to complement this chapter's details of laboratory waste regulation. There are many good compliance references to augment this book, and regulatory agencies should not be overlooked as another source of helpful information. Do not hesitate to seek legal advice when needed.

8.B CHEMICAL HAZARDOUS WASTE

8.B.1 In-Laboratory Hazard Reduction

The first and second tiers of waste management broadly describe methods of reducing quantity and level of hazard of laboratory waste. Hazard reduction is part of the broad theme of pollution prevention that is encouraged throughout this book. From a chemist's point of view, it is feasible to reduce the volume or the hazardous characteristics of many chemicals by conducting reactions and other hazard reduction procedures in the laboratory. It is becoming increasingly common to include such reactions as the final steps in an experimental sequence. Such procedures, as part of an academic or industrial experiment, usually involve small amounts of materials which can be handled easily and safely by laboratory personnel. Performing a hazard reduction procedure as part of an experiment has considerable economic advantages by eliminating the necessity to accumulate, handle, store, transport, and treat hazardous waste after the experiment. Furthermore, the laboratory professional who generates the potential waste often has the expertise and knowledge to safely handle the materials and perform hazard reduction procedures.

Conducting laboratory hazard reduction procedures for chemical hazardous waste makes most sense for hard-to-dispose-of waste, such as multihazardous waste, or for small or remote laboratories that generate very small quantities of easily treatable hazardous waste. In some cases, a simple procedure can make waste suitable for sewer disposal. When it can be done safely, knowledgeable laboratory staff may treat very small amounts of reactives that would otherwise pose a storage or transport risk. In some cases, waste is stabilized or encapsulated to enable safe storage and transport. More details can be found in section 8.B.6, below.

Keeping up-to-date chemical inventories can also reduce the in-laboratory hazards by simply reducing the quantity of hazardous material on-site. Ordering the smallest quantity of hazardous material required and reusing materials are also effective means of minimizing generation of hazardous waste.

Before beginning a detailed discussion of the handling of waste once it has been generated, it is important to understand the definition of waste, how it is characterized, and the regulations that govern it.

8.B.2 Characterization of Waste

Waste must be categorized as to its identity, constituents, and hazards so that it may be safely handled and managed. Categorization is necessary to determine a waste's regulatory status, hazardous waste ID number, and treatability group, and to determine its proper U.S. Department of Transportation (DOT) shipping name, and to meet other transport, treatment, and disposal requirements.

The great variety of laboratory waste makes waste categorization challenging. Transport and waste regulations are written for commercially available high-volume chemicals, which may make it difficult to categorize some laboratory chemicals, such as experimental or newly synthesized materials. Categorization procedures must account for the common laboratory waste management practices of placing small containers of waste chemicals into a larger overpack drum, and combining of many solvents and solutes into a single drum of flammable liquids.

There are several acceptable information sources for waste characterization, including the identity of the source or raw materials, in-laboratory test procedures (such as those described below), and analysis by an environmental laboratory. *Generator knowledge* can be used for waste characterization, such as the knowledge of waste characteristics and constituents by laboratory personnel who conducted the process, procedure, or experiment.

8.B.2.1 Characterization for Off-Site Management

When waste is to be shipped off-site for recycling, reclamation, treatment, or disposal, the waste characterization information needed depends on the waste management facility's requirements and its permit. Analytical methods have been established by the U.S. Environmental Protection Agency (EPA), and environmental laboratories that use EPA methods are often certified or accredited. Most of these methods are for commercially available chemicals, and so approved analytical procedures may not be available for some laboratory chemicals. It is important to work with your waste disposal firm to determine how laboratory waste is to be categorized. To avoid redundant analysis for recurring waste streams (e.g., chlorinated solvents, labpacks of organic solids), waste disposal firms and off-site facilities often establish a waste-stream profile. In some cases, detailed analytical information is not necessary if waste containers fall within the profile's hazard classification.

8.B.2.2 Identification Responsibilities of All Laboratory Personnel

Because proper management and disposal of laboratory waste requires information about its properties, it is very important that laboratory personnel accurately and completely identify and clearly label all chemical and waste containers in their laboratory, as well as maintain the integrity of source material labels. It is recommended that supplementary information be kept in a separate, readily available record (e.g., laboratory information system, lab notebook), especially for very small containers or collections. In academic laboratories where student turnover is frequent, identification is particularly important for the materials used or generated. This practice is as important for small quantities as it is for large quantities.

8.B.2.3 Characterization of Unknowns

Establishing the hazardous characteristics and evaluating the potential listing of clearly identified waste is usually quite simple. Unidentified materials (unknowns) present a problem, however, because recycling, treatment, and disposal facilities need to know characteristics and hazards to manage waste safely. All chemicals must be characterized sufficiently for safe transportation off-site.

Analysis of laboratory unknowns is expensive, especially if EPA methods must be used, or the presence of a constituent must be ruled out, and handling unknowns is risky due to the possible presence of unstable, reactive, or highly toxic chemicals or byproducts. Although expensive, some waste disposal firms offer on-site services to categorize unknown laboratory waste to prepare it for shipment to their treatment facility.

8.B.2.4 In-Laboratory Test Procedures for Unknowns

When the identity of the material is not known, simple in-laboratory test procedures can be carried out to determine the hazard class into which the material should be categorized. Because the generator may be able to supply some general information, it may be beneficial to carry out the test procedures before the materials are removed from the laboratory. Perform these tests only if they can be done safely, and only if they facilitate the characterization of the waste required

by your hazardous waste disposal firm. Understand that the following test procedures are only to provide additional information, and do not meet EPA regulatory requirements for waste analysis.

In general, precisely determining the molecular structure of the unknown material is not necessary. Hazard classification usually satisfies the regulatory requirements and those of the treatment disposal facility. However, it is important to establish which analytical data are required by the disposal facility.

Trained laboratory personnel who carry out the analytical procedures should be familiar with the characteristics of the waste and any necessary precautions. Because the hazards of the materials being tested are unknown, the use of proper personal protection and safety devices such as chemical hoods and shields is imperative. Older samples are particularly dangerous because they may have changed in composition, for example, through the formation of peroxides. (See Chapter 4, section 4.D.3.2, for information on the formation and identification of peroxides. See Chapter 6, section 6.G.3, for information on testing and disposal of peroxides.)

The following information is commonly required by treatment and disposal facilities before they agree to handle unknown materials:

- physical description,
- water reactivity,
- water solubility,
- pH,
- ignitability (flammability),
- presence of oxidizer,
- presence of sulfides or cyanides,
- presence of halogens,
- presence of radioactive materials,
- presence of biohazardous materials,
- presence of toxic constituents,
- presence of polychlorinated biphenyls (PCBs), and
- presence of high-odor compounds.

The following test procedures are readily accomplished by trained laboratory personnel. The overall sequence for testing is depicted in Figure 8.1 for liquid and solid materials.

- **Physical description.** Include the state of the material (solid, liquid), the color, and the consistency (for solids) or viscosity (for liquids). For liquid materials, describe the clarity of the solution (transparent, translucent, or opaque). If an unknown material is a bi- or tri-layered liquid, describe each layer separately, giving an approximate percentage of the total for each layer. After

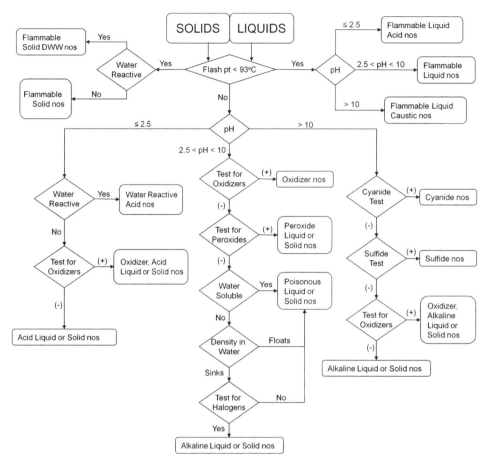

FIGURE 8.1   Flowchart for categorizing unknown chemicals for waste disposal. This decision tree shows the sequence of tests to be performed to determine the appropriate hazard category of an unknown chemical. DWW, dangerous when wet; nos, not otherwise specified. Following categorization, select a hazard reduction procedure (section 8.D) or disposal option (section 8.B.6).

taking appropriate safety precautions for handling the unknown, including the use of personal protection devices, remove a small sample for use in the following tests.

- **Water reactivity.** Carefully add a small quantity of the unknown to a few milliliters of water. Observe any changes, including heat evolution, gas evolution, and flame generation.
- **Water solubility.** Observe the solubility of the unknown in water. If it is an insoluble liquid, note whether it is less or more dense than water (i.e., does it float or sink?). Most nonhalogenated organic liquids are less dense than water.
- **pH.** Test the material with multirange pH paper. If the sample is water-soluble, test the pH of a 10% aqueous solution. Carrying out a neutralization titration may also be desirable or even required.
- **Ignitability (flammability).** Place a small sample of the material (<5 mL) in an aluminum test tray. Apply an ignition source, typically a propane torch, to the test sample for 0.5 second. If the material supports its own combustion, it is a flammable liquid with a flash point of less than 60 °C (140 °F). If the sample does not ignite, apply the ignition source again for 1 second. If the material burns, it is combustible. Combustible materials have a flash point between 60 and 93 °C (140 and 200 °F).
- **Presence of oxidizer.** Wet commercially available starch-iodide paper with 1 N hydrochloric acid, and place a small portion of the unknown on the wetted paper. A change in color of the paper to dark purple is a positive test for an oxidizer. The test can also be carried out by adding 0.1 to 0.2 g of sodium or potassium iodide to 1 mL of an acidic 10% solution of the unknown. Development of a yellow-brown color indicates an oxidizer. To test for hydroperoxides in water-insoluble organic solvents, dip the starch-iodine test paper into the solvent, and let it dry. Add a drop of water to the same section of the paper. Development of a dark color indicates the presence of hydroperoxides.
- **Presence of peroxides.** The following tests detect

most (but not all) peroxy compounds, including all hydroperoxides. Please take care. (See Chapter 4, section 4.D.3.2, and Chapter 6, section 6.G.3, for information on the hazards of peroxides.)

- ○ Peroxide test strips, which turn to an indicative color in the presence of peroxides, are available commercially. Note that these strips must be air dried until the solvent evaporates and exposed to moisture for proper operation.
- ○ Add 1 to 3 mL of the liquid to be tested to an equal volume of acetic acid, add a few drops of 5% aqueous potassium iodide solution, and shake. The appearance of a yellow to brown color indicates the presence of peroxides. Alternatively, addition of 1 mL of a freshly prepared 10% solution of potassium iodide to 10 mL of an organic liquid in a 25-mL glass cylinder produces a yellow color if peroxides are present.
- ○ Add 0.5 mL of the liquid to be tested to a mixture of 1 mL of 10% aqueous potassium iodide solution and 0.5 mL of dilute hydrochloric acid to which has been added a few drops of starch solution just prior to the test. The appearance of a blue or blue-black color within 1 minute indicates the presence of peroxides.

None of these tests should be applied to materials (such as metallic potassium) that may be contaminated with inorganic peroxides. (See Chapter 6, section 6.G.3, for more information about peroxide testing.)

- **Presence of sulfide.** Commercial test strips for the presence of sulfide are available, and their use is recommended. If the test strips are not available in the laboratory, the following test can be performed. **Warning: This test produces hazardous and odiferous vapors. Use only small quantities of solution for the test and use appropriate ventilation.** The test for inorganic sulfides is carried out only when the pH of an aqueous solution of the unknown is greater than 10. Add a few drops of concentrated hydrochloric acid to a sample of the unknown while holding a piece of commercial lead acetate paper, wet with distilled water, over the sample. Development of a brown-black color on the paper indicates generation of hydrogen sulfide.
- **Presence of cyanide.** Commercial test strips for the presence of cyanide are available, and their use is strongly recommended.
- **Presence of halogen.** Heat a piece of copper wire until red in a flame. Cool the wire in distilled or deionized water, and dip it into the unknown. Again heat the wire in the flame. The presence of halogen is indicated by a green color around the wire in the flame.

8.B.3 Regulated Chemical Hazardous Waste

An important question for planning within the laboratory is whether a waste is *regulated* as a hazardous waste, because regulated hazardous waste must be handled and disposed of in specific ways. This determination has important implications that can lead to significant differences in disposal cost. Regulatory definitions often differ from common definitions. EPA defines chemical hazardous waste under the Resource Conservation and Recovery Act of 1978 (RCRA 40 CFR Parts 260-272). EPA and RCRA establish the federal standards for chemical hazardous waste. The U.S. Nuclear Regulatory Commission (USNRC) defines radioactive waste. Hazardous biological waste is regulated less stringently under federal law, but its management is addressed in the Occupational Safety and Health Administration (OSHA) bloodborne pathogens standards and in *Biosafety in Microbiological and Biomedical Laboratories* (BMBL; HHS/CDC/NIH, 2007a).

Note that, although close attention must be paid to the regulatory definitions and procedures that govern the handling and disposal of waste, primary importance must be given to the safe and prudent handling of all laboratory wastes. Evaluate unregulated wastes and consider special handling if they pose occupational, environmental, or unknown risks.

Chemical waste that is regulated as "hazardous waste" is defined by EPA in either of two ways: (1) waste that has certain hazardous characteristics and (2) waste that is on certain lists of chemicals. The first category is based on properties of materials that should be familiar to all trained laboratory personnel. The second category comprises lists, established by EPA, of certain common hazardous chemicals and chemical wastes. These lists generally include materials that are widely used and recognized as hazardous. Chemicals are placed on these RCRA lists primarily on the basis of their toxicity. See below to determine if waste is hazardous or not.

Regardless of the regulatory definitions of hazard, understanding chemical characteristics that pose potential hazards is a fundamental part of the education and training of laboratory personnel. These characteristics may be derived from knowledge of the properties or precursors of the waste. The characteristics may also be established by specific tests cited in the regulations.

(Regulatory issues, specifically RCRA, are discussed further in Chapter 11, section 11.E.1.)

8.B.3.1 Definition of Characteristic Waste

According to federal law, the properties of chemical waste that pose hazards are as follows. Note that these definitions are unique, especially the definition of waste having the characteristic of toxicity.

1. **Ignitability.** Ignitable materials are defined as having one or more of the following characteristics:
   (a) liquids that have a flash point of less than 60 °C (140 °F) or some other characteristic that has the potential to cause fire;
   (b) materials other than liquids that are capable, under standard temperature and pressure, of causing fire by friction, adsorption of moisture, or spontaneous chemical changes and, when ignited, burn so vigorously and persistently that they create a hazard;
   (c) flammable compressed gases, including those that form flammable mixtures;
   (d) oxidizers that stimulate combustion of organic materials.
   Ignitable materials include most common organic solvents, gases such as hydrogen and hydrocarbons, and certain nitrate salts.
2. **Corrosivity.** Corrosive liquids have a pH ≤ 2 or pH ≥ 12.5 or corrode certain grades of steel. Most common laboratory acids and bases are corrosive. *Solid* corrosives, such as sodium hydroxide pellets and powders, are not legally considered by RCRA to be corrosive. However, trained laboratory personnel must recognize that such materials are extremely dangerous to skin and eyes and must be handled accordingly.
3. **Reactivity.** The reactivity classification includes substances that are unstable, react violently with water, detonate if exposed to some initiating source, or produce toxic gases. Alkali metals, peroxides and compounds that have peroxidized, and cyanide or sulfide compounds are classed as reactive.
4. **Toxicity.** Toxicity is established through the toxicity characteristic leaching procedure (TCLP) test, which measures the tendency of certain toxic materials to be leached (extracted) from the waste material under circumstances assumed to reproduce conditions of a landfill. The TCLP list includes a relatively small number of industrially important toxic chemicals and is based on the leachate concentration, above which a waste is considered hazardous. Failure to pass the TCLP results in classification of a material as a toxic waste. The TCLP test is primarily for solid materials; liquids are typically evaluated on a straight concentration basis. TCLP analyses are usually performed by environmental testing laboratories.

8.B.3.2 Definition of Listed Waste

A chemical waste that does not exhibit one of the above characteristics may still be regulated if it is a listed waste. Although EPA has developed several lists of hazardous waste, three regulatory lists are of most interest to trained laboratory personnel:

- **F list:** waste from nonspecific sources (e.g., spent solvents and process or reaction waste);
- **U list:** hazardous waste (e.g., toxic laboratory chemicals); and
- **P list:** acutely hazardous waste [e.g., highly toxic laboratory chemicals, that is, chemicals having a lethal dose (LD_{50}) of <50 mg/kg (oral, rat)].

Of the listed wastes, the most common for laboratories are the F wastes, which include many laboratory solvents. These include halogenated solvents (methylene chloride, tetrachloroethylene, and chlorinated fluorocarbons) and nonhalogenated solvents (xylene, acetone, ethyl acetate, ethyl benzene, ethyl ether, methyl isobutyl ketone, methanol, and *n*-butyl alcohol). Note that these are regulated under this listing only if they have been used (spent).

The other categories of listed waste common to laboratories are the U and P lists, which include many chemicals frequently found in laboratories. U and P lists pertain to

- waste chemicals that have not been used, because once used, the U or P listing does not apply;
- spills and spill cleanup material from U- or P-listed compounds; and
- rinsate from triple rinsing of empty containers of P compounds (described below), which is collected and handled as hazardous.

8.B.3.3 Determining the Regulatory Status of a Waste

The EPA regulations place the burden of determining whether a waste is regulated as hazardous and in what hazard classification it falls on the waste generator. Most laboratories rely on their EHS staff or their waste disposal firm to determine EPA and DOT regulatory categories (such as EPA ID numbers and transportation classes), as well as waste characterization information needed by the recycling, treatment, or disposal facility.

Testing is not necessarily required, and in most cases trained laboratory personnel are able to provide sufficient information about the waste to categorize it by general hazard categories. If the waste is not a common chemical with known characteristics, enough information about it must be supplied to satisfy the regulatory requirements and to ensure that it can be handled and disposed of safely. The information needed to characterize a waste also depends on the method of ultimate disposal. (See the discussion of disposal methods in sections 8.B.6 to 8.B.7, below.)

8.B.3.4 Empty Containers

The rules for disposal of empty hazardous waste containers, and cleaning the empty containers, are complex. A container or inner liner of a container that contained hazardous waste is "empty" under federal regulations if all waste has been removed by standard practice and no more than 2.5 cm (1 in.) of residue, or 3% by weight of containers less than 110 gal, remains. If the container held *acute* hazardous waste, triple rinsing or equivalent measures are required before the container is "empty" within the federal regulations. The rinsate must be collected and handled as acutely hazardous waste. "Empty" containers are no longer subject to federal regulation.

These are minimum standards. If empty containers are to be recycled or disposed of in the normal trash, it is recommended that labels be removed from empty hazardous waste containers, and that they be emptied as much as possible. Consider rinsing emptied containers with water or a detergent solution. Resulting rinsate from containers previously holding acutely hazardous waste are hazardous waste and must be disposed of accordingly. Rinsate resulting from cleaning of other hazardous waste containers is hazardous waste if it exhibits EPA's hazardous waste characteristics of ignitability, corrosivity, reactivity, or toxicity. It is prudent to follow these guidelines for disposing of empty containers of nonhazardous and nonregulated laboratory chemicals.

Properly cleaning containers as described above, and recycling or disposing of them with the normal trash, reduces costs as well as the volume of hazardous waste generated. Alternatively, some firms and institutions decide that it is more convenient to handle all empty chemical containers from laboratories as hazardous waste and dispose of them accordingly. This especially makes sense if the rinsate is hazardous.

8.B.4 Collection and Storage of Waste

8.B.4.1 Accumulation of Waste at the Location of Generation

Laboratory experiments generate a great variety of waste, including used disposable laboratory ware, filter media and similar materials, aqueous solutions, and hazardous and nonhazardous chemicals. As stated in the introduction to this chapter, begin no activity unless a plan for disposal of all waste, hazardous and nonhazardous, has been formulated.

The accumulation and temporary storage of waste in the laboratory is called *satellite accumulation*. The legal standards for satellite accumulation are included in this section; they are also good practices for the management of nonregulated waste. To ensure security and management oversight, chemical waste should be accumulated at or near the point of generation, and under control of laboratory personnel. Note that there is an optional alternative federal standard for the accumulation of waste within laboratories of colleges, universities, teaching hospitals, and certain nonprofit research facilities associated with colleges or universities. This is described in section 8.B.4.3, below.

Each category of waste has certain precautions and appropriate disposal methods. Below is a list of requirements and good practices for accumulating chemical waste in the laboratory:

- Collect hazardous or flammable waste solvents in an appropriate container pending transfer to the institution's central facility or satellite site for chemical waste handling or pickup by commercial disposal firm. Often, different kinds of waste are accumulated within a common container.
- Take care not to mix incompatible waste. This is a special concern with commingled waste solvents, which must be chemically compatible to ensure that heat generation, gas evolution, or another reaction does not occur. (See the discussion of commingling in section 8.B.4.2, below.) For example, waste solvents can usually be mixed for disposal, with due regard for the compatibility of the components.
- Keep wastes segregated by how they will be managed. For example, because nonhalogenated solvents are more suitable for fuel blending, many laboratories collect halogenated and nonhalogenated solvent wastes separately.
- Collect waste in dependable containers that are compatible with their contents. Keep containers closed except when adding or subtracting waste. Separate containers of incompatible materials physically or otherwise stored in a protective manner. (See Chapter 5, section 5.E.2, for storing chemicals according to their compatibility.)
- Use an appropriate container for the collection of liquid waste. Glass bottles are impervious to most chemicals but present a breakage hazard, and narrow-neck bottles are difficult to empty. The use of plastic (e.g., polyethylene jerry cans) or metal (galvanized or stainless steel) safety containers for the collection of liquid waste is strongly encouraged. Note that flame arresters in safety

containers can easily become plugged if there is sediment and may need to be cleaned occasionally. Do not store amines or corrosive materials in metal containers.

- Do not use galvanized steel safety cans for halogenated waste solvents because they tend to corrode and leak.

- As detailed below, clearly and securely label waste containers with their contents.

- Securely cap waste containers when not in immediate use. To minimize releases to the atmosphere, when a funnel is used either immediately reclose the container or use a capped waste funnel. Do not use the same funnel for containers containing incompatible waste types.

- Collect aqueous waste separately from organic solvent waste. Some laboratories may be served by a wastewater treatment facility that allows the disposal of aqueous waste to the sanitary sewer if it falls within a narrow range of acceptable waste types. Thus, solutions of nonhazardous salts or water-miscible organic materials may be acceptable in some localities. Solutions containing flammable or hazardous waste, even if water-miscible, are almost never allowed, and water-immiscible substances must never be put down the drain. Collect aqueous waste for nonsewer disposal in a container selected for resistance to corrosion. Do not use glass for aqueous waste if there is danger of freezing. Depending on the requirements of the disposal facility, adjustment of the pH of aqueous waste may be required. Such adjustment requires consideration of the possible consequences of the neutralization reaction that might take place: gas evolution, heat generation, or precipitation.

- Place solid chemical waste, such as reaction by-products or contaminated filter or chromatography media, in an appropriately labeled container to await disposal or pickup. Segregate unwanted reagents for disposal in their original containers, if possible. If original containers are used, labels should be intact and fully legible. Make every effort to use, share, or recycle unwanted reagents rather than commit them to disposal. (See Chapter 5, sections 5.D and 5.E, for a discussion of labeling alternatives.)

- Consider how to dispose of nonhazardous solid waste in laboratory trash or segregate it for recycling. Check the laboratory chemical safety summary, material safety data sheet, or other appropriate reference to determine toxicity. Consult institutional policy on nonhazardous solid waste disposal.

Trained laboratory personnel, who are most familiar with the waste and its generation, need to be actively involved in waste identification and management decisions, so that the waste is managed safely and efficiently. Often the appropriate time to decide to recycle or reuse surplus materials is shortly after the waste is generated, rather than when they are sent for disposal. Once combined with other waste materials, recycling or reuse may be more difficult. Evaluate all the costs and benefits of either decision at this time.

Safety considerations must be of primary concern. Store waste in clearly labeled containers in a designated location that does not interfere with normal laboratory operations. Ventilated storage may be appropriate. Use secondary containment such as trays, for spills or leakage from the primary containers. Many states require the use of secondary containment for wastes in satellite accumulation areas.

Federal regulations allow the indefinite accumulation of up to 55 gal of hazardous waste or 1 qt of acutely hazardous waste at or near the point of generation. However, prudence dictates that the quantities accumulated are consistent with good safety practices. Furthermore, satellite accumulation time must be consistent with the stability of the material. The general recommendation is that waste not be held for more than 1 year; some states specifically set this limit for satellite accumulation time. Within 3 days of the time that the amount of waste exceeds the 55-gal (or 1-qt) limit, manage it under the storage and accumulation time limits required at a central accumulation area, as described below.

Packaging and labeling are key parts of this initial in-laboratory operation. Label every container of hazardous waste with the material's identity and its hazard (e.g., flammable, corrosive) and the words "hazardous waste." Although the identity need not be a complete listing of all chemical constituents, knowledgeable laboratory professionals or waste handlers should be able to evaluate the hazard. However, when compatible wastes are collected in a common container, keep a list of the components to aid in later disposal decisions. Labeling must be clear and permanent. Although federal regulations do not require posting the date when satellite accumulation begins, some states do require this. The institution may suggest that this information be recorded as part of its chemical management plan.

8.B.4.2 Accumulation of Waste in a Central Area

The central accumulation area is an important component in the organization's chemical management plan. In addition to being the primary location where waste management occurs, it may also be the location where excess chemicals are held for possible redistribution. Along with the laboratory, the central accumulation area is often where hazard reduction of waste takes place through allowable on-site treatment processes.

The central accumulation area is often the appropriate place to accomplish considerable cost savings by commingling (i.e., combining) similar waste materials. This is the process where compatible wastes from various sources are combined prior to disposal. Commingling is particularly suitable for waste solvents because disposal of liquid in a 55-gal drum is generally much less expensive than disposal of the same volume of liquid in small containers. Because mixing waste requires transfer of waste between containers, the identity of all materials must be known and their compatibility understood. Although these procedures are very cost-effective, they require additional safety precautions, including the use of personal protective equipment and special and engineering controls. In addition to the facility needs described below, commingling areas require non-explosive electrical systems, grounding and bonding between floors and containers, nonsparking conductive floors and containers, and specialized ventilation systems. A walk-in fume hood is often used for both solvent commingling and the storage of commingling equipment. It is important to design the process to minimize lifting, awkward procedures that may cause injury, and the handling of heavy drums and equipment.

In some cases the disposal method and ultimate fate of the waste require that different wastes not be accumulated together. For example, if commingled waste contains significant amounts of halogenated solvents (usually >1%), disposing of the mixture can be considerably more expensive. In such cases segregation of halogenated and nonhalogenated solvents is economically favorable.

According to federal regulations, storage at a central accumulation area is normally limited to 90 days, although more time is allowed for small-quantity generators or other special situations (180 or 270 days). The count begins when the waste is brought to the central accumulation area from the laboratory or satellite accumulation area. *A special permit is required for storage beyond the above limits. Obtaining such a permit is usually too expensive and too time-consuming for most laboratory operations.* (See RCRA and Chapter 11, section 11.E.1, for more information.)

Store waste materials within a central accumulation area in appropriate and clearly labeled containers and separate them according to chemical compatibility as noted in Chapter 5, section 5.E.2. The label must include the accumulation start date and the words "hazardous waste."

Central accumulation areas should have fire suppression systems, ventilation, and dikes to avoid sewer contamination in case of a spill. Employees must be trained in correct handling of the materials as well as contingency planning and emergency response. The area should be secure, and employees should be encouraged to report any suspicious activity. Employees should know how to activate alarms, how to use fire extinguishers and other emergency response equipment, how to exit, and the location of the exterior assembly point. Be sure to document training and provide periodic refreshers.

Transportation of waste from laboratories (satellite accumulation areas) to the central accumulation area also requires specific attention to safety. Transport materials in appropriate and clearly labeled containers. Make provision for spill control in case of an accident during transportation and handling. Larger institutions are advised to have an internal tracking system to follow the movement of waste. If public roads are used to transport laboratory waste, additional DOT packaging, marking, labeling, and manifesting regulations may apply, as described below.

Final preparations for off-site disposal usually occur at the central accumulation area. Decisions on disposal options are best made here as larger quantities of waste are gathered. Identification of unknown materials not carried out within the laboratory must be completed at this point because unidentified waste cannot be shipped to a disposal site.

Your hazardous waste disposal firm is frequently involved with this phase of waste management. The decision of whether to involve a hazardous waste disposal firm, how, and when is largely based on logistics and economics. Table 8.1 describes the tasks involved in initiating off-site disposal and provides recommendations for what should be done in-house by staff and what should be contracted to professional service companies.

Laboratory waste typically leaves the generator's facility commingled in drums as compatible wastes or within a labpack. Labpacks are containers, often 55-gal drums, in which small containers of waste are packed with an absorbent material. Labpacks had been used as the principal method for disposing of laboratory waste within a landfill. However, recent landfill disposal restrictions severely limit landfill disposal of hazardous materials. Thus, the labpack has become principally a shipping container. The labpack is taken to a permitted treatment, storage, and disposal facility (TSDF), where it is either incinerated or unpacked and the contents redistributed for safe, efficient, and legal treatment and disposal.

If chemical hazardous waste is being accumulated for recycling (e.g., waste lead, solvents for redistillation), federal law requires 75% or more of these materials to be recycled or disposed of in each calendar year.

TABLE 8.1 Assignment of Tasks for Waste Handling

Task	Who Should Perform	Why?
Determine if waste is regulated as hazardous	Staff	Knowledge of the waste, liability considerations, and economics
Segregate according to hazard class	Staff	Economics and safety in storage
Determine if the material will be recycled or reused	Staff	Economics, knowledge of in-house requirements and capabilities
Commingling if appropriate	Staff	Economics, safety, liability, storage space; waste disposal firm could be consulted for advice
Determine appropriate disposal method	Staff and employees of the waste disposal firm	Waste disposal firm is aware of options for specific waste streams; staff should be involved because of liability and cost
Determine packing protocol for labpacks	Waste disposal firm	The waste disposal firm is aware of what is required by the treatment, storage, and disposal facility
Labpacking	Waste disposal firm	The waste disposal firm is generally required to do labpacking
Manifest preparation	Waste disposal firm; review by Staff	The waste disposal firm typically has more experience and will prepare the manifest; staff should be properly trained in how to review a manifest because of liability and cost considerations

This figure does not apply to surplus chemicals held for redistribution to other laboratories.

8.B.4.3 Special Regulations for Laboratories at Academic Institutions

Although laboratories are generally required to comply with the same regulations as industrial facilities, regulations promulgated in 2008 provide limited relief for academic institutions from some of the requirements associated with on-site management. When adopted by the state in which the laboratory is located, these alternative standards are available to colleges, universities, teaching hospitals, and certain nonprofit research facilities associated with colleges or universities. These standards are completely optional, at the discretion of the educational or research institution. To take advantage of these provisions, academic facilities must implement a performance-oriented laboratory management plan. This facility-specific plan must provide for seven required elements:

1. labeling standards,
2. container standards,
3. training requirements,
4. removal frequency of unwanted chemicals,
5. hazardous waste determinations,
6. laboratory cleanouts, and
7. prevention of emergencies.

The provisions allow academic facilities additional time to move waste from laboratories to central accumulation areas and additional time and flexibility in making waste determinations, and encourage laboratory cleanouts by providing relief from some time limits and generator classification provisions. However, these alternative standards require semiannual removal of all laboratory hazardous waste, whereas the standard satellite accumulation rule has no time limit for the accumulation of laboratory waste in unfilled containers smaller than 55 gal.

The academic waste rule does not apply in states with primacy over RCRA regulations until promulgated by the state. Check with your state environmental agency to see if the rule has been implemented. The entire text of the rule is available at the EPA Web site, www.epa.gov.

8.B.5 Disposal of Nonhazardous and Nonregulated Waste

Some nonregulated laboratory waste is hazardous and should be safely managed. There are more waste management options for nonregulated waste, especially with regard to hazard reduction procedures.

Some laboratories have policies that require all chemical waste to be handled as if it were regulated as hazardous. This recognizes the potential liabilities associated with misperceptions or the improper handling of nonregulated as well as regulated waste. For example, a trash hauler or landfill operator may become alarmed by a laboratory chemical container, even if it contains sucrose. Note that if different types of waste are comingled, though, then the mixture must be treated as hazardous waste, and the cost for disposal of the nonhazardous portion may increase. Also consider the possibility that a hazardous material may be improperly labeled or described as nonhazardous.

When safe and allowed by regulation, disposal of nonhazardous waste via the normal trash or sewer can substantially reduce disposal costs. Many state and local regulations restrict or prohibit the disposal of waste in municipal landfills or sewer systems, and so it is wise to check the rules and requirements of the local solid waste management authority and develop a list of materials that can be disposed of safely and legally in the normal trash. The common wastes usually not regulated as hazardous include certain salts (e.g., potassium chloride and sodium carbonate), many biochemicals, nutrients, and natural products (e.g., sugars and amino acids), and inert materials used in a laboratory (e.g., noncontaminated chromatography resins and gels). In some places, the laboratory's hazardous waste disposal firm may assist with disposal of nonregulated materials.

8.B.6 Treatment and Disposal Options

As described in the introduction to this chapter, the third tier of waste management entails reclamation and recycling of materials from the waste. These methods should be considered in conjunction with the fourth tier, disposal. Reclamation, recycling, and disposal methods for chemical hazardous waste are described in this section.

The question of what forms of treatment are allowed under federal regulations poses a dilemma for laboratory professionals. Federal regulations define treatment as "any method . . . designed to change the physical, chemical, or biological character or composition of any hazardous waste so as to neutralize such waste, or so as to recover energy or material resources from the waste, or so as to render the waste nonhazardous or less hazardous . . . ." In most cases, treatment requires a state or federal permit. The regulatory procedures and costs to obtain a permit for treatment are beyond the resources of most laboratories. Under federal law, laboratory treatment of chemical hazardous waste without a permit is allowed in the following instances:

- small-scale "treatment" that is part of a laboratory procedure, such as the last step of a chemical procedure;
- a state that allows "permit-by-rule," treatment, that is, by allowing categorical or blanket permitting of certain small-scale treatment methods;
- elementary acid-base neutralization; and
- treatment in the waste collection container (see section 8.D for regulatory information).

Of course, treatment restrictions apply only to chemical hazardous wastes that are regulated by EPA. Some biological toxins not listed by EPA can be easily

denatured without a permit by heat or an appropriate solvent. No permit is required to irretrievably mix small amounts of controlled substances (not regulated by EPA) into bulk waste flammable solvents. Because illegal waste treatment can lead to fines, it is most important that, before carrying out any processes that could be considered treatment, the responsible laboratory personnel or the institution's EHS office check with the local, state, or regional EPA to clarify its interpretation of the rules. Some states do allow small-scale treatment of waste, but many do not.

(Section 8.D, below, provides methods for small-scale treatment of common chemicals.)

To minimize costs and manage laboratory waste most efficiently, it is important to consider treatment and disposal options as early as possible, and plan ahead. For example, the method of waste collection impacts how waste will be stored, as well as its efficient transfer to a treatment or disposal facility. In addition to the hazard reduction procedures described above, laboratories utilize several treatment and disposal options because of the great variety of waste generated, and because each option (described below) has its own advantages for specific wastes, and so planning can be difficult. Although landfill disposal is not described separately below, it is often the disposal method for encapsulated waste, treatment residues, and ash from incineration. Note that disposal options change as technology and environmental concerns change. When feasible, waste minimization is always a best practice. (See Chapter 5, section 5.B, for step-by-step instructions on source reduction, and section 8.C, below, for general information on minimizing hazardous waste.)

8.B.6.1 Treatment and Recycling

There are various methods for physical and chemical treatment of hazardous wastes, as well as methods for recycling, reclamation, and recovery of valuable materials contained in the waste. These methods include neutralization, oxidation-reduction, distillation, digestion, encapsulation, and several forms of thermal treatment. While the expense and practicality of these technologies is largely based on the specific nature and volume of the material, treatment or recycling is preferable to incineration for some hazardous wastes. For example, high- and low-pH wastes may be neutralized, resulting in treatable wastewater and salts. Incineration of mercury and other toxic metals is restricted; recycling, recovery, or encapsulation is environmentally preferred. Filtration of aqueous-based wastes may also significantly decrease volumes and result in wastewaters suitable for treatment in a sewage treatment facility. Note that recycling and reclamation extend to reclamation of energy as well as materials, and flammable waste liquids from laboratory operations are almost universally consolidated and used in fuel blending operations, typically to power cement plants. These liquids may also be used as a fuel source for rotary kilns.

8.B.6.2 Disposal in the Sanitary Sewer

Disposal in the sewer system (down the drain) had been a common method of waste disposal until recent years. However, environmental concerns, the viability of publicly owned treatment works (POTW), and a changing disposal culture have changed that custom markedly. In fact, many industrial and academic laboratory facilities have completely eliminated sewer disposal. Most sewer disposal is controlled locally, and it is therefore advisable to consult with the POTW to determine what is allowed. Yet, if permitted by the sewer facility, it is often reasonable to consider disposal of some chemical waste materials in the sanitary sewer. These include substances that are water-soluble and those that do not violate the federal prohibitions on disposal of waste materials that interfere with POTW operations or pose a hazard.

Chemicals that may be permissible for sewer disposal include aqueous solutions that readily biodegrade and low-toxicity solutions of inorganic substances. When allowed by law, liquid laboratory wastes that are commonly disposed of in the sanitary sewer include spent buffer solutions, neutralized mineral acids and caustics, and very dilute aqueous solutions of water-soluble organic solvents (e.g., methanol, ethanol). After checking with authorities, some laboratories flush small amounts of water-soluble nontoxic solids into the sanitary sewer with excess water. Examples of potentially sewer-disposable solids include sodium or potassium chloride, nutrients, and other chemicals generally regarded as safe. Disposal of water-miscible flammable liquids in the sewer system is usually severely limited. Water-immiscible chemicals should never go down the drain.

Under federal, state, and local law, there are various exemptions, exclusions, effluent limits, and permitting requirements that may apply to laboratory wastewaters. For most labs, there are allowances for disposing of aqueous waste, rinsate, and certain hazardous and other laboratory wastes (within limits) via the sanitary sewer. Requirements vary by state, locale, and the individual laboratory's plumbing and sewer system, as well as other facility discharges and treatment systems that the laboratory is part of. Be aware that there are notification requirements for sewer discharges of any acute hazardous waste to a POTW (and more than 15 kg per month of other hazardous waste), and a one-time notification requirement for

discharges that fall within the federal domestic sewage exclusion.

In general, if laboratory wastes are discharged via a sanitary sewer to a POTW, follow the advice above to contact your POTW as to permitting and notification requirements and effluent limits. If not, contact your state water pollution control office to determine permitting and notification requirements and effluent limits.

Waste approved for drain disposal should be disposed of only in drains that flow to a POTW, never into a storm drain or septic system. Waste should be flushed with at least a 100-fold excess of water, and the facility's wastewater effluent should be checked periodically to ensure that concentration limits are not being exceeded.

8.B.6.3 Release to the Atmosphere

The release of vapors to the atmosphere, via, for example, open evaporation or laboratory chemical hood effluent, is not an acceptable disposal method. Apparatus for operations expected to release vapors should be equipped with appropriate trapping devices. Although laboratory emissions are not considered a major source under the Clean Air Act, deliberate disposal of materials via evaporation of vapors is strictly prohibited under RCRA.

Chemical hoods, the most common source of laboratory releases to the atmosphere, are designed as safety devices to transport vapors away from laboratory personnel, not as a routine means for volatile waste disposal. Units containing absorbent filters have been introduced into some laboratories, but have limited absorbing capacity. Redirection of hood vapors to a common trapping device can completely eliminate discharge into the atmosphere. (See Chapter 9, for more details.)

8.B.6.4 Incineration

Incineration is the most common disposal method for laboratory wastes. Incineration is normally performed in rotary kilns at high temperatures (1200–1400 °F). This technology provides for complete destruction of most organic materials and significantly reduces the volume of residual material which must be disposed of by landfill. However, it is an expensive option, generally requiring the use of significant volumes of fuel to generate the required temperatures. Also, some materials, such as mercury and mercury salts, may not be incinerated because of regulations and limitations of the destruction capability. (For information about reducing the use of mercury in laboratories, see Chapter 5, section 5.B.8.)

8.B.7 Monitoring Waste Services, Transport, and Off-Site Treatment and Disposal

The ultimate destination of waste is usually a treatment, recycling, and/or disposal facility, which is sometimes called a permitted TSDF. Here waste is treated (typically via chemical action or incineration), recycled, reclaimed, or disposed of in a landfill. Although the waste has left the generator's facility, *the generator retains the final responsibility for the long-term fate of the waste.* As explained below, it is important that the generator verify that the waste transporter and TSDF operate in a way that is safe, compliant, and environmentally sound, and minimizes long-term liability. The procedures for preparing and transporting the waste to such a facility are similar to those described above. (See section 8.B.3.)

Waste is rarely transported from a generator's site directly to the ultimate disposal facility. Because of the economics of transportation, only truckload quantities (usually at least 55- to 80-gal drums or a full tanker) are shipped directly. Most laboratory waste is transported to a transfer storage facility where it is consolidated with wastes from other generators and stored until truckload quantities are accumulated. Flammable solvents and other compatible materials are often blended to make them amenable for fuel recovery.

Generators are legally responsible for all aspects of hazardous waste management, including proper disposal, packaging, labeling, shipping, manifest preparation, recordkeeping, accumulation area operations and maintenance, accumulation time limit compliance, and contingency planning associated with hazardous waste management. However, generators may choose to handle some or all of these responsibilities using in-house personnel, or they may contract with professional waste management firms. The generator's responsibilities are summarized below:

- inspecting waste containers to ensure complete and legal labeling, and container dating when required;
- waste accumulation, and managing accumulation areas for safety, security, aisle width, and separation of incompatibles;
- inspecting waste containers to ensure that they are always closed and in good condition, including repackaging of leaking containers;
- preparation and updating of waste management and contingency plans;
- sampling and characterization of routine wastes;
- sampling and characterization of "unusual" or new wastes;
- preparation of transportation documents;

- identification of disposal sites; and
- performance of periodic waste management audits to ensure compliance with regulations.

Regardless of which activities a generator decides to conduct in-house, it is imperative that well-trained, qualified staff be available to conduct the waste management activities. It is also important that these persons be given the independence, authority, and resources to properly manage the facility's wastes and maintain regulatory compliance.

The selection of which activities to perform in-house and which services to handle through firms that specialize in waste disposal is dependent on the number, qualifications, and availability of in-house staff, organizational philosophy, and budgetary constraints. This is summarized in Table 8.1. It is very important to recognize that in the long term, it is the generator who bears the major liability to ensure proper handling and disposal of hazardous waste. Thus the choice of any outside waste disposal firm to participate in the process is extremely important.

When a generator designs or evaluates the effectiveness of its waste management program, it is important to know the types of outside services that are available and to determine if the use of such services is necessary or beneficial. Once a decision has been reached to hire a waste disposal firm for such services, it is important to know how to select, monitor, and work with such firms.

Waste disposal firms differ in the types of service they provide. Some firms furnish consulting services or directly provide transportation or disposal services. Other waste disposal firms provide both types of services, but usually specialize in one or the other.

Some waste disposal firms, referred to as brokers, generally provide few, if any, direct services. They coordinate, directly or through other specialty firms, the selection of disposal sites, acquisition of disposal approvals, and subsequent transportation and disposal. Brokers typically provide only limited direct consulting or waste management support services, which are often limited to waste sampling, packaging, preparation for shipment, and some regulatory compliance support.

Transportation services are designed to move the waste from the point of generation to the chosen TSDF. For laboratories, this typically involves transport of drums or other small containers by truck. For transport on public roads, federal regulations require laboratory waste to be contained in approved packages, such as drums approved by DOT. The packages must be marked and labeled according to DOT rules, and accompanied by a manifest, as described below. Preparers and drivers must receive DOT training. Drivers need a special license. In many cases, vehicles must be placarded. Generators depend on waste dispsosal firms and transportation services for compliance with these rules. Many transportation firms also provide waste characterization and preparation, disposal site selection, and disposal approval services. Transport services can be contracted directly by a generator or through a broker.

Waste disposal firms also include companies that operate permitted treatment facilities, disposal sites, and recycling/reclamation facilities. The appropriate TSDF should be selected by the generator, although often this decision is heavily influenced by the waste disposal firm, which usually has ties to selected TSDFs or will make recommendations. The generator should always be involved in this decision, because it can affect long-term liability as well as short-term cost.

8.B.7.1 Preparation for Off-Site Treatment or Disposal of Waste

Transportation and packaging requirements dictate how waste is prepared for shipment to a transfer, treatment, or disposal facility. Laboratory waste is often placed into DOT-compliant 1- to 55-gal overpacks. Reactives, gases, and highly incompatible and some highly toxic chemicals must be packaged separately.

Using labpacks is quite simple. As described above, small containers of compatible waste materials are placed in a larger container, usually a 30- to 55-gal drum, along with appropriate packing materials, as they are collected. When a drum is filled and ready for shipping, an inventory list of the contents of the labpack is prepared.

For certain similar and compatible liquid wastes, however, collecting containers of waste in a labpack is usually much more expensive than commingling (i.e., mixing) the wastes, partly because a 55-gal labpack only holds about 16 gal of waste in its individual containers. Most often, nonhalogenated solvents are ideally suited for commingling in bulk, because the best disposal strategies for flammable liquids are fuel blending or recycling. Commingling requires opening of containers and transferring their contents from the smaller laboratory containers to a larger drum. Furthermore, the containers should be rinsed before they are considered nonhazardous, and the rinsate must be treated as a hazardous waste.

Safety precautions and facility requirements for commingling are described above. See section 8.B.3.4, above, for information on rinsing and disposal of empty containers. Drums of commingled waste usually require an analysis or listing of their contents.

Commingling does not make sense if safe facilities are not available, or when insufficient amounts of compatible waste are generated within storage time limits.

8.B.7.2 Choice of Transporter and Disposal Facility

Because the long-term liability for the waste remains with the generator, it is imperative that the generator be thoroughly familiar with the experience and record of the transporter and TSDF. Economic factors alone should not govern choices, because the long-term consequences can be significant. The generator must obtain assurance, in terms of documentation, permits, records, insurance and liability coverage, and regulatory compliance history, that the chosen service provider is reliable.

There is often an advantage, particularly for smaller facilities, to contracting for all of the hazardous waste disposal operations. These include the packing and appropriate labeling of waste for off-site transportation and disposal, preparation of the shipping manifest, and arranging for the transporter and disposal facility. Again, *the liability remains with the generator,* and so the choice of such a waste disposal firm is critical.

In some states, Minnesota and Montana, for example, arrangements have been developed with local regulators to allow a large laboratory waste generator to handle the waste from very small laboratories such as those at small colleges and public schools. This plan results in informed assistance and cost savings for the smaller units. In Wisconsin, a statewide commercial contract that can be accessed by all state educational systems has been arranged. There is usually significant advantage to working with local and state agencies to develop acceptable plans for disposal methods that are environmentally and economically favorable for both large and small generators.

8.B.8 Liability Concerns

Generators are liable for proper disposal regardless of who is involved once waste leaves the facility. Long-term liability usually refers to Superfund liability. Superfund, the federal environmental policy initiative officially known as the Comprehensive Environmental Response, Compensation and Liability Act (CERCLA), provides EPA with broad authority to initiate the cleanup of hazardous waste sites. Originally funded primarily by taxes on chemicals and petroleum, the trust fund was fully depleted in 2003. All cleanups are now paid for by organizations found by EPA to have contributed to the contamination. Referred to as PRPs (potentially responsible parties), these organizations include four classes of parties:

- current owner or operator of the site [CERCLA § 107(a)(1)];
- owner or operator of a site at the time that disposal of a hazardous substance, pollutant, or contaminant occurred [CERCLA § 107(a)(2)];
- person who arranged for the disposal of a hazardous substance, pollutant, or contaminant at a site [CERCLA § 107(a)(3)]; and
- person who transported a hazardous substance, pollutant, or contaminant to a site; that transporter must have also selected that site for the disposal of the hazardous substances, pollutants, or contaminants [CERCLA § 107(a)(4)]. (See 42 U.S.C. § 9607.)

Sites may become Superfund sites through bankruptcy or other financial stress, illegal or improper disposal, act of nature (flooding, tornado, earthquake), or fire or explosion. For these reasons, generators should be concerned about the ultimate disposal of their wastes; many choose to pay higher initial costs to have wastes incinerated so as to reduce their long-term liability. Generators may perform other actions to ensure proper disposal, such as facility audits, credit checks, inspections, and/or regulatory compliance reviews.

8.B.9 Manifesting Hazardous Wastes

Hazardous wastes are shipped off-site using a special hazardous materials bill of lading known as a Uniform Hazardous Waste Manifest (see Figure 8.2). Transporters typically complete the manifest, but the shipper (generator) is legally responsible for its accuracy and completeness. Generators are also responsible for keeping a copy of the manifest, as well as the returned manifest that documents the waste's receipt at its destination. For this reason, it is important that generators have at least a basic understanding of the DOT and EPA requirements for proper completion of the form. Although instructions are provided with each manifest (printed on the back of the form), not all the information necessary to complete the manifest is provided.

Each container listed on the form must have a proper shipping name, hazard class, and EPA number. Shipping names are found in the Hazardous Materials Table (49 CFR § 172.101). For many chemical compounds, the name of the chemical is the proper shipping name. If the chemical name is not listed, then a generic hazard class (flammable liquid, corrosive solid, poisonous liquid, etc.) may be used. Once the proper shipping name has been identified, the columns to the right on the Hazardous Materials Table provide the hazard class, division, packaging requirements, and exemptions for the material.

EPA numbers must be provided for each line item on the manifest. These numbers are from the hazardous waste lists and characteristics associated with each waste. For instance, flammable liquids have a D001 hazardous waste ID number; corrosives are D002, and reactive wastes are D003. The toxic characteristics all have their own individual numbers. For example, arsenic wastes are D004. U- and P-listed chemicals each have their own number, beginning with either a U or a P, respectively. Should the waste carry more than one hazardous waste code, each should be listed on the manifest.

Other information on the manifest that must be provided (and completed correctly) are the generator's EPA ID number; 24-hour emergency response number; generator address; name, address, and EPA number of the transporter; name, address and EPA number of the designated TSDF; and the generator certification.

Each manifest that includes labpacked material should also include inventory sheets for each container within each labpack. These forms help to meet the regulatory requirement for description of hazardous materials for regulated transportation; they also provide the permitted TSDF with specific information regarding container contents. Other paperwork typically required for disposal includes a Land Disposal Restriction form, which provides information regarding proper disposal of each waste category on the manifest.

8.B.10 Records and Record Keeping

Records are needed both to meet regulatory requirements and to help monitor the success of the hazardous waste management program. Because the central accumulation area is usually the last place where waste is dealt with before it leaves the facility, it is often the most suitable place for ensuring that all appropriate and required records have been generated.

For regulatory purposes, the facility needs to keep records for on-site activities that include

- quantities and identification of waste generated and shipped;
- documentation of analyses of unknown materials if required;
- manifests for waste shipping as well as verification of disposal;
- inspection records, training records, contingency plans; and
- any other information required to ensure compliance to prevent long-term liability.

Although not required, a good practice is to request and keep certificates of final disposal or final disposi-

Please print or type. (Form designed for use on elite (12-pitch) typewriter.) Form Approved. OMB No. 2050-0039

UNIFORM HAZARDOUS WASTE MANIFEST

1. Generator ID Number | 2. Page 1 of | 3. Emergency Response Phone | **4. Manifest Tracking Number**

GENERATOR

5. Generator's Name and Mailing Address | Generator's Site Address (if different than mailing address)

Generator's Phone:

6. Transporter 1 Company Name | U.S. EPA ID Number

7. Transporter 2 Company Name | U.S. EPA ID Number

8. Designated Facility Name and Site Address | U.S. EPA ID Number

Facility's Phone:

9a. HM	9b. U.S. DOT Description (including Proper Shipping Name, Hazard Class, ID Number, and Packing Group (if any))	10. Containers No.	10. Containers Type	11. Total Quantity	12. Unit Wt./Vol.	13. Waste Codes
	1.					
	2.					
	3.					
	4.					

VOID

14. Special Handling Instructions and Additional Information

15. **GENERATOR'S/OFFEROR'S CERTIFICATION:** I hereby declare that the contents of this consignment are fully and accurately described above by the proper shipping name, and are classified, packaged, marked and labeled/placarded, and are in all respects in proper condition for transport according to applicable international and national governmental regulations. If export shipment and I am the Primary Exporter, I certify that the contents of this consignment conform to the terms of the attached EPA Acknowledgment of Consent.
I certify that the waste minimization statement identified in 40 CFR 262.27(a) (if I am a large quantity generator) or (b) (if I am a small quantity generator) is true.

Generator's/Offeror's Printed/Typed Name | Signature | Month Day Year

INT'L

16. International Shipments ☐ Import to U.S. ☐ Export from U.S. Port of entry/exit:
Transporter signature (for exports only): Date leaving U.S.:

TRANSPORTER

17. Transporter Acknowledgment of Receipt of Materials
Transporter 1 Printed/Typed Name | Signature | Month Day Year
Transporter 2 Printed/Typed Name | Signature | Month Day Year

DESIGNATED FACILITY

18. Discrepancy
18a. Discrepancy Indication Space ☐ Quantity ☐ Type ☐ Residue ☐ Partial Rejection ☐ Full Rejection
Manifest Reference Number:
18b. Alternate Facility (or Generator) | U.S. EPA ID Number
Facility's Phone:
18c. Signature of Alternate Facility (or Generator) | Month Day Year

19. Hazardous Waste Report Management Method Codes (i.e., codes for hazardous waste treatment, disposal, and recycling systems)
1. | 2. | 3. | 4.

20. Designated Facility Owner or Operator: Certification of receipt of hazardous materials covered by the manifest except as noted in Item 18a
Printed/Typed Name | Signature | Month Day Year

EPA Form 8700-22 (Rev. 3-05) Previous editions are obsolete.

DESIGNATED FACILITY TO DESTINATION STATE (IF REQUIRED)

FIGURE 8.2 Example of Uniform Hazardous Waste Manifest.

tion. Records of costs, internal tracking, and so forth can provide information on the success of the hazardous waste management program.

8.C MULTIHAZARDOUS WASTE

Multihazardous waste is a waste that presents any combination of chemical, radioactive, or biological hazards. This array of waste constituent hazards makes the management of multihazardous wastes difficult and complex. For example, low-level mixed waste (LLMW) is a multihazardous waste that contains both RCRA hazardous wastes that EPA regulates and low-level radioactive wastes (LLW or LLRW) that the USNRC regulates. The hazardous characteristics, treatment methods, and disposal requirements for these wastes are different and often incompatible. Other factors that further complicate the management of multihazardous wastes include a complex federal, state, and local regulatory framework; limited disposal options; and high disposal costs. Commercial treatment or disposal facilities for multihazardous waste from laboratories are scarce. There is little incentive for the development of a commercial market to treat and dispose of laboratory multihazardous waste because most of the waste that laboratories generate is unique to laboratories and small in volume. The management of multihazardous waste is particularly challenging for research laboratories where there are frequent changes in protocols, procedures, materials, and waste generating processes. These difficult and complex management issues can also make it difficult to promote and sustain prudent pollution prevention practices.

Medical, clinical, forensic, and environmental laboratories, and biomedical, biochemical, radiological, and other types of research laboratories generate multihazardous waste. Prudent management of these wastes is necessary to protect the health and safety of all laboratory personnel who handle, process, and store the waste for disposal, and to minimize the potential of harm to public health and the environment. A further objective of prudent management of multihazardous waste is to promote excellence in environmental stewardship. The Congress established a federal initiative for preventing or reducing pollution in the Pollution Prevention Act of 1990. This initiative can serve as a guide for developing prudent practices for managing multihazardous wastes.

The Pollution Prevention Act of 1990 established a national policy that emphasizes source reduction as the most desirable approach for preventing or reducing pollution. The policy created a new hierarchy for the management of hazardous wastes. The elements of that hierarchy are listed in order of priority and impor-

tance for accomplishing the objectives of the Pollution Prevention Act.

- **Source reduction.** Pollution should be prevented or reduced at the source whenever feasible.
- **Recycling.** Pollution that cannot be prevented should be recycled in an environmentally safe manner whenever feasible.
- **Treatment.** Pollution that cannot be prevented or recycled should be treated in an environmentally safe manner whenever feasible.
- **Disposal.** Disposal or other release into the environment should be employed only as a last resort and should be conducted in an environmentally safe manner.

Federal agencies are required to promote programs to advance this policy within their agencies and nationwide. The major research agencies within the federal government, and particularly the Department of Energy, National Institutes of Health (NIH), and EPA, are providing leadership in implementing the nation's pollution prevention policy, and are achieving results in source reduction. For example, NIH's low-level mixed waste (LLMW) minimization program demonstrated that a significant amount of mixed waste currently being generated can be reduced or eliminated. References found on the accompanying CD provide more detail about the source reduction and pollution prevention initiatives of those agencies, the achievements of which have encouraged academic research universities and corporate research facilities to focus their pollution prevention programs on source reduction. EHS programs at these institutions often share information on their Web sites regarding source reduction, recycling, treatment, and disposal.

Prudent waste management methods include a commitment by senior management to develop and support a waste minimization program. The program development should involve experienced laboratory personnel in planning waste minimization strategies and identifying source reduction options, such as incorporating pollution prevention goals into project proposals. Training of laboratory personnel to recognize opportunities for source reduction, reviewing research proposals to ensure adoption of available source reduction strategies, improving compliance with regulatory requirements, and institutional policy are among the new management initiatives at research institutions promoting pollution prevention. Multihazardous waste requires complex attention because of its combination of hazards and regulatory controls, as detailed in the following guidelines:

Assess the risk posed by the hazardous characteristics of the waste. A primary purpose of the risk assessment is to determine which hazardous constituent of the multihazardous waste presents the greatest risk. This knowledge can help identify source reduction and treatment possibilities to reduce the risk of the waste. An assessment that determines that a waste constituent does not present a significant risk may provide an opportunity for regulatory flexibility. For example, the USNRC or state authority may allow a licensee to manage a chemical–radioactive waste as a chemical waste without regard to radioactivity when the radioactive constituent concentration is less than what the USNRC specifies for an unrestricted area.

Minimize the hazardous constituents in the waste. Consider applying the waste minimization methods specific to each hazardous constituent of the waste. This strategy could result in reducing or eliminating one hazardous constituent from the waste stream and managing the waste as a single-hazard waste. For example, the substitution of nonignitable liquid scintillation fluid (LSF) for toluene-based LSF reduces a chemical–radioactive waste to a radioactive waste.

Determine options for managing the multihazardous waste. Waste management options include recycling, laboratory methods, management at institutional waste facilities, and treatment and disposal at commercial sites. Options can vary considerably between laboratories depending upon institutional capabilities and state and local laws. It may be appropriate to manage the waste in order of risk priority, from high to low risk. Options must be compatible with all hazards, and combinations of waste management methods may be limited by their order of application. Reject any combination or sequence of methods that may create an unreasonable risk to waste handlers or the environment, or that might increase the overall risk. If an option has a clear advantage in efficiency and safety, it should have highest priority. For example, if safe facilities are available on-site, hold short-half-life radioactive waste for decay before managing it as a chemical or biological waste. The EPA Final Rule on the storage, treatment, transportation, and disposal of low-level mixed waste will allow holding the waste for longer than 90 days.

Select a single management option when possible. Some waste management methods are appropriate for more than one waste hazard. Some multihazardous waste can be disposed of safely in the sanitary sewer when allowed by the local POTW.

8.C.1 *Chemical–Radioactive (Mixed) Waste*

LLMW is the most common form of multihazardous waste generated in laboratories and the most problem-

atic for waste management. "Mixed waste" is the regulatory term for multihazardous waste that exhibits both chemical and radioactive hazards (40 CFR § 266.210). Mixed wastes are defined by EPA as "wastes that contain a chemical hazardous waste component regulated under Subtitle C of RCRA and a radioactive component consisting of source, special nuclear, or byproduct material regulated under the Atomic Energy Act of 1946 (AEA)" (EPA, 1986). The complex challenge of managing waste controlled by two federal agencies was reduced by the EPA Final Rule on the storage, treatment, transportation, and disposal of low-level mixed waste of 2001 (40 CFR Part 266, Subpart N). The rule conditionally exempts the hazardous waste constituents of LLMW from RCRA during storage and treatment. This change provides more opportunities for treatment of the chemical constituents in LLMW, enabling disposal as a single hazard constituent LLRW. State regulations may continue to inhibit laboratory and on-site minimization, storage, and treatment of mixed waste. The rule applies only to LLMW that meets the specified conditions and is generated under a single USNRC license or USNRC Agreement State license. The rule also exempts LLMW and hazardous naturally occurring or accelerator-produced radioactive materials (NARM) waste from RCRA manifest, transportation, and disposal requirements that adhere to the specified conditions. Under this conditional exemption, the waste remains subject to manifest, transport, and disposal requirements under the USNRC (or USNRC Agreement State) regulations for LLW or eligible NARM. This flexibility allows on-site storage of LLMW for periods longer than 90 days. The management opportunity to treat the hazardous waste constituents of LLMW on-site can reduce the dependence on services provided by commercial treatment and disposal facilities.

Examples of laboratory mixed waste include

- used flammable (e.g., toluene) liquid scintillation cocktails,
- phenol–chloroform mixtures from extraction of nucleic acids from radiolabeled cell components,
- aqueous solutions containing radioactive material and chloroform that occur in solutions generated by the neutralization of radioactive trichloroacetic acid solutions,
- certain gel electrophoresis waste (e.g., methanol or acetic acid containing radionuclides), and
- lead contaminated with radioactive materials.

Mixed waste produced at university, clinical, and medical research laboratories is typically a mixture of a LLRW and chemical hazardous waste. Mixed waste from nuclear and energy research laboratories can include both low- and high-level (e.g., spent nuclear fuels) radioactive materials combined with chemical hazardous waste. Common laboratory waste management methods for radioactive constituents in waste include storage for decay and indefinite on-site storage, burial at a LLRW site, incineration, and sanitary sewer disposal. Disposal options for mixed waste are usually very expensive, and for many types of mixed waste, there are no management options other than indefinite storage on-site.

8.C.1.1 Minimization of Mixed Waste

Rigorous application of waste minimization principles can often solve the problems of managing mixed waste. Such efforts are most successful when scientists and EHS staff work together to evaluate laboratory processes. A successful collaborative minimization initiative undertaken by the NIH Mixed Waste Minimization Program demonstrated that the ultraviolet peroxidation treatment of aqueous mixed waste could reduce or eliminate a large portion of the mixed waste generated in the NIH research laboratories. This treatment method degrades hazardous organic compounds in high-volume aqueous mixed waste streams. The removal efficiency for a number of volatile and semi-volatile compounds is in excess of 99.99%. The treated waste can be discharged to the sanitary sewer (Rau, 1997).

Modifying laboratory processes, improving operations, or using substitute materials are approaches that can achieve minimization of mixed waste. Examples of these approaches include the following:

- Use 2.5-mL scintillation vials ("minivials") rather than 10-mL vials. Adapters are available for scintillation counters with 10-mL vial racks.
- Count phosphorus-32 (^{32}P) without scintillation fluid by the Cerenkov method on the tritium (^{3}H) setting of a liquid scintillation counter (approximately 40% efficiency); iodine-125 (^{125}I) can be counted without scintillation fluid in a gamma counter.
- Use microscale chemistry techniques.
- Eliminate the methanol/acetic acid (chemical) and radioactive mixed hazards in gel electrophoresis work by skipping the gel-fixing step if it is not required.
- Line lead containers with disposable plastic or use alternative shielding materials to prevent lead contamination by radioactivity.
- Reduce the volume of dry waste by compaction of contaminated waste gloves, absorbent pads, and glassware.

Some simple operational improvements can help minimize mixed waste. Purchase chemicals and radioactive materials in quantities necessary for a planned experiment to avoid creating surplus materials that may end up as waste. Establish procedures that will prevent commingling radioactive waste with noncontaminated materials and trash.

Consider substituting a less-hazardous constituent for either the chemical or the radioactive source of the mixed waste. The experimenter should use the minimum activity necessary and select the radionuclide with the most appropriate decay characteristics. Examples include the following:

- Use nonignitable scintillation fluid (e.g., phenylxylylethane, linear alkylbenzenes, and diisopropylnaphthalene) instead of flammable scintillation fluid (e.g., toluene, xylene, and pseudocumene). LSF that is sold as being "biodegradable" or "sewer disposable" is more appropriately labeled as "nonignitable" because biodegradability in the sanitary sewer can vary considerably with the local treatment facility.
- Use nonradioactive substitutes such as scintillation proximity assays for phosphorus-32 (^{32}P) or sulfur-35 (^{35}S) sequencing studies or ^{3}H cation assays, and enhanced chemiluminescence as a substitute for ^{32}P and ^{35}S DNA probe labeling and Southern blot analysis.
- Substitute enriched stable isotopes for radionuclides in some cases. Mass spectrometry (MS) techniques, such as inductively coupled plasma-MS, are beginning to rival the sensitivity of some counting methods. Examples include use of oxygen-18 (^{18}O) and deuterium (^{2}H) with mass spectrometry detection as substitutes for oxygen-19 (^{19}O) and ^{3}H.
- Substitution of shorter-half-life radionuclides such as ^{32}P ($t_{1/2}$ = 14 days) for phosphorus-33 (^{33}P) ($t_{1/2}$ = 25 days) in orthophosphate studies, or ^{33}P or ^{32}P for ^{35}S ($t_{1/2}$ = 87 days) in nucleotides and deoxynucleotides. In many uses, iodine-131 (^{131}I) ($t_{1/2}$ = 8 days) can be substituted for ^{125}I ($t_{1/2}$ = 60 days). Additional exposure precautions may be required.

8.C.1.2 Safe Storage of Mixed Waste

Store waste containing short-half-life radionuclides for decay prior to subsequent waste management processing and disposal. On-site decay-in-storage of LLW is very efficient and minimizes handling and transportation risks. Most institutions designate a room or facility equipped with good ventilation, effluent trapping, and fire suppression to contain and manage on-site decay-in-storage. Laboratory decay-in-storage space can provide safe storage for low-risk mixed wastes. Laboratory storage is not appropriate for storage of putrescent or reactive materials.

The specific USNRC requirements for decay-in-storage of radioactive waste are usually detailed in the institution's license. Decay-in-storage is usually limited to half-lives of less than 65 days. When the short-half-life radionuclides have decayed to background levels (the length of time depending on the initial radioactivity level but typically defined as a storage period of at least 10 half-lives), the chemical–radioactive waste can be managed as a chemical waste. After the decay period, USNRC licenses usually require that the mixed waste be surveyed for external radiation prior to releasing it to the chemical waste stream.

Storage of mixed waste for decay for more than 90 days may require the approval of the state chemical hazardous waste authority. In permitted storage facilities, storage may be limited to 1 year for some types of mixed waste. Workers should contact their institution's EHS staff or local hazardous waste agency to determine their regulatory status and requirements for storing mixed waste for decay.

8.C.1.3   Hazard Reduction of Mixed Waste

Chemical hazards can be reduced by carrying out various common chemical reactions with the waste in the laboratory. However, "treatment" of chemical hazardous waste has regulatory implications that must be considered. Many of the same considerations apply to treatment of mixed waste.

Nevertheless, there are still justifiable and legal reasons to carry out such operations in the laboratory when hazards can be minimized safely. Neutralization, oxidation, reduction, and various other chemical conversions as well as physical methods of separation and concentration can be applied prudently to many laboratory-scale mixed wastes. However, the dual character of the hazard, chemical and radioactive, requires that additional precautions be exercised. Treatment for the chemical hazard must not create a radioactivity risk for personnel or the environment. For example, vapors or aerosols from a reaction, distillation, or evaporation must not lead to escape of unsafe levels of radioactive materials into the atmosphere. Laboratory chemical hoods appropriate for such operations should be designed to trap any radioactive effluent. When mixed waste is made chemically safe for disposal into the sanitary sewer, the laboratory must ensure that the radioactivity hazard is below the standards set by the POTW. Several examples for reducing the hazard of mixed waste are described below:

- The worker can reduce the chemical hazard to a safe level and then handle the material as only a radioactive hazardous waste. Many low-level radiation materials can then be allowed to decay to a safe level, following which simple disposal is allowable.
- Some radioactive methanol–acetic acid solutions from gel electrophoresis can be recycled via distillation and the methanol reused. The solution is neutralized prior to distillation to protect the distillation equipment from corrosion and to reduce the level of methyl acetate formed during the process.
- The volume of waste phenol, chloroform, methanol, and water containing radionuclides can be reduced by separating the nonaqueous portion using a separatory funnel. After separation, the organic phase can be distilled to produce chloroform waste, which may contain levels of radioactivity below license limits for radioactive waste. The still bottom and aqueous phase must be handled as a mixed waste.
- High-performance liquid chromatography, used to purify radiolabeled proteins and lipids, can generate a waste radioactive solution of acetonitrile, water, methanol, acetic acid, and often a small amount of dimethylformamide. When the solution is distilled by rotary flash evaporation, the distillate of acetonitrile, methanol, and water is nonradioactive and can be handled as a chemical hazardous waste. The radioactive still bottom, containing 1 to 5% methanol and acetic acid, can usually be neutralized, diluted, and disposed of in the sanitary sewer.
- Aqueous solutions containing uranyl or thorium compounds can be evaporated to dryness and the residues disposed of as radioactive waste. Because of their toxicity, solidification may be necessary prior to burial at a LLRW site.
- Activated carbon, Molecular Sieves®, synthetic resins, and ion-exchange resins have been used with varying success in the separation of chemical and radioactive waste constituents. Activated carbon has been used to remove low concentrations of chloroform (less than 150 ppm) from aqueous mixed waste solutions. However, activated carbon is not suitable for high concentrations of phenol–chloroform or acetonitrile–water mixed waste. Amberlite® XAD resin, a series of Amberlite® polymeric absorbent resins used in chromatography, has been shown to be effective in removing the organic constituents from aqueous phenol, chloroform, and methanol solutions, leaving an aqueous solution that can be managed as

a radioactive waste. Chemical constituents can be separated from mixed waste by using supercritical fluid extraction (e.g., carbon dioxide), which is now available commercially.

- Surface contamination from radioactively contaminated lead can be removed by dipping the contaminated lead into a solution of 1 M hydrochloric acid. After rinsing the lead with water, it usually can be documented as nonradioactive. The acidic wash and rinse solutions contain radionuclides and lead and must be handled as mixed waste. However, decontaminating the lead results in a smaller mass of mixed waste and allows the decontaminated lead to be reused or recycled. Commercial rinse products are also available for this purpose.

- Incineration is advantageous as a treatment for many types of chemical–radioactive waste, especially those that contain toxic or flammable organic chemicals. Incineration can destroy oxidizable organic chemicals in the waste. To comply with radionuclide release limits, USNRC licensees need to control emissions and may need to restrict the incinerator's waste feed. Radioactive ash is typically managed as a radioactive waste. It is important to keep toxic metals (e.g., lead, mercury) out of the incinerable waste so that the ash is not chemically hazardous according to the TCLP test. On-site incineration minimizes handling and transportation risks; however, incineration of chemical waste is regulated by EPA and requires a permit, which is beyond the resources of most laboratory waste generators.

8.C.1.4   Commercial Disposal Services for Mixed Waste

Because of the great variety of laboratory mixed waste, it is often difficult to find a facility that can manage both the radioactive and the chemical hazards of the waste. In general, existing commercial disposal facilities are in business to manage mixed waste from the nuclear power industry, not waste from laboratories. Several commercial disposal facilities that accept mixed waste from off-site generators do exist in the United States. These sites have the capacity to manage LSF, halogenated organics, and other organic waste. Treatment capacity exists for stabilization, neutralization, decontamination/macroencapsulation of lead, and reduction of chromium waste.

In spite of this capacity, many types of laboratory mixed waste have no commercial repository. No commercial mixed-waste disposal facilities exist for waste contaminated with most toxic metals (such as mercury)

or for lead-contaminated oils. Commercial disposal capacity likewise does not exist for high concentrations of halogen-containing organics and other TCLP waste, such as waste that contains chloroform.

8.C.2   Chemical–Biological Waste

Medical, clinical, and biomedical research laboratories generate waste that contains potentially infectious materials and viable agents that are capable of causing human disease. Biohazardous wastes can include tissues and carcasses of experimental animals involved in infectious disease studies; cell cultures of infectious agents; contaminated sharps, gloves, gowns, glassware, and instruments; and blood and other clinical specimens. Biohazardous waste presents a hazard to persons who handle the waste within the generating facility. Waste decontamination is the treatment method to control or eliminate the exposure hazard prior to waste handling and disposal. The OSHA Occupational Exposure to Bloodborne Pathogens rule (29 CFR § 1910.1030) established federal requirements for the collection and containment of certain laboratory wastes that contain human blood or body fluids for the purpose of preventing exposure of personnel to bloodborne pathogens. This rule promotes the use of standard microbiological practices including safe practices for handling biohazardous wastes. The rule also requires the treatment of all contaminated waste from research laboratories handling human immunodeficiency virus (HIV) and hepatitis B virus (HBV) and other bloodborne pathogens by incineration or decontamination by a method known to destroy the pathogens within the waste materials. Federal regulations regarding transport and incineration may apply to the off-site management of nonlaboratory biohazardous waste, such as waste generated in medical or health care settings. Several states and local jurisdictions regulate the treatment and disposal of biohazardous wastes.

The Centers for Disease Control and Prevention (CDC) of the U.S. Department of Health and Human Services and the Animal and Plant Health Inspection Service of the U.S. Department of Agriculture promulgated rules under the Public Health Security and Bioterrorism Preparedness and Response Act of 2002 for the possession, use, and transfer of select agents and toxins. The rules require the destruction of select agents that are contaminants in any waste by validated laboratory decontamination methods, such as chemical decontamination or autoclave sterilization, before disposal.

Special procedures are required in disposing of multihazardous waste that includes both hazardous chemicals and materials contaminated with microorganisms that are potentially pathogenic. The purpose of the special procedures is to prevent the release of infectious agents to the environment. The procedures involve decontaminating the mixed hazardous chemical and biohazardous waste to eliminate the biohazardous characteristics of the waste prior to disposal. Autoclaving and chemical decontamination are the methods of choice for decontaminating biohazardous waste. Autoclaving volatile chemicals is not appropriate because this practice could cause the release of the chemical to the environment.

Disposal is most difficult for the very small amount of chemical–biological waste that is EPA-regulated as chemically hazardous or contains a chemical, such as lead, that is inappropriate for an animal or medical waste incinerator. Disposal of tissue specimens preserved in ethanol or another flammable solvent is also difficult. In most cases, storage of this waste is limited to 90 days and must be managed at an EPA-permitted chemical waste facility. However, few chemical waste facilities are prepared to handle waste that is putrescible, infectious, or biohazardous.

8.C.2.1 Disposal of Chemically Contaminated Animal Tissue

Animal carcasses and tissues that contain a toxic chemical may be the most prominent chemical–biological laboratory waste. Such waste includes biological specimens preserved in formalin and rodents that have been fed lead, mercury, or PCBs in toxicity studies. If storage of such putrescible waste is necessary, refrigeration is usually advisable. Infectious waste should be stored separately in a secure area.

Incineration, which destroys potential infectious agents, is the most appropriate disposal method for putrescible waste. Large research institutions are likely to have an on-site animal incinerator. Medical waste incineration is also available through commercial waste haulers.

(If animal or commercial incineration is unavailable, methods in section 8.C.3.3, below, may be adaptable to chemical–biological waste.)

8.C.2.2 Sewer Disposal of Chemical–Biological Liquids

Laboratories that manipulate infectious agents, blood, or body fluids may generate waste that is contaminated with these materials and toxic chemicals. In most cases, blood and body fluids that contain toxic chemicals can be disposed of safely in a sanitary sewer, which is designed to accept biological waste. Approval for such disposal should be requested from the local wastewater treatment works. Chemical concentrations

in such waste are typically low enough to be accepted by a local treatment works. OSHA recommends that a separate sink be used exclusively for disposal of human blood, body fluids, and infectious waste. It may be prudent to treat blood and body fluids with bleach (usually a 1:10 aqueous dilution of household bleach) prior to disposal in the sanitary sewer. Laboratory personnel should take care to prevent personal exposure while waste is being discharged into the sewer.

8.C.2.3 Disinfection and Autoclaving of Contaminated Labware

Contaminated labware may include cultures, stocks, petri plates, and other disposable laboratory items (e.g., gloves, pipettes, and tips). In many cases, the small quantities of infectious waste on labware can be disinfected safely with bleach or other chemical disinfectant (e.g., by soaking overnight). Once disinfected, the labware can be treated as a chemical waste. Laboratory personnel must check with the state or regional EPA office to determine if a treatment permit is required for chemical disinfection of chemical–biological waste.

Autoclaves can be used to steam-sterilize infectious waste but should be tested routinely for efficacy. Autoclaving does not require an EPA permit. Care must be taken because autoclaving of chemical–biological waste at 120 to 130 °C may result in the volatilization or release of the chemical constituent. Additional waste containment may be needed to minimize chemical releases, but it can interfere with steam penetration into the waste load and sterilization. Before autoclaving, evaluate the waste to verify that the heat and pressure of autoclaving do not create unsafe conditions.

Autoclaving waste containing flammable liquids may result in a fire or explosion. Note also that steam sterilization of waste that contains bleach may harm an autoclave. To autoclave voluminous chemical–biological waste streams, it may be appropriate to dedicate an autoclave room with ample ventilation and to restrict access.

8.C.2.4 Disposal of Chemically Contaminated Medical Waste and Sharps

Laboratories that work with human blood must adhere to OSHA's Standard for Occupational Exposure to Bloodborne Pathogens (29 CFR § 1910.1030), which requires waste containment, marking, and labeling. The OSHA standard also regulates waste disposal from laboratories that manipulate HIV or HBV. In general, such waste that has chemical contamination can be incinerated with other medical waste.

Waste hypodermic needles and other "sharps" (e.g.,

scalpels and razor blades) need to be contained in a puncture-resistant waste collection container. Sharps should be destroyed by incineration or by grinding as part of the disinfection treatment. Incineration of chemical- or drug-contaminated needles in a medical waste incinerator is appropriate if the waste is not an EPA-regulated chemical waste and if the chemical's toxicity or contamination is low. Needles and other sharps that are contaminated with toxic chemicals and infectious agents or blood can be autoclaved or disinfected on-site (see the precautions above), and then managed as a chemical waste. The waste container's biohazard symbol and markings should be defaced after autoclaving or disinfection to indicate that the waste has been sterilized. Noninfectious needles and sharps with high chemical toxicity or contamination are accepted by chemical incinerators.

Some biomedical research generates materials contaminated with blood and cytotoxic antineoplastic drugs or other highly potent drugs. Incineration of these materials as medical waste is appropriate. In some cases, chemical disinfection and treatment can be combined to destroy both infectious agents and the drug. Note that unemptied source containers of some drugs are EPA-listed hazardous waste and must be managed as a regulated chemical waste.

8.C.2.5 Minimization Methods for Chemical–Biological Waste

Waste minimization methods used for chemical waste can be used to reduce or eliminate the chemical hazard of chemical–biological waste. Some laboratories that generate biohazardous waste have replaced disposable items with reusable supplies, which are disinfected between uses.

For biological waste, waste minimization can be accomplished best through careful source separation of biological waste from other waste streams. When state guidelines for defining infectious waste do not exist, it is important for laboratories to define carefully those biological wastes that can be disposed of safely as noninfectious within the framework of the CDC and NIH guidelines (HHS/CDC/NIH, 2007a). Training workers to identify and separate biological waste will prevent its inadvertent mixing with other waste streams and normal trash.

8.C.3 Radioactive–Biological Waste

The management of radioactive–biological laboratory waste can be difficult because of limited on- and off-site disposal options. Basic principles for the management of radioactive–biological waste include the following:

- Risk associated with the waste should be assessed. It may be prudent to decontaminate highly biohazardous agents first to minimize handling risks. Appropriate containment, handling, and storage precautions should be taken prior to treatment.
- Radioactive–biological waste containing short-half-life radionuclides can be held for decay. After decay-in-storage, most USNRC licenses allow the waste to be managed as biological waste. If the waste supports the growth of an infectious agent that it contains, storage should be in a freezer to prevent the waste's infectious load from increasing.
- Refrigerated storage facilities or other preservation methods are necessary for putrescible waste.

8.C.3.1 Off-Site Management of Low-Level Radioactive Waste

Many laboratories do not have an on-site incinerator for radioactive–biological waste. Communities tend to oppose waste incinerators, and on-site incineration is prohibitively costly for some radioactive–biological waste generators. Even institutions that have incinerators must usually rely on off-site disposal for some of their radioactive waste. For radioactive putrescible waste, off-site disposal requires special packaging, storage, and transport considerations.

8.C.3.2 Disposal of Radioactive Animal Carcasses and Tissue

Waste radioactive animal carcasses and tissue generated from biomedical research typically pose no significant infectious hazard, but they are putrescible. USNRC regulations allow animal carcasses and tissue with less than 1.85 kBq/g of ^{3}H or ^{14}C to be disposed of without regard to radioactivity. Thus animal carcasses and tissue below this limit need not be managed as a radioactive–biological waste but only as a biological waste.

Animal tissue with higher levels of activity or other radionuclides must be managed as a radioactive waste. As with all putrescible waste, waste should be refrigerated, frozen, or otherwise preserved during accumulation, transport, and storage.

Although on-site incineration is the preferred method of managing radioactive animal carcasses and tissue, several alternatives exist. Alkaline digestion of animal carcasses containing ^{3}H, ^{14}C, and formaldehyde, followed by neutralization, results in an aqueous radioactive stream that can usually be disposed of in the sanitary sewer. The process uses 1 M potassium hydroxide at 300 °C and pressures up to 150 psi. Commercial units are available for this process. Radioactive animal carcasses may be accepted at a LLRW site when packed in lime.

Some institutions grind radioactive animal tissue for disposal in the sanitary sewer. USNRC requires that all such sewer-disposable waste be dispersible within the liquid effluent. Preventing contamination and exposure of waste handlers to dust or particles is an important safety measure in this operation.

Autoclaving of infectious animal carcasses is difficult because of the waste's high heat capacity and poor heat conductivity, and often unproductive because treated waste remains putrescible.

8.C.3.3 Disposal of Radioactive–Biological Contaminated Labware

Radioactive–biological contaminated labware (e.g., gloves and disposable laboratory articles) is generated from biomedical research using radioactive materials with infectious agents, blood, and body fluids. On-site incineration, autoclaving, and off-site disposal are the management options for this waste. Chemical decontamination (e.g., soaking in bleach) may be appropriate if it can be done without risking personal exposure, increasing waste volumes, or creating a waste that is difficult to handle (e.g., wet waste). After disinfection, radioactive–biological waste can be managed as radioactive waste.

Infectious waste and sharps containers that contain radionuclides can be autoclaved safely if the following precautions are satisfied:

- Monitor the air emissions of a test load to determine if the release of radioactive material is in compliance with USNRC license limits.
- Wipe-test the autoclave interior for surface contamination regularly.
- For ongoing treatment of this waste, dedicate an autoclave or autoclave room for this purpose. The room should have ample ventilation.
- Restrict access during autoclaving.
- Test the autoclave efficacy regularly using biological and chemical indicators.

Radioactive needles contaminated with infectious agents or blood should be autoclaved as described above, and then incinerated on-site or shipped to a LLRW site. To prevent injuries, it is important that hypodermic needles and other sharps be kept in waste containers that are puncture-resistant, leakproof, and closable from the point of discard through ultimate disposal. To prevent generation of radioactive aerosols, destruction of needles by grinding or a similar means is not recommended.

8.C.3.4 Sewer Disposal of Radioactive–Biological Liquids

Radioactive blood, body fluids, and other sewer-compatible liquids may be disposed of in the sanitary sewer if quantities are within USNRC license and treatment work limits. Precautions must be taken to prevent exposure of waste handlers. OSHA recommends that disposal of human blood and body fluids be done in a dedicated sink.

8.C.4 Chemical–Radioactive–Biological Waste

Chemical–radioactive–biological laboratory waste is the most difficult multihazardous waste to manage. The strategies for managing the various other types of multihazardous waste described above are generally applicable to chemical–radioactive–biological waste. For example, toxicological research sometimes generates animal tissue that contains a radioactively labeled toxic chemical. However, the chemical toxicity of such waste is commonly inconsequential, both legally and in relation to the waste's other characteristics. It could be appropriate to dispose of such animal tissue as a radioactive–biological waste, without regard to its low toxic chemical content.

Reduction or elimination of one of the waste hazards through waste management methods is often an efficient first step. Decay-in-storage is a simple, low-cost way to reduce the radioactivity hazard of a waste with short-lived radionuclides. After decay, most USNRC licenses allow the waste to be managed as a chemical–biological waste. Similarly, the use of a biological decontaminate can reduce a chemical–radioactive–biological waste to a chemical–radioactive waste. Autoclaves are readily available to most laboratories for destruction of infectious agents. Autoclaving or disinfection makes sense when any of the waste's characteristics (e.g., nutrient value) could support the growth of an infectious agent it contains and thus could increase the waste's risk.

Certain waste treatments reduce multiple hazards in one step. For example, incineration can destroy oxidizable organic chemicals and infectious agents, waste feed rates can be controlled to meet emission limits for volatile radionuclides, and radioactive ash can be disposed of as a dry radioactive waste. Likewise, some chemical treatment methods (e.g., those using bleach) both oxidize toxic chemicals and disinfect biological hazards. Such treatment could convert a chemical–radioactive–biological waste to a radioactive waste. The ultraviolet peroxidation treatment method may well demonstrate this capability.

8.D PROCEDURES FOR THE LABORATORY-SCALE TREATMENT OF SURPLUS AND WASTE CHEMICALS

As described above in section 8.B.5, there are many good reasons to perform in-laboratory-scale hazard reduction procedures. The pros and cons of many other waste management methods are discussed earlier in this book (see Chapter 5, section 5.B, and Chapter 6, section 6.B). The small-scale treatment, hazard reduction procedures, and deactivation of products (and byproducts) as part of the experiment plan make sense for certain wastes and certain situations at the level of the actual generator, the trained laboratory personnel. Beware that unless there is a significant reduction in risk by such action, there may be little benefit in carrying out a procedure that will simply produce another kind of waste with similar risks and challenges for disposal. Section 8.B.6 describes when federal law allows treatment of hazardous waste without permit. To recap, they are

- In certain states small-scale treatment is allowed within a laboratory, sometimes as part of a permit-by-rule allowance. Be sure to check with your state regulators.
- Treatment in an accumulation container is allowed.
- Elementary neutralization (see 8.D.1, below); the mixing of acidic and alkaline waste to form a salt solution, has long been encouraged as long as safety considerations are addressed. In particular, dilute solutions should be used to avoid rapid heat generation.
- Treatment is allowed as part of an experiment (or the last step) before it becomes a waste. Treatment of experimental byproducts assumes the material has not been declared a waste or handled in a wastelike manner. Such treatment cannot be performed anywhere other than the location where the byproduct was generated.

An explanation of the federal allowance to treat waste in an accumulation container has been published in the *Federal Register* (1986). For this allowance, the container must be kept closed except when adding or removing waste, and all standard time limits for accumulation and container management apply. Depending on the final disposition of treatment byproducts, federal Land Disposal Restrictions (40 CFR 268) treatability standards may apply.

To ensure compliance, be sure to check local and state regulations that may apply, and seek a legal review if any clarification is needed.

8.D.1 Treatment of Acids and Bases

Neutralization of acids and bases (corrosives) is generally exempt from a RCRA treatment permit. However, because the products of the reaction are often disposed of in the sanitary sewer, it is important to ensure that hazardous waste such as toxic metal ions is not a part of the effluent.

In most laboratories, both waste acids and waste bases are generated, and so it is most economical to collect them separately and then neutralize one with the other. If additional acid or base is required, sulfuric or hydrochloric acid and sodium or magnesium hydroxide, respectively, can be used.

Safety must be carefully considered before beginning any work. If the acid or base is highly concentrated, it is prudent to first dilute it with cold water (adding the acid or base to the water) to a concentration below 10%. Then the acid and base are mixed, and the additional water is slowly added when necessary to cool and dilute the neutralized product. The concentration of neutral salts disposed of in the sanitary sewer should generally be below 1%.

8.D.2 Treatment of Other Chemicals

The procedures listed below are for general use at the laboratory scale. Additional procedures can be found in the earlier editions of this book[1] and other books listed on the accompanying CD. See Tables 8.2 and 8.3 for a list of types of chemicals that have known treatment methods. Specific procedures for laboratory treatment are increasingly being included in the experimental sections of chemical journals and in publications such as *Organic Syntheses* and *Inorganic Syntheses*.

Safety must be the first consideration before undertaking any of the procedures suggested. Procedures presented in this book are intended to be carried out only by, or under the direct supervision of, a trained scientist or technologist who understands the chemistry and hazards involved. Appropriate personal protection should be used. (See Chapter 7, section 7.F, for information on protective equipment and Chapter 6 for more information about working with chemicals.) With the exception of neutralization, procedures are generally intended for application only in small quantities, that is, not more than a few hundred grams. *Because risks tend to increase exponentially with scale, larger quantities should be treated only in small batches unless a qualified chemist has demonstrated that the procedure can be scaled*

---

[1]*Prudent Practices for Disposal of Chemicals from Laboratories* (NRC, 1983); *Prudent Practices in the Laboratory: Handling and Disposal of Chemicals* (NRC, 1995).

TABLE 8.2 Classes and Functional Groupings of Organic Chemicals for Which There Are Existing Treatment Methods

Aldehydes	Hydroperoxides
Amines	Peroxides
Anhydrides	Sulfides
Halides	Thiols (mercaptans)

up safely. The generator must ensure that the procedure eliminates the regulated hazard before the products are disposed of as nonhazardous waste. In addition, if the procedure suggests disposal of the product into the sanitary sewer, this strategy must comply with local regulations.

TABLE 8.3 Classes and Functional Groupings of Inorganic Chemicals for Which There Are Existing Treatment Methods

Alkali Metals	Inorganic Peroxides and Hydroperoxides
Bromates	Iodates
Cations (precipitation to their hydroxides)	Metal azides
Chemicals in which neither the cation nor the anion presents a significant hazard	Metal catalysts
Chlorates	Metal hydrides
Chromates	Molybdates
Halides and acid halides of nonmetals	Periodates
Hypochlorites	Permanganates
Inorganic cyanides	Persulfates
Inorganic ions	Water-reactive metal halides

9 Laboratory Facilities

9.A INTRODUCTION   213

9.B GENERAL LABORATORY DESIGN CONSIDERATIONS   213
    9.B.1 Relationship Between Wet Laboratory Spaces and Other
        Spaces   213
        9.B.1.1 Relationship Between Laboratory and Office Spaces 213
    9.B.2 Open Laboratory Design   213
        9.B.2.1 Considerations for Open Laboratory Design   213
    9.B.3 Closed Laboratories and Access   214
    9.B.4 Equivalent Linear Feet of Workspace   215
    9.B.5 Laboratory Layout and Furnishing   215
        9.B.5.1 Adaptability   215
        9.B.5.2 Casework, Furnishings, and Fixtures   216
        9.B.5.3 Shared Spaces   216
        9.B.5.4 Flooring   216
        9.B.5.5 Doors, Windows, and Walls   216
    9.B.6 Noise and Vibration Issues   216
    9.B.7 Safety Equipment and Utilities   217
    9.B.8 Americans with Disability Act: Accessibility Issues Within the
        Laboratory   218
    9.B.9 Older Facilities   218

9.C LABORATORY VENTILATION   219
    9.C.1 Risk Assessment   219
    9.C.2 Laboratory Chemical Hoods   221
        9.C.2.1 Laboratory Chemical Hood Face Velocity   221
        9.C.2.2 Factors That Affect Laboratory Chemical Hood
            Performance   222
        9.C.2.3 Prevention of Intentional Release of Hazardous
            Substances into Chemical Hoods   222
        9.C.2.4 Laboratory Chemical Hood Performance Checks   222
        9.C.2.5 Housekeeping   223
        9.C.2.6 Sash Operation   223
        9.C.2.7 Constant Operation of Laboratory Chemical Hoods   224
        9.C.2.8 Testing and Verification   224
        9.C.2.9 Laboratory Chemical Hood Design and
            Construction   228
        9.C.2.10 Laboratory Chemical Hood Configurations   231
        9.C.2.11 Laboratory Chemical Hood Exhaust Treatment   234
    9.C.3 Other Local Exhaust Systems   236
        9.C.3.1 Elephant Trunks, Snorkels, or Extractors   237
        9.C.3.2 Slot Hoods   237
        9.C.3.3 Canopy Hoods   237
        9.C.3.4 Downdraft Hoods   237
        9.C.3.5 Clean Benches or Laminar Flow Hoods   239
        9.C.3.6 Ventilated Balance Enclosures   239
        9.C.3.7 Gas Cabinets   239

	9.C.3.8	Flammable-Liquid Storage Cabinets	239
	9.C.3.9	Benchtop Enclosers	240
9.C.4		General Laboratory Ventilation and Environmental Control Systems	240
	9.C.4.1	Constant Air Volume (CAV) Systems	241
	9.C.4.2	Variable Air Volume (VAV) Systems	241
9.C.5		Supply Systems	241
9.C.6		Exhaust Systems	241
	9.C.6.1	Individual Laboratory Chemical Hood Fans	241
	9.C.6.2	Manifolded (Common Header) Systems	241
	9.C.6.3	Hybrid Exhaust Systems	242
	9.C.6.4	Room Purge Systems	242
	9.C.6.5	Exhaust Stacks	242

| 9.D | ROOM PRESSURE CONTROL SYSTEMS | | 242 |

9.E	SPECIAL SYSTEMS		243
9.E.1	Gloveboxes		243
9.E.2	Clean Rooms		243
	9.E.2.1	Clean Room Classification	243
	9.E.2.2	Clean Room Protocols	244
	9.E.2.3	Laboratory Chemical Hoods and Laboratory Furniture in Clean Rooms	244
9.E.3	Environmental Rooms and Special Testing Laboratories		244
	9.E.3.1	Alternatives to Environmental Rooms	245
9.E.4	Biological Safety Cabinets and Biosafety Facilities		245
	9.E.4.1	Biosafety Cabinets	245
	9.E.4.2	Using a Biosafety Cabinet for Biological Materials	247
9.E.5	Nanoparticles and Nanomaterials		247
9.E.6	Explosion-Proof Chemical Hoods		248

| 9.F | MAINTENANCE OF VENTILATION SYSTEMS | | 248 |

9.G	VENTILATION SYSTEM MANAGEMENT PROGRAM		249
9.G.1	Design Criteria		249
9.G.2	Training Program		249
9.G.3	Inspection and Maintenance		250
9.G.4	Goals Performance Measurement		250
9.G.5	Commissioning		250

9.H	SAFETY AND SUSTAINABILITY		250
9.H.1	Low-Flow or High-Performance Laboratory Chemical Hoods		251
9.H.2	Automatic Sash Closers		251
9.H.3	Variable Air Volume (VAV) Systems with Setback Controls		252
9.H.4	Variable Air Volume Systems, Diversity Factors		252
9.H.5	Lower General Ventilation Rates		252
9.H.6	Laboratory Chemical Hood Alternatives		252
9.H.7	Retro-Commissioning		252
9.H.8	Components of Heating, Ventilation, and Air-Conditioning (HVAC)		252
9.H.9	How to Choose a Ventilation System		252

9.I	LABORATORY DECOMMISSIONING		253
9.I.1	Assessment		253
9.I.2	Removal, Cleaning, and Decontamination		253
9.I.3	Clearance		254

9.A INTRODUCTION

Trained laboratory personnel must understand how chemical laboratory facilities operate. Given the chance, they should provide input to the laboratory designers to ensure that the facilities meet the needs of the functions of the laboratory. Laboratory personnel need to understand the capabilities and limitations of the ventilation systems, environmental controls, laboratory chemical hoods, and other exhaust devices associated with such equipment and how to use them properly. To ensure safety and efficiency, the experimental work should be viewed in the context of the entire laboratory and its facilities.

9.B GENERAL LABORATORY DESIGN CONSIDERATIONS

9.B.1 Relationship Between Wet Laboratory Spaces and Other Spaces

Modern laboratories, particularly in academia, often have contiguous spaces that include wet laboratories, computer laboratories, instruments, write-up spaces, office areas, and other spaces with varying degrees of chemical use and hazards. Maintaining a positive safety culture and at the same time meeting the safety and comfort needs of laboratory personnel are challenging under these circumstances.

- Wherever possible, separate wet chemical areas or those with a higher degree of hazard from other areas with a physical barrier, such as a wall, divider, or control device. The objective is to protect the computer laboratory or otherwise low-hazard area from the risk of the higher hazard, and thus eliminate the need to use protective equipment in the low hazard area.
- When such areas cannot be physically separated, or where the risk cannot be eliminated completely, individuals working at the computer or in the write-up area need to evaluate what level of protection may be needed to control the risk of exposure to the hazards in the other areas. For example, all individuals in a computer laboratory must wear eye protection if there is a risk of eye injury from operations in a contiguous area.

9.B.1.1 Relationship Between Laboratory and Office Spaces

Almost all laboratory personnel require both laboratory and office support space. Their desire to be aware of procedures and to have a constant presence in the laboratory usually demands that office space be located near the laboratory. The need for personnel safety, evolutionary technology allowing for computer-based research and data monitoring outside of the laboratory, as well as a desire to foster better interaction between researchers has driven the offices outside the laboratory proper.

Locating all offices outside the laboratory environment allows for a safer workspace where food can be consumed, quiet work can be done, and more paper and books can be stored. Locating the office zone very close to or adjacent to the laboratory for easy access and communication is desirable.

Some laboratories have office spaces within research areas. In this design, it is best to have an obvious separation between the laboratory area and the office area using partitions or, at a minimum, aisle space, but preferably using a wall and a door that can be closed. Occupants should not have to walk through laboratory areas to exit from their office space. Visitors and students should not have to walk through laboratories to get to researchers' offices, because those persons do not have personal protective equipment (PPE). (See Vignette 9.1.)

9.B.2 Open Laboratory Design

Traditionally, laboratories were designed for individual research groups with walls separating the laboratories and support spaces. Group sizes ranged from 2 to 10 people, and most groups were completely self-contained, each with its own equipment and facilities (Figure 9.1).

Since the 1990s, the trend has been for researchers to collaborate in a cross-disciplinary nature; chemists, biologists, physicists, engineers, and computer scientists work together on a common goal. At the same time, laboratory designers have moved to open multiple-module laboratories that allow a wide variety of configurations for casework and equipment setups. These laboratories often support large or multiple teams and are configured with relocatable furnishings.

Even when not using a multidiscipline approach, many facilities have moved toward larger, more open laboratories with the belief that working in teams raises overall productivity, promote open communication, and facilitates resource sharing. Team sizes, in some disciplines, have risen and are frequently as high as 12 to 20 individuals.

9.B.2.1 Considerations for Open Laboratory Design

There are advantages and disadvantages to open laboratory design.

Advantages include

- visibility among researchers;
- better communication and collaboration;
- easy to share resources, including equipment, space, and support staff;
- flexibility for future needs because of open floor plan with adaptable furnishings;
- significant space savings compared with smaller, enclosed laboratories; and
- cost savings (first building/renovation costs and ongoing operating costs) compared with smaller, enclosed laboratories.

Disadvantages and limitations include

- for large spaces, challenging to balance the ventilation system;
- limitations to the size or placement of the labora-

tory (e.g., the floor of the building, the type of research) because of chemical storage code limitations for flammable and other materials;
- need for isolated spaces because of specific types of work being conducted, such as cell or tissue work where cross-contamination is an issue, use of certain radioactive materials, lasers, materials requiring special security measures, glass-washing facilities (see section 9.B.3 for more information);
- challenge of storing chemicals and supplies when there is a lack of natural spaces created by walls and other fixtures;
- noise from people and equipment may be higher than in a closed laboratory; and
- inability of some researchers to work effectively in an open laboratory environment.

Design teams should work with the research teams to find solutions that accommodate the needs of the researchers as much as possible. A combination of open laboratory spaces with smaller areas dedicated to special functions is often necessary.

9.B.3 Closed Laboratories and Access

Closed or separate laboratory spaces are often necessary for certain functions because of the nature of the operation, equipment needs, or security concerns. These areas may or may not be separated with a door. The need for a door and access control should be examined carefully for code requirements, safety protocol, and containment concerns.

The following issues should be considered:

- Do the exits require doors by code?
- Must the corridor walls, doors, and frames be fire-rated by code?
- Is containment of spills or smoke an issue that demands doors?
- Is noise an issue that demands separation and attenuation?
- Does the need for room air pressure control necessitate a door closing the laboratory space off from other areas?
- Does the work present a hazard that requires that access by untrained personnel be controlled?
- Do some materials or equipment present a security risk?
- Do the materials require compliance with biosafety guidelines?

Examples of operations or activities that may require separation from the main laboratory are in Table 9.1. The use of unusually hazardous materials may re-

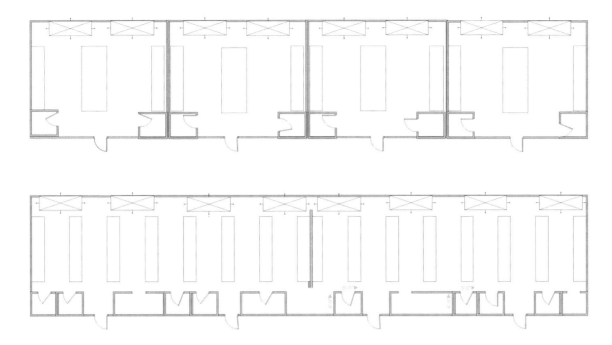

FIGURE 9.1 Open versus closed laboratory design. The top figure is an example of a typical closed laboratory design with four separate laboratories. The three walls separate the space and extend from floor to ceiling, with no shared spaces. The bottom figure is an example of an open laboratory in the same space. The wall extends from floor to ceiling but not from wall to wall (although in some designs, it could). Smaller working rooms with permanent or movable walls are set up for storage or activities that require closed spaces.

quire a dedicated area for such work to most efficiently manage security, safety, and environmental risk.

9.B.4 Equivalent Linear Feet of Workspace

When designing new laboratory spaces, consider the equivalent linear feet (ELF) of work surface within the laboratory. ELF can be divided into two categories: bench and equipment. Bench ELF is the required length of benchtop on which instruments can be set and where preparatory work takes place, as well as the length of laboratory chemical hoods. Equipment ELF includes the length of floor space for equipment that does not fit on a bench. Typically, every two laboratory personnel whose work mostly involves hazardous chemicals should have at least one chemical hood, and

these should be large enough to provide each person with a minimum of 3 linear ft, but it could be 8 ft or more depending on the planned activities and type of chemistry.

Typical chemistry laboratories are designed to provide from 28 to 30 ELF per person. Quality control, biology, and analytical laboratories range from 20 to 28 ELF per person. Quality control and production laboratories tend toward the low end of this range, whereas research laboratories are at or above the high end of the range. This number includes the support space outside the laboratory that is needed. These values can vary widely and must be addressed carefully for each project.

9.B.5 Laboratory Layout and Furnishing

9.B.5.1 Adaptability

The frequency of change in laboratory use has made it desirable to provide furnishings and services that can be moved and adapted quickly. Although some services and surfaces will be fixed elements in any laboratory, such as sinks and chemical hoods, there are several options available to meet the adaptable needs for various types of research.

Current design practice is to locate fixed elements

TABLE 9.1 Some Activities, Equipment, or Materials That May Require Separation from the Main Laboratory

Autoclaves	Animal Handling Areas
Darkrooms	Electron microscopes
Glasswashing facilities	High-powered lasers
Some radioactive materials	Tissue culture work
Exceptionally toxic materials	High-pressure equipment

such as laboratory chemical hoods and sinks at the perimeter of the laboratory, ensuring maximum mobility of interior equipment and furniture. Although fixed casework is common at the perimeters, moveable pieces are at the center to maximize flexibility. The central parts of the laboratory are configured with sturdy mobile carts, adjustable tables, and equipment racks.

Another trend for new laboratory buildings is to design interstitial spaces between the floors and to have all the utilities above the ceiling. The interstitial spaces are large enough to allow maintenance workers to access these utilities from above the ceiling for both routine servicing and to move plumbing and other utilities as research demands change.

Where interstitial spaces are not possible, overhead service carriers may be hung from the underside of the structural floor system. These service carriers may have quick connects to various utilities, such as local exhaust ventilation, computer cables, light fixtures, and electrical outlets.

9.B.5.2   Casework, Furnishings, and Fixtures

Casework should be durable and designed and constructed in a way that provides for long-term use, reuse, and relocation. Some materials may not hold up well to intensive chemistry or laboratory reconfiguration. Materials should be easy to clean and repair. For clean rooms, polypropylene or stainless steel may be preferable.

Work surfaces should be chemical resistant, smooth, and easy to clean. Benchwork areas should have knee space to allow for chairs near fixed instruments or for procedures requiring prolonged operation.

Work areas, including computers, should incorporate ergonomic features, such as adjustability, task lighting, and convenient equipment layout. Allow adequate space for ventilation and cooling of computers and other electronics.

Handwashing sinks for particularly hazardous materials may require elbow, foot, or electronic controls. Do not install more cupsinks than are needed. Unused sinks may develop dry traps that result in odor complaints.

9.B.5.3   Shared Spaces

Many facilities encourage sharing of some pieces of equipment. Locating the equipment in a space that is not defined as part of an individual's work zone facilitates sharing. Some examples of equipment that can be shared are in Table 9.2.

In an open laboratory setting, duplication of much of this equipment can be avoided. Often, if the equipment

is centrally located near a laboratory, it can be walled off to reduce noise.

The team needs to carefully address the need for alarms on specific pieces of equipment such as freezers and incubators that contain valuable samples.

Care must be taken, however, not to assume that sharing is always effective. There are certain pieces of equipment that must be dedicated to specific users.

9.B.5.4   Flooring

Wet laboratories should have chemically resistant covered flooring. Sheet goods are usually preferable to floor tiles, because floor tiles may loosen or degrade over time, particularly near laboratory chemical hoods and sinks. Rubberized materials or flooring with a small amount of grit may be more slip-resistant, which is desirable in chemical laboratories. Coved flooring that allows 4 to 8 in. of flooring material secured to the wall to form a wall base is also desirable.

Floors above areas with sensitive equipment, such as lasers, should be sealed to prevent leaks.

9.B.5.5   Doors, Windows, and Walls

Walls should be finished with material that is easy to clean and maintain. Fire code may require certain doors, frames, and walls to be fire-rated.

Doors should have view panels to prevent accidents caused by opening the door into a person on the other side and to allow individuals to see into the laboratory in case of an accident or injury. Doors should open in the direction of egress.

Laboratories should not have operable windows, particularly if there are chemical hoods or other local ventilation systems in the lab.

9.B.6   Noise and Vibration Issues

Many laboratories utilize equipment that may emit significant noise, require a stable structural environment, or both. During early planning stages, all equip-

TABLE 9.2   Examples of Equipment That Can Be Shared Between Researchers and Research Groups

Balances	Centrifuges
Gas chromatographs	High-performance liquid chromatographs
Ice machines	Incubators
Mass spectrometers	NMRs
Ovens	pH meters
Refrigerators/freezers	Weigh enclosures

ment should be discussed regarding any unique noise or vibration sensitivity in order to locate the equipment properly.

Large equipment such as centrifuges, shakers, and water baths often work best in separate equipment rooms. Pumps for older mass spectrometer units are both hot and noisy and are often located in either a small room or a hall. If in a closet, the area must have extra exhaust to remove heat, or else equipment may fail from overheating. With smaller and newer mass spectrometers, the pumps are often small and can fit into cabinets specifically designed for them. These pumps work especially well when water cooling is not required. Very few researchers need to hear their instrumentation running, but many want to see the equipment.

Another consideration crucial to equipment-intensive areas is the allowable vibration tolerance. Most analytical equipment such as NMRs, sensitive microscopes, mass spectrometers, and equipment utilizing light amplification (laser) require either vibration isolation tables or an area that is structurally designed to allow for very little vibration. Clarify the tolerance requirements with the user and equipment manufacturer during the equipment-programming phase, or early design process, so that the appropriate structure can be designed and the construction cost can be estimated more accurately.

9.B.7 *Safety Equipment and Utilities*

Each laboratory should have an adequate number and placement of safety showers, eyewash units, and fire extinguishers for its operations. (See Chapter 6, section 6.C.10, for more information.) The American National Standards Institute (ANSI) Z358.1-2004 standard provides guidance for safety shower and eyewash installation. The 2004 version recommends provision of tepid water, which can be complicated from an engineering standpoint. Although this standard does not address wastewater, most designers agree that emergency eyewash and shower units should be connected to drain piping. It is prudent to have floor drains near the units, preferably sloped to the drain to prevent excessive flooding and potential slip hazards. Consider choosing barrier-free safety showers and eyewash units that can accommodate individuals with disabilities. The maximum reach height for the activation control for safety showers is 48 in.

Sprinkler systems may be required by the building code and are almost always recommended. For areas with water-sensitive equipment or materials, consider preaction systems. Most dry or alternative systems do not function in a laboratory environment with chemical

hoods and other ventilation. There may be resistance to the idea of installing sprinkler systems in laboratories, particularly laboratories that use water-sensitive chemicals or equipment. The following facts may be helpful:

- Each sprinkler head is individually and directly activated by the heat of the fire, not by smoke or an alarm system. Thus, small fires are not likely to activate the sprinkler and moderate-size fires will likely activate only one or two heads. Indeed, more than 95% of fires are extinguished by one or two sprinkler heads.
- Statistics show that the sprinkler head failure rate is 1 in 16 million.
- In the event that the water from the sprinkler system reacts with water-sensitive materials, ensuing fires would be quenched once the reaction stopped. Damage is likely to be less severe than if a fire was not suppressed and was allowed to reach other flammable or combustible materials in the laboratory.
- Laboratory equipment, including lasers, is just as likely to be harmed by the fire as by the water. Without the sprinkler system, a fire that is large enough to activate the sprinkler system would result in response by the fire department. The sprinkler heads are designed to release water at a rate of 10–15 gallons per minute (gpm), whereas a firefighter's hose delivers 250–500 gpm.
- Dry chemical systems can seriously damage electronic and other laboratory equipment and are impractical in a building-wide system. Alternative agents are impractical because of the amount of space required for the cylinders and are most effective in rooms or areas that are sealed, which is not how laboratories are designed. These systems are most practical for an individual application, such as a piece of equipment or a "sealed" room.
- Locate utility shutoff switches outside or at the exit of the laboratory. The purpose of the switch is to shut down potentially hazardous operations quickly in the event of an emergency.
- Locate room purge buttons at the exits in laboratories with chemical hoods. For most laboratory buildings, activating the room purge button shuts down or minimizes supply air while increasing exhaust ventilation. In the event of a chemical spill, activating the purge system will help ventilate the resulting chemical vapors more quickly.
- Laboratories should have abundant electrical supply outlets to eliminate the need for extension cords and multiplug adapters. Place electrical panels in an accessible area not likely to be ob-

structed. Install ground-fault circuit interrupters near sinks and wet areas.

- Assess and provide for emergency power needs.
- Where possible, install chilled water loops for equipment requiring cooling. Chilled water loops save energy, water, and sewer costs.

9.B.8 Americans with Disability Act: Accessibility Issues Within the Laboratory

Title 1 of the Americans with Disabilities Act (ADA) of 1990 requires an employer to provide reasonable accommodation for qualified individuals with disabilities who are employees or applicants for employment, unless doing so would cause undue hardship. The design team and the owner are responsible for identifying what reasonable accommodations should and can be made to meet ADA guidelines or requirements.

In addition, some school systems and municipalities require a minimum number or percentage of accessible work areas in teaching laboratories. Accessible furniture, including laboratory chemical hoods, are readily available from most suppliers. The American Chemical Society has an excellent resource available online or in print, *Teaching Chemistry to Students with Disabilities: A Manual for High Schools, Colleges, and Graduate Programs* (ACS, 2001).

It is prudent to provide barrier-free safety showers and eyewash units for all laboratories. Figure 9.2 illustrates the specifications for barrier-free emergency equipment, according to ANSI 117.1-1992, "Accessible and Usable Building Facilities."

Additional accommodations will likely need to be made individually, depending on the special needs of the researcher. Partnering with the researcher, supervisor, and a laboratory safety professional will help determine the extent of the accommodations.

For wet laboratories, service animals should either have a place outside the lab or an area within the laboratory that is accessible without the animal having to traverse areas where chemicals or other hazardous materials could be present at floor level, including spills.

9.B.9 Older Facilities

Aging facilities can present multiple challenges. As materials of construction begin to degrade, the safety and environmental provisions of the facility often degrade as well. Although some equipment and materials may continue to function well for many years, modern alternatives may offer better safety and environmental sustainability features.

For older facilities, it is important to have a strong operations and maintenance program that monitors

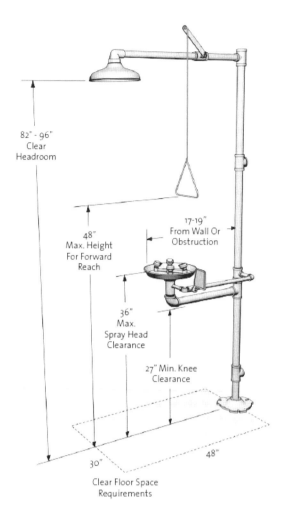

FIGURE 9.2 Specifications for barrier-free safety showers and eyewash units.

and maintains plumbing, ventilation, and structural components. Nonetheless, as individual laboratories or spaces are renovated for new uses or upgrades, there are opportunities for improving and modernizing building systems.

Depending on the location of the laboratory building, there may be requirements for bringing the entire building up to current building codes and standards once a certain percentage of the building is under renovation. These code requirements may include fire protection systems, accessibility, plumbing, ventilation, alarm systems, chemical storage restrictions, and egress issues.

With rising interest in energy conservation, there have been numerous studies and instances of retrocommissioning of laboratories. The focus is generally on the laboratory ventilation system, with the goal of managing airflow and temperature control to eliminate waste and reduce overall energy use. In "Laboratories for the 21st Century" the U.S. Environmental Protection Agency (EPA/DOE, 2006), reports that in most studied cases, retro-commissioning, when planned and executed well, resulted in reductions of at least 30% of overall facility energy use with a payback period of less than 3 years.

The typical retro-commissioning process proceeds in five major steps:

1. **Planning.** Bring facility and EHS staff, design engineers, and users together to discuss goals. Gather information about the current system, including the original plans, as-built plans, major alterations, and current function, including ventilation rates. Develop the retro-commissioning plan.
2. **Preinvestigation.** Verify all systems including the direct digital control or building automation systems, evaluate all components that affect energy use, and verify monitoring systems.
3. **Investigation.** Benchmark utility and energy use data, analyze trends, and test all equipment. Testing should include functional testing of chemical hoods and related components, including face velocity tests, containment tests, etc.
4. **Implementation.** Select which improvements will be made and prioritize them. Implement the improvements and test performance.
5. **Handoff.** Clearly document information and provide training to laboratory personnel and maintenance personnel.

Common conditions that lead to energy waste include

- overabundance of laboratory chemical hoods,
- laboratory chemical hoods with large bypass openings,
- dampers in fixed positions,
- overventilated laboratory spaces,
- excessive duct pressure,
- fans set to override position,
- fans that are no longer operating efficiently,
- constant volume systems with no setback for temperature or airflow when unoccupied, and
- high face velocities.

Whether retro-commissioning for energy efficiency or for safety, ensure that all stakeholders are involved in the process. Once the work is complete, continue to monitor efficiency and safety. It is important to include trained laboratory personnel in the feedback process. If systems are not used correctly or if they are bypassed, the retro-commissioning efficiency may deteriorate.

9.C LABORATORY VENTILATION

The laboratory ventilation system, whether it is the general ventilation, a chemical hood, or a specialized exhaust system, is a critical means to control airborne chemicals in the laboratory.

At a minimum, a well-designed laboratory ventilation system should include the following:

- Heating and cooling should be adequate for the comfort of laboratory occupants and operation of laboratory equipment.
- A differential should exist between the amount of air exhausted from the laboratory and the amount supplied to the laboratory to maintain a negative pressure between the laboratory and adjacent nonlaboratory spaces. This pressure differential prevents uncontrolled chemical vapors from leaving the laboratory. Clean rooms may require a slightly positive pressure differential. There should be separation between common spaces and the clean room to prevent migration of airborne contaminants.
- Exhaust ventilation devices should be appropriate to materials and operations in the laboratory.

Many devices are used to control emissions of hazardous materials in the laboratory. A risk assessment helps to determine the best choice for a particular operation or material (Table 9.3).

NOTE: Clean benches are **not** designed for use with hazardous materials. These are appropriate for use in work with materials that necessitate clean work conditions and should only be used for materials or chemicals that one could safety use on a benchtop.

9.C.1 Risk Assessment

For all materials, the objective is to keep airborne concentrations below established exposure limits (see Chapter 4, section 4.C.2.1). Where there is no established exposure limit, where mixtures are present, or where reactions may result in products that are not completely characterized, prudent practice keeps exposures ALARA (as low as reasonably achievable).

For chemicals, determine whether the material is flammable or reactive or if it poses a health hazard from inhalation. If no significant risk exists, the work does not likely require any special ventilation. If potential risk does exist, look at the physical properties of the chemical, specifically its vapor pressure and vapor density.

TABLE 9.3 Laboratory Engineering Controls for Personal Protection

Type of Ventilation	Typical Number of Air Changes or Face Velocity in Linear Feet per Minute (fpm) as Appropriate	Examples of Use
General laboratory ventilation	6–12 air changes/hour, depending on laboratory design and system operation	• Nonvolatile chemicals • Nonhazardous materials
Environmental rooms	0 air changes	• Materials that require special environmental controls • Nonhazardous amounts of flammable, toxic, or reactive chemicals.
Laboratory chemical hoods	10–15 air changes/minute or 60-100 fpm depending on hood type	• Flammable, toxic, or reactive materials • Products or mixtures with uncharacterized hazards
Unventilated storage cabinets	0 air changes	• Flammable liquids • Corrosives • Moderately toxic chemicals
Ventilated storage cabinets	1–2 air changes/minute	• Highly toxic, hazardous, or odiferous chemicals (if equipped with flame arrestors)
Recirculating biosafety cabinets	A1: 75 fpm A2: 100 fpm	• Biological materials • Nanoparticles, as of the date of publication
	B1: 100 fpm	• Biological materials • Nanoparticles, as of the date of publication • Minute amounts of volatile chemicals
Total exhaust biosafety cabinet	B2: 100 fpm	• Biological materials • Nanoparticles, as of the date of publication • Minute amounts of volatile chemicals
Glovebox	Varies from no change to very high rate of change, depending on the glovebox and the application	• Positive pressure for specialty environments • Negative pressure for highly toxic materials
Downdraft table	150–250 fpm depending on design	• Perfusions with paraformaldehyde, work with volatile, low to moderately hazardous materials with higher vapor density where access from more than one side is necessary
Elephant trunk	150–200 fpm at opening	• Local ventilation of a tabletop • Discharge from equipment such as a gas chromatograph
Canopy	N/A	• Ventilation of heat, steam, low or nontoxic materials with low vapor density
Ductless laboratory chemical hood	10–15 air changes/minute	• Materials that are compatible with the filtration system, in controlled quantities and under controlled conditions • Not suitable for particularly hazardous substances
Slot hood	Varies with application	• Local ventilation of higher density materials at the source, such as an acid bath
Ventilated balance enclosure	5–10 air changes/minute	• Weighing and initial dissolution of highly toxic or potent materials
Benchtop ventilated enclosures	Variable per the needs of the materials	• Benchtop equipment, such as rotovaps

Vapor pressure is usually measured in millimeters of mercury. A low vapor pressure (<10 mmHg) indicates that the chemical does not readily form vapors at room temperature. General laboratory ventilation or an alternative such as the elephant trunk or snorkel may be appropriate, unless the material is heated or in a higher temperature room that might promote vapor formation. High vapor pressure indicates that the material easily forms vapors and may require use of a ventilated enclosure, such as a chemical hood.

Vapor density is compared to that of air, which is 1. A chemical having a vapor density greater than 1 is heavier than air. If the vapors need to be controlled, a chemical hood or a ventilation device that draws air from below, such as a downdraft table or a slot hood or elephant trunk with the exhaust aimed low may be appropriate. Conversely, a chemical with a vapor density less than 1 is lighter than air. Besides a chemical hood, a ventilation device that draws air from above, such as an elephant trunk or snorkel with the exhaust positioned above the source, may work best.

For radioactive or biological materials, consider whether the operations might cause the materials to aerosolize or become airborne and whether inhalation poses a risk to health or the environment. Determine whether filtration or trapping is required or recommended.

For manipulating solid particulates, a chemical hood and similar equipment with higher airflow may be too turbulent. Weighing boxes or ventilated balance enclosures may be a better fit for such work.

For nanomaterials, a laboratory chemical hood might be too turbulent for manipulating the materials. Also, consider whether the exhaust containing these tiny particles should be filtered. Studies have shown that high-efficiency particulate air (HEPA) filters are very effective for nano-size particles. Containment tests for chemical hoods allow for a very minor amount of leakage into the breathing zone of the user. For chemical vapors, such an amount may be insignificant, but in the same volume of nanoparticles, the number of particles may be quite large, and biosafety cabinets, gloveboxes or filtering hoods would be better. (See section 9.E.5 for more information.)

More specialized ventilation systems, such as biosafety cabinets and gloveboxes, may be necessary to control specific types of hazards, as discussed later in this chapter.

9.C.2 Laboratory Chemical Hoods

Laboratory chemical hoods are the most important components used to protect laboratory personnel from exposure to hazardous chemicals and agents. Functionally, a standard chemical hood is a fire- and chemical-resistant enclosure with one opening (face) in the front with a movable window (sash) to allow user access to the interior. Large volumes of air are drawn through the face and out the top into an exhaust duct to contain and remove contaminants from the laboratory. Note that because a substantial amount of energy is required to supply tempered supply air to even a small hood, the use of hoods to store bottles of toxic or corrosive chemicals is a very wasteful practice, which can seriously impair the effectiveness of the hood as a local ventilation device. Thus, it is preferable to provide separate vented cabinets for the storage of toxic or corrosive chemicals. The amount of air exhausted by such cabinets is much less than that exhausted by a properly operating hood.

A well-designed hood, when properly installed and maintained, offers a substantial degree of protection to the user if it is used appropriately and its limitations are understood. Chemical hoods are the best choice, particularly when mixtures or uncharacterized products are present and any time there is a need to manage chemicals using the ALARA principle.

9.C.2.1 Laboratory Chemical Hood Face Velocity

The average velocity of air drawn through the face of the laboratory chemical hood is called the face velocity. The face velocity greatly influences the ability to contain hazardous substances, that is, its containment efficiency. Face velocities that are too low or too high reduce the containment efficiency.

Face velocity is only one indicator of hood performance and one should not rely on it as a sole basis for determining the containment ability of the chemical hood. There are no regulations that specify acceptable face velocity. Indeed, modern hood designs incorporate interior configurations that affect the airflow patterns and are effective at different ranges of face velocity.

For traditional chemical hoods, several professional organizations have recommended that the chemical hood maintain a face velocity between 80 and 100 feet per minute (fpm). Face velocities between 100 and 120 fpm have been recommended in the past for substances of very high toxicity or where outside influences adversely affect hood performance. However, energy costs to operate the chemical hood are directly proportional to the face velocity and there is no consistent evidence that the higher face velocity results in better containment. Face velocities approaching or exceeding 150 fpm should not be used; they may cause turbulence around the periphery of the sash opening and actually reduce the capture efficiency, and may reentrain settled particles into the air.

With the desire for more sustainable laboratory ventilation design, manufacturers are producing high-performance hoods, also known as low-flow hoods, that achieve the same level of containment as traditional ones, but at a lower face velocity. These chemical hoods are designed to operate at 60 or 80 fpm and in some cases even lower. (See section 9.C.2.9.3.6.)

Average face velocity is determined by measuring individual points across the plane of the sash opening and calculating their average. A more robust measure of containment uses tracer gases to provide quantitative data and smoke testing to visualize airflow patterns. ASHRAE/ANSI 110 testing is an example of this technique (see section 9.C.2.8 for more information). This type of testing should be conducted at the time the chemical hood is installed, when substantial changes are made to the ventilation system, including rebalancing and periodically as part of a recommissioning or maintenance program.

Once a chemical hood is tested and determined to be acceptable via the ASHRAE/ANSI 110 method or an equivalent means, the face velocity should be noted and used as the reference point for routine testing. Each chemical hood, laboratory, facility, or site must define the acceptable average face velocity, minimum acceptable point velocity, and maximum standard deviation of velocities, as well as when ASHRAE/ANSI 110 or visualization testing is required. These requirements should be incorporated into the laboratory's Chemical Hygiene Plan and ventilation system management plans (see section 9.H).

When first installed and balanced, a laboratory chemical hood must be subjected to the ASHRAE/ANSI 110 or equivalent test before it is commissioned. When multiple similar chemical hoods are installed at the same time, at least half should be tested, provided the design is standardized relative to location of doors and traffic, and to location and type of air supply diffusers.

9.C.2.2   Factors That Affect Laboratory Chemical Hood Performance

Tracer gas containment testing of chemical hoods reveals that air currents impinging on the face at a velocity exceeding 30 to 50% of the face velocity reduce the containment efficiency by causing turbulence and interfering with the laminar flow of the air entering the chemical hood. Thirty to fifty percent of a face velocity of 100 fpm, for example, is 30 to 50 fpm, which represents a *very* low velocity that can be produced in many ways. The rate of 20 fpm is considered to be still air because that is the velocity at which most people first begin to sense air movement.

9.C.2.2.1   Proximity to Traffic

Most people walk at approximately 250 fpm (approximately 3 mph [4.8 kph]) and as they walk, vortices exceeding 250 fpm form behind them. If a person walks in front of an open chemical hood, the vortices can overcome the face velocity and pull contaminants into the vortex, and into the laboratory. Therefore, laboratory chemical hoods should not be located on heavily traveled aisles, and those that are should be kept closed when not in use. Foot traffic near these chemical hoods should be avoided when work is being performed.

9.C.2.2.2   Proximity to Supply Air Diffusers

Air is supplied continuously to laboratories to replace the air exhausted through laboratory chemical hoods and other exhaust sources and to provide ventilation and temperature/humidity control. This air usually enters the laboratory through devices called supply air diffusers located in the ceiling. Velocities that exceed 800 fpm are frequently encountered at the face of these diffusers. If air currents from these diffusers reach the face of a chemical hood before they decay to 30 to 50% of the face velocity, they cause the same effect as air currents produced by a person walking in front of the chemical hood. Normally, the effect is not as pronounced as the traffic effect, but it occurs constantly, whereas the traffic effect is transient. Relocating the diffuser, replacing it with another type, or rebalancing the diffuser air volumes in the laboratory can alleviate this problem.

9.C.2.2.3   Proximity to Windows and Doors

Exterior windows with movable sashes are not recommended in laboratories. Wind blowing through the windows and high-velocity vortices caused when doors open can strip contaminants out of the chemical hoods and interfere with laboratory static pressure controls. Place hoods away from doors and heavy traffic aisles to reduce the chance of turbulence reducing the effectiveness of the hood.

9.C.2.3   Prevention of Intentional Release of Hazardous Substances into Chemical Hoods

Laboratory chemical hoods should be regarded as safety devices that can contain and exhaust toxic, offensive, or flammable materials that form as a result of laboratory procedures. Just as you should never flush laboratory waste down a drain, never intentionally send waste up the chemical hood. Do not use the chemical hood as a means of treating or disposing of chemical waste, including intentionally emptying hazardous gases from compressed gas cylinders or allowing waste solvent to evaporate.

For some operations, condensers, traps, and/or scrubbers are recommended or necessary to contain and collect vapors or dusts to prevent the release of harmful concentrations of hazardous materials from the chemical hood exhaust.

9.C.2.4   Laboratory Chemical Hood Performance Checks

When checking if laboratory chemical hoods are performing properly, observe the following guidelines:

- Evaluate each hood before initial use and on a regular basis (at least once a year) to visualize airflow and to verify that the face velocity meets the criteria specified for it in the laboratory's Chemical Hygiene Plan or laboratory ventilation plan.
- Verify the absence of excessive turbulence (see section 9.C.2.6, below).

- Make sure that a continuous performance monitoring device is present, and check it every time the chemical hood is used. (For further information, see section 9.C.2.8 on testing and verification.)

Box 9.1 provides a list of things to do to maximize chemical hood efficiency.

9.C.2.5 Housekeeping

Keep laboratory chemical hoods and adjacent work areas clean and free of debris at all times. Keep solid objects and materials (such as paper) from entering the exhaust ducts, because they can lodge in the ducts or fans and adversely affect their operation. The chemical hood will have better airflow across its work surface if it contains a minimal number of bottles, beakers, and laboratory apparatus; therefore, prudent practice keeps *unnecessary* equipment and glassware outside the chemical hood at all times and stores all chemicals in approved storage cans, containers, or cabinets. Furthermore, keep the workspace neat and clean in all laboratory operations, particularly those involving the use of chemical hoods, so that any procedure or experiment can be undertaken without the possibility of disturbing, or even destroying, what is being done.

9.C.2.6 Sash Operation

Except when adjustments to the apparatus are being made, keep the chemical hood closed, with vertical sashes down and horizontal sashes closed, to help prevent the spread of a fire, spill, or other hazard into the laboratory. Horizontal sliding sashes should not be removed. The face opening should be kept small to improve the overall performance of the hood. If the face velocity becomes excessive, the facility engineers should make adjustments or corrections.

For chemical hoods without face velocity controls (see section 9.C.4.1), the sash should be positioned to produce the recommended face velocity, which often occurs only over a limited range of sash positions. This range should be determined and marked during laboratory chemical hood testing. Do not raise the sash above the working height for which it has been tested to maintain adequate face velocity. Doing so may allow the release of contaminants from the chemical hood into the laboratory environment.

Chemical hood sashes may move vertically (sash moves up and down), horizontally (sash is divided in panes that move side to side to provide the opening to the hood interior), or a combination of both. Although both types of sash offer protection from the materials within the hood and help control or maintain airflow, consider the following:

BOX 9.1
Quick Guide for Maximizing Efficiency of Laboratory Chemical Hoods

Many factors can compromise the efficiency of chemical hood operation, and most are avoidable. Be aware of all behavior that can, in some way, modify the chemical hood and its capabilities. Always consider the following:

- Keep chemical fume hood exhaust fans on at all times.
- If possible, position the chemical hood sash so that work is performed by extending the arms under or around the sash, placing the head in front of the sash, and keeping the sash between the person and the chemical source. View the procedure through the sash, which acts as a primary barrier if a spill, splash, or explosion should occur.
- Avoid opening and closing the sash rapidly, and avoid swift arm and body movements in front of or inside the chemical hood. These actions may increase turbulence and reduce the containment efficiency.
- Place chemical sources and apparatus at least 6 in. behind the face. Paint a colored stripe or apply tape to the work surface 6 in. back from the face to serve as a reminder. Quantitative chemical hood containment tests reveal that the concentration of contaminant in the breathing zone can be 300 times higher from a source located at the front of the face than from a source placed at least 6 in. back. This concentration continues to decline as the source is moved farther toward the back.
- Place equipment as far to the back of the chemical hood as practical without blocking the bottom baffle.
- Separate and elevate each instrument by using blocks or racks; air should flow easily around all apparatus.
- Do not use large pieces of equipment in a chemical hood, because they tend to cause dead spaces in the airflow and reduce the efficiency.
- If a large piece of equipment emits fumes or heat outside a chemical hood, have a special-purpose hood designed and installed to ventilate that particular device. This method of ventilation is much more efficient than placing the equipment in a chemical fume hood, and it will consume much less air.
- Do not modify chemical hoods in any way that adversely affects performance. This includes adding, removing, or changing any of the components, such as baffles, sashes, airfoils, liners, and exhaust connections.
- Make sure all highly toxic or offensive vapors are scrubbed or adsorbed before the exit gases are released into the chemical hood exhaust system (see section 9.C.2.11.1 on chemical hood scrubbers).
- Keep the sash closed whenever the chemical hood is not actively in use or is unattended.

- Some experimentation requires the lab personnel to access equipment or materials toward the upper portion of the chemical hood. If the chemical hood is equipped with a vertical sash, it may be necessary to raise the sash completely in order to conduct the procedure.
  - The laboratory chemical hood must provide adequate containment at that sash height. Thus, the chemical hood must be tested in that position.
  - With the sash completely raised, it no longer provides a barrier between the chemical hood user and the materials within the hood.
  - If the only way to keep the sash in a fully raised position requires the use of a sash stop, the laboratory personnel may get into the habit of leaving the sash in this position, potentially reducing the safety and energy efficiency of the chemical hood.
- The standard operating position for the vertical sash may be comfortable for the majority of users. However, shorter laboratory personnel may find that this position does not provide an adequate barrier from the materials within the chemical hood and may need to adjust downward. Taller laboratory personnel may need to raise the sash more in order to comfortably work in the chemical hood.

For chemical hoods with horizontal sashes, the intended operating configuration is to open the panes in such a way that at least one pane is between both arms, providing a barrier between the user and the contents of the chemical hood. In addition,

- Do not remove panes. Permanently removing panes may decrease the safety afforded by the sash barrier and negatively affect containment and waste energy.
- Working with all panes moved to one side or through an opening in the center of the laboratory chemical hood provides no barrier between the user and the materials within the chemical hood. The chemical hood is not intended to be used in this configuration.

Sash panes should be equal width with a maximum of 15 in. (375 mm) to accommodate use of the sash pane as a protective barrier with operator arm on either side.

Conventional glass or plastic sashes are not designed to provide explosion protection per ANSI/NFPA (ANSI, 2004; NFPA, 2004). Sash panes and viewing panes constructed of composite material (safety glass backed by polycarbonate, with the safety glass toward

the explosion hazard) are recommended for chemical hoods used when there is the possibility of explosion or violent overpressurization (e.g., hydrogenation, perchloric acid).

For all laboratory chemical hoods, **the sash should be kept closed when the hood is not actively attended.** Lowering or closing the sash not only provides additional personal protection but also results in significant energy conservation. Some chemical hoods may be equipped with automatic sash-positioning systems with counterweighting or electronic controls (see section 9.H.2).

9.C.2.7 Constant Operation of Laboratory Chemical Hoods

Although turning laboratory chemical hoods off when not in use saves energy, keeping them on at all times is safer, especially if they are connected directly to a single fan. Because most laboratory facilities are under negative pressure, air may be drawn backward through the nonoperating fan, down the duct, and into the laboratory unless an ultralow-leakage backdraft damper is used in the duct. If the air is cold, it may freeze liquids in the hood. The ducts are rarely insulated; therefore, condensation and ice may form in cold weather. When the chemical hood is turned on again and the duct temperature rises, the ice will melt, and water will run down the ductwork, drip into the hood, and possibly react with chemicals in the hood.

Chemical hoods connected to a common exhaust manifold offer the advantage that the main exhaust system is rarely shut down. Hence, positive ventilation is available on the system at all times. In a constant air volume (CAV) system (see section 9.C.4.1), install shutoff dampers to each chemical hood, allowing passage of enough air to prevent fumes from leaking into the laboratory when the sash is closed. Prudent practice allows 10 to 20% of the full volume of flow to be drawn through the laboratory chemical hood in the off position to prevent excessive corrosion.

Some laboratory chemical hoods on variable air volume (VAV) systems (see section 9.C.4.2) have automatic setback controls that adjust the airflow to a lower face velocity when not in use. The setback may be triggered by occupancy sensors, a light switch, or a timer or a completely lowered sash. Understand what triggers the setback and ensure that the chemical hood is not used for hazardous operations when in setback mode.

Some chemical hoods do have on/off switches and may be turned off for energy conservation reasons. They should only be turned off when they are empty of hazardous materials. An example of an acceptable operation would be a teaching laboratory where the empty chemical hoods are turned off when the laboratory is not in use.

9.C.2.8 Testing and Verification

The OSHA lab standard includes a provision regarding laboratory chemical hoods, including a requirement for some type of continuous monitoring device on each chemical hood to allow the user to verify performance and routine testing of the hood. It does not specify a test protocol.

Laboratory chemical hoods should be tested at least as follows:

- containment test by manufacturer;
- containment test after installation and prior to initial use (commissioning);
- annual or more frequent face velocity and airflow visualization;
- performance test any time a potential problem is reported; and
- containment test after significant changes to the ventilation system, including rebalancing or recommissioning.

9.C.2.8.1 Initial Testing

All laboratory chemical hoods should be tested before they leave the manufacturer according to ANSI/ASHRAE Standard 110-1995 or equivalent, Methods of Testing Performance of Laboratory Fume Hoods (ANSI, 1995). They should pass the low- and high-volume smoke challenges with no leakage or flow reversals and have a control level of 0.05 ppm or less on the tracer gas test. It is highly recommended that chemical hoods be retested by trained personnel after installation in their final location, using ANSI/ASHRAE 110-1995 or equivalent testing. The control level of tracer gas for an "as installed" or "as used" test via the ANSI/ASHRAE 110-1995 method should not exceed 0.1 ppm.

The ANSI/ASHRAE 110-1995 test is the most practical way to determine chemical hood capture efficiency quantitatively. The test includes several components, which may be used together or separately, including face velocity testing, flow visualization, face velocity controller response testing, and tracer gas containment testing. These tests are much more accurate than face velocity and smoke testing alone. Respectively, ASHRAE and ANSI found that 28% or 38% of chemical hoods tested using this method did not meet the pass criteria, even though face velocity testing alone found them to be in an acceptable face velocity range.

Performance should be evaluated against the design specifications for uniform airflow across the chemical hood face as well as for the total exhaust air volume.

Equally important is the evaluation of operator exposure. The first step in the evaluation of hood performance is the use of a smoke tube or similar device to determine that the laboratory chemical hood is on and exhausting air. The second step is to measure the velocity of the airflow at the face of the hood. The third step is to determine the uniformity of air delivery to the hood face by making a series of face velocity measurements taken in a grid pattern.

Leak testing is normally conducted using a mannequin equipped with sensors for the test gas. As an alternative, a person wearing the sensors or collectors may follow a sequence of movements to simulate common activities, such as transferring chemicals. It is most accurate to perform the in-place tests with the chemical hood at least partially loaded with common materials (e.g., chemical containers filled with water, equipment normally used in the chemical hood), in order to be more representative of operating conditions.

For the ASHRAE 110-1995 leak testing, the method calls for a release rate for the test gas of 4 liters per minute (Lpm), but suggests that higher rates may be used. One-liter per minute release rate approximates pouring a volatile solvent from one beaker to another. Eight liters per minute approximates boiling water on a 500-W hot plate. The 4-Lpm rate is an intermediate of these two conditions. If there is a possibility that the chemical hood will be used for volatile materials under heating conditions, consider a higher release rate of up to 8 Lpm for worst-case conditions.

The total volume of air exhausted by a laboratory chemical hood is the sum of the face volume (average face velocity times face area of the hood) plus air leakage, which averages about 5 to 15% of the face volume. If the laboratory chemical hood and the general ventilating system are properly designed, face velocities in the range of the design criteria will provide a laminar flow of air over the work surface and sides of the hood. Higher face velocities (150 fpm or more), which exhaust the general laboratory air at a greater rate, waste energy and are likely to degrade hood performance by creating air turbulence at the face and within the chemical hood, causing vapors to spill out into the laboratory (Figure 9.3).

An additional method for containment testing is the EN 14175, which is the standard adopted by the European Union and replaces several other procedures that were in place for individual countries. Parts 3 (Type tests) and 4 (On-site tests) of this standard address methods for "as manufactured" and "as installed/used" systems, respectively.

9.C.2.8.2 Routine Testing

At least annually, the following test procedures should be conducted for all chemical hoods:

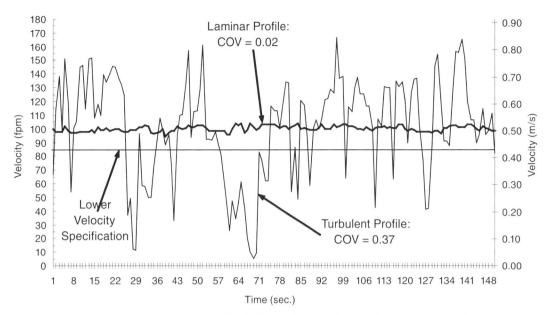

FIGURE 9.3 Laminar versus turbulent velocity profile. Velocity data are from a single traverse point on two separate hoods. The light line represents a hood where supply air interference caused large variations in velocity, a "typical" turbulent profile. Eddy currents and flow reversals caused by a turbulent airflow pattern may cause spillage and leakage of contaminants from the hood into the laboratory environment. In contrast, the bold line represents a hood having an almost ideal velocity profile, indicative of a laminar airflow pattern. The coefficient of variation (COV) is used as a predictor of the level of turbulence experienced at the face of a hood. A high COV indicates a turbulent air profile and most likely is a strong indicator of poor containment; a low COV indicates a laminar flow profile and likely good containment.
SOURCE: Maupins and Hitchings (1998). Reprinted by permission of Taylor & Francis.

- Analyze face velocity using the method and criteria described in section 9.C.2.8.4.
- Visualize airflow using smoke tubes, bombs, or fog generators.
- Verify that continuous flow monitoring devices are working properly.
- Verify that other controls, including automatic sash positioners, alarm systems, etc. are functioning properly.
- Check the sash to ensure that it is in good condition, moves easily, is unobstructed, and has adequate clarity to see inside the laboratory chemical hood.
- Ensure that the laboratory chemical hood is being used as intended (e.g., no evidence of perchloric acid in a chemical hood not designed for it, not using it as a chemical storage device).
- Note any conditions that could affect laboratory chemical hood performance, such as large equipment, excessive storage, etc.
- Take corrective actions where necessary and retest.

Provide information and test results to the chemical hood users and/or supervisors. Document the results in order to maintain a log showing the history of chemical hood performance.

9.C.2.8.3   Additional Testing

Laboratory personnel should request a chemical hood performance evaluation any time there is a change in any aspect of the ventilation system. Thus, changes in the total volume of supply air, changes in the locations of supply air diffusers, or the addition of other auxiliary local ventilation devices (e.g., more chemical hoods, vented cabinets, and snorkels) all call for reevaluation of the performance of all chemical hoods in the laboratory.

9.C.2.8.4   Face Velocity Testing

Visually divide the face opening of a laboratory chemical hood into an imaginary grid, with each grid space being approximately 1 ft^2 in area. Using an anemometer, velometer, or similar device, take a measurement at the center of each grid space. Face velocity readings should be integrated for at least 10 seconds (20 is preferable) because of the fluctuations in flow. The measured velocity will likely fluctuate for several seconds; record the reading once it has stabilized. Calculate the average of the velocity for every grid space. The resulting number is the average face velocity. Analyze the results to determine if any one measurement is 20% or more above or below the average. Such readings indicate the possibility of turbulent or nonlaminar airflow. Smoke tests will help confirm whether this is problematic.

Traditional handheld instruments are subject to probe movement and positioning errors as well as reading errors owing to the optimistic bias of the investigator. Also, the traditional method yields only a snapshot of the velocity data, and no measure of variation over time is possible. To overcome this limitation, take velocity data while using a velocity transducer connected to a data acquisition system and read continuously by a computer for approximately 30 seconds at each traverse point. If the transducer is fixed in place, using a ring stand or similar apparatus, and is properly positioned and oriented, this method overcomes the errors and drawbacks associated with the traditional method. The variation in data for a traverse point can be used as an indicator of turbulence, an important additional performance indicator that has been almost completely overlooked in the past.

If the standard deviation of the average velocity profile at each point exceeds 20% of the mean, or the average standard deviation of velocities at each traverse point (turbulence) exceeds 15% of the mean face velocity, corrections should be made by adjusting the interior baffles and, if necessary, by altering the path of the supply air flowing into the room (see Figure 9.4). Most laboratory chemical hoods are equipped with a baffle that has movable slot openings at both the top and the bottom, which should be moved until the airflow is essentially uniform. Larger chemical hoods may require additional slots in the baffle to achieve uniform airflow across the face. These adjustments should be made by an experienced laboratory ventilation engineer or technician using proper instrumentation.

9.C.2.8.5 Testing Criteria

Prior to the initial tests, determine the acceptance criteria for the ANSI/ASHRAE 110-1995 leak test, face velocity (based on the results of the ANSI/ASHRAE testing and the design of the laboratory chemical hood), and visual airflow tests.

One important factor to consider is acceptable sash position. It is common to set the acceptance criteria as an acceptable level of containment and/or face velocity range at the standard operating position of the sash, often 18 in. However, one must understand how the chemical hood will be used to determine the range of

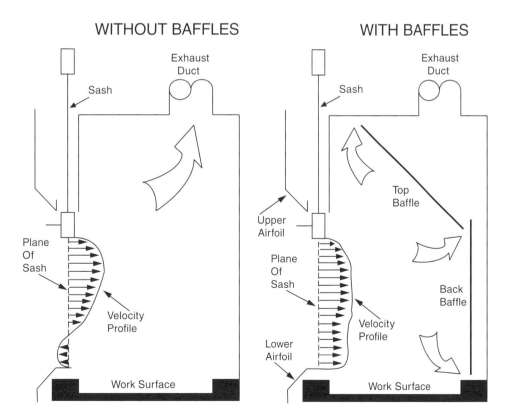

FIGURE 9.4 Effect of baffles on face velocity profile in a laboratory chemical hood.

sash positions needed. For example, if the users will need to sometimes use the hood with vertical sash fully open, then the test criteria should be for 100% sash opening.

It may be prudent to set the acceptance criteria with the sash 100% open and 80% open, ensuring adequate containment at both of these positions.

9.C.2.8.6   Instrumentation

Anemometers and other instruments used to measure face velocity must be accurate in order to supply meaningful data. Instruments should be calibrated at least once a year and the calibration should be National Institute of Standards and Technology traceable.

9.C.2.8.7   Additional Exposure Monitoring

If there is any concern that a laboratory chemical hood or other ventilation device may not provide enough protection to the trained laboratory personnel, it is prudent to measure worker exposure while the hood is being used for its intended purpose. By conducting personal air-sampling using traditional industrial hygiene techniques, worker exposure (both excursion peak and time-weighted average) can be measured. The criterion for evaluating the hood should be the desired performance (i.e., does the hood contain vapors and gases at the desired worker-exposure level?). A sufficient number of measurements should be made to define a statistically significant maximum exposure based on worst-case operating conditions. Direct-reading instruments may be available for determining the short-term concentration excursions that may occur in chemical hood use.

9.C.2.9   Laboratory Chemical Hood Design and Construction

When specifying a laboratory chemical hood for use in a particular activity, laboratory personnel should be aware of the design features. Assistance from an industrial hygienist, ventilation engineer, or laboratory consultant is recommended when deciding to purchase a chemical hood.

9.C.2.9.1   General Design Recommendations

Construct laboratory chemical hoods and the associated exhaust ducts of nonflammable materials. Equip them with vertical, horizontal, or combination vertical/horizontal sashes that can be closed. For the glass within the sash, use either laminated safety glass that is at least 7/32-in. thick or other equally safe material that will not shatter if there is an explosion inside. Locate the utility control valves, electrical receptacles, and other fixtures outside the chemical hood to minimize the need to reach within the chemical hood proper.

Other specifications regarding the construction materials, plumbing requirements, and interior design vary, depending on the intended use. (See Chapter 7, sections 7.C.1.1 and 7.C.1.2.) Information regarding the minimum flow rate through hoods can be found in ANSI Z9.5.

Although chemical hoods are most commonly used to control concentrations of toxic vapors, they can also serve to dilute and exhaust flammable vapors. Although theoretically possible, it is extremely unlikely (even under worst-case scenarios) that the concentration of flammable vapors will reach the lower explosive limit (LEL) in the exhaust duct. However, somewhere between the source and the exhaust outlet of the chemical hood, the concentration will pass through the upper explosive limit and the LEL before being fully diluted at the outlet. Both the designer and the user should recognize this hazard and eliminate possible sources of ignition within the chemical hood and its ductwork if there is a potential for explosion. The use of duct sprinklers or other suppression methods in laboratory hood ductwork is not necessary or desirable.

9.C.2.9.2   Special Design Features

Since the invention of the chemical hood, two major improvements have been made in the design—airfoils and baffles. Include both features on any new purchases.

Airfoils built into the bottom and sides of the sash opening significantly reduce boundary turbulence and improve capture performance. Fit new chemical hoods with airfoils and retrofit any hoods without airfoils

When air is drawn through a laboratory chemical hood without a baffle (see Figure 9.4), most of the air is drawn through the upper part of the opening, producing an uneven velocity distribution across the face opening. All chemical hoods should have baffles. When baffles are installed, the velocity distribution is greatly improved. Adjustable baffles can improve hood performance and are desirable if the adjustments are made by an experienced industrial hygienist, consultant, or technician.

9.C.2.9.3   Laboratory Chemical Hood Airflow Types

The first chemical hoods were simply boxes that were open on one side and connected to an exhaust duct. Since they were first introduced, many variations on this basic design have been made. Six of the major variants in airflow design are listed below with their characteristics. Conventional laboratory chemical hoods are the most common and include benchtop, distillation, and walk-in hoods of the CAV, CAV bypass, nonbypass, and VAV, with or without airfoils. Auxiliary air hoods and ductless chemical hoods are not considered conventional and are used less often.

Trained laboratory personnel should know what kind they are using and what its advantages and limitations are. In general, the initial cost of a CAV system may be less than VAV, but the life-cycle cost of the VAV will almost always be lower than a CAV system.

9.C.2.9.3.1 *Constant Air Volume Laboratory Chemical Hoods*

A CAV chemical hood draws a constant exhaust volume regardless of sash position. Because the volume is constant, the face velocity varies inversely with the sash position. The laboratory chemical hood volume should be adjusted to achieve the proper face velocity at the desired working height of the sash, and the chemical hood should be operated at this height. (See section 9.C.4.)

9.C.2.9.3.2 *Constant Air Volume Nonbypass Laboratory Chemical Hoods*

A nonbypass chemical hood has only one major opening through which the air may pass, that is, the sash opening. The airflow pattern is shown in Figure 9.5. A CAV nonbypass chemical hood has the undesirable characteristic of producing very large face velocities at small sash openings. As the sash is lowered, face velocities may exceed 1,000 fpm near the bottom. Face velocities are limited by the leakage through cracks and under the airfoil and by the increasing pressure drop as the sash is closed.

A common misconception is that the volume of air exhausted by this type of chemical hood decreases when the sash is closed. Although the pressure drop increases slightly as the sash is closed, no appreciable change in volume occurs. All chemical hoods should be closed when not in use, because they provide a primary barrier to the spread of a fire or chemical release.

Many trained laboratory personnel are reluctant to close their CAV nonbypass chemical hoods because of the increase in air velocity and noise that occurs when the sash is lowered. This high-velocity air jet sweeping over the work surface often disturbs gravimetric measurements, causes undesired cooling of heated vessels and glassware, and can blow sample trays, gloves, and paper towels to the back of the laboratory chemical hood, where they may be drawn into the exhaust system. Exercise care to prevent materials from entering the exhaust system where they can lodge in the ductwork, reducing airflow, or can be conveyed through the system and drawn into the exhaust fan and damage the fan or cause sparks.

Because of numerous operational problems with the design of nonbypass hoods, their installation in new

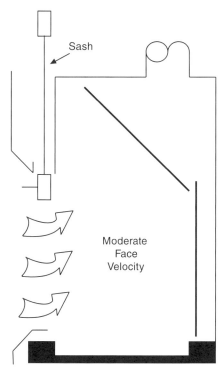

SASH FULLY OPEN

Sash

Moderate Face Velocity

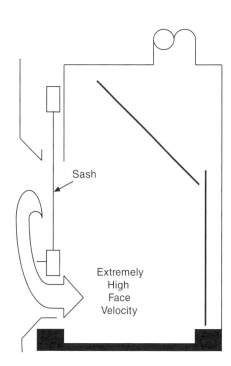

SASH PARTLY OPEN

Sash

Extremely High Face Velocity

FIGURE 9.5 Effect of sash placement on airflow in a nonbypass laboratory chemical hood.

SASH FULLY OPEN

SASH PARTLY OPEN

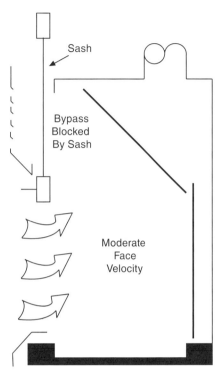
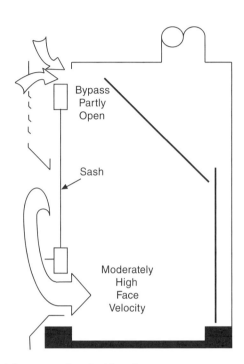

FIGURE 9.6 Effect of sash placement on airflow in a bypass laboratory chemical hood.

facilities is discouraged. If present in existing facilities, their replacement should be considered. In many instances, the cost of replacement can be recouped from the resulting reduction in energy costs.

9.C.2.9.3.3 Bypass Laboratory Chemical Hoods

A bypass chemical hood is shown in Figure 9.6. It is similar to the nonbypass design but has an opening above the sash through which air may pass at low sash positions. Because the opening is usually 20 to 30% of the maximum open area of the sash, this hood will still exhibit the increasing velocity characteristic of the nonbypass chemical hood as the sash is lowered. But the face velocity stops increasing as the sash is lowered to the position where the bypass opening is exposed by the falling sash. The terminal face velocity of these types of hoods depends on the bypass area but is usually in the range of 300 to 500 fpm—significantly higher than the recommended operating face velocity. Therefore, the air volume for bypass laboratory chemical hoods should also be adjusted to achieve the desired face velocity at the desired sash height, and the hood should be operated at this position. This arrangement is usually found in combination with a vertical sash, because this is the simplest arrangement for opening the bypass. Varieties are available for horizontal sashes, but the bypass mechanisms are

complicated and may cause maintenance problems. For a well-designed bypass hood, the face velocity will stay relatively constant until open about 12 in., then increases rapidly.

9.C.2.9.3.4 Variable Air Volume Laboratory Chemical Hoods

A VAV chemical hood, also known as a constant velocity hood, is one that has been fitted with a face velocity control, which varies the amount of air exhausted from the chemical hood in response to the sash opening to maintain a constant face velocity. In addition to providing an acceptable face velocity over a relatively large sash opening (compared to a CAV hood), VAV hoods also provide significant energy savings by reducing the flow rate when it is closed. These types of hoods are usually of the nonbypass design to reduce air volume (see below). Even though the face velocity responds to the position of the sash, the face velocity may drop off as the sash height increases, depending on the design. As a result, there is a maximum sash height above which the chemical hood becomes less effective.

9.C.2.9.3.5 Auxiliary Air Laboratory Chemical Hoods

Quantitative tracer gas testing of many auxiliary air laboratory chemical hoods has revealed that, even when adjusted properly and with the supply air properly conditioned, significantly higher personnel exposure to the materials used may occur than with conventional (non-auxiliary air) chemical hoods. They should not be purchased for new installations, and existing ones should be replaced or modified to eliminate the supply air feature. This feature causes a disturbance of the velocity profile and leakage of fumes into the personnel breathing zone.

The auxiliary air chemical hood was developed in the 1970s primarily to reduce laboratory energy consumption and is a combination of a bypass hood and a supply air diffuser located at the top of the sash. They were intended to introduce unconditioned or tempered air, as much as 70% of the air exhausted, directly to the front of the chemical hood. Ideally, this unconditioned air bypasses the laboratory and significantly reduces air-conditioning and heating costs. In practice, however, many problems are caused by introducing unconditioned or slightly conditioned air above the sash, all of which may produce a loss of containment.

9.C.2.9.3.6   *Low-Flow or High-Performance Laboratory Chemical Hoods*

With rising energy costs and high interest in sustainable laboratory design, manufacturers are producing low-flow, "high-performance" hoods that are able to meet the performance criteria of the ANSI/ASHRAE 110-1995 tests at a lower face velocity. They tend to be deeper than the traditional laboratory chemical hood and some have altered air front airfoils, internal or external auxiliary air devices, and/or automatic baffle controls. There are other design differences from a traditional chemical hood; thus, it is usually not possible to simply reduce the flow of a traditional hood to a lower face velocity and expect it to meet the same performance criteria as these specially designed hoods.

Like any other chemical hood, the design criteria and limitations need to be fully understood before one is selected for the laboratory. For example, if the chemical hood is designed to meet performance criteria at a sash height of 18 in., but users must operate it at a sash height of 24 in., the hood may not be effective at 24 in., creating a potentially hazardous situation.

Reviews by users have been mixed. For best results, be sure that the engineers, trained laboratory personnel, and the vendors understand how the chemical hoods are intended to be used. Their design and function continue to improve.

9.C.2.9.3.7   *Ductless Laboratory Chemical Hoods*

Ductless laboratory chemical hoods are ventilated enclosures that have their own fan, which draws air out and through filters and ultimately recirculates it into the laboratory. The filters are designed to trap vapors generated in the chemical hood and exhaust clean air back into the laboratory. They frequently use activated carbon filters, HEPA filters, or a combination of the two. Newer filter materials on the market claim that they capture a larger variety of chemicals.

These ventilated enclosures do not necessarily achieve the same level of capture and containment as a chemical hood. Unlike a conventional laboratory chemical hood, it is not possible to conduct tracer gas studies to measure containment even with the newer technology ductless hoods. Because the collection efficiency of the filters decreases over time, the filters must be monitored and replaced routinely. Depending on the materials and the laboratory environment, chemicals can desorb from the filter and reenter the laboratory over time. They do not control fire hazards and National Fire Protection Association standard 45 states "Ductless chemical fume hoods that pass air from the hood interior through an absorption filter and then discharge the air into the laboratory are only applicable for use with nuisance vapors and dusts that do not present a fire or toxicity hazard" (NFPA, 2004).

Ductless chemical hoods have *extremely* limited applications and should be used *only* where the hazard is very low, where the access to the hood and the chemicals used in it are carefully controlled, and under the supervision of a laboratory supervisor who is familiar with its serious limitations. If these limitations cannot be accommodated, do not use this type of device.

The benefits of recirculating chemical hoods are that they are much more energy efficient than a ducted chemical hood and they do not require a ventilation system that relies on a fan on the roof or upper levels. Some urban buildings retrofitted with laboratories on lower floors, buildings with limitations on the ventilation system or laboratories with minor chemical use have successfully used these ductless hoods, under the limited conditions cited above and with rigorous filter maintenance programs. They can also be used for control of particulate material where a chemical hood or even Class 1 or 2 biosafety cabinets provide too much turbulent air (see section 9.E.4.1).

To determine whether recirculating hoods are appropriate, an industrial hygienist or safety professional should conduct a risk assessment that includes

- an analysis of the chemicals that will be used, the hazards they pose, and the materials they generate as byproducts;
- the frequency and duration of use of these chemicals; and

• the nature of the materials that must be controlled compared to the filter media provided with the recirculating hood.

Individuals using recirculating hoods need training on the use and limitations of the recirculating hood. Each ductless chemical hood should have signage explaining the limitations, how to detect whether the filter media are working, and the filter maintenance schedule.

9.C.2.10 Laboratory Chemical Hood Configurations

9.C.2.10.1 Benchtop Laboratory Chemical Hoods

As the name implies, a benchtop chemical hood sits on a laboratory bench with the work surface at bench height. It can be of the CAV or VAV variety and can have a bypass or nonbypass design. The sash can be a vertical-rising or a horizontal-sliding type or a combination of the two. Normally, the work surface is dished or has a raised lip around the periphery to contain spills. Sinks in chemical hoods are not recommended because they encourage laboratory personnel to dispose of chemicals in them. If they must be used, to drain cooling water from a condenser, for instance, they should be fitted with a standpipe to prevent

chemical spills from entering the drain. The condenser water drain can be run into the standpipe. Spills will be caught in the cupsink by the standpipe for later cleanup and disposal. A lip on the cupsink could be used as an alternative to a standpipe to prevent spills from getting into the sink. A typical benchtop chemical hood is shown in Figure 9.7.

9.C.2.10.2 Distillation (Knee-High or Low-Boy) Chemical Fume Hoods

The distillation hood is similar to the benchtop hood except that the work surface is closer to the floor to allow more vertical space inside for tall apparatuses such as distillation columns. A typical distillation hood is shown in Figure 9.8.

9.C.2.10.3 Walk-In Laboratory Chemical Hoods

A walk-in hood stands on the floor of the laboratory and is used for very tall or large apparatus. The sash can be either horizontal or double- or triple-hung vertical. These hoods are usually of the nonbypass type. The word "walk-in" is a misnomer; one should never actually walk into a chemical hood when it is operating and contains hazardous chemicals. Once past the plane of the sash, the personnel are inside with the chemicals. If the personnel are required to enter the hood during operations where hazardous chemicals are present,

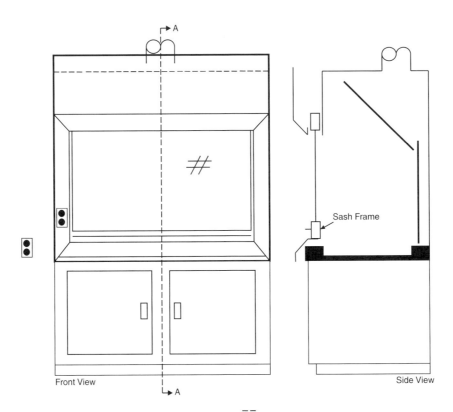

Front View

Side View

Sash Frame

FIGURE 9.7 Diagram of a typical benchtop laboratory chemical hood.

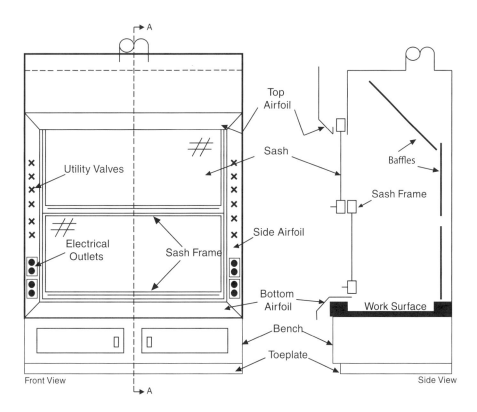

FIGURE 9.8 Diagram of a typical distillation hood.

they should wear PPE appropriate for the hazard. It may include respirators, chemical splash goggles, rubber gloves, boots, suits, and self-contained breathing apparatus. A typical walk-in chemical hood is shown in Figure 9.9.

9.C.2.10.4 California Laboratory Chemical Fume Hoods and Ventilated Enclosures

The California chemical fume hood is a ventilated enclosure with a movable sash on more than one side. They are usually accessed through a horizontal sliding sash from the front and rear. They may also have a sash on the ends. Because their configuration precludes the use of baffles and airfoils, they may not provide a suitable face velocity distribution across their many openings.

A ventilated enclosure is any site-fabricated chemical hood designed primarily for containing processes such as scale-up or pilot plant equipment. Most do not have baffles or airfoils, and most designs have not had the rigorous testing and design refinement that conventional mass-produced chemical hoods enjoy. Working at the opening of the devices, even when the plane of the opening has not been broken, may expose personnel to higher concentrations of hazardous materials than if a conventional hood were used.

9.C.2.10.5 Perchloric Acid Laboratory Chemical Hoods

The perchloric acid laboratory chemical hood, with its associated ductwork, exhaust fan, and support systems, is designed especially for use with perchloric acid and other materials that can deposit shock-sensitive crystalline materials in the hood and exhaust system. These materials become pyrophoric when they dry or dehydrate (see also Chapter 6, section 6.G.6). Special water spray systems are used to wash down all interior surfaces of the hood, duct, fan, and stack, and special drains are necessary to handle the effluent from the washdown. The liner and work surface are usually stainless steel with welded seams. Perchloric acid hoods have drains in their work surface. Water spray heads are usually installed in the top, behind the baffles, and in the interior. The water spray should be turned on whenever perchloric acid is being heated in the chemical fume hood. The ductwork should be fabricated of plastic, glass, or stainless steel and fitted with spray heads approximately every 10 ft on vertical runs and at each change in direction. The fan and stack should be fabricated of plastic, fiberglass, or stainless steel. Welded or flanged and gasketed fittings to provide airtight and watertight connections are recommended. Avoid horizontal runs because they

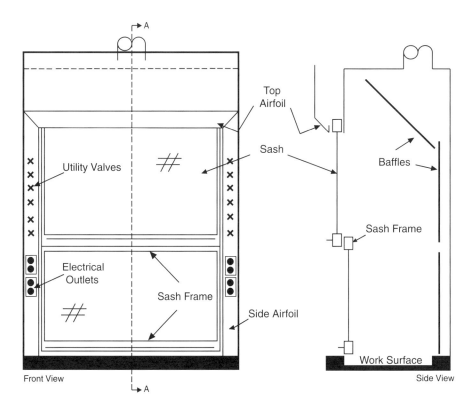

FIGURE 9.9 Diagram of a typical walk-in laboratory chemical hood.

inhibit drainage, and the spray action is not as effective on the top and sides of the duct. Any washdown piping, which is located outside must be protected from freezing. A drain and waste valve on the water supply piping that allows it to drain when not in use is helpful. Route the drain lines carefully to prevent the creation of traps that retain water. Write special operating procedures to cover the washdown procedure for these types of hoods. The exhaust from a perchloric acid hood should not be manifolded with that from other types of chemical hoods.

9.C.2.10.6   Radioisotope Laboratory Chemical Hoods

Design chemical hoods used for work with radioactive sources or materials so that they can be decontaminated completely on a regular basis. A usual feature is a one-piece stainless steel welded liner with smooth curved corners that can be cleaned easily and completely. The superstructure of radioisotope hoods is usually made stronger than that of a conventional hood to support lead bricks and other shielding that may be required. Special treatment of the exhaust from radioisotope hoods may be required by government regulations to prevent the release of radioactive material into the environment. This treatment usually involves the use of HEPA filters (see section 9.C.4.2).

Another practical way to handle radioactive materi-

als that require special exhaust treatment is to use a containment chamber within a traditional chemical hood. Several safety supply companies offer portable disposable glovebag containment chambers with sufficient space to conduct the work and then dispose of them in accordance with applicable nuclear regulatory standards.

9.C.2.10.7   Clean Room Laboratory Chemical Hoods

Chemical hoods in clean rooms are generally no different than traditional chemical hoods, except that they are usually made of polypropylene or thermoplastics. Some have hinged sashes rather than sliding sashes. Most require separate chemical hoods for acid work and solvent work.

Polypropylene hoods burn easily, melt quickly, and may become fully involved in a fire. There are fire-retardant polypropylene and other thermoplastics available, but they cost more. Alternatively, an automatic fire extinguisher may be installed inside.

9.C.2.11   Laboratory Chemical Hood Exhaust Treatment

Until recently, treatment of laboratory chemical hood exhausts has been limited. Because effluent quantities and concentrations are relatively low compared to those of other industrial air emission sources, their removal is technologically challenging. And the chemistry for a given chemical hood effluent can be difficult to predict and may change over time.

Nevertheless, legislation and regulations increasingly recognize that certain materials in laboratory chemical hoods may be sufficiently hazardous that they can no longer be expelled directly into the air. Therefore, the practice of removing these materials from exhaust streams will become increasingly more prevalent.

9.C.2.11.1 Laboratory Chemical Hood Scrubbers and Contaminant Removal Systems

A number of technologies are evolving for treating chemical hood exhaust by means of scrubbers and containment removal systems. Whenever possible, experiments involving toxic materials should be designed so that they are collected in traps or scrubbers rather than released. If for some reason collection is impossible, HEPA filters are recommended for highly toxic particulates. Liquid scrubbers may also be used to remove particulates, vapors, and gases from the exhaust system. None of these methods, however, is completely effective, and all trade an air pollution problem for a solid or liquid waste disposal problem. Incineration may be the ultimate method for destroying combustible compounds in exhaust air, but adequate temperature and dwell time are required to ensure complete combustion.

Incinerators require considerable capital to build and energy to operate; hence, other methods should be studied before resorting to their use. Determine the optimal system for collecting or destroying toxic materials in exhaust air on a case-by-case basis. Treatment of exhaust air should be considered only if it is not practical to pass the gases or vapors through a scrubber or adsorption train before they enter the exhaust airstream. Also, if an exhaust system treatment device is added to an existing chemical hood, carefully evaluate the impact on the fan and other exhaust system components. These devices require significant additional energy to overcome the pressure drop they add to the system. (See also Chapter 8, section 8.B.6.1.)

9.C.2.11.2 Liquid Scrubbers

A laboratory chemical hood scrubber is a laboratory-scale version of a typical packed-bed liquid scrubber used for industrial air pollution control. Figure 9.10 shows a schematic of a typical chemical hood scrubber.

Contaminated air from the chemical hood enters the unit and passes through the packed-bed, liquid spray section, and mist eliminator and into the exhaust system for release up the stack. The air and the

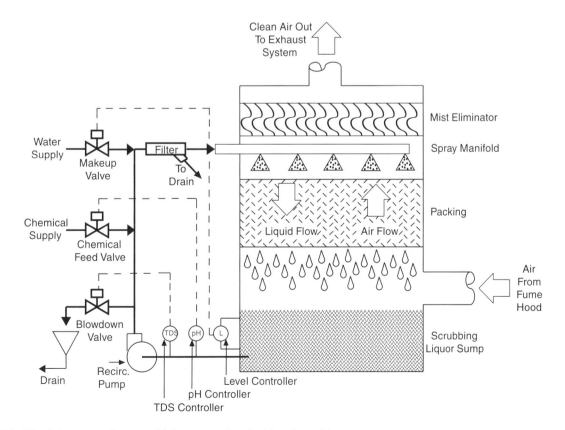

FIGURE 9.10 Schematic of a typical laboratory chemical hood scrubber.

scrubbing liquor pass in a countercurrent fashion for efficient gas-liquid contact. The scrubbing liquor is recirculated from the sump and back to the top of the system using a pump. Water-soluble gases, vapors, and aerosols are dissolved into the scrubbing liquor. Particulates are also captured quite effectively by this type of scrubber. Removal efficiencies for most water-soluble acid- and base-laden airstreams are usually between 95 and 98%.

Scrubber units are typically configured vertically and are located next to the chemical hood as shown in Figure 9.10. They are also produced in a top-mount version, in which the packing, spray manifold, and mist eliminator sections are located on top of the chemical hood and the sump and liquid-handling portion are underneath for a compact arrangement taking up no more floor area than the hood itself. Most hoods do not require a scrubber unit, assuming the exhaust stack is designed properly and chemical quantities of volatile materials are low.

9.C.2.11.3   Other Gas-Phase Filters

Another basic type of gas-phase filtration is available for chemical hoods in addition to liquid scrubbers. These are inert adsorbents and chemically active adsorbents. The inert variety includes activated carbon, activated alumina, and molecular sieves. These substances typically come in bulk form for use in a deep bed and are available also as cartridges and as panels for use in housings similar to particulate filter housings. They are usually manufactured in the form of beads, but they may take many forms. The beads are porous and have extremely large surface areas with sites onto which gas and vapor molecules are trapped or adsorbed as they pass through. Chemically active adsorbents are simply inert adsorbents impregnated with a strong oxidizer, such as potassium permanganate (purple media), which reacts with and destroys the organic vapors. Although there are other oxidizers targeted to specific compounds, the permanganates are the most popular. Adsorbents can handle hundreds of compounds, including most volatile organic components but also have an affinity for harmless species such as water vapor.

As the air passes through the adsorbent bed, gases are removed in a section of the bed. (For this discussion, gas means gases and vapors.) As the bed loads with gases, and if the adsorbent is not regenerated or replaced, eventually contaminants will break through the end of the bed. After breakthrough occurs, gases will pass through the bed at higher and higher concentrations at a steady state until the upstream and downstream levels are almost identical. To prevent breakthrough, the adsorbent must be either changed or regenerated on a regular basis. Downstream monitor-ing to detect breakthrough or sampling of the media to determine the remaining capacity of the bed should be performed regularly.

An undesirable characteristic of these types of scrubbers is that if high concentrations of organics or hydrocarbons are carried into the bed, as would occur if a liquid were spilled inside the hood, a large exotherm occurs in the reaction zone of the bed. This exotherm may cause a fire in the scrubber. Place these scrubbers and other downstream devices such as particulate filters in locations where the effects of a fire would be minimized. Fires can start in these devices at surprisingly low temperatures because of the catalytic action of the adsorbent matrix. Therefore, use and operate such devices with care.

9.C.2.11.4   High-Efficiency Filters

Air from laboratory chemical hoods and biological safety cabinets (BSCs) in which some radioactive or biologically active particulates are used should be properly filtered to remove these agents and prevent their release into the atmosphere. Other hazardous particulates may require this type of treatment as well. The most popular method of removal is a HEPA filter. These HEPA filters trap 99.97% of all particulates greater than 0.3 μm in diameter and may be just as effective with smaller particle sizes. Studies have shown that HEPA filters can be quite effective at trapping nanoparticles, due to Brownian motion and electrostatic capture. Before any filtration system is installed, a risk assessment should be performed to determine the need and the appropriate level of filtration required.

Ultra-low penetration air (ULPA) filters are an alternative to HEPA filters. These filters are 99.9995% efficient in removing particles greater than 0.12 μm. However, ULPA filters are more expensive than HEPA filters, and they increase the system static pressure. Note that any system designed to provide protection against radioactive particles can be expected to be effective against nanoparticles, and studies have confirmed that HEPA filters provide sufficient capture for nanoparticles (HHS/CDC/NIOSH, 2009a) making ULPA unnecessary.

These systems must be specified, purchased, and installed so that the filters can be changed without exposing the personnel or the environment to the agents trapped in the filter. Sterilizing the filter bank is prudent before changing filters that may contain etiologic agents.

The bag-in, bag-out method of replacing filters is a popular way to prevent personnel exposure. This method separates the contaminated filter and housing from the personnel and the environment by using a special plastic barrier bag and special procedures to prevent exposure to or release of the hazardous agent.

9.C.2.11.5   Thermal Oxidizers and Incinerators

Thermal oxidizers and incinerators are extremely expensive to purchase, install, operate, and maintain. However, they are one of the most effective methods of handling toxic and etiologic agents. The operational aspects of these devices are beyond the scope of this book. Also, their application to chemical hoods has historically been rare. When considering this method of pollution control, call an expert for assistance.

9.C.3   Other Local Exhaust Systems

Many types of laboratory equipment and apparatus that generate vapors and gases should not be used in a conventional laboratory chemical hood. Some examples are gas chromatographs, atomic absorption spectrophotometers, mixers, vacuum pumps, and ovens. If the vapors or gases emitted by these types of equipment are hazardous or noxious, or if it is undesirable to release them into the laboratory because of odor or heat, contain and remove them using local exhaust equipment. Local capture equipment and systems should be designed only by an experienced engineer or industrial hygienist. Also, users of these devices must have appropriate training.

Whether the emission source is a vacuum-pump discharge vent, a gas chromatograph exit port, or the top of a fractional distillation column, the local exhaust requirements are similar. The total airflow should be high enough to transport the volume of gases or vapors being emitted, and the capture velocity should be sufficient to collect the gases or vapors.

Despite limitations, specific ventilation capture systems provide effective control of emissions of toxic vapors or dusts if installed and used correctly and, in some cases, can result in energy savings. A separate dedicated exhaust system is recommended. Do not attach the capture system to an existing laboratory chemical hood duct unless fan capacity is increased and airflow to both hoods is properly balanced. One important consideration is the effect that such added local exhaust systems will have on the ventilation for the rest of the laboratory. Each additional capture hood will be a new exhaust port in the laboratory and will compete with the existing exhaust sources for air supply.

Downdraft ventilation has been used effectively to contain dusts and other dense particulates and high concentrations of heavy vapors that, because of their density, tend to fall. Such systems require special engineering considerations to ensure that the particulates are transported in the airstream. Here again, consult a ventilation engineer or industrial hygienist if this type of system is deemed suitable for a particular laboratory operation.

9.C.3.1   Elephant Trunks, Snorkels, or Extractors

An elephant trunk, or snorkel, is a piece of flexible duct or hose connected to an exhaust system. To capture contaminants effectively, it must be closer than approximately one-half a diameter of the hood from the end of the hose. An elephant trunk is particularly effective for capturing discharges from gas chromatographs, pipe nipples, and pieces of tubing if the hose is placed directly on top of the discharge with the end of the discharge protruding to the hose. Note that unless the intake for the snorkel is placed very close to the point source, it will be susceptible to inefficient capture. Newer designs mount the intake on an articulated arm, which tends to make the systems more effective and convenient to use. (See Figure 9.11.) The volume flow rate of the hose must be at least 110 to 150% of the flow rate of the discharge.

The face velocity for a snorkel or elephant trunk is usually 150–200 fpm. The velocity and the capture efficiency drop sharply with distance from the intake. As a result, efficient capture of contaminants is generally adequate when the discharge source is 2 in. away, but inadequate if it is 3 in. away. In cases where there is a question about efficacy of capture, perform a smoke test to determine if the flow rate is adequate (ACGIH, 2004).

9.C.3.2   Slot Hoods

Slot hoods are local exhaust ventilation hoods specially designed to capture contaminants generated according to a specific rate, distance in front of the hood, and release velocity for specific ambient airflow. In general, if designed properly, these hoods are more effective and operate using much less air than either elephant trunks or canopy hoods. To be effective, however, the geometry, flow rate, and static pressure must all be correct.

Typical slot hoods are shown in Figure 9.12. Each type has different capture characteristics and applications. If laboratory personnel believe that one of these devices is necessary, a qualified ventilation engineer should design the hood and exhaust system.

9.C.3.3   Canopy Hoods

The canopy hood is not only the most common local exhaust system but also probably the most misunderstood piece of industrial ventilation equipment. Industrial ventilation experts estimate that as many as 95% of the canopy hoods in use (other than in homes and restaurants) are misapplied *and* ineffective. The capture range of a canopy hood is extremely limited, and a large volume of air is needed for it to operate

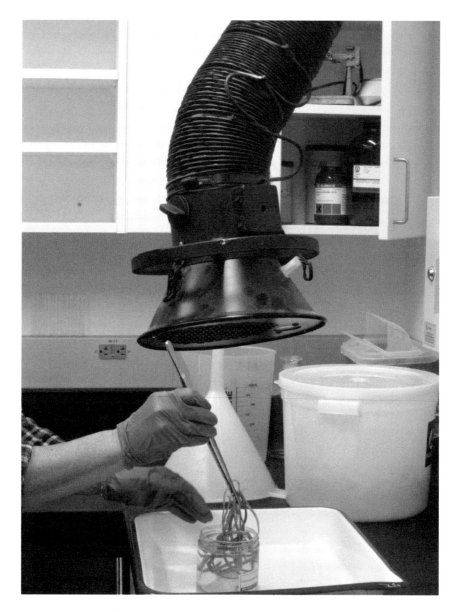

FIGURE 9.11   Fume extractor or snorkel.

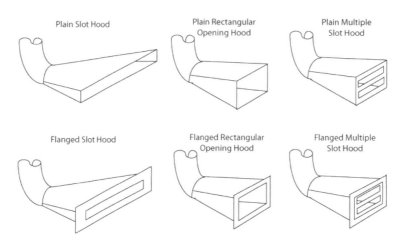

FIGURE 9.12   Diagrams of typical slot hoods.

effectively. Thus, a canopy hood works best when thermal or buoyant forces exist that move the contaminant up to the hood capture zone (a few inches below the opening). However, because canopy hoods are generally placed well above a contaminant source so that laboratory personnel can operate underneath them, they draw contaminants past the breathing zone and into the exhaust system. If a canopy hood exists in a laboratory, use it only for nonhazardous service, such as capturing heated air or water vapor from ovens or autoclaves. For design advice, consult the American Conference of Governmental Industrial Hygienists ventilation manual (ACGIH, 2004) and ANSI Z9.2.

9.C.3.4   Downdraft Hoods

Downdraft hoods or necropsy tables are specially designed work areas with ventilation slots on the sides of the work area. This type of system is useful for animal perfusions, gross anatomy laboratories, and other uses of chemicals where there is a need to have full access over and around the materials (which would be obstructed by the three sides of a chemical hood) and the chemicals in use have vapor densities that are heavier than air.

9.C.3.5   Clean Benches or Laminar Flow Hoods

A clean bench or laminar flow hood resembles a chemical hood but is not intended to provide protection to the user. A clean bench is generally closed on three sides and either is fully open in the front or has a partial opening. Some have hinged or sliding sashes. On the top or back of the clean bench, HEPA filters pull room air through the filters and pass that air across the work surface, providing clean air. The clean bench is for product protection, not personal protection, and is not connected to the ventilation system. Mark such equipment "not for use with hazardous materials" to remind laboratory personnel not to use anything in it that they would not use on the benchtop.

9.C.3.6   Ventilated Balance Enclosures

Ventilated balance enclosures are commonly used in laboratories to weigh toxic particulates. These devices are installed with different specifications for face velocity than the standard laboratory chemical hood and are well suited for locating sensitive balances that might be disturbed if placed in a laboratory chemical hood. The average face velocity is specified at 75 fpm plus or minus 10 fpm (0.40 ± 0.05 m/s). Individual face velocity at each grid point should be within a tolerance of plus or minus 20 fpm (0.10 m/s). Ventilated balance

enclosures are typically equipped with HEPA filters to remove hazardous particulates captured within the device prior to exhaust. They can be either the recirculating type or 100% exhausted to the exterior.

Housings for ventilated balance enclosures are generally constructed of minimum 3/8-in.-thick (10-mm-thick) clear acrylic. Edges of the vertical sides are beveled, rounded, or otherwise aerodynamically designed to reduce turbulence at the perimeter of the face. Ventilated balance enclosures consist of an integrated dished base that facilitates cleaning at the interface of the vertical and horizontal surfaces. Airfoil sills have an ergonomic radius on the front edge. Sash configuration consists of a hinged single sash pane for cabinet widths and provides a full, clear, and unobstructed side-to-side view of the entire cabinet interior. Sash openings are usually at a fixed height of 8 to 12 in. (200 to 300 mm) above the work surface.

9.C.3.7   Gas Cabinets

Whenever possible, minimize use of highly toxic or hazardous gases and restrict them to lecture bottles that are placed on stands and used within the confines of a chemical hood.

Use and store containers for highly toxic or hazardous gases, such as diborane, phosgene, or arsine, that are too large to be used within a chemical hood in ventilated gas cabinets. In the event of a leak or rupture, a gas cabinet prevents the gas from contaminating the laboratory. Consult the standards developed by SEMI for specific, recommended exhaust rates for gas cabinets.

Connect gas cabinets to laboratory exhaust ventilation using metal ductwork, rather than flexible tubing, because such tubing is more apt to develop leaks. Use coaxial tubing for delivering gas from the cylinder to the apparatus. Coaxial tubing consists of an internal tube containing the toxic gas, inside another tube. Nitrogen, which is maintained at a pressure higher than the delivery pressure of the toxic gas, is between the two sets of tubing, ensuring that, in the event of a leak in the inner tubing, the gas will not leak into the room.

9.C.3.8   Flammable-Liquid Storage Cabinets

Store flammable and combustible liquids only in approved flammable-liquid storage cabinets, not in a chemical hood, on the bench, or in an unapproved storage cabinet. These cabinets are designed to prevent the temperature inside the cabinet from rising quickly in the event of a fire directly outside of the cabinet. These cabinets may be ventilated or unventilated. Ventilating flammable-liquid storage cabinets is a matter for debate. One view is that all such cabinets should be vented by using an approved exhaust system, because it reduces the concentration of flammable vapors below the LEL inside the cabinet. A properly designed cabinet ventilation system does this under most circumstances and results in a situation in which no fuel is rich enough in vapor to support combustion. However, with liquid in the cabinet and a source of fresh air provided by the ventilation system, all that is needed is an ignition source. The other view is that in most circumstances flammable-liquid storage cabinets should not be ventilated.

Both opinions are valid, depending on the conditions. Ventilation is prudent when the liquids stored in the cabinet are highly toxic or extremely odoriferous. Particularly odoriferous substances such as mercaptans have such a low odor threshold that even with meticulous housekeeping the odors persist; and, ventilation may be desired. Local authorities may have specific regulations regarding the need for ventilation within the fire cabinet.

If a ventilated flammable-liquid storage cabinet is used under a chemical hood, do not vent it into the chemical hood above it. It should have a separate exhaust duct connected to the exhaust system. Fires occur most frequently in chemical hoods and may propagate into a flammable-liquid storage cabinet that is directly vented into it.

If a specially designed flammable storage cabinet ventilation system is installed, use an Air Movement Control Association C-type spark-resistant fan and an explosion-proof motor. Most fractional horsepower fans commonly used for this purpose do not meet this criterion and should not be used. If the building has a common laboratory chemical hood exhaust system, hook a flammable-liquid storage cabinet up to it for ventilation.

9.C.3.9   Benchtop Enclosures

Many laboratory ventilation system manufacturers offer ventilated enclosures that can be sized to fit equipment that would normally be placed in a chemical hood, such as rotovaps and microwave ovens. They can be made of metal or plastic and could have doors or sashes for access. The velocity of air will vary depending on the material being ventilated. The enclosure may be fitted with a filtration system for nanomaterials. By placing larger equipment in a ventilated enclosure rather than a hood, the amount of space in the hood in maximized and smaller hoods may be acceptable, resulting in energy and space savings.

9.C.4   General Laboratory Ventilation and Environmental Control Systems

General ventilation systems control the quantity and quality of the air supplied to and exhausted from the laboratory. The general ventilation system should ensure that the air is continuously replaced so that concentrations of odoriferous or toxic substances do not increase during the workday and are not recirculated from laboratory to laboratory.

Exhaust systems fall into two main categories: general and specific. General systems serve the whole laboratory and include devices such as chemical hoods and snorkels, as codes and good design practices allow. Specific systems serve isotope hoods, perchloric acid hoods, or other high-hazard sources that require isolation from the general laboratory exhaust systems.

General laboratory ventilation is typically set to provide 6 to 12 room air changes per hour. However, there is no specific requirement for ventilation rates. More airflow may be required to cool laboratories with high internal heat loads, such as those with analytical equipment, or to service laboratories with large specific exhaust system requirements or those with high densities of chemical hoods or other local exhaust ventilation devices. The ACGIH industrial ventilation manual states that "'Air changes per hour' or 'air changes per minute' is a poor basis for ventilation criteria where environmental control of hazards, heat and/or odors is required. The required ventilation depends on the problem, not the size of the room in which it occurs" (ACGIH, 2004). Where dilution ventilation will be the primary means of controlling exposure, the ventilation rate is dependent upon the materials in use. Standard industrial hygiene calculations may help to determine the required rate. Computational fluid dynamics models are often utilized to determine minimal rates when the lab is occupied and unoccupied.

Air should always flow from the offices, corridors, and support spaces into the laboratories. Exhaust all air from chemical laboratories outdoors and do not recirculate it. Thus, the air pressure in chemical laboratories should be negative with respect to the rest of the building unless the laboratory is also a clean room (see section 9.E.2). The outside air intakes for a laboratory building should be in a location that reduces the possibility of reentrainment of laboratory exhaust or contaminants from other sources such as waste disposal areas and loading docks.

Although the supply system provides dilution of toxic gases, vapors, aerosols, and dust, it gives only modest protection, especially if these impurities are released into the laboratory in any significant quantity. Perform operations that release these toxins, such as running reactions, heating or evaporating solvents, and transfer of chemicals from one container to another, in a laboratory chemical hood where possible. Vent laboratory apparatus that may discharge toxic vapors, such as vacuum pump exhausts, gas chromatograph exit ports, liquid chromatographs, and distillation columns to an exhaust device such as an elephant trunk.

The steady increase in the cost of energy, coupled with a greater awareness of the risks associated with the use of chemicals in the laboratory, has caused a conflict between the desire to minimize the costs of heating, cooling, humidifying, and dehumidifying laboratory air and the need to provide laboratory personnel with adequate ventilation. However, cost considerations should never take precedence over ensuring that personnel are protected from hazardous concentrations of airborne toxic substances.

9.C.4.1 Constant Air Volume Systems

CAV air systems assume constant exhaust and supply airflow rates throughout the laboratory. Although such systems are the easiest to design, and sometimes are the easiest to operate, they have significant drawbacks due to their high energy consumption and limited flexibility. Classical CAV design assumes that all chemical hoods operate 24 hours per day, 7 days per week, and at constant maximum volume. Adding, changing, or removing chemical hoods or other exhaust sources for CAV systems requires rebalancing the entire system to accommodate the changes. Most CAV systems in operation today are unbalanced and operate under significant negative pressure. These conditions are caused by the inherent inflexibility of this design type, coupled with the addition of chemical hoods not originally in the plan.

9.C.4.2 Variable Air Volume Systems

VAV systems are based on laboratory chemical hoods with face velocity controls. As users operate the chemical hoods, the exhaust volume from the laboratory changes and the supply air volume must adapt to maintain a volume balance and room pressure control. Consult an experienced laboratory ventilation engineer to design these systems, because the systems and controls are complex and must be designed, sized, and matched to operate effectively together.

VAV systems provide many opportunities for increased safety and energy conservation that cannot be accomplished with a CAV system. Individual laboratory chemical hoods, groups, or all chemical hoods on the same system can be adjusted to a lower airflow when not in use through timers, occupancy sensors, or other means (see section 9.H.3). Similarly, exhaust

may be automatically increased to purge the room in the event of a spill or release (see section 9.C.6.4).

9.C.5 Supply Systems

Well-designed laboratory air supply systems approach the ideal condition of laminar airflow, directing clean incoming air over laboratory personnel and sweeping contaminated air away from their breathing zone. Ventilation systems with well-designed diffusers that optimize complete mixing may also be satisfactory. Usually, several carefully selected supply-air diffusers are used in the laboratory. Ceiling plenums with perforated ceiling tiles have been used with some success, but can be difficult to design and maintain properly. In cases where high airflow rates are required, fabric diffusers may be a better option.

9.C.6 Exhaust Systems

9.C.6.1 Individual Laboratory Chemical Hood Fans

In some exhaust systems, particularly where there are just a few hoods in a building, each chemical hood has its own exhaust fan. This arrangement has both advantages and disadvantages.

Advantages include the following:

- The possibility of cross-contamination from one laboratory chemical hood discharge to another is eliminated.
- The potential to treat individual chemical hood exhaust (as opposed to treating all chemical hood exhaust) is excellent.
- A fan failure will affect only one chemical hood.

Disadvantages include the following:

- There is no way to dilute the laboratory chemical hood effluent before release.
- Providing redundancy and emergency power for this arrangement is difficult and expensive.
- The potential to use diversity (see section 9.C.6.2) is limited, as is the potential to use VAV controls.
- The potential to recover heat from individual fans is almost nonexistent.
- The maintenance requirement for these systems is considerable, because they contain many pieces of equipment and have many roof penetrations, which can cause leakage problems.
- The mechanical (shaft) space requirements, initial cost, and operating cost are higher than for alternative systems, such as manifolded systems.

9.C.6.2 Manifolded (Common Header) Systems

For compatible exhaust streams, providing a common manifolded exhaust system is an attractive design alternative to individual laboratory chemical hood fans. This design is chosen increasingly for new laboratory buildings and is compatible with VAV systems.

Manifolded VAV systems also allow design engineers to take advantage of diversity. Simply stated, diversity is an estimate of the actual expected peak airflow rate expressed as a percentage of the total exhaust capacity. The rationale is that it is not reasonable to expect that all chemical hoods would have all sashes open and laboratory personnel actively working at the same time. Thus, the exhaust system is designed for a maximum load of a lower percentage of the exhaust capacity, rather than 100%, which results in both installation cost savings and ongoing energy savings. If an aggressive diversity rate is desired, the customer may need to indemnify the designers. Engineering firms may be reluctant to design for aggressive diversity because of the potential for liability if the system capacity is later deemed inadequate. Additionally, understanding building usage patterns is critical to diversity calculation. For example, because of consistency in laboratory practices from day to day, an industrial research lab may be able to accommodate a much greater diversity than a chemistry instructional laboratory.

Manifolded systems have the following advantages and disadvantages:

- The potential for mixing and dilution of high concentrations of contaminants from a single chemical hood by the air exhaust from all the other chemical hoods on the system is excellent.
- The cross-contamination potential from one hood to another is minimal.
- The potential to provide redundancy of exhaust fans and emergency power to these systems is excellent.
- Conversely, the effects of a fan failure are widespread and serious; hence, redundancy is required in most cases.
- The potential to take advantage of VAV diversity and flow variation is also excellent, as is the ability to oversize the system for future expansion and flexibility.
- The ability to treat individual exhausts is retained by using new in-line liquid scrubber technologies.
- The maintenance, operating, and initial costs of these systems are all lower than for individual chemical hood fan systems, and these systems require fewer roof penetrations.
- The heat recovery potential for these systems is maximized by collecting all the exhaust sources into a common duct.

Some municipalities have adopted the International Mechanical Codes, including IMC 510, which poses some restrictions on manifolding exhaust ductwork. Check with your code compliance officials to determine how this affects your buildings. Most of this code includes exemptions for laboratory applications.

9.C.6.3 Hybrid Exhaust Systems

Certain types of laboratory chemical hoods and exhaust sources, such as perchloric acid hoods, should not be manifolded with other types of chemical hood exhausts. In large buildings where the designer wishes to take advantage of the benefits of manifolded exhaust systems but wishes to isolate a few exhaust streams, a combination, or hybrid, of these two types of systems is usually the most prudent and cost-effective alternative.

9.C.6.4 Room Purge Systems

A room purge system is useful in the event of a spill or release in the laboratory. These systems may be found in areas with NMR spectrometers or other large sources of cryogenic gases as well as other areas where a spill or release may cause an asphyxiation hazard. The system works when laboratory personnel depress the purge button located at the entrance to the laboratory. Air in the laboratory ceases to be recirculated and the exhaust is maximized with the goal of removing the dangerous materials. The button can be pulled to its original position to reset the ventilation system to normal operation or may require a key for reset. Note, however, that these systems should not be considered foolproof. In any case where there is a threat of asphyxiation, it is best to evacuate the area and follow emergency protocols as mandated by your organization. In areas with documented asphyxiation risks, flow restrictors should be put in place where feasible.

9.C.6.5 Exhaust Stacks

Proper stack design and placement are an extremely important aspect of good exhaust system design. Recirculation of contaminated air from the chemical hood exhaust system into the fresh air supply of the facility or adjacent facilities may occur if stacks are not provided or if they are not designed properly to force the contaminated exhaust air up and into the prevailing wind stream. Stack design should take into account building aerodynamics, local terrain, nearby structures, and local meteorological information. Consult an experienced laboratory consultant with expertise in atmospheric dispersion to design exhaust stacks for a laboratory facility. Typical exhaust discharge velocity is recommended to be 3,000–4,000 fpm, but velocities can be lower with manifolded systems because of the benefit of dilution.

9.D ROOM PRESSURE CONTROL SYSTEMS

Laboratories and clean rooms usually require that a differential pressure be maintained between them and adjoining nonlaboratory spaces. This requirement for a pressure differential may come from code considerations or from the intended use of the space. For example, NFPA 45 states that "air pressure in the laboratory work areas shall be negative with respect to corridors and non-laboratory areas of the laboratory unit . . ." (NFPA, 2004). This rule helps prevent the migration of fire, smoke, and chemical releases from the laboratory space. Laboratories containing radiation hazards or biohazards may also be required by government agencies to maintain a negative pressure to contain these hazards. Clean rooms, on the other hand, are normally operated at a positive static pressure to prevent infiltration of particulates. (See sections 9.E.2 and 9.E.3, below, for further information.)

9.E SPECIAL SYSTEMS

9.E.1 Gloveboxes

Unlike a chemical hood, gloveboxes are fully enclosed and are under negative or positive pressure. Gloveboxes are usually small units that have multiple openings in which arm-length rubber gloves are mounted. The operator works inside the box by using these gloves. Construction materials vary widely, depending on the intended use. Clear plastic is frequently used, because it allows visibility of the work area and is easily cleaned.

A glovebox operating under negative pressure is generally used for highly toxic materials, when a chemical hood might not offer adequate protection. A rule of thumb is that a chemical hood offers protection for up to 10,000 times the immediately dangerous concentration of a chemical. The airflow through the glovebox is relatively low, and the exhaust usually must be filtered or scrubbed before it is released into the exhaust system. Nanoparticles can also be used in a glovebox. A cautionary statement pertaining to devices such as gloveboxes: Because these devices are designed with very low airflow rates, the rate of contaminant dilution is minimal. Therefore, to ensure adequate protection to laboratory personnel, these devices must be routinely tested for leaks to ensure that enclosure integrity is sufficient. If leakage is detected, the source of contaminant release must be identified and repaired prior to any further work.

A glovebox operating under positive pressure may be used for experiments that require protection from moisture or oxygen or a high-purity inert atmosphere. Usually, the chamber is pressurized with argon or nitrogen. If this type of glovebox is to be used with hazardous chemicals, test the glovebox for leaks before each use. A method to monitor the integrity of the system (such as a shutoff valve or a pressure gauge designed into it) is required.

9.E.2 Clean Rooms

Clean rooms are special laboratories or workspaces in which large volumes of air are supplied through HEPA filters to reduce the particulates present in the room, in order to protect research materials. As nanotechnology becomes more prevalent in scientific research across many disciplines, clean rooms are becoming more and more in demand, not just in pharmaceutical, microbiological, optical, and microelectronics laboratories. Special construction materials and techniques, air-handling equipment, filters, garments, and procedures are required, depending on the cleanliness level of the facility. Consult a laboratory expert in clean room operation before a clean room is designed, built, or worked in.

9.E.2.1 Clean Room Classification

Clean room classifications refer to the number of particles larger than 0.5 $\mu m/ft^3$ of volume. Unfiltered ambient air has approximately 500,000 to 1,000,000 particles/ft^3.

Many laboratories still use the U.S. FED STD 209E classification system (GSA, 1992; see also ACGIH, 1998), although it was officially cancelled by the General Services Administration of the U.S. Department of Commerce on November 29, 2001 (http://www.iest.org/). This system denotes Class 10,000 to Class 1, with the class number referring to the maximum number of particles larger than 0.5 $\mu m/ft^3$ of air. Thus, a Class 1000 clean room allows no more than 1,000 particles larger than 0.5 $\mu m/ft^3$ of air (Table 9.4). Although not as widely used in the United States, the International Organization for Standardization (ISO) 14644-1 clean room standard classification system is becoming more common. This classification uses a logarithm of the maximum number of particles larger than 0.1 $\mu m/m3$ of volume. Thus, an ISO Class 3 clean room has a maximum of $10^3 = 1,000$ particles of 0.1 $\mu m/m^3$ of air (Table 9.5).

9.E.2.2 Clean Room Protocols

The main objective of a clean room is to protect the materials and equipment from particulates. Whereas

TABLE 9.4 US FED STD 209E Clean Room Classification

| Class | Maximum Particles/ft³ | | | | | ISO Equivalent |
	>0.1 μm	>0.2 μm	>0.3 μm	>0.5 μm	>5 μm	
1	35	7	3	1	–	ISO 3
10	350	75	30	10	–	ISO 4
100	–	750	300	100	–	ISO 5
1,000	–	–	–	1,000	7	ISO 6
10,000	–	–	–	10,000	70	ISO 7
100,000	–	–	–	100,000	700	ISO 8

SOURCE : ANSI/IEST/ISO 14644-1:1999.

most laboratories maintain negative airflow with respect to adjacent nonlaboratory areas, clean rooms may be slightly positive. Thus, it is important to ensure that hazardous materials are stored in ventilated cabinets and work with volatile hazardous materials is done with proper ventilation.

Depending on the clean room level, laboratory personnel may need to follow special protocols to minimize generation of particulates, including some or all of the following:

- Wear special clothing ranging from shoe covers-only to shoe covers and special laboratory coats to fully encapsulating bunny suits with head cover and beard cover.
- Use an air shower before entering the clean room.
- Keep personal items out of the clean room.
- Use only specially made notebooks and paper in the clean room; no felt-tip pens (except permanent markers).
- Avoid bringing wood-pulp-based products into the clean room, such as magazines, books, regular tissues, and regular paper.
- Do not bring styrofoam or powders or any products that may produce dusts or aerosols into the clean room.

9.E.2.3 Laboratory Chemical Hoods and Laboratory Furniture in Clean Rooms

Laboratory chemical hoods and laboratory furniture in clean rooms must be easy to clean and not subject to rust or chalking. Most prefer not to use materials with painted surfaces, which may chalk or peel over time, or wood products that may form wood dusts. Stainless steel and thermoplastics are the most common materials.

Polypropylene chemical hoods are commonplace in clean rooms. The main concern is that this material burns and melts very easily. In the event of a fire, a polypropylene chemical hood may become fully involved. For this reason, it is prudent to choose either a fire-retardant polypropylene or another thermoplastic or to install an automatic fire extinguisher within the hood.

For nanomaterials, consider whether a chemical hood might be too turbulent for manipulating the materials. A biosafety cabinet, a ventilated enclosure with HEPA filtration, or a glovebox may be better alternatives. (See section 9.E.5 for more information.)

TABLE 9.5 ISO Classification of Air Cleanliness for Clean Rooms

| Class | Maximum Particles/m³ | | | | | | FED STD 209E Equivalent |
	>0.1 μm	>0.2 μm	>0.3 μm	>0.5 μm	>1 μm	>5 μm	
ISO 1	10	2	—	—	—	—	
ISO 2	100	24	10	4	—	—	
ISO 3	1,000	237	102	35	8	—	Class 1
ISO 4	10,000	23,700	1,020	352	83	—	Class 10
ISO 5	100,000	237,000	10,200	3,520	832	29	Class 100
ISO 6	1,000,0000	—	102,000	35,200	8,320	293	Class 1000
ISO 7	—	—	—	352,000	83,200	2930	Class 10,000
ISO 8	—	—	—	3,520,000	832,000	29,300	Class 100,000
ISO 9	—	—	—	35,200,000	8,320,000	293,000	Room air

SOURCE: ACGIH (1998). Copyright 1998. Reprinted with permission.

9.E.3 Environmental Rooms and Special Testing Laboratories

Environmental rooms, either refrigeration cold rooms or warm rooms, for growth of organisms and cells, are designed and built to be closed air circulation systems. Thus, the release of any toxic substance into these rooms poses potential dangers. Their contained atmosphere creates significant potential for the formation of aerosols and for cross-contamination of research projects. Control for these problems by preventing the release of aerosols or gases into the room. Special ventilation systems can be designed, but they will almost always degrade the temperature and humidity stability of the room. Special environmentally controlled cabinets are available to condition or store smaller quantities of materials at a much lower cost than in an environmental room.

Because environmental rooms have contained atmospheres, personnel who work inside them must be able to escape rapidly. Doors for these rooms should have magnetic latches (preferable) or breakaway handles to allow easy escape. These rooms should have emergency lighting so that a person will not be confined in the dark if the main power fails. Because these rooms are often missed when evaluating building alarm systems, be sure that the fire alarm or other alarm systems are audible and/or visible from inside the room.

As is the case for other refrigerators, do not use volatile flammable solvents in cold rooms (see Chapter 7, section 7.C.3). The exposed motors for the circulation fans can serve as a source of ignition and initiate an explosion.

Avoid the use of volatile acids in these rooms, because such acids can corrode the cooling coils in the refrigeration system, which can lead to leaks of refrigerants. Also avoid other asphyxiants such as nitrogen gas in enclosed spaces. Oxygen monitors and flammable gas detectors are recommended when the possibility of a low oxygen or flammable atmosphere exists in the room.

Box 9.2 provides some basic guidelines for working in environmental rooms.

9.E.3.1 Alternatives to Environmental Rooms

Shaker boxes may be a viable alternative to environmental rooms. These boxes come in a variety of shapes and sizes and may be stackable. They use less electricity, take up much less space, and have just as much control over the environment.

A shaker box is a sealed cabinet with a pull-out work surface. The user may control the environment within the cabinet, including the temperature, humidity, carbon dioxide level, lighting, and vibration. Shaker boxes

may be used as incubators or for cooling, giving a full range of options.

9.E.4 Biological Safety Cabinets and Biosafety Facilities

BSCs are common containment and protection devices used in laboratories working with biological agents. BSCs and other facilities in which viable

---

**BOX 9.2
Quick Guide for Working in Environmental Rooms**

Mold growth can cause problems for an experiment and affect personnel health. To avoid mold:

- Report any leaks or condensation to maintenance personnel for repair.
- Clean up spills immediately. Mold thrives on organic material.
- Do not keep papers or cardboard in the room. If such materials are needed, keep them in plastic bags.
- Do not use wood. Replace wood shelving with plastic or metal.
- Clean all surfaces with a hospital-grade disinfectant.

Be wary of using flammable materials in this room:

- There may be sources of ignition in the room, including fan motors.
- Do not store flammable liquids in the room.

Ventilation is limited:

- Chemical vapors may accumulate. Do not use materials that require local ventilation. Even materials that normally may be used on a benchtop may pose a risk in a closed environment. Do a full risk assessment.
- Limit the use of compressed gases in the event that they may displace oxygen and cause an oxygen-deficient atmosphere.
- Do not store dry ice or liquid nitrogen in an environmental room, because sublimation of the carbon dioxide may displace the air in the room, creating an asphyxiation hazard.

Do not store foods in an environmental room:

- Do not store alcoholic or nonalcoholic beverages.
- Foods may absorb chemical vapors. Do not store any food in these rooms.

organisms are handled require special construction and operating procedures to protect laboratory personnel and the environment. Conventional chemical hoods should never be used to contain biological hazards. *Biosafety in Microbiological and Biomedical Laboratories* (HHS/CDC/NIH, 2007a), *Primary Containment for Biohazards: Selection, Installation, and Use of Biological Safety Cabinets* ((HHS/CDC/NIH, 2007b), and *Biosafety in the Laboratory: Prudent Practices for the Handling and Disposal of Infectious Materials* (NRC, 1989) give detailed information on this subject.

9.E.4.1 Biosafety Cabinets

A biosafety cabinet is specially designed and constructed to offer protection to the laboratory personnel and clean filtered air to the materials within the workspace. A biosafety cabinet may also be effective for controlling nanoparticles.

The three classes of biosafety cabinets for work with biological agents are briefly described below. For more information, see the guide *Primary Containment for Bio-*

hazards: Selection, Installation, and Use of Biological Safety Cabinets (HHS/CDC/NIH, 2007b).

- A Class I biosafety cabinet does not provide a clean work environment but does provide some protection to the user. Like a chemical hood, it draws air through the face of the cabinet away from the user, across the work surface, through a set of HEPA filters, and back into the laboratory.

- A Class II biosafety cabinet (Type A1, A2, B1, or B2) provides a clean work environment and protection to the user. Internal supply air passes through a HEPA filter in a downward laminar flow across the work surface, preventing cross-contamination. It works by drawing room air around laboratory personnel through slots in the work surface at the front of the cabinet, offering user protection. Air also is exhausted through a grill along the back of the cabinet and is either recirculated through HEPA filters to the internal workspace or passes through another set of filters to be exhausted to the room or through ductwork and out of the building. (See Figure 9.13.)

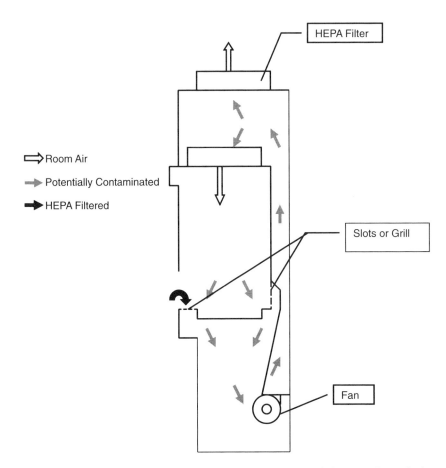

FIGURE 9.13 Example of a Class II biosafety cabinet. Room air passes around the user through the grill at the front of the cabinet. Filtered air passes into the cabinet over the materials, providing a clean environment for the materials in the cabinet. Potentially contaminated air moves through the grill and slots, across the cabinet, and passes through HEPA filters.

- A Class III biosafety cabinet provides maximum protection to laboratory personnel and the working environment. This type of cabinet is a glovebox with HEPA filter exhaust.

A biosafety cabinet is generally not suited for work with hazardous chemicals. Most biosafety cabinets exhaust the contaminated air through HEPA filters back into the laboratory. This type of filter will not contain most hazardous materials, particularly gases, fumes, or vapors. Even when connected to the laboratory exhaust system, a ducted biosafety cabinet may not provide enough containment for work with hazardous chemicals. For field testing of biosafety cabinets, consult NSF/ANSI Standard 49-2009.

Some Class II biosafety cabinets may be connected to the laboratory exhaust system and may be touted as a combination biosafety cabinet and chemical hood. However, even when ducted, a biosafety cabinet may not provide adequate containment for work with hazardous materials.

Table 9.6 provides an overview of the characteristics of different types of biosafety cabinets.

9.E.4.2 Using a Biosafety Cabinet for Biological Materials

The following protocol should be followed when using a biosafety cabinet for work with biological materials:

- Turn the cabinet on at least 10–15 minutes prior to use, if the cabinet is not left running. Verify the cabinet is operating properly and has been certified within the dates recommended by your institution.
- Disinfect work surface with 70% alcohol or other suitable disinfectant.
- Place items into the cabinet so that they can be worked with efficiently without unnecessary disruption of the airflow, working with materials from the clean to the dirty side.
- Wear appropriate PPE. At a minimum, this will include a buttoned laboratory coat and gloves.
- Adjust the working height of the stool or stand so that the worker's face is above the front opening.
- Delay manipulation of materials for approximately 1 minute after placing the hands/arms inside the cabinet.
- Minimize the frequency of moving hands in and out of the cabinet.
- Do not disturb the airflow by covering any of the grill or slots with materials.
- Work at a moderate pace to prevent airflow disruption that occurs with rapid movements.
- Wipe the bottom and sides of the cabinet surfaces with disinfectant when work is completed.

Unlike a chemical hood, a biosafety cabinet contains filters that must be changed on a regular basis. The biosafety cabinet must be decontaminated before replacing the filters and then recertified for use. Check

TABLE 9.6 Comparison of Biosafety Cabinet Characteristics

BSC Class	Face Velocity	Airflow Pattern	Applications	
			Nonvolatile Toxic Chemicals and Radionuclides	Volatile Toxic Chemicals and Radionuclides
I	75	In at front through HEPA to the outside or into the room through HEPA	Yes	When exhausted outdoors[a,b]
II, A1	75	70% recirculated to the cabinet work area through HEPA; 30% balance can be exhausted through HEPA back into the room or to outside through a canopy unit; plenums are under negative pressure	Yes (minute amounts)	No
II, B1	100	30% recirculated, 70% exhausted; exhaust cabinet air must pass through a dedicated duct to the outside through a HEPA filter	Yes	Yes (minute amounts)[a,b]
II, B2	100	No recirculation; total exhaust to the outside through a HEPA filter	Yes	Yes (small amounts)[a,b]
II, A2	100	Similar to II, A1, but has 100 fpm intake air velocity and plenums are under negative pressure to room; exhaust air can be ducted to outside through a canopy unit	Yes	When exhausted outdoors (formerly "B3") (minute amounts)[a,b]
III	N/A	Supply air is HEPA filtered; exhaust air passes through two HEPA filters in series and is exhausted to the outside via a hard connection	Yes	Yes (small amounts)[a,b]

[a]Installation may require a special duct to the outside, an in-line charcoal filter, and a sparkproof (explosion-proof) motor and other electrical components in the cabinet. Discharge of a Class I or Class II, Type A2 cabinet into a room should not occur if volatile chemicals are used.

[b]In no instance should the chemical concentration approach the lower explosion limits of the compounds.

SOURCE: HHS/CDC/NIH (2007b); NSF/ANSI Standard 49-2009.

with your institutional biosafety officer for required frequency.

9.E.5 Nanoparticles and Nanomaterials

Engineering control techniques such as source enclosure (i.e., isolating the generation source from the worker) and local exhaust ventilation systems should be effective for capturing airborne nanomaterials, based on what is known of nanomaterial motion and behavior in air.

Though traditional chemical hoods may be used for research on nanoscale particles and materials, some researchers find it challenging to work with nanoparticles in hoods operating with a 100-fpm face velocity because of turbulent airflow. In addition, limited studies demonstrate that chemical hoods that operate at a 100-fpm face velocity, even those that pass the ANSI/ASRAE containment tests, may allow nanoparticles to escape in quantities that may pose a risk to health or the environment (Ellenbecker and Tsai, 2008). This is similar to the experience of pharmaceutical companies handling dry powder formulations research. Lower-flow, reduced-turbulence hoods may be warranted. Even at lower face velocities, dispersion of particles may result in loss of materials or contamination of surfaces or both. This active area of research should be carefully monitored by anyone working with nanoparticles in a laboratory.

Because the effect of nanomaterials on the environment is still a topic of research and debate, prudent practice ensures that they do not disperse into the environment through the ventilation system. HEPA filters, which are 99.99% efficient at removing 0.3-µm and larger particles, are also very effective in trapping nanoscale particles. Some vendors offer ULPA filters, which are 99.9995% efficient at removing 0.12-µm and larger particles. Although ULPA filters are more efficient than HEPA filters, HEPA filters are generally acceptable for nanoparticle work. HEPA filters should be properly seated in well-designed filter housings.

Ionizers that are placed in either a chemical hood or a cabinet or are integrated into a cabinet can help minimize dispersion of nanomaterials, reducing loss of materials and keeping the work surfaces cleaner. Exercise caution working with explosive or highly flammable chemicals near an ionizer.

Stainless steel is much easier to clean and may show areas where materials have dispersed. Enclosures with stainless steel work surfaces are good for nanomaterial work but are not necessary.

There are several alternatives for controlling nanomaterials in the laboratory, and many ventilation vendors are working on systems specifically designed for nanoparticles.

- A low-flow enclosure or chemical hood equipped with a HEPA filter on the exhaust side is effective at reducing turbulence, preventing nanomaterials from being released into the environment through the exhaust system, and providing good containment for both nanomaterials and hazardous chemicals. For laboratories that can only provide one type of containment, this is a good alternative.
- A negative-pressure glovebox is effective.
- Class I biosafety cabinets that exhaust air through HEPA filters into the room or those that are hard-ducted to the outdoors may provide good containment for nanoparticles. Class II biosafety cabinets that exhaust air through HEPA filters back into the room or those that are hard-ducted to the outdoors may be a good choice. A glovebox provides a high level of protection and can be equipped with HEPA filtration.

Some vendors have produced other alternatives, most of which are Class I biosafety cabinets equipped with an ionizer near the front edge.

For laboratories with both hazardous chemicals and nanoparticle work, one strategy is to handle the nanoparticles in a Class I or II biosafety cabinet or a low-flow enclosure (see above), transfer the particles into solution, and then continue work in a laboratory chemical hood.

Do not use horizontal laminar-flow hoods (clean benches) that direct a flow of HEPA-filtered air into the user's face for any operations involving hazardous materials or engineered nanomaterials.

9.E.6 Explosion-Proof Chemical Hoods

For operations involving materials that could explode, protection aimed at preventing ignition and containing an explosion may be necessary. The sash should be composed of a composite material of safety glass backed by polycarbonate, with the safety glass on the interior side of the sash. In addition, all components of the hood, including the electrical supply, lighting, etc., must be explosion-proof.

9.F MAINTENANCE OF VENTILATION SYSTEMS

Even the best-engineered and most carefully installed ventilation system requires routine maintenance. Blocked or plugged air intakes and exhausts, as well as control system calibration and operation, alter the performance of the total ventilation system. Filters become loaded, belts loosen, bearings require lubrication, motors need attention, ducts corrode, and minor components fail. These malfunctions, individually or collectively, affect overall ventilation performance. Some laboratory ventilation systems have become so complex that prudent practice requires a special team of facilities staff dedicated to the maintenance of the system.

Inspect and maintain facility-related environmental controls and safety systems, including chemical hoods and room pressure controls, fire and smoke alarms, and special alarms and monitors for gases, on a regular basis.

Evaluate each laboratory periodically for the quality and quantity of its general ventilation and anytime a change is made, either to the general ventilation system for the building or to some aspect of local ventilation within the laboratory. The size of a room and its geometry, coupled with the velocity and volume of supply air, determine its air patterns. Airflow paths into and within a room can be determined by observing smoke patterns. Convenient sources of smoke for this purpose are the commercial smoke tubes available from local safety and laboratory supply companies. If the general laboratory ventilation is satisfactory, the movement of supply air from corridors and other diffusers into the laboratory and out through laboratory chemical hoods and other exhaust sources should be relatively uniform. There should be no areas where air remains static or areas that have unusually high airflow velocities. If stagnant areas are found, consult a ventilation engineer, and make appropriate changes to supply or exhaust sources to correct the deficiencies.

The number of air changes per hour within a laboratory can be estimated by dividing the total volume of the laboratory (in cubic feet) by the rate at which exhaust air is removed (in cubic feet per minute) and multiplying the total by 60. For each exhaust port (e.g., laboratory chemical hoods), the product of the face area (in square feet) and the average face velocity (in linear feet per minute) gives the exhaust rate for that source (in cubic feet per minute). The sum of these rates for all exhaust sources yields the total rate at which air is exhausted from the laboratory. The rate at which air is exhausted from the laboratory should equal the rate at which supply air is introduced into the room. Thus, decreasing the flow rate of supply air (perhaps to conserve energy) decreases the number of air changes per hour in the laboratory, the face velocities of the chemical hoods, and the capture velocities of all other local ventilation systems.

Airflows are usually measured with thermal anemometers or velometers. These instruments are available from safety supply companies or laboratory supply houses. The proper calibration and use of these instruments and the evaluation of the data are a separate discipline. Consult an industrial hygienist or a ventilation engineer whenever serious ventilation problems are suspected or when decisions on appropriate changes to a ventilation system are needed to achieve a proper balance of supply and exhaust air.

All ventilation systems should have a device that readily permits the user to monitor whether the total system and its essential components are functioning properly. Manometer, pressure gauges, and other devices that measure the static pressure in the air ducts are sometimes used to reduce the need to manually measure airflow. Determine the need for and the type of monitoring device on a case-by-case basis. If the substance of interest has excellent warning properties and the consequence of overexposure is minimal, the system will need less stringent control than if the substance is highly toxic or has poor warning properties.

9.G VENTILATION SYSTEM MANAGEMENT PROGRAM

The laboratory ventilation system is one of the most important aspects of laboratory safety and, at the same time, is likely to be the highest consumer of energy in the laboratory building. Managing all facets of the ventilation system is crucial to maximize safety and energy conservation.

The AIHA/ANSI Z9.5-2003 Laboratory Ventilation Standard provides an outline for a ventilation management program and recommends appointing a responsible person to oversee the program. The Leadership in Energy and Environmental Design (LEED) Green Building Rating System™ of the U.S. Green Building Council uses this model as a consideration in its certification system for rating laboratory buildings.

Overall, there are four main aspects of a ventilation system management program: design criteria, training for laboratory personnel, system maintenance, and performance measurement.

9.G.1 Design Criteria

The institution should determine the criteria to use for all new installations of chemical hoods and other ventilation systems. This might include

- testing criteria as installed (e.g., all or a representative sampling of the hoods must pass ANSI/ASHRAE 110-1995 containment testing as installed);
- chemical hood design criteria (e.g., face velocity criteria at specific sash height and sash design);
- types of continuous monitoring systems preferred or required (e.g., face velocity reading, magnehelic gauge);
- acceptable diversity factors;
- energy conservation strategies;

- alarm systems;
- type of duct work;
- noise criteria;
- preference for VAV systems (designing one extra fan into each system); and
- backup power.

9.G.2 Training Program

No matter how well a system is designed or maintained, no matter what lengths an institution has gone to for the sake of safety and energy conservation, if laboratory personnel do not use the equipment properly, individual users can defeat these efforts with their own behaviors.

Laboratory personnel who insist on working at the edge of the laboratory chemical hood, raise the sash above its maximum operating height, defeat alarms, disable sash closures, do not move an elephant trunk close to the source, block baffles, use loose materials in the chemical hood and clog the ductwork, leave the sash open when not working at the chemical hood, fail to report that a filter needs to be changed reduce safety and sustainability efforts. Sometimes, these actions are due to lack of consideration; sometimes personnel may simply not understand the implications.

All laboratory personnel should receive training that includes

- how to use the ventilation equipment,
- consequences of improper use,
- what to do in the event of system failure,
- what to do in the event of a power outage,
- special considerations or rules for the equipment,
- significance of signage and postings.

Training may be one-on-one, classroom, Web-based, or whatever format fits the culture of the institution and the needs of the laboratory.

Many laboratories, particularly academic research laboratories, experience high turnover rates. Good signage and postings complement training and act as constant reminders (Figure 9.14).

Consider the following types of signs and postings:

- sash position for laboratory chemical hoods,
- telltales (ribbons or similar materials on chemical hood sashes with a key to good performance),
- meaning of any audible or visual alarms,
- function of occupancy sensors (e.g., setback mode tied to light switch),
- downtimes if the system has a setback mode that is on a timer, and
- reminder to lower the sash when not in active use.

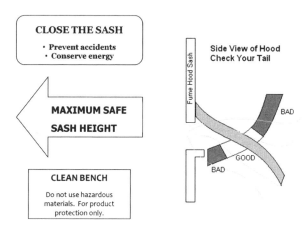

FIGURE 9.14 Examples of postings for laboratory chemical hoods. Clockwise from top left: reminder to close the chemical hood sash, guide to checking the telltale ribbon taped to the sash of the chemical hood, reminder that a clean bench is not for hazardous chemicals, indicator showing the safe maximum sash height.

9.G.3 Inspection and Maintenance

Maintenance is key to a ventilation system management program. The program should describe the elements of the inspection and maintenance program, including

- designation of who conducts inspections and how often;
- how inspections are recorded;
- inspection criteria for laboratory chemical hoods including
  - face velocity testing—equipment used, history,
  - how recorded,
  - how posted on the chemical hood, and
  - will maximum sash height be marked and how;
- criteria for working on roofs and around stacks;
- fan maintenance schedule;
- VAV system maintenance schedule;
- alarms and controls maintenance schedule; and
- schedule for recommissioning the ventilation system.

9.G.4 Goals Performance Measurement

The old adage that "you can't manage what you don't measure" rings true too with the ventilation management program. At least annually, evaluate the effectiveness of the program, including

- energy use and savings,
- emission issues,

- trends in chemical hood performance (signs of deterioration, etc.), and
- review of the life cycle of the ventilation system.

9.G.5 Commissioning

When a new ventilation system in installed, new components are installed, or any significant change to the ventilation system occurs, consider hiring a commissioning agent with experience with laboratory facilities. An outside commissioning agent will ensure that the system meets the criteria you have selected, note any design errors, handle problems, and facilitate testing, installation, etc. In-house staff or hired consultants will continue to maintain the equipment, but the startup issues can be overwhelming. Ensure that those who will be using and maintaining the system receive training.

9.H SAFETY AND SUSTAINABILITY

Cost considerations should never take precedence over ensuring that laboratory personnel are protected from hazardous concentrations of airborne toxic substances. That sentiment bears repeating. However, since the 1980s, the chemical hood has become a fixture in a laboratory, sometimes whether it was needed or not. Many laboratory research buildings have several chemical hoods that remain unused, even as thousands of cubic feet of conditioned air passes through them every minute. In a typical laboratory building containing office space, meeting space, and laboratories, the labs constitute one-sixth of the floor space, yet consume a third of the energy.

One suburban university that is relatively typical of a research campus conducted a study of the origin of its carbon inventory and determined that 37% was from laboratory buildings, which constitute 15% of the total building area on campus (see Figure 9.15).

Typically, at any one time, fewer than half the hoods in a given laboratory are in active use. Chemical hoods are excellent, but they are not the only solution for reducing exposure to a safe level. Where laboratory chemical hoods are needed, the amount of energy they consume can be reduced. (See Vignette 9.2.)

Several options for energy conservation have been presented in previous sections of this chapter. More technologies are being developed and become available every year. Each deserves attention and scrutiny before using them in a research laboratory environment.

This section focuses on sustainability with respect to ventilation, but sustainability can be supported in other areas through water conservation, following appropriate waste disposal techniques, considering the principles of green chemistry when performing

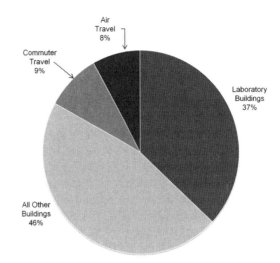

FIGURE 9.15 Carbon inventory of a research university campus.

research, and investigating ways to reduce the energy needs of the building.

9.H.1 Low-Flow or High-Performance Laboratory Chemical Hoods

Low-flow or high-performance hoods operate at a lower face velocity and save energy by reducing the amount of conditioned air that passes through them. They tend to be more expensive than traditional chemical hoods, but the energy savings generally result in a quick payback. They are deeper than a traditional chemical hood and may not occupy the same space in a retrofit situation. See section 9.C.2.9.3.6 for more information.

**VIGNETTE 9.2
Sustainability considerations in laboratory ventilation design**

In the initial design discussions for an academic research laboratory, the principal investigator called for six 8-ft chemical hoods plus two ventilated Class II biosafety cabinets. After discussions about how this equipment was to be used and the operations of the laboratory, the EHS staff and the engineers suggested alternatives, including ventilated equipment enclosures and snorkels. These changes resulted in significant savings in first costs, space, operating costs, and energy consumption, while better fitting the needs of the researchers.

9.H.2   Automatic Sash Closers

For most laboratory chemical hoods, especially those on VAV systems, when the sash is closed, they draw much less air, resulting in significant energy savings. Laboratory personnel do forget to close the sash or find it cumbersome to keep closing the sash every time they step away.

Modern automatic sash closers have a sensor technology that uses a proximity or motion detector to sense when there is no one in front of the chemical hood. The sensor has a timer that can be adjusted to a set time period; after that time, if no one appears to be working at the hood, the system gently closes the sash. Like a garage door closer, there is usually a sensor at the bottom edge of the sash, such that if anything, even a pipette, crosses the plane of the sash, the sash will stop closing to avoid breaking or bumping whatever is below the sash.

Some sash designs include counterweights that automatically lower the sash to a set level when the laboratory personnel step away. The sash does not close completely but does lower substantially.

Automatic sash closers can result in significant cost savings and add to the safety of laboratory personnel by keeping a barrier between the materials in the chemical hood and personnel and materials in the laboratory.

9.H.3   Variable Air Volume Systems with Setback Controls

Most chemical hoods are used only a portion of the day. An advantage of a VAV system is that individual chemical hoods or an entire system can be adjusted to a setback mode, a low flow that maintains negative pressure but conserves energy.

The setback mode may be activated in a number of ways, such as:

- a timer for an individual chemical hood or an entire system where work schedules are predictable;
- occupancy sensors, set back when sensors indicate that the laboratory or the chemical hood is not in use;
- sash position, set back when the sash is fully closed, especially useful in conjunction with automatic sash closers; and
- light switch, set back when lights are turned off, indicating that the laboratory is unoccupied.

9.H.4   Variable Air Volume Systems, Diversity Factors

Another advantage of a VAV system with manifolded exhaust is that the system could be designed for just a portion of that maximum airflow, rather than for a system that handles 100% of the hoods it serves. The rationale is that it is extremely unlikely that all the chemical hoods would be operating with the sash open at the same time. The diversity factor is the maximum percentage of airflow ever needed at once.

By designing the system to handle a smaller number of chemical hoods, the system takes advantage of smaller ductwork and fewer fans, resulting in both first-cost savings and ongoing energy cost savings. Prudent practice adds at least one extra fan to the system both for maintenance reasons (always able to have one fan down) and for future growth.

9.H.5   Lower General Ventilation Rates

As discussed in section 9.C.4, many laboratories have a minimum of 6 to 12 air changes per hour. Some laboratories have been able to lower these rates based on the materials and operations in the laboratory. Consultants experienced in computational fluid dynamics modeling are able to take information about the chemicals and processes and the ventilation system and predict how a lower air change rate might affect laboratory air quality.

Some laboratories have installed active chemical monitoring systems that sample for and provide real-time measurements of carbon dioxide and specific chemicals, adjusting the airflow in the room as needed to maintain an acceptable air quality. Limitations do exist for this method, but it may be useful in some situations.

9.H.6   Laboratory Chemical Hood Alternatives

The laboratory chemical hood is a fabulous engineering control, but it is not the only one. Perform a risk assessment and consider the other alternatives. Many of the alternatives will result in lower energy usage without compromising safety.

9.H.7   Retro-Commissioning

For facilities with ventilation systems that were not designed for energy efficiency, consider whether it makes sense to replace all or parts of the system with newer, more efficient alternatives. Retro-commissioning a laboratory ventilation system can result in large energy savings and a safer ventilation system and may have a relatively short payback period. See section 9.B.9 for additional information.

9.H.8 Components of Heating, Ventilation, and Air-Conditioning (HVAC)

There are many technologies aimed at energy conservation for ventilation systems. Examples include chilled beams for cooling labs and offices, reheat systems that cool or heat within zones rather than for all labs on the system, and enthalpy wheels for retaining latent and sensible heat, just to name a few.

Technologies continue to improve and new ideas are being tested constantly. The following resources, mostly available online, may be useful in identifying and evaluating these systems:

- EPA Laboratories in the 21st Century (Labs 21) (http://www.labs21century.gov/),
- US Green Building Council's LEED (http://www.usgbc.org), and
- ASHRAE Laboratory Design Guide (http://ateam.lbl.gov/).

9.H.9 How to Choose a Ventilation System

There is no one choice that is right for every laboratory. The designers, the laboratory users, and the facilities staff must discuss the possibilities. EHS professionals and laboratory managers are helpful in these discussions as well. The individuals who decide which systems to install must understand the needs of the users, and the users must understand how the systems work, the capabilities and limitations of the systems, and what to expect from them. The facilities staff must understand how the systems need to be maintained, and those who are choosing the system need to know whether there is in-house expertise to maintain them.

Check local, state, and federal codes and regulations before choosing a new system. Only a few actual regulations cover ventilation systems, but more and more municipalities are adopting international building and mechanical codes. These codes impose limitations on manifolding ductwork and may require detection or sprinklers within ducts.

When considering a new technology, benchmarking is usually helpful. Find someone who is using a similar system and discuss their experience. Ask for samples. Visit laboratories that use similar products. Find the systems that work best for your applications. Continue communications between the users and the installers and the maintenance staff to ensure that the systems are working as intended.

Remember that even if all the chemical hoods are removed, ventilation is still needed in the laboratory.

9.I LABORATORY DECOMMISSIONING

A laboratory must be properly decommissioned prior to changing its use. Among other steps, decommissioning entails decontamination and the removal of hazards to ensure the safety of future occupants and others who may enter the space. Decommissioning must be done prior to renovation, even if the space is to be reused as a laboratory. Because laboratory operations differ, it is appropriate to decommission a laboratory whenever there is a significant change in occupancy. Areas outside of the laboratory, such as ventilation ductwork, coldrooms, hallway freezers and common storage areas, should also be decommissioned if they are concurrently subject to a significant change in use or occupancy. Decommissioning must also be done prior to the demolition of a laboratory.

Before decommissioning begins it is important to establish a level of cleanliness that meets the regulatory and institutional safety standards for the next occupancy. Detailed radiological assessment and decontamination guidelines are available in the Multi-Agency Radiation Survey and Site Investigation Manual (MARSSIM), available from the Nuclear Regulatory Commission and other government agencies (EPA/USNRC/DOE/DOD, 2000). Although a helpful Laboratory Decommissioning Standard is available from the American National Standards Institute (ANSI Z9.11, 2008), there are few standards for an acceptable level of residual chemical contamination. Even when environmental cleanup standards exist, it may be difficult to apply them to laboratory decommissioning.

Be sure to document the assessment, decontamination and removal activities, and to issue a final clearance statement. A Laboratory Closeout Checklist is included on the disc that accompanies this book. It may be appropriate to prepare a written Decommissioning Plan.

9.I.1 Assessment

The first step in laboratory decommissioning is to assess any hazards that may remain in the space. Review the known or likely historic uses of the space, as well as records of spills and accidents, laboratory manuals and notebooks, and published papers of research conducted in the lab. Ask former occupants what hazardous materials they used and if they know of any contaminated areas.

The assessment of radiological hazards is relatively straightforward and requires standard methods for handheld survey meters and wipe tests for removable contamination. Because it is easy to do, a radiological survey should be done unless it can be assured that no radioactive material had been used in the space.

Because many chemicals require a unique protocol for sampling and analysis, a chemical contamination assessment usually requires that the potential contaminants be well-defined. A field sampling plan should describe how wipe tests will be taken, the wetting solvent used, the protocol for grid sampling (or other sampling scheme), necessary analytical sensitivity, and the methodology that will be used to evaluate the results.

9.1.2 Removal, Cleaning, and Decontamination

The second step in decommissioning is to remove all hazards from the space. Be sure that all chemicals, radioactive materials, and biologicals have been removed from use and storage areas, including refrigerators and freezers. Movable equipment should be appropriately cleaned and/or disinfected, and removed from the lab.

Residual perchloric acid and mercury contamination are common concerns for laboratory decommissioning. If perchloric acid was used outside of a hood designed for that purpose, hoods and ductwork can become contaminated with explosive metal perchlorates. (See section 9.C.2.10.5 for information about the hazards of perchloric acid in laboratory hoods and ventilation.)

Mercury is used in most laboratories, and mercury spills are common. Unless it is certain that no mercury was used, laboratory decommissioning should include testing of floors, sinks, cupboards, and molding around furniture and walls. Be sure to check and clean sink p-traps. Visual inspection alone is inadequate as historic spills may reach beneath floor tiles and furniture, and behind walls. As described in the ANSI Laboratory Decommissioning Standard, modern mercury testing utilizes a portable atomic absorption spectrophotometer with a sensitivity of 2 ng/m^3. Decommissioning clearance levels consider the U.S. Agency for Toxic Substances and Disease Registry's Minimal Risk Level (MRL) of 200 ng/m^3 for non-occupationally exposed individuals. Chapter 6, section 6.C.10.8, includes information on dealing with mercury contamination. Additional mercury testing may be necessary as furniture, floors, walls, and plumbing are removed during renovation.

After hazardous materials and movable equipment have been removed, areas known to be contaminated (e.g., stained floors and cupboards) should be cleaned appropriately, or destructively removed and disposed of. Chemical decontamination can be done using appropriate surfactant soaps, solvents, neutralizing agents, or other cleaners.

Unless is it known that no biological materials were used in the space, the furniture, equipment, and other surfaces should be cleaned with an appropriate disinfectant. Sophisticated biological decontamination technologies are available for areas where high-risk pathogens have been used.

As a precautionary measure, it is a standard practice to remove dusts and other settled particulates via a thorough final wet-cleaning of floors, vertical surfaces and furniture using commercial cleaning products.

9.1.3 Clearance

Final tests or survey results can be used to verify decontamination. In some cases regulatory authorities allow permanent marking of a porous floor or wall where a radioactive material or chemical has penetrated deeply, and destructive removal is impractical prior to the building's demolition. When removal, decontamination, and cleaning meet planned decommissioning standards, a final area clearance statement can be issued, and renovation, demolition, or the new occupancy can commence.

10 Laboratory Security

10.A	INTRODUCTION	256
10.B	SECURITY BASICS	256
	10.B.1 Physical and Electronic Security	256
	10.B.1.1 Door Locks	257
	10.B.1.2 Video Surveillance	258
	10.B.1.3 Other Systems	258
	10.B.2 Operational Security	258
	10.B.3 Information Security	258
	10.B.3.1 Backup Systems	259
	10.B.3.2 Confidential or Sensitive Information	259
10.C	SYSTEMS INTEGRATION	259
10.D	DUAL-USE HAZARD OF LABORATORY MATERIALS	259
10.E	LABORATORY SECURITY REQUIREMENTS	260
	10.E.1 Biological Materials and Infectious Agents	260
	10.E.2 Research Animals	260
	10.E.3 Radioactive Materials and Radiation-Producing Equipment	261
	10.E.4 Chemicals	261
	10.E.4.1 Drug Enforcement Agency Chemicals	261
	10.E.4.2 DHS Chemicals of Interest (COI)	261
10.F	SECURITY VULNERABILITY ASSESSMENT	261
10.G	DUAL-USE SECURITY	262
10.H	SECURITY PLANS	262
	10.H.1 Levels of Security	263
	10.H.1.1 Normal (Security Level 1)	263
	10.H.1.2 Elevated (Security Level 2)	263
	10.H.1.3 High (Security Level 3)	264
	10.H.2 Managing Security	264
	10.H.3 Training	264

10.A INTRODUCTION

The world has become more security conscious, and that awareness extends to laboratories. New guidelines and approaches, driven by legislation and regulation—to say nothing of common sense—are promulgated every year. A laboratory security system is put in place to mitigate a number of risks and is complementary to existing laboratory security policies. In very broad terms, laboratory safety keeps people safe from chemicals, and laboratory security keeps chemicals safe from people. This chapter is intended to provide the reader with an overview of laboratory security concerns and to raise awareness of the issue. Risks to laboratory security include

- theft or diversion of chemicals, biologicals, and radioactive or proprietary materials (such materials could be stolen from the laboratory, diverted or intercepted in transit between supplier and laboratory, at a loading dock, or at a stockroom, and then sold or used, directly or as precursors, in weapons or manufacture of illicit substances);
- theft or diversion of mission-critical or high-value equipment;
- threats from activist groups;
- intentional release of, or exposure to, hazardous materials;
- sabotage or vandalism of chemicals or high-value equipment;
- loss or release of sensitive information; and
- rogue work or unauthorized laboratory experimentation.

The type and extent of the security system needed depend on several factors, including

- known and recognized threats gleaned from the experience of other laboratories, institutions, or firms;
- history of theft, sabotage, vandalism, or violence directed at or near the laboratory, institution, or firm;
- presence of valuable or desirable materials, equipment, technology, or information;
- intelligence regarding groups or individuals who pose a general threat to the discipline or a specific threat to the institution;
- regulatory requirements or guidance;
- concerns regarding information security; and
- the culture and mission of the institution.

A good laboratory security system should, among other things, increase overall safety for laboratory personnel and the public, improve emergency preparedness by assisting with preplanning, and lower the organization's liability.

10.B SECURITY BASICS

There are four integrated domains to consider when improving security of a facility:

- physical or architectural security—doors, walls, fences, locks, barriers, controlled roof access, and cables and locks on equipment;
- electronic security—access control systems, alarm systems, password protection procedures, and video surveillance systems;
- operational security—sign-in sheets or logs, control of keys and access cards, authorization procedures, background checks, and security guards; and
- information security—passwords, backup systems, shredding of sensitive information.

These domains are complementary, and each should be considered when devising security protocols. Any security system should incorporate redundancy to prevent failure in the event of power loss or other environmental changes.

Security systems should help

- **detect** a security breach, or a potential security breach, including intrusion or theft;
- **delay** criminal activity by imposing multiple layered barriers of increasing stringency or "hardening" in the form of personnel and access controls; and
- **respond** to a security breach or an attempt to breach security.

10.B.1 Physical and Electronic Security

There are many systems available for physical and electronic laboratory security. The choice and implementation depends on the level of security needed and resources available. The following sections provide some examples, although new technologies are always under development.

The concept of concentric circles of protection, as shown in Figure 10.1, is useful when considering a laboratory's physical security. Physical and electronic security begins at the perimeter of the building and becomes increasingly more stringent as one moves toward the interior area (e.g., at the intervention zone), where sensitive material, equipment, or technology reside. Note that although physical measures are

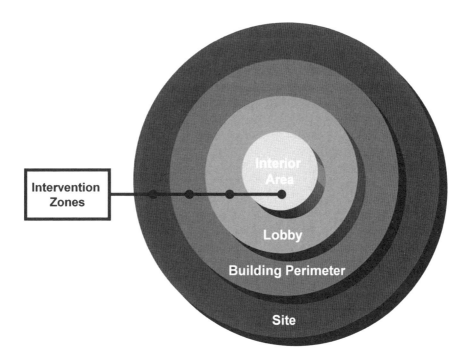

FIGURE 10.1 Concentric circles of physical protection.

implemented in the intervention zones, electronic and operational security measures are implemented only under certain conditions, depending on need.

10.B.1.1 Door Locks

Within a laboratory, perhaps the most obvious form of security is the door lock. There are many choices available, including

- Traditional locks with regular keys (which are subject to duplication, loss, theft, and failure to return after access) should no longer be utilized in areas where dual-use materials are located.
- Traditional locks with keys marked "Do Not Duplicate" have the same drawbacks as above, but may be less likely to be duplicated.
- Cipher locks with an alpha or numeric keypad may be vulnerable to thieves who are able to deduce the access code from the appearance of the keys. Access codes should be changed from the factory default when the lock is installed.
- High-security cores are difficult to break into and to duplicate.
- Card access (dip locks) traditionally have datalogging capabilities that allow those with access

to security records to identify which cards were used to gain access.
- Card access (swipe cards). These provide a transaction record and can be programmed for different levels and times of access.
- Key fobs or card access (proximity card readers) have the same benefits as swipe cards, but there is no requirement to place the card physically in the reader.
- Biometric readers offer a high level of security but are expensive and require more intensive maintenance.

Each of these systems requires training, management, and maintenance, whether it is a key inventory system or controls for card access. Of course, the system is only as effective as the users allow it to be. Users should be trained to not hold doors open for others, and that everyone needs to use their key to pass through an access point. Unauthorized personnel should not be allowed to enter the laboratory, and if there is any question, laboratory personnel should be instructed to call security for guidance. The organization should ensure that there is a program in place to collect keys or revoke card access to the laboratory before a person leaves the workplace.

10.B.1.2 Video Surveillance

Video surveillance systems are often used to supplement locks for documenting access and may be continuously monitored by security personnel. Recordings of relevant video may be reviewed after an incident.

When implementing a video surveillance system, document the purpose and ensure that personnel understand the objectives. Video surveillance may be used to

- prevent crime by recognizing unusual activity in real time, which requires staff dedicated to watching the camera output and is most effective when the presence of individuals alone is suspicious;
- validate entry authorization by verifying the identity of the worker; and
- verify identity of unauthorized personnel after unauthorized access.

Video surveillance cameras should be located to provide a clear image of people in the area, particularly those entering or exiting. They are not as useful in the work area itself unless suspicious behavior is obvious.

If video is recorded, a system of storage and documentation is needed. Establish the duration of recording retention, the media used, and the need for permanent archiving. Create a procedure to quickly find, maintain, and duplicate critical recordings if an incident occurs.

No matter the objective of the video surveillance system, it is crucial to establish a policy and procedure for using it and for reviewing recordings. Involve human resources and legal personnel in the policy-making process. For example, if the video surveillance system is designed to record unauthorized entry, it may not be allowable by the institution to use it to track worker productivity. Clarify under what circumstances the information may be viewed, and by whom.

10.B.1.3 Other Systems

There are many other methods of implementing physical and electronic security, ranging from simple to sophisticated, which can be employed for crime deterrence, recognition, or investigation. A few examples include

- glass-break alarms for windows and doors,
- intrusion alarms,
- hardware to prevent tampering with window and/or door locks,
- lighting of areas where people may enter a secure area,
- bushes and other barriers to reduce visibility of sensitive areas from outside the building,
- locks on roof access doors,
- walls that extend from the floor to the structural ceiling,
- tamper-resistant door jambs,
- blinds on windows,
- locks and cables on equipment to prevent easy removal,
- badges or other forms of identification, and
- sign-in logs.

10.B.2 Operational Security

Operational security is responsible for the people within the laboratory. A security system is only as strong as the individuals who support it, and thus, among the goals of an operational security system are to increase awareness of security risks and protocols, to provide authorization for people who need access to a given area or material, and to provide security training.

Though far from comprehensive, elements of operational security include

- screening full- and part-time personnel before providing access to sensitive materials or information;
- providing ID badges;
- working to increase the situational awareness of laboratory personnel (e.g., knowing who is in the laboratory, identifying suspicious activity);
- encouraging the reporting of suspicious behavior, theft, or vandalism;
- restricting off-hour access to laboratories;
- providing entry logs at building and laboratory access points; and
- inspecting and inventorying materials removed from the laboratory.

10.B.3 Information Security

Information and data security can be as critical as security of equipment and materials. Loss of data and computer systems from sabotage, viruses, or other means can be devastating for a laboratory.

The issue of dual use applies to information as well as laboratory materials. Over the years, several examples of cybersecurity breaches have led to loss of sensitive information. A detailed description of a laboratory procedure may find its way into the public domain, creating a new resource for those with illicit intentions, or simply depriving the researchers of recognition for their work.

Most institutions and firms have information security policies and procedures and information technology support staff who can help implement security systems. Laboratory managers and personnel should be familiar with and follow their protocols.

10.B.3.1 Backup Systems

Develop and institute a plan for backing up data on a regular basis with backup media off-site, in fire-safe storage, or at a central facility (e.g., the institution's information technology facility).

10.B.3.2 Confidential or Sensitive Information

Assess the type of data produced by the laboratory, department, or group. Laboratories that possess chemicals of interest (COI) and are covered by the Chemical Facilities Anti-Terrorism Standards (CFATS) are subject to U.S. Department of Homeland Security (DHS) requirements for Chemical-terrorism Vulnerability Information (CVI). CVI may not be openly shared. It includes data and results from an inventory assessment called a Top-Screen (see section 10.E.4.2), the facility's DHS Security Vulnerability Assessment and Site Security Plan (e.g., procedures and physical safeguards), as well as training and incident records, and drill information.

Other data may fit into the following categories:

- public, shared freely with anyone;
- internal, shared freely within the institution;
- department, shared only within the department;
- laboratory, shared only in the laboratory; or
- confidential,[1] shared only with those directly involved with the data or on a need-to-know basis.

If the laboratory produces private, sensitive, or proprietary data,

- Provide training to those with access to this information, stressing the importance of confidentiality. Review any procedures for releasing such information outside the laboratory or group.
- Consider a written and signed confidentiality agreement for those with access to such information.
- Keep passwords confidential. Do not store or write them in an obvious place.
- Change passwords routinely.
- Safeguard keys, access cards, or other physical security tools.
- Before discarding materials that contain sensitive information, render them unusable by shredding them, or by erasing magnetic tape.

- Report any known or suspected breaches in security immediately.
- Establish policies and procedures for the storage of proprietary information on hard drives or portable storage media and for the removal of proprietary information from the laboratory or secure area.

Many services and programs are available to protect data from viruses and similar threats as well as high levels of security. Refer to the institution's information technology group or an outside consultant.

10.C SYSTEMS INTEGRATION

Since events such as the attacks on the World Trade Center, institutions and firms have steadily improved their security systems for personal as well as institutional protection. They have incorporated more rigorous planning, staffing, training, and command systems and have implemented emergency communications protocols, drills, background checks, card access systems, video surveillance, and other measures. What's more, many colleges and universities, to say nothing of commercial institutions, have engaged their own sworn and armed on-site police force.

Security is not new, at least for some laboratories. For years, secure management of controlled substances and denatured alcohol has been required by law; however, global events have raised the stakes for these laboratories as well as for those that were not previously concerned about security. It is not enough to implement a laboratory security system; it is imperative that such a system protect the laboratory and also be compatible, consistent, and integrated smoothly with the overarching systems in the institution. The institution is responsible for the general security atmosphere, and laboratory systems focus on residual and specialized security risks.

Moreover, the security plan should identify protocols, policies, and responsible parties, clearly delineating response to security issues. This includes coordination of institution and laboratory personnel and coordination of internal and external responders, including local police and fire departments.

10.D DUAL-USE HAZARD OF LABORATORY MATERIALS

In addition to inadvertent misuse of chemicals, it is apparent that chemicals can also be misused intentionally, for example, as precursors of illicit narcotics. Much of the recent focus on security in research and teaching laboratories pertains to "dual use" materials. Dual-use or multiple-use materials are materials that have both a bona fide use in scientific research and education, but

---

[1]The term "confidential" may have special meaning for some operations and funding resources. Use care in choosing terminology for sensitive information. In the event of an inspection by a government agency or association providing information or funding, there may be expectations related to the use of these terms. Classified information is often defined further as confidential, secret, or top secret.

also can be used for criminal or terrorist activities. For example, common chemical substances that are easily removed from the laboratory without notice or readily purchased, such as acetone and hydrogen peroxide, can be converted to highly explosive or otherwise hazardous products. Although certain dual-use materials can be obtained from hair salons, hardware stores, and the like, laboratories are also a source, and security should be considered.

Dual-use biological agents include live pathogens and biological toxins that have a realistic potential to be used for terrorism (e.g., anthrax). There are national as well as international regulations to address the risk of dual use, such as import and export controls. Firms and institutions may wish to integrate their facility dual-use controls with both levels of regulation.

Terrorist Web sites have suggested that their operatives can pose as students to gain access to university laboratories and remove hazardous chemical, biological, or radiological agents. However, meaningful quantities of some dual-use chemicals can also be found outside the laboratory in situations that are less secure than laboratories. As a result, the acquisition and dual use of laboratory chemicals is a real possibility, especially utilizing chemicals that can pose a high risk in relatively small laboratory quantities.

Although there is no comprehensive list of dual-use chemicals, DHS has developed a list of COI because of concern about dual use. (See section 10.E.4.2 for more information.) In addition to known warfare agents, such as nitrogen mustard and sarin (which are difficult to acquire or synthesize in makeshift laboratories), more common laboratory reagents, such as ammonia, chlorine, phosgene, cyanogen chloride, sodium cyanide, and sodium azide are considered dual-use compounds. These substances can cause human injury—either directly or after acidification—that is relatively resistant to medical treatment (Shea and Gottron, 2004), and therefore could be sought by terrorists gaining access to laboratory facilities. Alternatively, a research laboratory could be used for the illicit synthesis of terror substances.

Objective evaluation of the utility of a given chemical to terrorists might underestimate the true risk posed by malicious intent. For example, osmium tetroxide, which is highly toxic in pure solid form and in solution, has been judged to be a poor choice for terrorists to use, because of its high cost, its rapid evaporation, and the fact that an explosion would convert it to harmless products. Nonetheless, osmium tetroxide poisoning was suspected to be the intended means of a thwarted terror attack in the vicinity of London, England (Kosal, 2006). One cannot assume terrorists will follow the same logical path or practical considerations as an individual who is trained in laboratory sciences.

10.E LABORATORY SECURITY REQUIREMENTS

For most laboratories, there are a few general security requirements; however, most security measures are based on an assessment of the vulnerabilities and needs of an individual laboratory or institution. For some materials or operations, regulations or strict guidance documents specify the type or level of security.

10.E.1 Biological Materials and Infectious Agents

Certain biological agents, including viruses, bacteria, fungi, and their genetic elements, are considered dual-use materials because of their potential for use by terrorists to harm human health. Biological materials pose a unique problem because these materials can replicate; thus, theft of even small amounts is significant.

In the United States, these dual-use biological materials are called Select Agents and Toxins, and their laboratory use is regulated by the Centers for Disease Control and Prevention (CDC) and the U.S. Department of Agriculture's Animal and Plant Health Inspection Service (APHIS). Individuals planning to use Select Agents and Toxins are required to perform a security risk assessment (i.e., a detailed background check) to determine whether they are permitted to work with the materials. There are additional requirements for laboratory security, and the CDC or APHIS will conduct periodic inspections to assess compliance.

In addition, federal guidance from the National Institutes of Health (NIH) addresses the management of dual-use risks from gene synthesis, synthetic biology, and certain experiments. The publication *Biosafety in Microbiology and Biomedical Laboratories* (BMBL; HHS/CDC/NIH, 2007a) includes guidance for security of biological materials, based on a risk assessment method described in the document. For institutions that receive NIH funding, compliance with the BMBL is a grant requirement for recombinant DNA research.

10.E.2 Research Animals

Animal research is the focus of numerous animal rights organizations, including some that have engaged in malicious behavior. Vivarium security is critical for the safety of animals and researchers. The Association for Assessment and Accreditation of Laboratory Animal Care International provides guidance for security of laboratory animals and research facilities.

10.E.3 Radioactive Materials and Radiation-Producing Equipment

In most laboratories, the quantity, isotope, and characteristics of radioactive materials used for research or teaching do not pose a serious dual-use risk. However, any radioactive materials can be perceived as a risk by the community.

In the United States, use of radioactive materials is regulated by the U.S. Nuclear Regulatory Commission (USNRC) or USNRC-authorized state agencies. Compulsory guidelines for security are included in the requirements for licensing and use of these materials. Specific USNRC security requirements typically vary depending on the risk of the material.

10.E.4 Chemicals

Chemical security is garnering increasing attention from regulators. Most regulations that require specific security measures are aimed at facilities with large stores of materials—such as production facilities—rather than laboratory-scale quantities. However, federal, state, and local regulatory agencies are increasingly applying standards to chemical laboratories.

10.E.4.1 Drug Enforcement Agency Chemicals

Illicit drugs and their precursors pose a theft risk because of their resale (street) value. The U.S. Drug Enforcement Agency (DEA) has strict rules about procurement, inventory, use, disposal, and security of these chemicals. A person using materials regulated by DEA must obtain a user license or work under the direction of a person with such a license. The materials must be secured, with the level of security needed dependent on the classification of the material.

Laboratories in which DEA-regulated materials are used must keep an inventory log that documents the quantity and date that any amount of material is removed, as well as a signature or other record to identify who removed the material. Once a DEA-regulated material has expired or is ready for disposal, it must be either destroyed or returned to the manufacturer or distributor. Destruction must render the material unusable and unidentifiable as the original agent and must be done by a person designated by the licensed user and witnessed by at least two people, one of whom, preferably, is a law enforcement officer. The destroyed

materials must be disposed of in accordance with applicable laws (see Chapter 9 for disposal details).

10.E.4.2 DHS Chemicals of Interest (COI)

DHS has promulgated regulations that apply to chemical facilities, including laboratories, with the purpose of keeping dual-use chemicals out of the possession and control of terrorists. The Chemical Facility Anti-Terrorism Standards are concerned with the following types of chemicals:

- EPA Risk Management Plan chemicals,
- highly toxic gases,
- chemical weapons convention chemicals,
- explosives, and
- precursors of the above chemicals.

In the DHS process for determination of risk, all laboratory facilities are expected to survey their entire facility (including nonlaboratory areas) for the presence of COI and compare their inventory to the threshold screening quantities established in the standard. If the facility meets or exceeds the threshold quantity for any chemical of interest, the facility must report the inventory by completing an assessment document called "Top-Screen."

Upon receiving a completed Top-Screen, the facility is required to conduct a security vulnerability assessment. There are four risk tiers, with tier 1 for facilities posing the greatest risk and tier 4 posing the least risk. Based on the results of the assessment and the risk tier, the facility is expected to develop and implement an approved site security plan. There are also requirements for information security and training provisions under this rule.

As of the time of publication, DHS was continuing to develop rules and guidance for chemical facilities, including laboratories.

10.F SECURITY VULNERABILITY ASSESSMENT

Whether or not the security of a laboratory material is regulated by a government agency, it is prudent to assess risk. A security vulnerability assessment (SVA) is used to catalog potential security risks to the laboratory and the magnitude of possible threats. It begins with a walk-through of the laboratory, building, and building perimeter, and includes discussion with laboratory staff pertaining to the chemicals, equipment, procedures, and data that they use or produce. The SVA process will also assess the adequacy of the systems already in place and help determine the security planning needs for the laboratory, building, or department.

There are a number of ways to conduct an SVA. DHS has developed an SVA protocol for higher risk facilities, which may include laboratories if threshold amounts of COI are present. Completion of this SVA is mandatory for facilities that DHS has classified into a risk tier (see section 10.E.4.2). The DHS SVA is available on its Web site for use by any facility, even those not regulated by DHS.

Many states have adopted SVAs for their critical infrastructure, which often includes colleges, universities, and other facilities with research or pilot laboratories. Several professional organizations have also developed SVA checklists, such as the one by the American Chemical Society Committee on Chemical Safety, which is available on the CD that accompanies this book.

The following is a partial list of issues to review as part of an SVA:

- existing threats, based on the history of the institution (e.g., theft of laboratory materials, sabotage, data security breaches, protests);
- the attractiveness of the institution as a target, and the potential impact of an incident;
- chemicals, biological agents, radioactive materials, or other laboratory equipment or materials with dual-use potential (see section 10.D);
- sensitive data or computerized systems;
- animal care facilities;
- infrastructure vulnerabilities (e.g., accessible power lines, poor lighting);
- security systems in place (e.g., access control, cameras, intrusion detection);
- access controls for laboratory personnel (e.g., background checks, authorization procedures, badges, key controls, escorted access);
- institutional procedures and culture (e.g., tailgating, open laboratories, no questioning of visitors);
- security plans in place; and
- training and awareness of laboratory personnel.

Where the perceived risk is high, institutions should consider contracting a laboratory security consultant to conduct the SVA with input and feedback from security, safety, and laboratory staff.

10.G DUAL-USE SECURITY

When assessing security needs, determine whether laboratories possess materials, equipment, or technologies that have the potential for dual use, such as Select Agents or COI. Whether or not security regulations apply, take prudent steps to reduce the risk of theft or use for terrorist activity.

- Maintain inventory records of dual-use materials.
- Limit the number of laboratory personnel who have access to dual-use agents.
- Provide easy access to a means of emergency communication, in case of a security breach or a threat from within or outside. Consider adding repeaters, or bidirectional signal amplifiers, so that someone with a cell phone can make an emergency phone call from within the secure area.
- Periodically and carefully review laboratory access controls to areas where dual-use agents are used or stored.
- Maintain a log of who has gained access to areas where dual-use materials are used or stored.
- Develop a formal policy prohibiting use of laboratory facilities or materials without the consent of the principal investigator or laboratory supervisor.
- Monitor and authorize specific use of these materials.
- Remain alert and aware of the possibility of removal of any chemicals for illicit purposes. Report such activity to the head of security.
- Train all laboratory personnel who have access to these substances, including a discussion of the security risks of dual-use materials.

As appropriate, address these steps in the SVA and ensure that the security plans adequately provide for the issues these steps address.

10.H SECURITY PLANS

The SVA findings provide a list of risks, needs, and options for improvement (i.e., materials and laboratories in need of security measures beyond a lock and key). There is no template that can apply to every laboratory security plan, because several factors make each organization unique, including building architecture, building use (e.g., mixed use with classrooms, offices, or meeting rooms), organizational culture, and so on.

DHS provides guidance on the planning process in its Risk-Based Performance Standard for chemical security. These guidelines were prepared for dual-use materials that pose high or unusual risks. Recognizing that facilities need "the flexibility to choose the most cost-effective method for achieving a satisfactory level of security based on their risk profile" (DHS, 2008), this guidance provides an outline of elements that should be considered for any laboratory security plan:

- Identify the leadership structure for security issues.
- Secure the assets identified in the vulnerability assessment in a manner that prevents access by unauthorized individuals.

- Deter cyber sabotage, including unauthorized on-site or remote access to critical process controls.
- Prevent diversion using secure shipping, receiving, and storage of target materials.
- Detect theft or diversion of target materials through inventory controls.
- Establish a process for personnel surety, such as background checks, of laboratory personnel, visitors, and others with access to the laboratory.
- Screen and control access to the facility using identification badges, electronic access controls, and security personnel. Check individuals to ensure individuals do not bring harmful materials into the laboratory.
- Train laboratory personnel on the security measures, response, and importance of compliance with security procedures.
- Deter and delay a security breach through the use of multiple security layers and the physical security measures discussed below. Deterrents add time between the detection of a breach and the successful act (i.e., theft or release), which allows more time for responders to prevent the act.
- Monitor (detect) the security of those assets, such that a security breach would be noticed, and (for high-risk materials) would prompt an immediate response by laboratory or security personnel.
- Maintain monitoring, communication, and warning systems.
- Develop and implement response plans for security breaches, and exercise those plans.
- Investigate and track reports of security-related incidents. Document the incident reports, including findings and mitigation.
- Report significant incidents involving chemical security to local law enforcement.
- Maintain records of compliance with the security plan.
- Establish information-sharing and communication networks with associations and government agencies that regularly evaluate and categorize threats relevant to the laboratory or laboratory personnel. Develop a multilevel security plan that identifies appropriate security processes, procedures, and systems for normal security operations and increasing levels of security for periods of higher risk.

DHS also recommends that security plans address the security of the site perimeter and institute vehicle checks. These elements may be appropriate where laboratories are located within an industrial facility, but may be impractical at a medical, research, or educational facility.

Background checks are important for individuals working with dual-use or high-security materials, but it can be challenging to make them complete and accurate. Criminal background checks sometimes include only local crimes, rather than those committed in other areas, or vice versa. However, potential problems can be identified by noting gaps in job history and verifying employment and education background information provided by the applicant. It is often very difficult to get good background information for people who have lived, worked, or been educated in a foreign country.

10.H.1 Levels of Security

When developing a security plan, it is important to establish levels of security that correspond to the security needs of a particular laboratory or portion of a laboratory. These needs will also be influenced by the mission of the organization. For example, in many universities, research laboratories are housed in the same building as instructional classrooms. In those cases, strong access controls to the building are not practical, and would likely cause consternation on campus. Establishing security levels facilitates the review of security needs for a laboratory, ensures consistency in the application of security principles, and integrates the specific measures described above.

The following is one example of a management system for laboratory security, which illustrates how an institution or firm might set three security levels based on operations and materials.

10.H.1.1   Normal (Security Level 1)

In this example, a laboratory characterized as Security Level 1 (see Table 10.1) poses low risk for extraordinary chemical, biological, or radioactive hazards. Loss to theft, malicious pranks, or sabotage would have minimal impact to operations, health, or safety.

10.H.1.2   Elevated (Security Level 2)

A laboratory characterized as Security Level 2 (see Table 10.2) poses moderate risk for potential chemical, biological, or radioactive hazards. The laboratory may contain equipment or material that could be misused or threaten the public. Loss to theft, malicious pranks, or sabotage would have moderately serious health

TABLE 10.1   Security Features for Security Level 1

Physical	• Lockable doors and windows
Operational	• Lock doors when not occupied • Ensure all laboratory personnel receive security awareness training • Control access to keys, use judgment in providing keys to visitors

TABLE 10.2 Security Features for Security Level 2

Physical	• Lockable doors, windows, and other passageways • Door locks with high-security cores • Separate from public areas • Hardened doors, frames, and locks • Perimeter walls extending from the floor to the ceiling (prevent access from one area to the other over a drop ceiling)
Operational	• Secure doors, windows, and passageways when not occupied • Ensure all laboratory personnel receive security awareness training • Escort visitors and contractors, consider an entry log
Electronic	• Access control system recommended • Intrusion alarm recommended where sabotage, theft, or diversion is a concern

TABLE 10.3 Security Features for Security Level 3

Physical	• Lockable doors, windows, and other passageways • Door locks with high-security cores • Separate from public areas • Hardened doors, frames, and locks • Perimeter walls extending from the floor to the ceiling (prevent access from one area to the other over a drop ceiling) • Double-door vestibule entry
Operational	• Secure doors, windows, and passageways when not occupied • Ensure all laboratory personnel receive security awareness training • Escort and log in visitors and contractors • Lock doors, windows, and passageways at all times • Inspect items carried into or removed from the laboratory • Have an inventory system is in place for materials of concern. • Perform background checks on individuals with direct access to the materials of concern or within the control zone.
Electronic	• Access control system that records the transaction history of all authorized individuals • Biometric personal verification technology recommended • Intrusion alarm system • Closed-circuit television cameras for entrance and exit points, materials storage, and special equipment

and safety impact, and be detrimental to the research programs and the reputation of the institution.

10.H.1.3   High (Security Level 3)

A laboratory characterized as Security Level 3 (see Table 10.3) in this example can pose serious or potentially lethal biological, chemical, or radioactive risks to students, employees, or the environment. Equipment or material loss to theft, malicious pranks, or sabotage would have serious health and safety impacts and consequences to the research programs, the facilities, and the reputation of the institution.

10.H.2   Managing Security

As noted above, any security plan, no matter what level of security is needed, should identify a person or group responsible for the overall plan. The person or group managing the program should have at least basic security knowledge, understand the risks and vulnerabilities, and should be provided sufficient resources, responsibility, and authority.

10.H.3   Training

Security should be an integral part of the laboratory safety program. Ensure all personnel are trained in security issues, in addition to safety issues. Although safety and security are two different things, there are many overlaps between measures used to increase security and those used to increase safety, including

- minimizing the use of hazardous and precursor chemicals, which reduces health, safety, and potential security risks;
- minimizing the supply of hazardous materials on-site;
- restricting access to only those who need to use the material and understand the hazards from both a chemical standpoint and a security standpoint; and
- knowing what to do in an emergency or security breach, and how to recognize threats.

Ensure that all personnel understand the security measures in place and how to use them. No matter how complex a system may be, the weakest link tends to be personnel. For example, even the best access control system may not prevent laboratory personnel from granting an unauthorized individual access to a sensitive area.

11 Safety Laws and Standards Pertinent to Laboratories

11.A	INTRODUCTION	267
	11.A.1 Making Safety Laws and Their Rationale	267
	11.A.2 OSHA and Laboratories	268
	11.A.2.1 OSHA Enforcement and State OSHA Laws	268
	11.A.2.2 The General Duty Clause and "Nonlaboratory" OSHA Standards	268
	11.A.2.3 Laboratory Standard Versus Hazard Communication Standard	268
	11.A.2.4 PELs, TLVs, and RELs	269
	11.A.3 Understanding Other Laboratory Safety Requirements	269
11.B	REGULATION OF LABORATORY DESIGN AND CONSTRUCTION	272
11.C	REGULATION OF CHEMICALS USED IN LABORATORIES	273
	11.C.1 OSHA Standards for Specific Chemicals	273
	11.C.2 The OSHA Laboratory Standard	273
	11.C.2.1 The Chemical Hygiene Plan	274
	11.C.2.2 Particularly Hazardous Substances	274
	11.C.3 Chemical Facility Anti-Terrorism Standards	275
	11.C.4 Regulations Covering Polychlorinated Biphenyls (PCBs)	275
11.D	REGULATION OF BIOHAZARDS AND RADIOACTIVE MATERIALS USED IN LABORATORIES	276
11.E	ENVIRONMENTAL REGULATIONS PERTAINING TO LABORATORIES	276
	11.E.1 Management of Chemical Hazardous Waste	276
	11.E.1.1 Definitions and Types of Hazardous Waste Generators	276
	11.E.1.2 Implications of EPA's Definition of *On-Site*	276
	11.E.1.3 Minimum Requirements for Generators	277
	11.E.1.4 RCRA Waste Minimization Requirements	277
	11.E.1.5 Transportation of Chemicals and Hazardous Waste	277
	11.E.2 Management of Radioactive and Biohazardous Waste	278
	11.E.3 Discharges to the Sewer	278
	11.E.4 Air Emissions from Laboratories	278
11.F	SHIPPING, EXPORT, AND IMPORT OF LABORATORY MATERIALS	278
	11.F.1 General Shipping Regulations	278
	11.F.2 EPA Requirements for Chemical Export and Import	279
	11.F.2.1 TSCA Research and Development Exemption	279
	11.F.2.2 TSCA Record-Keeping Requirements for R&D Laboratories	279
	11.F.2.3 Chemical Exports from R&D Laboratories	279

        11.F.2.4   TSCA Requirements for Other Chemical
                   Shipments   280
        11.F.2.5   Chemical Imports from R&D Laboratories   280
        11.F.2.6   Nanomaterials Under TSCA   280
    11.F.3   Requirements for Biological Export and Import   280
    11.F.4   Other Export Regulations   281

11.G   LABORATORY ACCIDENTS, SPILLS, RELEASES, AND
       INCIDENTS   281
    11.G.1   Laboratory Injuries and Illnesses   281
    11.G.2   Planning for Chemical Emergencies   281
    11.G.3   Notification Requirements for Spills, Releases, and Other
             Emergencies   281
    11.G.4   Emergency Training and Response   281

11.A INTRODUCTION

There are a number of federal, state, and local laws, regulations, ordinances, and standards that pertain to the laboratory activities and conditions that affect the environment, health, and safety. These are reviewed briefly in this chapter. For safety laws and standards described in detail elsewhere in this book, this chapter will refer to that section.

Laws, rules, regulations, and ordinances are created and enforced by federal, state, and local governments. International regulations apply to air and marine transport of laboratory materials. Safety standards and codes are created by nongovernmental bodies, but are important to know because they may be required by a law (by reference), as condition of occupancy, by your insurance company, by an accrediting body, or as a widely accepted industry standard. In some cases, following a safety guideline is a condition of receiving a research grant.

Please note that this chapter is not meant to be a compliance guide. This chapter only provides an overview of certain laws. Further, this chapter mostly focuses on federal requirements. State and local requirements may be more stringent, so be sure to check to determine the specific rules that apply.

11.A.1 Making Safety Laws and Their Rationale

Organizations that handle chemicals in laboratories should participate in the regulatory process so that regulators will understand the impact that proposed rules can have on the laboratory environment. The best way to provide input to this process is through dialogue with the regulators, which can take place directly or in collaboration with the institution's environmental health and safety (EHS) or governmental relations office. Also, professional associations, such as the American Chemical Society (ACS), the American Industrial Hygiene Association (AIHA), the American Conference of Governmental Industrial Hygienists (ACGIH), and the American Institute of Chemical Engineers (AIChE), as well as trade associations such as the American Chemistry Council (ACC) and the Campus Health Safety and Environmental Management Association (CHSEMA), regularly comment on proposed regulations, especially proposed federal regulations (which, by law, require solicitation of comment from interested parties). Participation in the regulatory process through such groups is encouraged.

A brief description of the federal legislative and regulatory processes may be helpful. Laws are a product of legislative activity. Legislation is usually proposed by senators and representatives to achieve a desired result, for example, improved employee safety or environmental protection. Proposed laws are often known by their Senate or House file numbers, for example, S.xxx or H.R.xxx. Copies of proposed laws can be obtained by visiting thomas.loc.gov, the Web site for the legislative search engine at the Library of Congress, or by requesting them from local offices of House or Senate members. Sponsors of proposed legislation are open to comment from the public. Once a law is passed, it is known by its Public Law number, for example, P.L. 94-580, Resource Conservation and Recovery Act (RCRA). It is published in the *United States Code* and is referenced by title and section number; 42 USC § 6901 et seq. is the citation for RCRA.

When a law is passed, it is assigned to an administrative unit (agency or department) for development of rules and regulations that will implement the purpose of the legislation. The major federal agencies involved in regulation of laboratory chemicals are the U.S. Occupational Safety and Health Administration (OSHA), the U.S. Environmental Protection Agency (EPA), the U.S. Drug Enforcement Agency, the U.S. Department of Homeland Security (DHS), and the U.S. Department of Transportation (DOT). Proposed regulations are published in the *Federal Register,* a daily publication of federal agency activities. Typically, a public comment period and perhaps public hearings are specified, during which all affected parties have an opportunity to present their support for or concerns with the regulations as proposed. This is the second significant opportunity for involvement in the regulatory process. Final rules are published in the *Federal Register* and in the *Code of Federal Regulations* (CFR), which is updated annually to include all changes during the previous year. Rules in the CFR are referenced by title and part number; for example, 40 CFR Parts 260–272 is the citation for RCRA's hazardous waste rules.

It is helpful to understand the rationale that underlies EHS laws and regulations. These laws reflect congressional, state, and local legislative concerns about worker safety, human health, and the environment, and enjoy strong public support.

Regulations and compliance with them is complicated by the fact that it is a virtual impossibility for EHS regulators to weigh *every* risk precisely. To attempt chemical-by-chemical regulation of the thousands of known, and unknown, chemicals would be so onerous and time-consuming as to leave many serious hazards unregulated. Consequently, regulators attempt to strike a balance by regulating classes of hazards and risks.

Those managing and working in laboratories should also recognize that violation of EHS laws and regulations not only may pose unnecessary risks to those in the laboratory and the surrounding community, but also can result in significant civil penalties (at publication of this book, some laws allow maximum fines of more than $30,000 per day per violation), as well as criminal penalties. Violations can erode community confidence in an institution's seriousness of purpose in safeguarding the environment and complying with the law. Prudent practice requires not only scientific prudence, but also prudent behavior in terms of preventing the risks of noncompliance, adverse publicity, and damage to public trust and an institution's community support.

11.A.2 OSHA and Laboratories

It is important to understand the relationships between the regulations and standards that mediate laboratory activities. The OSHA Laboratory Standard (Occupational Exposure to Hazardous Chemicals in Laboratories, 29 CFR § 1910.1450) is the primary regulation, but laboratory personnel and EHS staff should understand its relationship to the hazard communication standard. In addition, the general duty clause is often invoked, and OSHA standards not written specifically for laboratories may also apply. Laboratory personnel also need to know the relationship between OSHA's permissible exposure limits (PELs), ACGIH threshold limit values (TLVs), and the National Institute of Occupational Safety and Health (NIOSH) recommended exposure limits (RELs).

11.A.2.1 OSHA Enforcement and State OSHA Laws

Enforcement of OSHA standards (such as the Laboratory Standard), may be a shared responsibility of the federal government and of state occupational safety and health programs. Under Section 18 of the Occupational Safety and Health Act, individual states may be authorized by federal OSHA to administer the act if they adopt a plan for development and enforcement of standards that is at least as effective as the federal standards. These states are known as "state-plan" states. In states that do not administer their own occupational safety and health programs, federal OSHA is the regulator, covering all nonpublic employers. State-plan states have generally included public employees in their regulatory approach. What this means is that a given institution may be subject to (1) the federal Laboratory Standard, enforced by federal OSHA; (2) a state laboratory standard, enforced by state OSHA; or (3) if a public institution is not subject to OSHA regulation, state public institution health and safety regulations enforced by a state agency. The EHS office at each institution should have a copy of the applicable standard.

11.A.2.2 The General Duty Clause and "Nonlaboratory" OSHA Standards

Another important point to understand about OSHA and laboratories is that although the Laboratory Standard supersedes existing OSHA *health* standards, other OSHA rules on topics not specifically addressed in the standard remain applicable. The so-called general duty clause of the Occupational Safety and Health Act, which requires an employer to "furnish to each of his employees . . . a place of employment . . . free from recognized hazards that are likely to cause death or serious physical harm . . ." and requires an employee to "comply with occupational safety and health standards and all rules . . . issued pursuant to this chapter which are applicable to his own actions and conduct" continues to be applicable and, indeed, is one of the most commonly cited sections in cases of alleged OSHA violations.

11.A.2.3 Laboratory Standard Versus Hazard Communication Standard

As noted above, the Laboratory Standard is intended, with limited exceptions, to be the primary OSHA standard governing employees who routinely work in laboratories. The Hazard Communication Standard, on the other hand, applies to all nonlaboratory operations "where chemicals are either used, distributed or are produced for use or distribution."

The obvious difficulty is that workers in maintenance shops, even if in a laboratory building, would be covered by the Hazard Communication Standard, not the Laboratory Standard. The requirements of the Hazard Communication Standard are, in certain respects, more demanding than those of the Laboratory Standard. For example, the Hazard Communication Standard requires that *each* container of hazardous chemicals used by the employee be labeled clearly with the identity of the chemical and appropriate hazard warnings, whereas the Laboratory Standard requires only that employers "ensure that labels on incoming containers of hazardous chemicals are not removed or defaced."

The Hazard Communication Standard further requires that copies of material safety data sheets (MSDSs) for *each* hazardous chemical be readily accessible to employees, whereas the Laboratory Standard requires only that employers "maintain MSDSs that are received with incoming shipments, and ensure that they are readily accessible. . . ."

Custodial and maintenance staff who service the laboratory continue to be governed by the Hazard Communication Standard and other OSHA standards, which set forth the information, training, and health and safety protections required to be provided to non-laboratory employees.

Many organizations, faced with the difficulty of designing EHS programs that meet both the requirements of the Laboratory Standard and the requirements of the Hazard Communication Standard, have opted to follow the requirements of the Hazard Communication Standard for all workplaces, laboratory and nonlaboratory, while additionally adopting and implementing the Chemical Hygiene Plan requirements of the Laboratory Standard as they apply to laboratories. Careful comparison of the two standards should be made when designing an EHS program.

11.A.2.4 PELs, TLVs, and RELs

OSHA has developed PELs for chemicals. These are enforceable regulatory limits for the air concentration of individual substances to which a worker may be exposed. Many PELs are based on TLVs, which are nonregulatory exposure limits prepared by ACGIH using existing published, peer-reviewed scientific literature. Quoting the TLV booklet (ACGIH, 2009), "The TLVs . . . represent conditions under which ACGIH believes that nearly all workers may be repeatedly exposed without adverse health effects. They are not fine lines between safe and dangerous exposures, nor are they a relative index of toxicology." PELs and TLVs are average concentrations for a normal 8-hour workday and a 40-hour workweek. This time-weighted average (TWA) approach to evaluating airborne contaminant exposure means that some periods of the day may have higher or lower exposures than others, reflecting the variability in most work with chemicals.

For a small number of compounds, both OSHA and ACGIH have also established a short-term exposure limit (STEL), a concentration considered safe for no more than four 15-minute periods a day. STELs are published only for compounds where toxic effects have been reported from high-level, short-duration exposures in humans or animals. In addition, both groups have also established ceiling limits for some compounds (indicated by a "C" preceding the TLV or PEL value). The ceiling limit is the concentration that should not be exceeded during any time portion of exposure. For compounds that include neither a STEL nor a C notation, a limit on the upper level of exposure should still be imposed. According to the TLV booklet, "Excursions in worker exposure levels may exceed 3 times the TLV-TWA for no more than a total of 30 minutes during a work day, and under no circumstances should they exceed 5 times the TLV-TWA, provided that the TLV-TWA is not exceeded."

The action level (AL) is an OSHA regulatory concept applied to only a few substances. The AL is also an exposure limit for airborne concentration (lower than its associated PEL) that, if exceeded, requires certain additional protective measures to be implemented, such as additional confirmatory exposure monitoring, training, or medical surveillance. Although personal exposures in research laboratory environments are generally controlled well below all of these limits by the use of local exhaust devices and room air change rates, laboratories working with any of the chemicals covered by an OSHA substance-specific standard must be aware of the applicable regulatory provisions and implement them.

RELs are additional exposure values that are developed by the National Institute for Occupational Safety and Health (NIOSH). Like TLVs, RELs are not legal standards but are science-based recommendations that do not need to take into account feasibility, financial impact, or other consequences of their use. As a result, RELs and TLVs are generally more conservative (i.e., lower, more protective) than OSHA's limits.

11.A.3 Understanding Other Laboratory Safety Requirements

These rules are vast, complex, and intricate in their details and interrelationships. As noted above, the application and specifics of federal laws vary from state to state, local jurisdictions, and among federal regulatory agency regional offices. Further, there is a great variety of state and local laws, and so requirements depend on the laboratory's location. State and local laws are not covered here, and so specific requirements may vary from the general information provided here. Where available, an EHS officer who is familiar with the details of these rules can act as a resource for scientists. Smaller organizations can seek advice directly from their counsel, insurance provider, regulatory agencies, EHS professionals at other organizations, or consultants.

Table 11.1 lists safety laws that pertain to laboratories, along with their associated regulations. This table is not comprehensive. As noted previously, a detailed explanation of these requirements, and all the nonregulatory safety standards that apply to laboratories, is beyond the scope of this book. Laboratory safety standards that are among the most relevant are those published by the American Industrial Hygiene Association, American National Standards Institute (ANSI; e.g., laboratory decommissioning standard), Clinical and Laboratory Standards Institute (e.g., clinical laboratory waste management), College of American Pathologists, International Association for Assessment and Accreditation of Laboratory Animal Care, and the National Council on Radiation Protection and Measurement (e.g., radiation exposure, waste manage-

TABLE 11.1 Federal Safety Laws and Regulations That Pertain to Laboratories

Law or Regulation	Citation	Purpose	Comments
Regulation of Chemicals Used in Laboratories			
Occupational Safety and Health Act (OSHA)	29 USC § 651 et seq.	Worker protection	
General duty clause	29 USC § 654(5)(a) and (b)	Assurance of workplace free from recognized hazards that are causing or likely to cause serious physical harm	Foundation enforceable requirement in absence of a specific standard
Occupational Exposure to Hazardous Chemicals in Laboratories (Laboratory Standard)	29 CFR § 1910.1450	Laboratory worker protection from exposure to hazardous chemicals	Requires a chemical hygiene plan. Title 29 rules are written and enforced by OSHA
Hazard Communication Standard	29 CFR § 1910.1200	General worker protection from chemical use	Requires labeling and material safety data sheets (MSDSs)
Air contaminants	29 CFR §§ 1910.1000–1910.1050	Standards for exposure to hazardous chemicals	See section 11.C.1 for chemical-specific regulations pertinent in laboratories
Hazardous materials	29 CFR §§ 1910.101–1910.111	Protection against hazards of compressed gases, flammable and combustible liquids, explosives, anhydrous ammonia	See also Uniform Fire Code and National Fire Protection Association standards
OSHA Respiratory Protection Standard	29 CFR § 1910.134	When respiratory protection is required; how to fit and use respirators; and medical review	
Personal protective equipment	29 CFR §§ 1910.132–1910.138	Head, hand, foot, eye, face, and respiratory tract protection	See also American National Standards Institute standards
Control of hazardous energy (Lock out/Tag out)	29 CFR § 1910.147	Worker protection from electrical and other stored energy hazards	
Machinery and machine guarding	29 CFR §§ 1910.211–1910.219	Worker protection from mechanical hazards	
Controlled substances	21 CFR §§ 1300-1399	Requires licenses and controls for the purchase, use, and possession of controlled substances, illicit drugs, and certain drug precursors	Enforced by the Drug Enforcement Agency
Chemical Facility Anti-Terrorism Standards (CFATS) with Appendix	6 CFR Part 27	Establishes risk-based performance standards for the security of chemical facilities	Appendix A of the regulation contains list of chemicals of interest and their threshold quantities
Toxic Substances Control Act (TSCA) Polychlorinated biphenyls (PCBs)	40 CFR Part 761	Prohibition against PCBs in manufacturing, processing, distribution in commerce, and certain uses	Permits certain limited laboratory use of PCBs
Permit and excise tax for purchase of 190- and 200-proof ethanol	27 CFR Part 211	Control of the sale of ethanol	Enforced by the U.S. Bureau of Alcohol, Tobacco, and Firearms
Regulation of Biohazards and Radioactive Material Used in Laboratories			
Occupational exposure to bloodborne pathogens	29 CFR § 1910.1030	Worker protection from exposure to bloodborne pathogens	
Select agents and toxins	42 CFR Part 73	Establishes the requirements for possession, use, and transfer of select agents and toxins.	Select agents are biological agents that are a terror risk. Rules are administered by the U.S. Centers for Disease Control and Prevention and the U.S. Animal and Plant Health Inspection Service

TABLE 11.1 Continued

Law or Regulation	Citation	Purpose	Comments
Atomic Energy Act Energy Reorganization Act	42 USC § 2073 et seq. 42 USC § 5841 et seq.	Establish standards for protection against radiation hazards	See also OSHA, Ionizing Radiation
Standards for Protection Against Radiation; Licenses	10 CFR Part 20 10 CFR Parts 30–35	Establish exposure limits and license conditions	Title 10 rules are written and enforced by Nuclear Regulatory Commission
Notices, Instructions, and Reports to Workers; Inspections	10 CFR Part 19	Workplace information that must be posted where radiation or radioactive materials are present	
Environmental Regulations Pertaining to Laboratories			
Resource Conservation and Recovery Act (RCRA)	42 USC § 6901 et seq.	Protection of human health and environment	
Hazardous waste management	40 CFR Parts 260–272	"Cradle-to-grave" control of chemical waste	Subpart K of 40 CFR Part 262 is an opt-in rule specific to laboratories in academia. Title 40 rules are written and enforced by EPA
Clean Air Act (CAA)	42 USC § 7401 et seq.	Protection of air quality and human health	
CAA Amendments of 1990	42 USC § 7409 et seq.	Expansion of air quality protection	Requires development of specific rules for laboratories
National Emission Standards for Hazardous Air Pollutants	40 CFR Part 82	Control of air pollutant emissions	
Montreal Protocol for Protection of Stratospheric Ozone	40 CFR Part 82	Control of emission of ozone-depleting compounds	Severely limits use of certain chlorofluorocarbons
Federal Water Pollution Control Act	33 USC § 1251 et seq.	Improvement and protection of water quality	
Criteria and standards for the National Pollutant Discharge Elimination System (NPDES)	40 CFR Part 125	Control of discharge to public waters	
General pretreatment regulations for existing and new sources of pollution	40 CFR Part 403	Control of discharge of pollutants to public treatment works	Implemented by local sewer authorities
Shipping, Export, and Import of Laboratory Materials			
Hazardous Materials Transportation Act	48 USC § 1801 et seq.	Control of movement of hazardous materials	
Hazardous material regulations	49 CFR Parts 100–199	Regulation of packaging, labeling, placarding, and transporting	Standards of the International Air Transport Agency apply to chemicals shipped by air. Title 49 rules are written and enforced by DOT.
Hazardous materials training requirements	49 CFR §§ 172.700–172.704	Assurance of training for all persons involved in transportation of hazardous materials	Also known as HM126F
TSCA	15 USC § 2601 et seq.	Requires testing and necessary restrictions on use of certain chemical substances	Collection and development of information on chemicals
Reporting and recordkeeping requirements	40 CFR Part 704	One provision exempts users of small quantities solely for research and development (R&D)	Must follow R&D exemption requirements
Significant adverse reaction	40 CFR Part 717	Record of new allegation that chemical substances or mixture caused significant adverse effect for health or the environment	TSCA § 8(c)

continued

TABLE 11.1 Continued

Law or Regulation	Citation	Purpose	Comments
Technically qualified individual (TQI)	40 CFR § 720.3(ee)	Definition of TQI by background; understanding of risks, responsibilities, and legal requirements	Follow TQI requirements with R&D
TSCA exemption for R&D	40 CFR § 720.36	Exemption for R&D from PMN if chemical substance not on TSCA inventory or is manufactured or imported only in small quantities solely for R&D	Follow R&D exemption requirements including labeling and MSDS information
Exports of samples, chemicals, biologicals, other materials, and laboratory equipment	15 CFR Parts 730–774	Regulates shipments of certain chemicals and other research materials out the United States	These rules are administered by the U.S. Department of Commerce; other export regulations may apply

Regulation of Laboratory Injuries, Accidents, and Spills

Recording and reporting occupational injuries and illnesses	29 CFR Part 1904	Standards for employee reporting and recordkeeping	
Employee emergency plans and fire prevention plans	29 CFR § 1910.38	Requirements for written emergency and fire prevention plans	
Medical services and first aid	29 CFR § 1910.151	Provision of medical services, first-aid equipment, and facilities for quick drenching and flushing of eyes	
Superfund Amendments and Reauthorization Act (SARA)	42 USC § 9601 et seq. 42 USC § 11000 et seq. 40 CFR Part 370 (§ 311–312) 40 CFR Part 372 (§ 313)	Planning for emergencies and reporting of hazardous materials	Title III, also known as Community Right-to-Know Act
Emergency planning and notification	40 CFR Part 355	Requirements for reporting of extremely hazardous materials and unplanned releases	Applies to all chemical users
Hazardous Waste Operations and Emergency Response	29 CFR § 1910.120 40 CFR Part 311	Worker protection during hazardous waste cleanup	Applies to state and local government employees not covered by OSHA

Other Laboratory Regulations and Standards

Americans with Disabilities Act	28 CFR Part 36	Standards for making workplace accommodations for students and employees with disabilities	
Access to employee exposure and medical records	29 CFR § 1910.20	Employee and privacy and other rights; employer responsibilities	
Occupational noise exposure	29 CFR § 1910.95	Standards for noise, monitoring and medical surveillance	

ment). (See Chapter 10, section 10.E, for an explanation of laboratory security requirements.)

Two laws that have perhaps the most impact on laboratories are the Occupational Safety and Health Administration's *Occupational Exposure to Hazardous Chemicals in Laboratories* (the OSHA Laboratory Standard) and RCRA, under which EPA regulates chemical hazardous waste. Because of its importance, the text of the OSHA Laboratory Standard is reprinted in Appen-

dix A. Laboratory workers and managers should read and understand these regulations.

11.B REGULATION OF LABORATORY DESIGN AND CONSTRUCTION

Laboratory design, construction, and renovation are regulated mainly by state and local laws that incorporate, by reference, generally accepted standard practices set out in various uniform codes, such as the International Building Code (IBC), the International Fire Code (IFC), and the National Fire Protection Association standards.[1] For laboratory buildings where hazardous chemicals are stored or used, detailed requirements usually cover spill control, drainage, containment, ventilation, emergency power, special controls for hazardous gases, fire prevention, and building height. Some localities have initiated regulations aimed at increasing efficiency and sustainability in building design. These may become more common in the future, and laboratory designers may wish to consider these issues when planning new construction.

Building and fire codes also apply after construction has been completed. These codes are typically enforced by the fire authority having jurisdiction—usually the local fire marshal. As explained in Chapter 6, sections 6.F.5 and 6.F.7 these codes describe how flammables, reactives, and gases must be stored, and limit their quantities in fire control areas.

In addition, OSHA standards affect some key laboratory design and construction issues, for example, eyewashes, safety showers, and special ventilation requirements. Other consensus standards prepared by organizations such as ANSI and the American Society of Heating, Refrigeration, and Air Conditioning Engineers are relevant to laboratory design. It is not uncommon for various codes and consensus standards to be incorporated into state or federal regulations.

11.C REGULATION OF CHEMICALS USED IN LABORATORIES

OSHA and EPA regulation of chemical use in laboratories is described below. The laboratory use of controlled substances, regulated by the U.S. Drug Enforcement Agency, is described in Chapter 10, section 10.E.4.1. Select agent toxins are regulated by the Centers for Disease Control and Prevention (CDC) and the U.S. Department of Agriculture Animal and Plant Health Inspection Service (APHIS).

11.C.1 OSHA Standards for Specific Chemicals

OSHA has developed comprehensive standards for several chemicals, which are listed in Table 11.2. To prevent exposure to personnel, these standards cover all aspects of the use of these chemicals. These standards are above those required by the Laboratory Standard and, in some cases, may require special signs, medi-

TABLE 11.2 Chemicals Covered by Specific OSHA Standards

1001 Asbestos
1002 Coal tar pitch volatiles
1003 4-Nitrobiphenyl (and 12 related carcinogens)
1004 α-Naphthylamine
1006 Methyl chloromethyl ether
1007 3,3'-Dichlorobenzidine (and its salts)
1008 bis-Chloromethyl ether
1009 β-Naphthylamine
1010 Benzidine
1011 4-Aminodiphenyl
1012 Ethyleneimine
1013 β-Propiolactone
1014 2-acetylaminofluorene
1015 4-Dimethylaminoazobenzene
1016 N-Nitrosodimethylamine
1017 Vinyl chloride
1018 Inorganic arsenic
1025 Lead
1026 Chromium VI
1027 Cadmium
1028 Benzene
1044 1,2-Dibromo-3-chloropropane
1045 Acrylonitrile
1047 Ethylene oxide
1048 Formaldehyde
1050 Methylenedianiline
1051 1,3-Butadiene
1052 Methylene chloride

Each standard is in 29 CFR § 1910.XXXX, where XXXX is the section number that precedes the chemical name:

cal surveillance, and routine air monitoring of your workplace. For more information, see 29 CFR Part 1910 as well as in specific standards following section 1910.1000, such as the vinyl chloride standard, 29 CFR § 1910.1017, which prohibits direct contact with liquid vinyl chloride.

Other OSHA standards setting forth PELs apply to the extent that they require limiting exposures to below the PEL, and, where the PEL or AL is routinely exceeded, the Laboratory Standard's provisions (described below) require exposure monitoring and medical surveillance (see Appendix A, sections (d) and (g)).

If you use these chemicals routinely, even for short periods of time, you should have your workplace evaluated by your EHS officer to ensure that your work practices and engineering controls are sufficient to keep your exposures below the OSHA-specified limits. Because of their common use in campus laboratories, the specific standards for formaldehyde (used as formalin for preservation of tissue samples), benzene, and ethylene oxide are of particular concern.

11.C.2 The OSHA Laboratory Standard

In 1990, OSHA promulgated its Laboratory Standard (Occupational Exposure to Hazardous Chemicals in

---

[1]In 2003, the Building Officials and Code Administrators International, Inc., the International Conference of Building Officials, and the Southern Building Code Congress International, Inc. formed the International Code Council. This body now publishes both the IFC and the IBC among other documents.

Laboratories, 29 CFR § 1910.1450; see Appendix A). In brief, the OSHA Laboratory Standard requires organizations to

1. Keep laboratory personnel exposures to chemicals below OSHA's PELs.
2. Write a Chemical Hygiene Plan.
3. Designate a Chemical Hygiene Officer to implement the plan.
4. Train and inform new laboratory personnel of
   - the OSHA Laboratory Standard,
   - the Chemical Hygiene Plan and its details,
   - OSHA's PELs,
   - the signs and symptoms of exposure to hazardous chemicals,
   - MSDSs,
   - *Prudent Practices in the Laboratory*,
   - methods to detect the presence of hazardous chemicals,
   - the physical and health hazards of the chemicals, and
   - measures to protect laboratory personnel from chemical hazards.
5. In certain circumstances, provide laboratory personnel access to medical consultations and examinations.
6. Keep labels of supplied chemicals intact.
7. Maintain the MSDSs for all your supplied chemicals.
8. For chemical substances developed in your laboratory, train laboratory personnel as described above.
9. Use respirators properly.

It is important to understand that the OSHA PELs and substance-specific standards do not include all hazardous chemicals. It is the laboratory manager's responsibility under the Laboratory Standard and its general duty clause to apply scientific knowledge in safeguarding workers against risks, even though there may be no specifically applicable OSHA standard. In circumstances where exposure limits are exceeded or where work with particularly hazardous substances is conducted, laboratories must keep records of exposure monitoring and medical surveillance.

The Laboratory Standard refers to the National Research Council's *Prudent Practices for Handling Hazardous Chemicals in Laboratories* (NRC, 1981) as "nonmandatory . . . guidance to assist employers in the development of the Chemical Hygiene Plan."

One of the most common Laboratory Standard OSHA citations has been for failure to have a Chemical Hygiene Plan or for missing an element in the plan. Another commonly cited violation is failure to meet the employee information and training requirements of the Laboratory Standard.

11.C.2.1 The Chemical Hygiene Plan

The centerpiece of the Laboratory Standard is the Chemical Hygiene Plan. This is a written plan developed by employers. It has the following major elements:

- employee information and training about the hazards of chemicals in the work area, including how to detect their presence or release, work practices and how to use protective equipment, and emergency response procedures;
- circumstances under which a particular laboratory operation requires prior approval from the employer;
- standard operating procedures for work with hazardous chemicals;
- criteria for use of control measures, such as engineering controls or personal protection equipment;
- measures to ensure proper operation of fume hoods and other protective equipment;
- provisions for additional employee protection for work with "select carcinogens" (as defined in the Laboratory Standard) and for reproductive toxins or substances that have a high degree of acute toxicity;
- provisions for medical consultations and examinations for employees; and
- designation of a Chemical Hygiene Officer.

Section 2.B of Chapter 2 describes additional elements, not required by law, that may be added to a Chemical Hygiene Plan.

Some firms and institutions have developed a single generic Chemical Hygiene Plan for the entire organization. To be most effective, however, the plan should include detailed protections that are specific to each laboratory, project, experiment, procedure, and worker. Laboratory-specific plans allow considerable flexibility in achieving the performance-based goals of the Laboratory Standard. Model Chemical Hygiene Plans are available from your state OSHA consultation service or the American Chemical Society.

11.C.2.2 Particularly Hazardous Substances

There are special provisions in the Laboratory Standard regarding work with "particularly hazardous substances," a term that includes "select carcinogens," "reproductive toxins," and "substances with a high degree of acute toxicity."

- A *select carcinogen* is defined in the standard as any substance (1) regulated by OSHA as a carcinogen; (2) listed as "known to be a carcinogen" in the *Report on Carcinogens* published by the National Toxicology Program (HHS/CDC/NTP, 1995); (3) listed under Group 1 ("carcinogenic to humans") by the *International Agency for Research on Cancer (IARC) Monographs*; or (4) *in certain cases*, listed in either Group 2A or 2B by IARC or under the category "reasonably anticipated to be carcinogens" by NTP. A category (4) substance is considered a select carcinogen only if it causes statistically significant tumor incidence in experimental animals in accordance with any of the following criteria: (1) after inhalation exposure of 6 to 7 hours per day, 5 days per week, for a significant portion of a lifetime to dosages of less than $10\,mg/m^3$; (2) after repeated skin application of less than $300\,mg/kg$ of body weight per week; or (3) after oral dosages of less than $50\,mg/kg$ of body weight per day.
- "*Reproductive toxins*" are defined as those chemicals that affect reproductive capabilities, including chromosomal damage (mutations) and effects on fetuses (teratogenesis).
- *Chemicals with a "high degree of acute toxicity"* are highly toxic noncarcinogenic or highly volatile toxic materials that may be fatal or cause damage to target organs as a result of a single exposure or exposures of short duration. Examples include hydrogen sulfide, nitrogen dioxide, hydrogen cyanide, and methylmercury.

Although "select carcinogens" are specifically identified through reference to other publications, "reproductive toxins" and chemicals with a "high degree of acute toxicity" are not specified further, which has made it difficult to apply these categories. Some organizations have chosen to adopt the OSHA Hazard Communication Standard definition of "highly toxic" ($LD_{50} < 50\,mg/kg$ oral dose) as a workable definition of high degree of acute toxicity. There is little agreement on how to determine reproductive toxins.

The OSHA-mandated special provisions for work with carcinogens, reproductive toxins, and substances that have a high degree of acute toxicity include consideration of "designated areas," use of containment devices, special handling of contaminated waste, and decontamination procedures. The OSHA requirement is for evaluation, assessment, and implementation of these special controls, when appropriate. These special provisions are to be included in the Chemical Hygiene Plan.

11.C.3 Chemical Facility Anti-Terrorism Standards

In 2007, Congress authorized DHS to "establish risk-based performance standards for security chemical facilities." In response, DHS issued the Chemical Facility Anti-Terrorism Standards (CFATS). According to the agency, the standards identify high-risk facilities based on the likelihood of an attack, the consequences of an attack, and the threat of an attack based on the intent and capability of an adversary. The standards are concerned with

- EPA Risk Management Plan chemicals,
- highly toxic gases,
- chemical weapons convention chemicals, and
- explosives.

The specific "Chemicals of Interest" are listed in Appendix A of the CFATS rule. (See Chapter 10, section 10.E.4.2 for examples.) The standard applies to any institution that meets or exceeds the threshold quantity established for these chemicals. All facilities, including those with laboratories, are expected to survey their site for the presence of the chemicals of interest and compare the inventory to the threshold screening quantities established in Appendix A of the standard. If the facility meets or exceeds the threshold quantity for any chemical of interest, the facility must report the inventory to DHS.

On the basis of the report, chemical facilities are categorized into risk-based tiers. Each facility is assigned a tier ranging from tier 1 (highest risk) to tier 4 (lowest risk). Facilities that fall into risk tiers 1–3 must prepare a security vulnerability assessment (SVA) to identify facility security vulnerabilities, and develop and implement site security plans. Should a facility fall into tier 4, circumstances may allow for submission of alternate security programs in lieu of an SVA, a site security plan, or both.

For more information about SVAs and CFATS, see Chapter 10, sections 10.F and 10.E.4.2.

11.C.4 Regulations Covering Polychlorinated Biphenyls (PCBs)

Regulations pursuant to the Toxic Substances Control Act (TSCA) apply to the use of PCBs and monochlorobiphenyls in laboratories. Although the rules except the use of "small quantities for research and development" and use "as an immersion oil in microscopy," researchers contemplating work with PCBs (including environmental studies with PCB-contaminated media) should consult their institution's EHS officer because of the stringency of these regulations.

11.D REGULATION OF BIOHAZARDS AND RADIOACTIVE MATERIALS USED IN LABORATORIES

As explained in Chapter 4, sections 4.H, and Chapter 6, section 6.E.2, most radioactive materials that are used in laboratories are regulated by the U.S. Nuclear Regulatory Commission (USNRC). Rules most pertinent to laboratories are in Title 10 of the *Code of Federal Regulations*, Parts 20 and 30. The USNRC licenses the use of radioactive materials. Many institutions and firms obtain a broadscope license from the NRC, which provides flexibility but requires an institutional Radiation Safety Officer and Radiation Safety Committee.

As explained in Chapter 4, section 4.G, and Chapter 6, section 6.E.1, the most widely accepted standards for using biohazards in laboratories can be found in *Biosafety in Microbiological and Biomedical Laboratories* (BMBL; HHS/CDC/NIH, 2007a). The Foreword explains that, "the BMBL remains an advisory document recommending best practices for the safe conduct of work in biomedical and clinical laboratories, from a biosafety perspective and is not intended as a regulatory document." However, many accrediting bodies, grant-making organizations, and state regulators expect laboratories that use biohazards to adhere to the BMBL.

Select agents are regulated by CDC and the Department of Agriculture's APHIS.

11.E ENVIRONMENTAL REGULATIONS PERTAINING TO LABORATORIES

Federal and state environmental regulations apply to laboratory waste, air emissions, and discharges to the sewer. Of these, EPA's rules for chemical hazardous waste may be the most demanding.

11.E.1 Management of Chemical Hazardous Waste

Chapter 8 covers the regulation of chemical hazardous waste in laboratories, while this section covers the regulation of that waste at an institutional level.

RCRA was enacted by Congress in 1976 to address the problem of improper management of hazardous waste. Subtitle C of that Act established a system for controlling hazardous waste from generation to disposal, often referred to as "cradle to grave." Under RCRA, EPA is given great responsibilities in promulgating detailed regulations governing the generation, transport, treatment, storage, and disposal of hazardous (chemical) waste. RCRA and EPA regulations apply to laboratories that use chemicals.

11.E.1.1 Definitions and Types of Hazardous Waste Generators

A generator is any firm or institution whose processes and actions create hazardous waste. There are three categories of generator:

1. **Large-quantity generators** are those whose facilities generate 1,000 kg or more per month (about four 55-gal drums of hazardous waste) or over 1 kg of "acutely hazardous waste" per month. By this measure, most large research organizations, including the larger universities, are large-quantity generators.
2. **Generators of more than 100 but less than 1,000 kg of hazardous waste per month,** and less than 1 kg of "acutely hazardous waste" per month (and accumulate less than 1 kg at any one time). This category may not accumulate more than 6,000 kg at any one time.
3. **Conditionally exempt small-quantity generators of 100 kg or less of hazardous waste per month** and less than 1 kg of "acutely hazardous waste." The special requirements applicable to conditionally exempt small-quantity generators can be found in 40 CFR § 261.5.

11.E.1.2 Implications of EPA's Definition of On-Site

Federal and state definitions of on-site have bearing on the generation category of each site, and how laboratory hazardous waste is transported and managed. This is particularly important for firms, colleges, universities, and other organizations that are transected by public roads.

"Individual generation site" is defined by RCRA regulation as a contiguous site at or on which hazardous waste is generated. A firm or institution located in one geographic area may be viewed as a single generator with a single EPA generator identification number or, if it is transected by public roads, may be viewed as multiple generator sites requiring multiple EPA generator identification numbers. Multisite facilities are required to have separate EPA identification numbers for each site.

Note that each individual laboratory generating waste is not itself a RCRA "generator," but instead is part of the "generator" site. Each laboratory therefore must comply with the requirements applicable to the site's generator category.

RCRA defines "on-site," as "the same or geographically contiguous property which may be divided by public or private right-of-way, provided the entrance and exit between the properties is at a crossroads intersection, and access is by *crossing as opposed to going along* [emphasis added] the right-of-way."

The significance of this definition is that, with one exception, hazardous waste that is being transported on public roads can be sent only to a permitted treatment, storage, and disposal facility (TSDF). The exception [in 40 CFR § 262.20 (f)] explains that this restriction does "not apply to the transport of hazardous wastes on a public or private right-of-way within or along the border of contiguous property under the control of the same person, even if such contiguous property is divided by a public or private right-of-way."

In all other cases, hazardous waste cannot be transported on public roads to an unpermitted holding facility, even if the public road and the receiving location are within the boundaries of an institution.

11.E.1.3 *Minimum Requirements for Generators*

Generators must obtain an EPA identification number, prepare the waste for transport, follow accumulation and storage requirements, manifest hazardous waste, and adhere to detailed record-keeping and reporting requirements. At most firms and institutions, hazardous waste is shipped off-site, treated, stored, and disposed of at commercial EPA-permitted TSDFs. Note that generators producing more than 1 kg in a calendar month of "acute hazardous waste" (see above) are subject to full regulation under RCRA as a large-quantity generator.

Although conditionally exempt small-quantity generators are partially exempt from these requirements, they must still

- identify their waste to determine whether it is hazardous,
- not accumulate more than 1,000 kg of hazardous waste, and
- ensure that the waste is sent to a permitted TSDF or a recycling facility.

Note that state laws may differ. For example, some states regulate all generators of hazardous waste with no exemptions, and some states regulate chemical wastes that are not included in RCRA (e.g., used oil, as hazardous waste).

See Chapter 8, section 8.B.4 for a detailed explanation of hazardous waste collection and storage requirements.

11.E.1.4 *RCRA Waste Minimization Requirements*

Generators are required to certify on the manifest accompanying off-site shipment of waste that they have a waste minimization program. Guidelines for a waste minimization program are available from EPA. By signing the manifest, the generator is certifying the following:

Large-Quantity Generators: "I have a program in place to reduce the volume and toxicity of waste generated to the degree I have determined to be economically practicable and I have selected the practicable method of treatment, storage, or disposal currently available to me which minimizes the present and future threat to human health and the environment."

Small-Quantity Generators: "I have made a good faith effort to minimize my waste generation and select the best waste management method that is available to me and that I can afford."[2]

11.E.1.5 *Transportation of Chemicals and Hazardous Waste*

For organizations whose laboratory operations are at a single site, transportation within that site is not regulated, as long as that transport involves no travel along public ways. Most organizations, however, have developed policies for on-site transport covering labeling, segregation of incompatibles, containment and double containment, and other necessary safeguards to prevent accidental release to the environment or injury to persons during transportation.

As with hazardous materials, *off-site* transportation of hazardous waste is regulated by DOT in accordance with the Hazardous Materials Transportation Uniform Safety Act. These regulations apply not only to those who actually transport, but also to those who *initiate or receive* hazardous waste shipments. DOT regulations applicable to transport of laboratory chemicals include those governing packaging, labeling, marking, placarding, and reporting of discharges. Those who prepare hazardous materials for transportation must also meet certain training requirements.

Under the DOT Materials of Trade exception, facilities may transport their own chemicals to another facility owned by the same organization under certain conditions. This exemption also applies to transport for the purpose of chemical demonstrations, such as at a local high school or as part of a special event. All chemicals must be properly packaged in DOT-specification containers. Hazardous waste may not be transported as a Material of Trade.

As explained in Chapter 8, section 8.B.7, EPA's RCRA rules include additional requirements for transporta-

---

[2]From 40 CFR § 262.27(a).

tion of hazardous waste. Detail of the many requirements for transporting hazardous waste is beyond the scope of this book.

11.E.2 Management of Radioactive and Biohazardous Waste

Disposal of low-level radioactive waste from laboratories is governed by USNRC rules in Title 10 of the *Code of Federal Regulations*, Parts 20 and 30, as well as conditions specified in institutional licenses. Short-half-life radwaste is typically held for decay in storage, and then disposed of without regard to its radioactivity.

Federal laws that regulate laboratory biohazardous and infectious waste are limited. Most important are the OSHA bloodborne pathogen standard, DOT rules for transporting biomedical waste, and EPA medical waste incineration rules. The OSHA bloodborne pathogen standard addresses the collection and management of needles, blades, and other sharps. Most states regulate the treatment and disposal of laboratory biohazardous waste; consult your state laws for specific requirements.

11.E.3 Discharges to the Sewer

Contact your local publicly owned treatment works (POTW) for rules on discharges to the sanitary sewer. Your POTW is the best source for information about limits and prohibitions for the discharge of laboratory wastewaters that contain chemicals, biologicals, or radioactive materials. Federal rules exist that pertain to the discharge of hazardous waste and radioactive materials, but those limits are usually incorporated in POTW ordinances.

11.E.4 Air Emissions from Laboratories

The Clean Air Act (CAA) regulates emissions into the air. Laboratories should be aware of the regulations that control stratospheric-ozone-depleting substances. The list of such substances can be found in 40 CFR Part 82, Appendixes A and B to Subpart A. The list includes as "Class I" substances most common freons, carbon tetrachloride, and methyl chloroform.

Under the CAA, EPA also sets national emission standards for hazardous air pollutants (NESHAPs). NESHAPs for radionuclides and sterilants have been established, and these may apply to some laboratories. EPA has not established emission standards for volatile organic compounds or other emissions from laboratory operations, nor has EPA established a special source category for research or laboratory facilities. However,

some states have set emission limits that apply to laboratories, or require permits for laboratory hoods. Check with your state environmental agency to determine if there are specific air emission requirements for your laboratory.

11.F SHIPPING, EXPORT, AND IMPORT OF LABORATORY MATERIALS

Shipping, export, and import laws strictly regulate the domestic and international transport of an extensive list of laboratory materials, including many chemicals, vaccines, genetic elements, microbiological agents, radioactive materials, and a wide array of research equipment, technologies, and supplies. Many items that are not perceived to be particularly hazardous, valuable, or uncommon are nevertheless subject to export control laws and shipping regulations. Export and import laws may require special licenses or permits prior to leaving or entering the United States.

Regulated activities include conveying laboratory materials via

- shipments and mailings using the U.S. Post Office and other mail couriers;
- receiving or sending regulated materials by any method of transport;
- shipments to (exporting) or from (importing) a foreign country;
- transporting any amount of regulated material in a commercial aircraft, whether on your person or in carry-on luggage or checked luggage.

The many laws for shipping laboratory materials are described below, including regulations from the U.S. Department of Commerce (DOC), DOT, and EPA. Although these rules are described individually, please note that several regulations often apply to a single shipment.

In addition to these requirements, your institution may have entered into a Material Transfer Agreement, which controls any transfer of the research materials from your institution to another.

See Chapter 5, section 5.F for practical, nonlegal information about shipping laboratory materials.

11.F.1 General Shipping Regulations

Regulations on the transportation of hazardous materials are aimed at ensuring that the public and the workers in the transportation chain are protected from exposure to potentially hazardous materials being transported. Protection is achieved through the following requirements:

- rigorous packaging that will withstand rough handling and contain all liquid material within the package without leakage to the outside,
- appropriate labeling of the package to alert the workers in the transportation chain to the hazardous contents within,
- documentation of the hazardous contents within the package and emergency contact information in the event of an emergency with the package, and
- training of workers in the transportation chain to familiarize them with the hazardous contents so as to be able to respond to emergency situations.

DOT is the national authority that regulates the shipment and transport of hazardous material. DOT regulations governing hazardous materials transport are detailed in Title 49 of the *Code of Federal Regulations*, Parts 171–178.

Technical Instructions for the Safe Transport of Dangerous Goods by Air, published by the International Civil Aviation Organization (ICAO), are the legally binding international regulations. Annually, the International Air Transport Association (IATA) publishes *Dangerous Goods Regulations* (DGR) that incorporates the ICAO provisions and may add further restrictions. The ICAO rules apply on all international flights. For national flights (i.e., flights within one country), national civil aviation authorities apply national legislation. This is normally based on the ICAO provisions, but may incorporate variations. State and operator variations are published in the ICAO Technical Instructions and in the IATA DGR.

11.F.2 EPA Requirements for Chemical Export and Import

TSCA, administered by EPA, was established to ensure that the human health and environmental effects of chemical substances are identified and properly controlled prior to placing these materials into commerce.

Chemical substances regulated by TSCA include, "Any organic or inorganic substances of a particular molecular identity including any combination of such substances occurring, in whole or in part, as a result of chemical reaction or occurring in nature and any element or uncombined radical." Chemical substances not regulated or excluded by TSCA include pesticides regulated by the Federal Insecticide, Fungicide, and Rodenticide Act; tobacco and tobacco products regulated by the Bureau of Alcohol, Tobacco, Firearms, and Explosives; radioactive materials regulated by the USNRC; and foods, food additives, drugs, and cosmetics or devices regulated by the Food and Drug Administration.

11.F.2.1 TSCA Research and Development Exemption

TSCA includes a "Research and Development (R&D) Exemption" which greatly reduces requirements for laboratories, but does not eliminate them. Under the R&D Exemption, laboratory chemicals are exempted from many TSCA requirements if they are

- imported, manufactured, or used in small quantities ("not greater than reasonably necessary for such (R&D) purposes"); and
- solely for purposes of noncommercial scientific experimentation, analysis, or research, and
- under the supervision of a technically qualified individual.

To maintain this exemption status, laboratories engaged in R&D must keep records of allegations of adverse reactions and discovery of substantial risk. Also, chemical imports need to be certified in writing, and certain chemical exports require notification of the receiving countries.

11.F.2.2 TSCA Record-Keeping Requirements for R&D Laboratories

For R&D laboratories, TSCA is primarily an administrative, records-intensive program. Establish a TSCA compliance file to log significant adverse effects, file reports of substantial risks, and document imports and exports.

Under TSCA § 8(c), laboratories are required to keep records of allegations of significant adverse effects from R&D chemicals. For example, laboratories must create and maintain records of allegations of, for example, a skin rash, allergic reaction, or respiratory effect that may be attributable to exposure to an R&D chemical.

Laboratories must also document the discovery of any significant risks to human health or the environment potentially associated with R&D substances. Include this report in your TSCA compliance file.

Your file should also contain TSCA import and export certifications, as described below. Copies of the written notifications provided (i.e., the letter, the MSDS, and copies of all labels affixed to sample containers) should be maintained in a file for 5 years.

11.F.2.3 Chemical Exports from R&D Laboratories

Laboratories must complete and send to EPA a TSCA Export Notification Form prior to the exportation of chemical substances listed in the EPA's Chemicals on Reporting Rule (CORR) Database. Sample forms, the CORR Database, and EPA submission instructions are available via the Web (search for "TSCA Export Certification Form," "EPA CORR Database" and "EPA TSCA Exports").

Copies of this form must be kept in lab records for 3 years.

An export notification is not required for R&D chemicals not listed in the CORR Database.

11.F.2.4   TSCA Requirements for Other Chemical Shipments

For shipments of R&D chemicals to locations within the United States, TSCA requires that laboratories

- Label the containers, shipping containers, and shipping papers with "This material is not listed on the TSCA Inventory. It should be used for research and development purposes only under the direct supervision of a technically qualified individual."
- Prepare and include an MSDS for the substance. This MSDS should evaluate and communicate risks of the substance. On the "composition, information on ingredients," section of the MSDS, indicate, "This material is for R&D evaluation only. It can only be used for R&D evaluations until PMN review by EPA is completed. If this material is used in plants or non-R&D locations for R&D evaluation, its use must be supervised by a technically qualified individual. Review all sections of this MSDS prior to use." Alternatively, this information may be included on the shipment form.

If an R&D-exempt chemical is transferred to a pilot plant or manufacturing plant, see EPA rules for the additional requirements.

11.F.2.5   Chemical Imports from R&D Laboratories

Laboratories must complete the TSCA Import Certification Form for all R&D samples and chemicals received from a foreign country. There are no exceptions to this requirement.

Sample forms are available via the Web (search for "TSCA Import Certification Form"), chemical supply vendors, or customs brokers. Unless the imported chemical is excluded (see above), check "Positive Certification" on the Form. Provide this form to the mail or express delivery service or customs broker prior to the import date. Keep a copy in your TSCA compliance file for 3 years.

11.F.2.6   Nanomaterials Under TSCA

In January 2008, EPA issued "TSCA Inventory Status of Nanoscale Substances—General Approach," which describes the agency's perspective on whether nanomaterials are required to be registered under TSCA. EPA uses "molecular identity" to determine if a chemical substance is new. Substances are said to have different molecular identities if they

- have different molecular formulas,
- have the same molecular formulas but different atom connectivities,
- have the same molecular formulas and atom connectivities but different spatial arrangements of atoms,
- have the same types of atoms but have different crystal lattices,
- are different allotropes of the same element, or
- have different isotopes of the same elements.

Differences in physical characteristics such as particle size and shape are not considered part of a substance's molecular identity. Thus, EPA states, "a nanoscale substance that has the same molecular identity as a substance listed on the TSCA Inventory . . . is considered an existing chemical, i.e., the nanoscale and non-nanoscale forms are considered the same chemical substances because they have the same molecular identity" (EPA, 2008).

Regulatory controls on nanomaterials will likely change as the field develops, and it is important for researchers and organizations to monitor this area. To assist with future regulatory questions regarding nanomaterials, EPA has created the Nanoscale Materials Stewardship Program. Those who work with nanomaterials should also be aware of international efforts through the International Organization for Standardization and others to develop standards, testing, health and safety practices, etc. and may affect future regulations.

11.F.3   Requirements for Biological Export and Import

Laboratories that export or import infectious substances, related biological substances, and/or materials that may contain infectious substances should be aware of the following regulatory programs:

- Infectious Substance (human pathogens) Import Permit Program (U.S. Department of Health and Human Services, U.S. Public Health Service, CDC, 42 CFR Part 71);

- Animal Pathogens and Related Biological Materials Import Permit Program (USDA APHIS);
- Importing a Plant Pathogen or Plant Product (USDA/APHIS Plant Protection and Quarantine PPQ, 7 CFR Part 330); and
- U.S. Fish and Wildlife and CITES Endangered Species Permits (Convention on International Trade in Endangered Species).

11.F.4 Other Export Regulations

Scientists who ship or carry a research material oversees may be subject to the export licensing requirements of DOC.

DOC's Export Administration Regulations (EAR) require licenses for the export of a wide variety of research materials. These materials (including chemicals and laboratory equipment) are classified and assigned an Export Control Classification Number. (On the Web, search "ECCN List" for examples of regulated exports.)

The type of material, the destination, and the proposed recipient are all subject to approval by the DOC Bureau of Industry and Security, who issues (or may deny) the license. If any of these elements (material, destination, and recipient) are under the control of the EAR, then an export license will be required.

11.G LABORATORY ACCIDENTS, SPILLS, RELEASES, AND INCIDENTS

Chapter 3 describes laboratory emergency planning and response. This section describes legal requirements for incidents that may occur in a laboratory.

11.G.1 Laboratory Injuries and Illnesses

Immediately report all laboratory injuries and illnesses to your firm or institution's appropriate office (e.g., Workers' Compensation, Risk Management, EHS), even if consultation with a medical professional is not deemed necessary. OSHA requires tracking and reporting of workplace injuries and accidents. State workers' compensation laws detail procedures, provisions, employer responsibilities, and employee rights when dealing with workplace injuries and medical care. Your EHS officer should also be informed of any near misses, spills, releases, accidents, and incidents so that they can be investigated and safety problems are corrected.

OSHA standards and workers' compensation laws apply only to "employees" of laboratory facilities. Unpaid students are not employees within the scope of the Occupational Safety and Health Act, but both moral and legal considerations suggest that colleges and universities provide the same protections to students as are provided to all employees regularly working in the laboratory.

11.G.2 Planning for Chemical Emergencies

Title III of the Superfund Amendments and Reauthorization Act (SARA Title III) was passed in 1986 to facilitate planning for chemical emergencies. One provision of the law requires that any institution with an EPA-listed "extremely hazardous substance" on-site in greater than its "threshold planning quantities" must notify emergency response authorities. The quantity limits are based on the total quantity of the hazardous chemical present at the facility rather than in an individual laboratory.

SARA Title III also requires facilities that use hazardous chemicals to submit copies of the MSDSs used in their operations and report inventories of hazardous chemicals. Research and clinical laboratories are exempt from these requirements because the law defines "hazardous chemical" to exclude any chemical, "to the extent it is used in a research laboratory or hospital or other medical facility under the direct supervision of a technically qualified individual." Note that some states require chemical inventories or release notification for laboratories regardless of SARA exemptions.

11.G.3 Notification Requirements for Spills, Releases, and Other Emergencies

SARA Title III also requires that accidental releases be reported to emergency planning authorities. This emergency notification requirement applies to all facilities, including research laboratories, hospitals, and other medical facilities. A firm or institution must notify state and community authorities in the event of a release into the environment of a "hazardous substance" or an "extremely hazardous substance" in excess of EPA-established "reportable quantities."

Also be sure to determine the additional emergency reporting requirements of your state and locale.

11.G.4 Emergency Training and Response

OSHA's standard for hazardous waste operations and emergency response (29 CFR § 1910.120) establishes criteria for training, worker protection, and cleanup of spills and releases to the environment. This standard is an excellent reference for planning your response to laboratory spills and releases. This standard must be followed by spill response contractors and fire departments when they respond to a laboratory emergency involving hazardous materials.

In most cases, however, the immediate, simple

cleanup of a spill by laboratory staff is not subject to this requirement. According to 29 CFR § 1910.120(a)(3), "Responses to incidental releases of hazardous substances where the substance can be absorbed, neutralized, or otherwise controlled at the time of release by employees in the immediate release area, or by maintenance personnel are not considered to be emergency responses within the scope of this standard." That section goes on to say, "Responses to releases of hazardous substances where there is no potential safety or health hazard (i.e., fire, explosion, or chemical expo-

sure) are not considered to be emergency responses." It is important that facilities have a clear understanding of the circumstances under which employees are expected to respond to incidents, and train employees to be able to identify the difference between a routine incidental release and an emergency requiring outside assistance.

OSHA's bloodborne pathogen standard describes the necessary precautions for cleaning a spill of human blood or body fluids.

Bibliography

ACGIH (American Conference of Governmental Industrial Hygienists). 1998. *Industrial Ventilation, A Manual of Recommended Practice*, 23rd ed. Cincinnati, OH: ACGIH.

————. 2004. *Industrial Ventilation: A Manual of Recommended Practice*, 25th ed. Cincinnati, OH: ACGIH.

————. 2008a. *Control Banding: Issues and Opportunities: A Report of the ACGIH Exposure Control Banding Task Force.* Publication No. 08-001. Cincinnati, OH: ACGIH.

————. 2008b. *Documentation of the Threshold Limit Values and Biological Exposure Indices.* Cincinnati, OH: ACGIH.

————. 2009. *2009 TLVs and BEIs: Based on the Documentation of the Threshold Limit Values for Chemical Substances and Physical Agents & Biological Exposure Indices.* Cincinnati, OH: ACGIH.

ACS (American Chemical Society). 1991. *Design of Safe Chemical Laboratories: Suggested References*, 2nd ed. Washington, DC: ACS.

————. 2001. *Teaching Chemistry to Students with Disabilities: A Manual for High Schools, Colleges, and Graduate Programs*, 4th ed. Washington, DC: ACS. Available at http://portal.acs.org/preview/fileFetch/C/CTP_005073/pdf/CTP_005073.pdf. Accessed June 9, 2010.

————. 2003. *Safety in Academic Chemistry Laboratories*, 7th ed. Washington, DC: ACS.

Alaimo, R. J. 2001. *Handbook of Chemical Health and Safety.* New York: Oxford University Press.

American Industrial Hygiene Association. 2007. *Guidance for Conducting Control Banding Analyses.* Fairfax, VA: American Industrial Hygiene Association.

ANSI (American National Standards Institute). 1992. *ANSI 117.1-1992: Accessible and Usable Building Facilities:* New York: ANSI.

————. 1995. *ANSI/ASHRAE 110-1995: Methods of Testing Performance of Laboratory Fume Hoods.* New York: ANSI.

————. 2004. *ANSI Z358.1-2004: Emergency Eyewash and Shower Equipment.* New York: ANSI.

————. 2007. *ANSI Z136.1-2007: Safe Use of Lasers:* New York: ANSI.

Anastas, P. T., and J. C. Warner. 1998. *Green Chemistry: Theory and Practice.* New York: Oxford University Press.

ASME (American Society of Mechanical Engineers). 1992. Boiler and Pressure Vessel Code. Section VII, B31.1. PTC 25.3, NQA-1. Fairfield, NJ: ASME Press.

ASTM International. 2007a. ASTM E77-07: Standard Test Method for Inspection and Verification of Thermometers. West Conshohocken, PA: ASTM International.

————. 2007b. ASTM E 2535-07: Standard Guide for Handling Unbound Engineered Nanoscale Particles in Occupational Settings. West Conshohocken, Pa.: ASTM International.

Backus, B. D., S. Dowdy, K. Boschert, T. Richards, and M. Becker-Hapak. 2001. Safety guidance for laboratory personnel working with *trans*-activating transduction (TAT) protein transduction domains. *Journal of Chemical Health and Safety* 8(2): 5–11.

Beyler, R. E., and V. K. Myers. 1982. What every chemist should know about teratogens. *Journal of Chemical Education* 59(9): 759–763.

Bingham, E., B. Cohrssen, and C. H. Powell. 2001. *Patty's Industrial Toxicology*, 5th ed. Hoboken, NJ: John Wiley and Sons.

Bollinger, N. 2004. *NIOSH Respirator Selection Logic.* Publication No. 2005-100. Cincinnati, OH: National Institute for Occupational Safety and Health. Available at http://www.cdc.gov/niosh/docs/2005-100/pdfs/05-100.pdf. Accessed March 17, 2010.

Bracker, A. L., T. F. Morse, and N. J. Simcox. 2009. Training health and safety committees to use control banding: Lessons learned and opportunities for the United States. *Journal of Occupational and Environmental Hygiene* 6(5): 307–314.

Bretherick, L., ed. 1986. *Hazards in the Chemical Laboratory*, 4th ed. Cambridge, UK: Royal Society of Chemistry.

Bruker BioSpin GmbH. 2008a. *General Safety Considerations for the Installation and Operation of Superconducting Magnet Systems.* Rheinstetten, Germany: Bruker.

————. 2008b. *Site Planning Guide for Superconducting NMR Systems.* Rheinstetten, Germany: Bruker.

Carson, P. A., and C. J. Mumford. 2002. *Hazardous Chemicals Handbook*, 2nd ed. Woburn, MA.: Elsevier.

Cember, H., and T. Johnson. 2008. *Introduction to Health Physics*, 4th ed. New York: McGraw-Hill.

Code of Federal Regulations. 1998. Title 21, Parts 1300 to End. Washington, DC: U.S. Government Printing Office.

DHS (U.S. Department of Homeland Security). 2008. *Risk-Based Performance Standards Guidance. Chemical Facility Anti-Terrorism Standards*, Version 2.4, October. Washington, DC: DHS.

Dillon, K. J. 1994. *Apprentice to Paracelsus: My Search for the Secrets of Healing.* McLean, VA: Scientia Press.

DOE (U.S. Department of Energy). 1994. *Integrated Safety Management System Manual DOE P 450.4.* Washington, DC: U.S. Government Printing Office. Available at http://www.hss.energy.gov/healthsafety/ism/. Accessed September 21, 2009.

————. 2006. *Safety Management System Manual DOE M 450.4-1.* Washington, DC: U.S. Government Printing Office. Available at http://www.hss.energy.gov/healthsafety/ism/. Accessed September 21, 2009.

————. 2008. *Nanoscale Science Research Centers: Approach to Nanomaterial ES&H*, Rev. 3a, May. Available at http://www.er.doe.gov/bes/DOE_NSRC_Approach_to_Nanomaterial_ESH.pdf. Accessed Nov. 23, 2009.

————. 2009. *The Safe Handling of Unbound Engineered Nanoparticles.* DOE N 456.1. Washington, DC: DOE. Available at https://www.directives.doe.gov/pdfs/doe/doetext/neword/456/n4561.pdf. Accessed November 23, 2009.

Dou, D., E. N. Duesler, and R. T. Paine. 1999. Synthesis and structural characterization of new polyphosphorus ring and cage compounds. *Inorganic Chemistry* 38(4): 788–793.

Doxsee, K. M., and J. E. Huchison. 2003. *Green Organic Chemistry Strategies, Tools, and Laboratory Experiments*, 1st ed. London: Thomson/Brooks-Cole.

Earley, M. W., J. S. Sargent, J. V. Sheehan, and E. W. Buss. 2008. *NFPA 70: National Electrical Code Handbook.* Quincy, MA: National Fire Protection Association.

Ellenbecker, M. J., and C. S. Tsai. 2008. Nanoparticle Occupational and Environmental Health and Safety: the CHN Experience. Presentation to the Committee on Prudent Practices in the Laboratory: An Update, Washington, DC, October 30.

Ellis, R. T. 2001. Commissioning: Getting it right. *Engineered Systems* 1(Jan.). Available online at http://findarticles.com/p/articles/mi_m0BPR/is_1_18/ai_69759100/?tag=content;col1. Accessed May 22, 2010.

EPA (U.S. Environmental Protection Agency).1986. *Understanding the Small Quantity Generator Hazardous Waste Rules: A Handbook for Small Business.* EPA 530-SW-86-019. U.S. EPA, Washington, D.C.

————. 1996. *Profiles and Management Options for EPA Laboratory Generated Mixed Waste.* EPA 402-R-96-015. Washington, DC: EPA.

————. 2002. *Environmental Management Guide for Small Laboratories.* EPA 233-B-00-001. Washington, DC: EPA. Available at nepis.epa.gov/Exe/ZyPURL.cgi?Dockey=200092JP.txt. Accessed September 19, 2009.

————. 2008. *TSCA Inventory Status of Nanoscale Substances—General Approach.* Available at http://www.epa.gov/oppt/nano/nmsp-inventorypaper.pdf. Accessed June 11, 2010.

EPA/DOE (U.S. Environmental Protection Agency and U.S. Department of Energy). 2006. *Retro-Commissioning Laboratories for Energy Efficiency.* Laboratories for the 21st Century Technical Bulletin. Washington, DC: EPA. Available at http://labs21century.gov/pdf/bulletin_retrocx_508.pdf. Accessed June 9, 2010.

EPA/USNRC/DOE/DOD (U.S. Environmental Protection Agency/U.S. Nuclear Regulatory Commission, U.S. Department of Energy, U.S. Department of Defense). 2000. Multi-Agency Radiation Survey and Site Investigation Manual, Rev. 1, NUREG-1575. Available at http://www.nrc.gov/reading-rm/doc-collections/nuregs/.

Federal Register. Investigation Manual, Rev. 1, NURGE-1575. Available at http://wbanrc.gov:8080/res/view_content.jsp. March 24, 1986. 51: 10168.

Ford, M., K. A. Delaney, L. Ling, and T. Erickson. 2001. *Clinical Toxicology,* 1st ed. Philadelphia: Saunders.

Foster, B. L. 2005a. Mercury thermometer replacements in chemistry laboratories at West Virginia University. *Journal of Chemical Education* 82: 269.

Foster, B. L. 2005b. The chemical inventory management system in academia. *Journal of Chemical Health & Safety* 12(5): 21–25.

Foster, B. L., and M. E. Cournoyer. 2005. The use of microwave ovens with flammable liquids. *Journal of Chemical Health & Safety* 12(4): 27–32.

Gosselin, R. E, R. P. Smith, and H. C. Hodge. 1984. *Clinical Toxicology of Commercial Product.* Baltimore: Williams and Wilkins.

GSA (General Services Administration). 1992. FED-STD209E: Airborne Particulate Cleanliness Classes in Cleanrooms and Clean Zones, September 11. Washington, DC: GSA.

Hamstead, A. 1964. Destroying peroxides of isopropyl ether. *Journal of Industrial and Engineering Chemistry* 56(6): 37.

Hashimoto, H., T. Goto, N. Nakachi, H. Suzuki, T. Takebayashi, and K. Mori. 2007. Evaluation of the control banding method—Comparison with measurement-based comprehensive risk assessment. *Journal of Occupational Health* 49(6): 482–492.

Hathaway, G. J., and N. H. Proctor. 2004. *Proctor and Hughes' Chemical Hazards of the Workplace,* 5th ed. Hoboken, NJ: John Wiley and Sons.

HHS/CDC/NIH (U.S. Department of Health and Human Services, Centers for Disease Control and Prevention, and National Institutes of Health). 2007. *Primary Containment for Biohazards: Selection, Installation, and Use of Biological Safety Cabinets,* 3rd ed. L. C. Chosewood and D. E. Wilson, eds. Available at http://www.cdc.gov/biosafety/publications/primary_containment_for_biohazards.pdf. Accessed June 10, 2010.

————. 2009. *Biosafety in Microbiological and Biomedical Laboratories,* 5th ed., L. C. Chosewood and D. E. Wilson, eds. Washington, DC: U.S. Government Printing Office. Available at http://www.cdc.gov/od/OHS/biosfty/bmbl5/BMBL_5th_Edition.pdf. Accessed March 16, 2010.

HHS/CDC/NIOSH (U.S. Department of Health and Human Services, Centers for Disease Control and Prevention, National Institute for Occupational Safety and Health). 1981. *Occupational Health Guidelines for Chemical Hazards,* F. W. Mackison, R. S. Stricoff, and L. J. Partridge, eds. NIOSH Publication No. 81-123 I CDC/NIOSH. Available at http://www.cdc.gov/niosh/docs/81-123/. Accessed March 17, 2010.

————. 1995. *Occupational Health Guidelines for Chemical Hazards Supplement IV-OHG.* Publication No. 95-121. Washington, DC: U.S. Government Printing Office. Available at http://www.cdc.gov/niosh/docs/95-121/. Accessed June 11, 2010.

————. 1996. *The Effects of Workplace Hazards on Male Reproductive Health.* NIOSH Publication No. 96-132. Washington, DC: U.S. Government Printing Office. Available at http://www.cdc.gov/niosh/malrepro.html. Accessed March 16, 2010.

————. 1997. *Preventing Allergic Reactions to Natural Rubber Latex in the Workplace.* NIOSH Publication No. 97-135. Cincinnati, OH: NIOSH. Available at http://www.cdc.gov/Niosh/pdfs/97-135sum.pdf. Accessed March 17, 2010.

————. 1999. *The Effects of Workplace Hazards on Female Reproductive Health.* Publication No. 99-104. Washington, DC: U.S. Government Printing Office. Available at http://www.cdc.gov/niosh/docs/99-104/#workers. Accessed March 16, 2010.

————. 2005. *Contact Lens Use in a Chemical Environment.* NIOSH Publication No. 2005-139, Current Intelligence Bulletin 59. Available at http://www.cdc.gov/niosh/docs/2005-139/. Accessed June 4, 2010.

————. 2007. *NIOSH Pocket Guide to Chemical Hazards.* Publication No. 2005-149. Washington, DC: U.S. Government Printing Office.

————. 2009a. *Approaches to Safe Nanotechnology: Managing the Health and Safety Concerns Associated with Engineered Nanomaterials.* NIOSH Publication No. 2009-125. Available at http://www.cdc.gov/niosh/docs/2009-125/pdfs/2009-125.pdf. Accessed June 1, 2009.

————. 2009b. *Qualitative Risk Characterization and Management of Occupational Hazards: Control Banding (CB).* NIOSH Publication No. 2009-152. Available at http://www.cdc.gov/niosh/docs/2009-152. Accessed March 17, 2010.

HHS/CDC/NTP (U.S. Department of Health and Human Services, Centers for Disease Control and Prevention, and National Toxicology Program). 2005. *Annual Report on Carcinogens,* 11th ed. Summary from the National Toxicology Program. Research Triangle Park, NC: Public Information Office.

HHS/PHS/NIOSH (U.S. Department of Health and Human Services, Public Health Service, National Institute for Occupational Safety and Health). Registry of Toxic Effects of Chemical Substances. Revised annually. Washington, DC: U.S. Government Printing Office.

Hileman, B. 1993. Health effects of electromagnetic fields remain unresolved. *Chemical & Engineering News* 71(45): 15–29.

ICNIRP (International Commission on Non-ionizing Radiation Protection). 2009. Guidelines on limits of exposure to static magnetic field. *Health Physics* 96(4): 504–514.

Institute of Occupational Medicine. 2004. *Nanoparticles: An Occupational Hygiene Review.* Research Report 274 prepared for the Health and Safety Executive. Available at http://www.hse.gov.uk/research/rrpdf/rr274.pdf. Accessed June 11, 2010.

IPCS (International Programme on Chemical Safety). 2009. *International Chemical Safety Cards.* Available at www.cdc.gov/niosh. Accessed September 15, 2009.

Izzo, R. M. 2002. Care and use of liquid-in-glass thermometers. *Lab Safety and Environmental Management,* June.

Jackson, H. L., W. B. McCormack, C. S. Rondestvedt, K. C. Smeltz, and I. E. Viele. 1970. Control of peroxidizable compounds. *Journal of Chemical Education* 47(3): A175.

Karn, B. 2008. Prudent Practice for Nanomaterials. Presentation to the Committee on Prudent Practices in the Laboratory: An Update, Washington, DC, October 30.

Kelly, R. J. 1996. Review of safety guidelines for peroxidizable organic chemicals. *Journal of Chemical Health & Safety* 3(5): 28–36.

Khattak, S., G. K-Moghtader, K. McMartin, M. Barrera, D. Kennedy, and G. Koren. 1999. Pregnancy outcome following gestational exposure to organic solvents. *Journal of the American Medical Association* 281: 1106–1109.

Klaassen, C. D. 2007. *Casarett and Doull's Toxicology: The Basic Science of Poisons*, 7th ed. New York: McGraw-Hill.

Kosal, M. E. 2006. Near term threats of chemical weapons terrorism. *Strategic Insights* 5(6). Available at http://www.nps.edu/Academics/centers/ccc/publications/OnlineJournal/2006/Jul/kosalJul06.html. Accessed September 19, 2009.

Lee, E. G., M. Harper, R. B. Bowen, and J. Slaven. 2009. Evaluation of COSHH essentials: Methylene chloride, isopropanol, and acetone exposures in a small printing plant. *Annals of Occupational Hygiene* 53(5): 463–474.

Lewis, R. J., Sr. 1991. *Reproductively Active Chemicals: A Reference Guide*. New York: Van Nostrand Reinhold.

———. 2004. *Sax's Dangerous Properties of Industrial Materials*, 11th ed. Hoboken, NJ: John Wiley and Sons.

Mallinkrodt Baker, Inc. 2008. Mercury (CAS Number 7439-97-6). Material Safety Data Sheet #M1599, August 20. Phillipsburg, NJ: Mallinckrodt Baker, Inc.

Marquart, H., H. Heussen, M. Le Feber, D. Noy, E. Tielemans, J. Schinkel, J. West, and D. Van Der Schaaf. 2008. "Stoffenmanager," a Web-based control banding tool using an exposure process model. *Annals of Occupational Hygiene* 52(6): 429–441.

Maupins, K., and D. Hitchings. 1998. Reducing employee exposure potential using the ANSI/ASHRAE 110 *Method of Testing Performance of Laboratory Fume Hoods* as a diagnostic tool. *American Industrial Hygiene Association Journal* 59(2): 133–138.

McKenzie, L. C., J. E. Thompson, R, Sullivan, and J. E. Hutchison. 2004. Green chemical processing in the teaching laboratory: A convenient liquid CO_2 extraction of natural products. *Green Chemistry* 6: 355–358.

National Ag Safety Database. Glove Guide. Available at www.nasd.org. Accessed July 8, 2008.

National Committee on Clinical Laboratory Standards. 2002. *Clinical Laboratory Waste Management; Approved Guideline*, 2nd ed. NCCLS Document No. GP5-A2. Wayne, PA: NCCLS.

NCRP (National Council on Radiation Protection and Measurements). 2003. *Management Techniques for Laboratories and Other Small Institutional Generators to Minimize Off-Site Disposal of Low-Level Radioactive Waste*. NCRP Report No. 143. Bethesda, MD: NCRP.

NFPA (National Fire Protection Association). 2001. *Fire Protection Guide to Hazardous Materials*, 13th ed. Quincy, MA: NFPA.

———. 2004. *NFPA 45: Standard on Fire Protection for Laboratories Using Chemicals*. Quincy, MA: NFPA.

———. 2007. *NFPA 704: Standard System for the Identification of the Hazards of Materials for Emergency Response*. Quincy, MA: NFPA.

———. 2008. *NFPA 70: National Electrical Code*. Quincy, MA: NFPA.

———. 2010. *NFPA 13: Standard System for the Installation of Sprinkler Systems*. Quincy, MA: NFPA.

NIH (National Institutes of Health). 2011. *NIH Guidelines for Research Involving Recombinant DNA Molecules*. Available at http://oba.od.nih.gov/. Accessed June 1, 2010.

NRC (National Research Council). 1981. *Prudent Practices for Handling Hazardous Chemicals in Laboratories*. Washington, DC: National Academy Press.

———. 1983. *Prudent Practices for Disposal of Chemicals from Laboratories*. Washington, DC: National Academy Press.

———. 1989. *Biosafety in the Laboratory—Prudent Practices for Handling and Disposal of Infectious Materials*. Washington, DC: National Academy Press. Available at http://www.nap.edu/openbook.php?isbn=0309039754. Accessed June 5, 2010.

———. 1995. *Prudent Practices in the Laboratory: Handling and Disposal of Chemicals*. Washington, DC: National Academy Press.

Organic Syntheses, Coll. 5 (1973): 351; 41 (1961): 16. Available at http://www.orgsyn.org/orgsyn/orgsyn/prepcontent.asp?prep=cv5p0351. Accessed May 30, 2010.

Paik, S. Y., D. M. Zalk, and P. Swuste. 2008. Application of a pilot control banding tool for risk level assessment and control of nanoparticle exposures. *Annals of Occupational Hygiene* 52(6): 419–428.

Patnaik, P. A. 2007. *A Comprehensive Guide to the Hazardous Properties of Chemical Substances*, 3rd ed. Hoboken, NJ: John Wiley and Sons.

Peacock, R. N. 1993. Safety and health considerations related to vacuum gauging. *Journal of Vacuum Science & Technology A: Vacuum, Surfaces, and Films* 11(4): 1627–1630.

Pine, S. H. 1994. Safety lessons from an earthquake zone. *Journal of Chemical Health & Safety* 1(July/August): 10.

Pohanish, R. P. 2008. *Sittig's Handbook of Toxic and Hazardous Chemicals and Carcinogens*, 5th ed. Norwich, CT: William Andrew.

Pohanish, R., and S. Greene. 2003. *Wiley Guide to Chemical Incompatibilities*, 2nd ed. New York: John Wiley and Sons.

Porter, R. L., P. A. Lobo, and C. M. Sliepcevich. 1956. Design and construction of barricades. *Industrial & Engineering Chemistry* 48(5): 841–845. Available at http://pubs.acs.org/doi/abs/10.1021/ie50557a021. Accessed April 22, 2010.

Purchase, R., Ed. 1994. *The Laboratory Environment*. Special Publication No. 136. Cambridge, UK: Royal Society of Chemistry.

Ramsey, H., and W. H. J. Breazeale, Jr. 1998. Contact lenses. *Chemical & Engineering News* 76(22): 6.

Rau, E. H. 1997. Trends in management and minimization of biomedical mixed waste at the National Institutes of Health. *Technology Journal of the Franklin Institute* 334(A): 397–419.

———. 2005. Management of multihazardous wastes from high containment laboratories. Pp 103–133 in *Anthology of Biosafety VIII. Evolving Issues in Containment*, J. Y. Richmond, ed. Mundelein, IL: American Biological Safety Association.

Rau, E. H., R. J. Alaimo, P. C. Ashbrook, S. M. Austin, N. Borenstein, N. R. Evans, H. M. French, R. W. Gilpin, J. Hughes, Jr., S. J. Hummel, A. P. Jacobsohn, C. Y. Lee, S. Merkle, T. Radzinski, R. Sloane, K. D. Wagner, and L. E. Weaner. 2000. Minimization and management of wastes from biomedical research. *Environmental Health Perspectives* 108(6): 953–977.

Reinhardt, P. A., K. L. Leonard, and P. C. Ashbrook, eds. 1996. *Pollution Prevention and Waste Minimization in Laboratories*. Boca Raton, FL: CRC Press.

Schenck, J. F. 2000. Safety of strong, static magnetic fields. *Journal of Magnetic Resonance Imaging* 12(1): 2–18. Available at http://www3.interscience.wiley.com/journal/72512386/abstract. Accessed April 16, 2010.

Schwindeman, J. A., C. J. Woltermann, and R. J. Letchford. 2002. Safe handling of organolithium compounds in the laboratory. *Journal of Chemical Health & Safety* 9(3): 6.

Shapiro, J. 2002. *Radiation Protection: A Guide for Scientists, Regulators, and Physicians*, 4th ed. Cambridge, MA: Harvard University Press.

Shea, D. A., and F. Gottron. 2004. *Small-Scale Terrorist Attacks Using Chemical and Biological Agents: An Assessment Framework and Preliminary Comparisons*. Congressional Research Service Report for Congress No. RL32391. Available at http://fpc.state.gov/documents/organization/33629.pdf. Accessed June 11, 2010.

Shepard, T. H., and R. J. Lemire. 2004. *Catalog of Teratogenic Agents*, 11th ed. Baltimore: Johns Hopkins University Press.

———. 2007. *Catalog of Teratogenic Agents*, 12th ed. Baltimore: Johns Hopkins University Press.

Smith, D. T. 1964. Shields and barricades for chemical laboratory operations. *Journal of Chemical Education* 41(7): A520. Available online at http://pubs.acs.org/doi/abs/10.1021/ed041pA520. Accessed April 22, 2010.

Steere, N. V. 1964. Safety in the chemical laboratory. *Journal of Chemical Education* 41(8): A575–A579.

Swift E. H., and W. P. Schaefer. 1961. *Qualitative Elemental Analysis.* San Francisco: Freeman.

Task Force on Laboratory Waste Management. 1993. *Less Is Better: Laboratory Chemical Management for Waste Reduction*, 2nd ed.Washington, DC: American Chemical Society. Available at http://portal.acs.org/portal/PublicWebSite/about/governance/committees/chemicalsafety/publications/WPCP_012290. Accessed June 2, 2010.

Tuma, L. D. 1991. Identification of process hazards using thermal analytical techniques. *Thermochimica Acta* 192: 121–128.

United Nations Economic Commission for Europe. 2009. *Globally Harmonized System of Classification and Labelling of Chemicals (GHS)*, 3rd ed. New York and Geneva: United Nations.

Urben, P. G., ed. 2007. *Bretherick's Handbook of Reactive Chemical Hazards*, 7th ed. Oxford, UK: Academic Press.

Urben, P., and L. Bretherick. 1999. *Bretherick's Handbook of Reactive Chemical Hazards*, 6th ed. Oxford, UK: Butterworth-Heinemann.

U.S. Department of Labor. Occupational Health and Safety Administration. Occupational exposure to hazardous chemicals in laboratories. (29 CFR §§ 1910.1450–1990). Available at http://www.osha.gov/pls/oshaweb/owadisp.show_document?p_table=STANDARDS&p_id=10106 Accessed September 21, 2009.

————. Safety and Health Management e-Tool—Hazard Prevention and Control. Available at http://www.osha.gov/SLTC/etools/safetyhealth/comp3.html Accessed September 21, 2009.

Waters, T. R., V. Putz-Anderson, and A. Garg. 1994. *Applications Manual for the Revised NIOSH Lifting Equation.* NIOSH Publication No. 94-110. Cincinnati, OH: National Institute for Occupational Safety and Health. Available at http://www.cdc.gov/niosh/docs/94-110/. Accessed May 22, 2010.

Yarnell, A. 2002. Putting safety first. *Chemical & Engineering News* 80(20): 43. Available at http://pubs.acs.org/isubscribe/journals/cen/80/i20/html/8020sci2.html. Accessed April 28, 2010.

Yaws, C. L., and W. Braker. 2001. *Matheson Gas Data Book*, 7th ed. Parsippany, NJ: Matheson Tri-Gas and McGraw-Hill.

Young, J. A., ed. 1991. *Improving Safety in the Chemical Laboratory: A Practical Guide*, 2nd ed. New York: John Wiley and Sons.

Zalk, D. M., and D. I. Nelson. 2008. History and evolution of control banding: A review. *Journal of Occupational and Environmental Hygiene* 5(5): 330–346.

APPENDIXES

Appendix A: OSHA Laboratory Standard

29 CFR 1910.1450—Occupational Exposure to Hazardous Chemicals in Laboratories

(a) Scope and application.

(1) This section shall apply to all employers engaged in the laboratory use of hazardous chemicals as defined below.

(2) Where this section applies, it shall supersede, for laboratories, the requirements of all other OSHA health standards in 29 CFR part 1910, subpart Z, except as follows:

(i) For any OSHA health standard, only the requirement to limit employee exposure to the specific permissible exposure limit shall apply for laboratories, unless that particular standard states otherwise or unless the conditions of paragraph (a)(2)(iii) of this section apply.

(ii) Prohibition of eye and skin contact where specified by any OSHA health standard shall be observed.

(iii) Where the action level (or in the absence of an action level, the permissible exposure limit) is routinely exceeded for an OSHA regulated substance with exposure monitoring and medical surveillance requirements paragraphs (d) and (g)(1)(ii) of this section shall apply.

(3) This section shall not apply to:

(i) Uses of hazardous chemicals which do not meet the definition of laboratory use, and in such cases, the employer shall comply with the relevant standard in 29 CFR part 1910, subpart Z, even if such use occurs in a laboratory.

(ii) Laboratory uses of hazardous chemicals which provide no potential for employee exposure. Examples of such conditions might include:

(A) Procedures using chemically-impregnated test media such as Dip-and-Read tests where a reagent strip is dipped into the specimen to be tested and the results are interpreted by comparing the color reaction to a color chart supplied by the manufacturer of the test strip; and

(B) Commercially prepared kits such as those used in performing pregnancy tests in which all of the reagents needed to conduct the test are contained in the kit.

(b) Definitions—"Action level" means a concentration designated in 29 CFR part 1910 for a specific substance, calculated as an eight (8)-hour time-weighted average, which initiates certain required activities such as exposure monitoring and medical surveillance.

"Assistant Secretary" means the Assistant Secretary of Labor for Occupational Safety and Health, U.S. Department of Labor, or designee. "Carcinogen" (see "select carcinogen").

"Chemical Hygiene Officer" means an employee who is designated by the employer, and who is qualified by training or experience, to provide technical guidance in the development and implementation of the provisions of the Chemical Hygiene Plan. This definition is not intended to place limitations on the position description or job classification that the designated individual shall hold within the employer's organizational structure.

"Chemical Hygiene Plan" means a written program developed and implemented by the employer which sets forth procedures, equipment, personal protective equipment and work practices that (i) are capable of protecting employees from the health hazards presented by hazardous chemicals used in that particular workplace and (ii) meets the requirements of paragraph (e) of this section. "Combustible liquid" means any liquid having a flashpoint at or above 100 deg. F (37.8 deg. C), but below 200 deg. F (93.3 deg. C), except any mixture having components with flashpoints of 200 deg. F (93.3 deg. C), or higher, the total volume of which make up 99 percent or more of the total volume of the mixture.

"Compressed gas" means: (i) A gas or mixture of gases having, in a container, an absolute pressure exceeding 40 psi at 70 deg. F (21.1 deg. C); or (ii) A gas or mixture of gases having, in a container, an absolute pressure exceeding 104 psi at 130 deg. F (54.4 deg. C) regardless of the pressure at 70 deg. F (21.1 deg. C); or (iii) A liquid having a vapor pressure exceeding 40 psi at 100 deg. F (37.8 deg. C) as determined by ASTM D-323-72.

"Designated area" means an area which may be used for work with "select carcinogens," reproductive toxins or substances which have a high degree of acute toxicity. A designated area may be the entire laboratory, such as a laboratory hood.

"Emergency" means any occurrence such as, but not limited to, equipment failure, rupture of containers or failure of control equipment which results in an uncontrolled release of a hazardous chemical into the workplace.

"Employee" means an individual employed in a laboratory workplace who may be exposed to hazardous chemicals in the course of his or her assignments.

"Explosive" means a chemical that causes a sudden, almost instantaneous release of pressure, gas, and heat when subjected to sudden shock, pressure, or high temperature.

"Flammable" means a chemical that falls into one of the following categories:

(i) "Aerosol, flammable" means an aerosol that, when tested by the method described in 16 CFR 1500.45, yields a flame protection exceeding 18 inches at full valve opening, or a flashback (a flame extending back to the valve) at any degree of valve opening;

(ii) "Gas, flammable" means: (A) A gas that, at ambient temperature and pressure, forms a flammable mixture with air at a concentration of 13 percent by volume or less; or (B) A gas that, at ambient temperature and pressure, forms a range of flammable mixtures with air wider than 12 percent by volume, regardless of the lower limit.

(iii) "Liquid, flammable" means any liquid having a flashpoint below 100 deg F (37.8 deg. C), except any mixture having components with flashpoints of 100 deg. C) or higher, the total of which makes up 99 percent or more of the total volume of the mixture.

(iv) "Solid, flammable" means a solid, other than a blasting agent or explosive as defined in 1910.109(a), that is liable to cause fire through friction, absorption of moisture, spontaneous chemical change, or retained heat from manufacturing or processing, or which can be ignited readily and when ignited burns so vigorously and persistently as to create a serious hazard. A chemical shall be considered to be a flammable solid if, when tested by the method described in 16 CFR 1500.44, it ignites and burns with a self-sustained flame at a rate greater than one-tenth of an inch per second along its major axis.

"Flashpoint" means the minimum temperature at which a liquid gives off a vapor in sufficient concentration to ignite when tested as follows:

(i) Tagliabue Closed Tester (See American National Standard Method of Test for Flash Point by Tag Closed Tester, Z11.24-1979 (ASTM D 56-79))- for liquids with a viscosity of less than 45 Saybolt Universal Seconds (SUS) at 100 deg. F (37.8 deg. C), that do not contain suspended solids and do not have a tendency to form a surface film under test; or

(ii) Pensky-Martens Closed Tester (See American National Standard Method of Test for Flashpoint by Pensky-Martens Closed Tester, Z11.7-1979 (ASTM D 93-79))—for liquids with a viscosity equal to or greater than 45 SUS at 100 deg. F (37.8 deg. C), or that contain suspended solids, or that have a tendency to form a surface film under test; or

(iii) Setaflash Closed Tester (see American National Standard Method of Test for Flash Point by Setaflash Closed Tester (ASTM D 3278-78)). Organic peroxides, which undergo autoaccelerating thermal decomposition, are excluded from any of the flashpoint determination methods specified above.

"Hazardous chemical" means a chemical for which there is statistically significant evidence based on at least one study conducted in accordance with established scientific principles that acute or chronic health effects may occur in exposed employees. The term "health hazard" includes chemicals which are carcinogens, toxic or highly toxic agents, reproductive toxins, irritants, corrosives, sensitizers, hepatotoxins, nephrotoxins, neurotoxins, agents which act on the hematopoietic systems, and agents which damage the lungs, skin, eyes, or mucous membranes. Appendices A and B of the Hazard Communication Standard (29 CFR 1910.1200) provide further guidance in defining the scope of health hazards and determining whether or not a chemical is to be considered hazardous for purposes of this standard.

"Laboratory" means a facility where the "laboratory use of hazardous chemicals" occurs. It is a workplace where relatively small quantities of hazardous chemicals are used on a non-production basis.

"Laboratory scale" means work with substances in which the containers used for reactions, transfers, and other handling of substances are designed to be easily and safely manipulated by one person.

"Laboratory scale" excludes those workplaces whose function is to produce commercial quantities of materials.

"Laboratory-type hood" means a device located in a laboratory, enclosure on five sides with a movable sash or fixed partial enclosed on the remaining side; constructed and maintained to draw air from the laboratory and to prevent or minimize the escape of air contaminants into the laboratory; and allows chemical manipulations to be conducted in the enclosure without insertion of any portion of the employee's body other than hands and arms. Walk-in hoods with adjustable sashes meet the above definition provided that the sashes are adjusted during use so that the airflow and the exhaust of air contaminants are not compromised and employees do not work inside the enclosure during the release of airborne hazardous chemicals.

"Laboratory use of hazardous chemicals" means handling or use of such chemicals in which all of the following conditions are met:

(i) Chemical manipulations are carried out on a "laboratory scale;"

(ii) Multiple chemical procedures or chemicals are used;

(iii) The procedures involved are not part of a production process, nor in any way simulate a production process; and

(iv) "Protective laboratory practices and equipment" are available and in common use to minimize the potential for employee exposure to hazardous chemicals.

"Medical consultation" means a consultation which takes place between an employee and a licensed phy-

sician for the purpose of determining what medical examinations or procedures, if any, are appropriate in cases where a significant exposure to a hazardous chemical may have taken place.

"Organic peroxide" means an organic compound that contains the bivalent —O—O— structure and which may be considered to be a structural derivative of hydrogen peroxide where one or both of the hydrogen atoms have been replaced by an organic radical.

"Oxidizer" means a chemical other than a blasting agent or explosive as defined in 1910.109(a), that initiates or promotes combustion in other materials, thereby causing fire either of itself or through the release of oxygen or other gases.

"Physical hazard" means a chemical for which there is scientifically valid evidence that it is a combustible liquid, a compressed gas, explosive, flammable, an organic peroxide, an oxidizer pyrophoric, unstable (reactive) or water-reactive.

"Protective laboratory practices and equipment" means those laboratory procedures, practices and equipment accepted by laboratory health and safety experts as effective, or that the employer can show to be effective, in minimizing the potential for employee exposure to hazardous chemicals.

"Reproductive toxins" means chemicals which affect the reproductive chemicals which affect the reproductive capabilities including chromosomal damage (mutations) and effects on fetuses (teratogenesis).

"Select carcinogen" means any substance which meets one of the following criteria:

(i) It is regulated by OSHA as a carcinogen; or

(ii) It is listed under the category, "known to be carcinogens," in the Annual Report on Carcinogens published by the National Toxicology Program (NTP) (latest edition); or

(iii) It is listed under Group 1 ("carcinogenic to humans") by the International Agency for Research on Cancer Monographs (IARC) (latest editions); or

(iv) It is listed in either Group 2A or 2B by IARC or under the category, "reasonably anticipated to be carcinogens" by NTP, and causes statistically significant tumor incidence in experimental animals in accordance with any of the following criteria: (A) After inhalation exposure of 6-7 hours per day, 5 days per week, for a significant portion of a lifetime to dosages of less than 10 mg/m(3); (B) After repeated skin application of less than 300 (mg/kg of body weight) per week; or (C) After oral dosages of less than 50 mg/kg of body weight per day.

"Unstable (reactive)" means a chemical which in the pure state, or as produced or transported, will vigorously polymerize, decompose, condense, or will become self-reactive under conditions of shocks, pressure or temperature.

"Water-reactive" means a chemical that reacts with water to release a gas that is either flammable or presents a health hazard.

(c) Permissible exposure limits. For laboratory uses of OSHA regulated substances, the employer shall assure that laboratory employees' exposures to such substances do not exceed the permissible exposure limits specified in 29 CFR part 1910, subpart Z.

(d) Employee exposure determination

(1) Initial monitoring. The employer shall measure the employee's exposure to any substance regulated by a standard which requires monitoring if there is reason to believe that exposure levels for that substance routinely exceed the action level (or in the absence of an action level, the PEL).

(2) Periodic monitoring. If the initial monitoring prescribed by paragraph (d)(1) of this section discloses employee exposure over the action level (or in the absence of an action level, the PEL), the employer shall immediately comply with the exposure monitoring provisions of the relevant standard.

(3) Termination of monitoring. Monitoring may be terminated in accordance with the relevant standard.

(4) Employee notification of monitoring results. The employer shall, within 15 working days after the receipt of any monitoring results, notify the employee of these results in writing either individually or by posting results in an appropriate location that is accessible to employees.

(e) Chemical hygiene plan—General. (Appendix A of this section is non-mandatory but provides guidance to assist employers in the development of the Chemical Hygiene Plan.)

(1) Where hazardous chemicals as defined by this standard are used in the workplace, the employer shall develop and carry out the provisions of a written Chemical Hygiene Plan which is:

(i) Capable of protecting employees from health hazards associated with hazardous chemicals in that laboratory and

(ii) Capable of keeping exposures below the limits specified in paragraph (c) of this section.

(2) The Chemical Hygiene Plan shall be readily available to employees, employee representatives and, upon request, to the Assistant Secretary.

(3) The Chemical Hygiene Plan shall include each of the following elements and shall indicate specific measures that the employer will take to ensure laboratory employee protection:

(i) Standard operating procedures relevant to safety and health considerations to be followed when laboratory work involves the use of hazardous chemicals;

(ii) Criteria that the employer will use to determine and implement control measures to reduce employee exposure to hazardous chemicals including engineering controls, the use of personal protective equipment and hygiene practices; particular attention shall be given to the selection of control measures for chemicals that are known to be extremely hazardous;

(iii) A requirement that fume hoods and other protective equipment are functioning properly and specific measures that shall be taken to ensure proper and adequate performance of such equipment;

(iv) Provisions for employee information and training as prescribed in paragraph (f) of this section;

(v) The circumstances under which a particular laboratory operation, procedure or activity shall require prior approval from the employer or the employer's designee before implementation;

(vi) Provisions for medical consultation and medical examinations in accordance with paragraph (g) of this section;

(vii) Designation of personnel responsible for implementation of the Chemical Hygiene Plan including the assignment of a Chemical Hygiene Officer, and, if appropriate, establishment of a Chemical Hygiene Committee; and

(viii) Provisions for additional employee protection for work with particularly hazardous substances. These include "select carcinogens," reproductive toxins and substances which have a high degree of acute toxicity. Specific consideration shall be given to the following provisions which shall be included where appropriate:

(A) Establishment of a designated area;

(B) Use of containment devices such as fume hoods or glove boxes;

(C) Procedures for safe removal of contaminated waste; and

(D) Decontamination procedures.

(4) The employer shall review and evaluate the effectiveness of the Chemical Hygiene Plan at least annually and update it as necessary.

(f) Employee information and training.

(1) The employer shall provide employees with information and training to ensure that they are apprised of the hazards of chemicals present in their work area.

(2) Such information shall be provided at the time of an employee's initial assignment to a work area where hazardous chemicals are present and prior to assignments involving new exposure situations. The frequency of refresher information and training shall be determined by the employer.

(3) Information. Employees shall be informed of:

(i) The contents of this standard and its

appendices which shall be made available to employees;

(ii) the location and availability of the employer's Chemical Hygiene Plan;

(iii) The permissible exposure limits for OSHA regulated substances or recommended exposure limits for other hazardous chemicals where there is no applicable OSHA standard;

(iv) Signs and symptoms associated with exposures to hazardous chemicals used in the laboratory; and

(v) The location and availability of known reference material on the hazards, safe handling, storage and disposal of hazardous chemicals found in the laboratory including, but not limited to, Material Safety Data Sheets received from the chemical supplier.

(4) Training.

(i) Employee training shall include:

(A) Methods and observations that may be used to detect the presence or release of a hazardous chemical (such as monitoring conducted by the employer, continuous monitoring devices, visual appearance or odor of hazardous chemicals when being released, etc.);

(B) The physical and health hazards of chemicals in the work area; and

(g) Medical consultation and medical examinations.

(1) The employer shall provide all employees who work with hazardous chemicals an opportunity to receive medical attention, including any follow-up examinations which the examining physician determines to be necessary, under the following circumstances:

(i) Whenever an employee develops signs or symptoms associated with a hazardous chemical to which the employee may have been exposed in the laboratory, the employee shall be provided an opportunity to receive an appropriate medical examination.

(ii) Where exposure monitoring reveals an

exposure level routinely above the action level (or in the absence of an action level, the PEL) for an OSHA regulated substance for which there are exposure monitoring and medical surveillance requirements, medical surveillance shall be established for the affected employee as prescribed by the particular standard.

(iii) Whenever an event takes place in the work area such as a spill, leak, explosion or other occurrence resulting in the likelihood of a hazardous exposure, the affected employee shall be provided an opportunity for a medical consultation. Such consultation shall be for the purpose of determining the need for a medical examination.

(2) All medical examinations and consultations shall be performed by or under the direct supervision of a licensed physician and shall be provided without cost to the employee, without loss of pay and at a reasonable time and place.

(3) Information provided to the physician. The employer shall provide the following information to the physician:

(i) The identity of the hazardous chemical(s) to which the employee may have been exposed;

(ii) A description of the conditions under which the exposure occurred including quantitative exposure data, if available; and

(iii) A description of the signs and symptoms of exposure that the employee is experiencing, if any.

(4) Physician's written opinion.

(i) For examination or consultation required under this standard, the employer shall obtain a written opinion from the examining physician which shall include the following:

(A) Any recommendation for further medical follow-up;

(B) The results of the medical examination and any associated tests;

(C) Any medical condition which may

be revealed in the course of the examination which may place the employee at increased risk as a result of exposure to a hazardous workplace; and

(D) A statement that the employee has been informed by the physician of the results of the consultation or medical examination and any medical condition that may require further examination or treatment.

(ii) The written opinion shall not reveal specific findings of diagnoses unrelated to occupational exposure.

(h) Hazard identification.

(1) With respect to labels and material safety data sheets:

(i) Employers shall ensure that labels on incoming containers of hazardous chemicals are not removed or defaced.

(ii) Employers shall maintain any material safety data sheets that are received with incoming shipments of hazardous chemicals, and ensure that they are readily accessible to laboratory employees.

(2) The following provisions shall apply to chemical substances developed in the laboratory:

(i) If the composition of the chemical substance which is produced exclusively for the laboratory's use is known, the employer shall determine if it is a hazardous chemical as defined in paragraph (b) of this section. If the chemical is determined to be hazardous, the employer shall provide appropriate training as required under paragraph (f) of this section.

(ii) If the chemical produced is a byproduct whose composition is not known, the employer shall assume that the substance is hazardous and shall implement paragraph (e) of this section.

(iii) If the chemical substance is produced for another user outside of the laboratory, the employer shall comply with the Hazard Communication Standard (29 CFR 1910.120) including the requirements for preparation of material safety data sheets and labeling.

(i) Use of respirators. Where the use of respirators is necessary to maintain exposure below permissible exposure limits, the employer shall provide, at no cost to the employee, the proper respiratory equipment. Respirators shall be selected and used in accordance with the requirements of 29 CFR 1910.134.

(j) Record-keeping.

(1) The employer shall establish and maintain for each employee an accurate record of any measurements taken to monitor employee exposures and any medical consultation and examinations including tests or written opinions required by this standard.

(2) The employer shall assure that such records are kept, transferred, and made available in accordance with 29 CFR 1910.20.

(k) Dates.

(1) Effective date. This section shall become effective May 1, 1990.

(2) Start-up dates.

(i) Employers shall have developed and implemented a written Chemical Hygiene Plan no later than January 31, 1991.

(ii) Paragraph (a)(2) of this section shall not take effect until the employer has developed and implemented a written Chemical Hygiene Plan.

(1) Appendices. The information contained in the appendices is not intended, by itself, to create any additional obligations not otherwise imposed or to detract from any existing obligation.

Appendix A To 1910.1450—National Research Council Recommendations Concerning Chemical Hygiene In Laboratories (Non-Mandatory)

Table Of Contents

Foreword

Corresponding Sections of the Standard and This Appendix

A. General Principles

1. Minimize All Chemical Exposures
2. Avoid Underestimation of Risk
3. Provide Adequate Ventilation
4. Institute a Chemical Hygiene Program
5. Observe the PELs and TLVs

B. Responsibilities

1. Chief Executive Officer
2. Supervisor of Administrative Unit
3. Chemical Hygiene Officer
4. Laboratory Supervisor
5. Project Director
6. Laboratory Worker

C. The Laboratory Facility

1. Design
2. Maintenance
3. Usage
4. Ventilation

D. Components of the Chemical Hygiene Plan

1. Basic Rules and Procedures
2. Chemical Procurement, Distribution, and Storage
3. Environmental Monitoring
4. Housekeeping, Maintenance and Inspections
5. Medical Program
6. Personal Protective Apparel and Equipment
7. Records
8. Signs and Labels
9. Spills and Accidents
10. Training and Information
11. Waste Disposal

E. General Procedures for Working with Chemicals

1. General Rules for All Laboratory Work with Chemicals
2. Allergens and Embryotoxins
3. Chemicals of Moderate Chronic or High Acute Toxicity
4. Chemicals of High Chronic Toxicity
5. Animal Work with Chemicals of High Chronic Toxicity

F. Safety Recommendations

G. Material Safety Data Sheets

Foreword

As guidance for each employer's development of an appropriate laboratory Chemical Hygiene Plan, the following non-mandatory recommendations are provided. They were extracted from "Prudent Practices for Handling Hazardous Chemicals in Laboratories" (referred to below as "Prudent Practices"), which was published in 1981 by the National Research Council and is available from the National Academy Press, 2101 Constitution Ave., NW, Washington DC 20418. "Prudent Practices" is cited because of its wide distribution and acceptance and because of its preparation by members of the laboratory community through the sponsorship of the National Research Council. However, none of the recommendations given here will modify any requirements of the laboratory standard. This appendix merely presents pertinent recommendations from "Prudent Practices," organized into a form convenient for quick reference during operation of a laboratory facility and during development and application of a Chemical Hygiene Plan. Users of this appendix should consult "Prudent Practices" for a more extended presentation and justification for each recommendation.

"Prudent Practices" deals with both safety and chemical hazards while the laboratory standard is concerned primarily with chemical hazards. Therefore, only those recommendations directed primarily toward control of toxic exposures are cited in this appendix, with the term "chemical hygiene" being substituted for the word "safety." However, since conditions producing or threatening physical injury often pose toxic risks as well, page references concerning major categories of safety hazards in the laboratory are given in section F. The recommendations from "Prudent Practices" have been paraphrased, combined, or otherwise reorganized, and headings have been added. However, their sense has not been changed.

Corresponding Sections of the Standard and This Appendix

The following table is given for the convenience of those who are developing a Chemical Hygiene Plan which will satisfy the requirements of paragraph (e) of the standard. It indicates those sections of this appendix which are most pertinent to each of the sections of paragraph (e) and related paragraphs.

Paragraph and topic in laboratory standard appendix section	Relevant
(e)(3)(i) Standard operating procedures for handling toxic chemicals.	C, D, E

Paragraph and topic in laboratory standard appendix section	Relevant
(e)(3)(ii) Criteria to be used for implementation of measures to reduce exposures.	D
(e)(3)(iii) Fume hood performance.	C4b
(e)(3)(iv) Employee information and training (including emergency procedures).	D10, D9
(e)(3)(v) Requirements for prior approval of laboratory activities.	E2b, E4b
(e)(3)(vi) Medical consultation and medical examinations.	D5, E4f
(e)(3)(vii) Chemical hygiene responsibilities.	B
(e)(3)(viii) Special precautions for work with particularly hazardous substances.	E2, E3, E4

In this appendix, those recommendations directed primarily at administrators and supervisors are given in sections A-D. Those recommendations of primary concern to employees who are actually handling laboratory chemicals are given in section E. (References to page numbers in "Prudent Practices" are given in parentheses.)

A. General Principles for Work with Laboratory Chemicals

In addition to the more detailed recommendations listed below in sections B-E, "Prudent Practices" expresses certain general principles, including the following:

1. It is prudent to minimize all chemical exposures. Because few laboratory chemicals are without hazards, general precautions for handling all laboratory chemicals should be adopted, rather than specific guidelines for particular chemicals (2,10). Skin contact with chemicals should be avoided as a cardinal rule (198).

2. Avoid underestimation of risk. Even for substances of no known significant hazard, exposure should be minimized; for work with substances which present special hazards, special precautions should be taken (10, 37, 38). One should assume that any mixture will be more toxic than its most toxic component (30, 103) and that all substances of unknown toxicity are toxic (3, 34).

3. Provide adequate ventilation. The best way to prevent exposure to airborne substances is to prevent their escape into the working atmosphere by use of hoods and other ventilation devices (32, 198).

4. Institute a chemical hygiene program. A mandatory chemical hygiene program designed

to minimize exposures is needed; it should be a regular, continuing effort, not merely a standby or short-term activity (6,11). Its recommendations should be followed in academic teaching laboratories as well as by full-time laboratory workers (13).

5. Observe the PELs, TLVs. The Permissible Exposure Limits of OSHA and the Threshold Limit Values of the American Conference of Governmental Industrial Hygienists should not be exceeded (13).

B. Chemical Hygiene Responsibilities

Responsibility for chemical hygiene rests at all levels (6, 11, 21) including the:

1. Chief executive officer, who has ultimate responsibility for chemical hygiene within the institution and must, with other administrators, provide continuing support for institutional chemical hygiene (7, 11).

2. Supervisor of the department or other administrative unit, who is responsible for chemical hygiene in that unit (7).

3. Chemical hygiene officer(s), whose appointment is essential (7) and who must:

(a) Work with administrators and other employees to develop and

(b) Monitor procurement, use, and disposal of chemicals used in the lab (8);

(c) See that appropriate audits are maintained (8);

(d) Help project directors develop precautions and adequate facilities (10);

(e) Know the current legal requirements concerning regulated substances (50); and

(f) Seek ways to improve the chemical hygiene program (8, 11).

4. Laboratory supervisor, who has overall responsibility for chemical hygiene in the laboratory (21) including responsibility to:

(a) Ensure that workers know and follow the chemical hygiene rules, that protective equipment is available and in working order, and that appropriate training has been provided (21, 22);

(b) Provide regular, formal chemical hygiene and housekeeping inspections including routine inspections of emergency equipment (21, 171);

(c) Know the current legal requirements concerning regulated substances (50, 231);

(d) Determine the required levels of protective apparel and equipment (156, 160, 162); and

(e) Ensure that facilities and training for use of any material being ordered are adequate (215).

5. Project director or director of other specific operation, who has primary responsibility for chemical hygiene procedures for that operation (7).

6. Laboratory worker, who is responsible for:

(a) Planning and conducting each operation in accordance with the institutional chemical hygiene procedures (7, 21, 22, 230); and

(b) Developing good personal chemical hygiene habits (22).

C. The Laboratory Facility

1. Design. The laboratory facility should have:

(a) An appropriate general ventilation system (see C4 below) with air intakes and exhausts located so as to avoid intake of contaminated air (194);

(b) Adequate, well-ventilated stockrooms/storerooms (218, 219);

(c) Laboratory hoods and sinks (12, 162);

(d) Other safety equipment including eyewash fountains and drench showers (162, 169); and

(e) Arrangements for waste disposal (12, 240).

2. Maintenance. Chemical-hygiene-related equipment (hoods, incinerator, etc.) should undergo continual appraisal and be modified if inadequate (11, 12).

3. Usage. The work conducted (10) and its scale (12) must be appropriate to the physical facilities available and, especially, to the quality of ventilation (13).

4. Ventilation—

(a) General laboratory ventilation. This system should: Provide a source of air for breathing and for input to local ventilation devices (199); it should not be relied on for protection from toxic substances released into the laboratory (198); ensure that laboratory air is continually replaced, preventing increase of air concentrations of toxic substances during the working day (194); direct air flow into the laboratory from non-laboratory areas and out to the exterior of the building (194).

(b) Hoods. A laboratory hood with 2.5 linear feet of hood space per person should be provided for every 2 workers if they spend most of their time working with chemicals (199); each hood should have a continuous monitoring device to allow convenient confirmation of adequate hood performance before use (200, 209). If this is not possible, work with substances of unknown toxicity should be avoided (13) or other types of local ventilation devices should be provided (199). See pp. 201-206 for a discussion of hood design, construction, and evaluation.

(c)   Other local ventilation devices. Ventilated storage cabinets, canopy hoods, snorkels, etc. should be provided as needed (199). Each canopy hood and snorkel should have a separate exhaust duct (207).

(d)   Special ventilation areas. Exhaust air from glove boxes and isolation rooms should be passed through scrubbers or other treatment before release into the regular exhaust system (208). Cold rooms and warm rooms should have provisions for rapid escape and for escape in the event of electrical failure (209).

(e)   Modifications. Any alteration of the ventilation system should be made only if thorough testing indicates that worker protection from airborne toxic substances will continue to be adequate (12, 193, 204).

(f)   Performance. Rate: 4-12 room air changes/hour is normally adequate general ventilation if local exhaust systems such as hoods are used as the primary method of control (194).

(g)   Quality. General air flow should not be turbulent and should be relatively uniform throughout the laboratory, with no high velocity or static areas (194, 195); airflow into and within the hood should not be excessively turbulent (200); hood face velocity should be adequate (typically 60-100 lfm) (200, 204).

(h)   Evaluation. Quality and quantity of ventilation should be evaluated on installation (202), regularly monitored (at least every 3 months) (6, 12, 14, 195), and reevaluated whenever a change in local ventilation devices is made (12, 195, 207). See pp. 195-198 for methods of evaluation and for calculation of estimated airborne contaminant concentrations.

D.   Components of the Chemical Hygiene Plan

1.   Basic Rules and Procedures (Recommendations for these are given in section E, below.)

2.   Chemical Procurement, Distribution, and Storage
(a)   Procurement. Before a substance is received,

information on proper handling, storage, and disposal should be known to those who will be involved (215, 216). No container should be accepted without an adequate identifying label (216). Preferably, all substances should be received in a central location (216).

(b)   Stockrooms/storerooms. Toxic substances should be segregated in a well-identified area with local exhaust ventilation (221). Chemicals which are highly toxic (227) or other chemicals whose containers have been opened should be in unbreakable secondary containers (219). Stored chemicals should be examined periodically (at least annually) for replacement, deterioration, and container integrity (218-19). Stockrooms/storerooms should not be used as preparation or repackaging areas, should be open during normal working hours, and should be controlled by one person (219).

(c)   Distribution. When chemicals are hand carried, the container should be placed in an outside container or bucket. Freight-only elevators should be used if possible (223).

(d)   Laboratory storage. Amounts permitted should be as small as practical. Storage on bench tops and in hoods is inadvisable. Exposure to heat or direct sunlight should be avoided. Periodic inventories should be conducted, with unneeded items being discarded or returned to the storeroom/stockroom (225-6, 229).

3.   Environmental Monitoring
Regular instrumental monitoring of airborne concentrations is not usually justified or practical in laboratories but may be appropriate when testing or redesigning hoods or other ventilation devices (12) or when a highly toxic substance is stored or used regularly (e.g., 3 times/week) (13).

4.   Housekeeping, Maintenance, and Inspections
(a)   Cleaning. Floors should be cleaned regularly (24).
(b)   Inspections. Formal housekeeping and chemical hygiene inspections should be held at least quarterly (6, 21) for units which have frequent personnel changes and semiannually for others; informal inspections should be continual (21).
(c)   Maintenance. Eye wash fountains should be inspected at intervals of not less than 3 months (6). Respirators for routine use should be inspected periodically by the laboratory supervisor (169). Other safety equipment should be inspected regularly (e.g., every 3-6 months) (6, 24, 171).

Procedures to prevent restarting of out-of-service equipment should be established (25).

(d) Passageways. Stairways and hallways should not be used as storage areas (24). Access to exits, emergency equipment, and utility controls should never be blocked (24).

5. Medical Program

(a) Compliance with regulations. Regular medical surveillance should be established to the extent required by regulations (12).

(b) Routine surveillance. Anyone whose work involves regular and frequent handling of toxicologically significant quantities of a chemical should consult a qualified physician to determine on an individual basis whether a regular schedule of medical surveillance is desirable (11, 50).

(c) First aid. Personnel trained in first aid should be available during working hours and an emergency room with medical personnel should be nearby (173). See pp. 176-178 for description of some emergency first aid procedures.

6. Protective Apparel and Equipment
These should include for each laboratory:

(a) Protective apparel compatible with the required degree of protection for substances being handled (158-161);

(b) An easily accessible drench-type safety shower (162, 169);

(c) An eyewash fountain (162)

(d) A fire extinguisher (162-164);

(e) Respiratory protection (164-9), fire alarm and telephone for emergency use (162) should be available nearby; and (f) Other items designated by the laboratory supervisor (156, 160).

7. Records

(a) Accident records should be written and retained (174).

(b) Chemical Hygiene Plan records should document that the facilities and precautions were compatible with current knowledge and regulations (7).

(c) Inventory and usage records for high-risk substances should be kept as specified in section E3e below.

(d) Medical records should be retained by the institution in accordance with the requirements of state and federal regulations (12).

8. Signs and Labels
Prominent signs and labels of the following types should be posted:

(a) Emergency telephone numbers of emergency personnel/facilities, supervisors, and laboratory workers (28);

(b) Identity labels, showing contents of containers (including waste receptacles) and associated hazards (27, 48);

(c) Location signs for safety showers, eyewash stations, other safety and first aid equipment, exits (27) and areas where food and beverage consumption and storage are permitted (24); and

(d) Warnings at areas or equipment where special or unusual hazards exist (27).

9. Spills and Accidents

(a) A written emergency plan should be established and communicated to all personnel; it should include procedures for ventilation failure (200), evacuation, medical care, reporting, and drills (172).

(b) There should be an alarm system to alert people in all parts of the facility including isolation areas such as cold rooms (172).

(c) A spill control policy should be developed and should include consideration of prevention, containment, cleanup, and reporting (175).

(d) All accidents or near accidents should be carefully analyzed with the results distributed to all who might benefit (8, 28).

10. Information and Training Program

(a) Aim: To assure that all individuals at risk are adequately informed about the work in the laboratory, its risks, and what to do if an accident occurs (5, 15).

(b) Emergency and Personal Protection Training: Every laboratory worker should know the location and proper use of available protective apparel and equipment (154, 169). Some of the full-time personnel of the laboratory should be trained in the proper use of emergency equipment and procedures (6). Such training as well as first aid instruction should be available to (154) and encouraged for (176) everyone who might need it.

(c) Receiving and stockroom/storeroom personnel should know about hazards, handling equipment, protective apparel, and relevant regulations (217).

(d) Frequency of Training: The training and education program should be a regular, continuing activity-not simply an annual presentation (15).

(e) Literature/Consultation: Literature and consulting advice concerning chemical hygiene should be readily available to laboratory personnel, who should be encouraged to use these information resources (14).

11. Waste Disposal Program

(a)   Aim: To assure that minimal harm to people, other organisms, and the environment will result from the disposal of waste laboratory chemicals (5).

(b)   Content (14, 232, 233, 240): The waste disposal program should specify how waste is to be collected, segregated, stored, and transported and include consideration of what materials can be incinerated. Transport from the institution must be in accordance with DOT regulations (244).

(c)   Discarding Chemical Stocks: Unlabeled containers of chemicals and solutions should undergo prompt disposal; if partially used, they should not be opened (24, 27). Before a worker's employment in the laboratory ends, chemicals for which that person was responsible should be discarded or returned to storage (226).

(d)   Frequency of Disposal: Waste should be removed from laboratories to a central waste storage area at least once per week and from the central waste storage area at regular intervals (14).

(e)   Method of Disposal: Incineration in an environmentally acceptable manner is the most practical disposal method for combustible laboratory waste (14, 238, 241). Indiscriminate disposal by pouring waste chemicals down the drain (14,231,242) or adding them to mixed refuse for landfill burial is unacceptable (14). Hoods should not be used as a means of disposal for volatile chemicals (40, 200). Disposal by recycling (233, 243) or chemical decontamination (40, 230) should be used when possible.

E.   Basic Rules and Procedures for Working with Chemicals

The Chemical Hygiene Plan should require that laboratory workers know and follow its rules and procedures. In addition to the procedures of the sub programs mentioned above, these should include the rules listed below.

1.   General Rules

The following should be used for essentially all laboratory work with chemicals:

(a)   Accidents and Spills—Eye Contact: Promptly flush eyes with water for a prolonged period (15 minutes) and seek medical attention (33, 172).

Ingestion: Encourage the victim to drink large amounts of water (178).

Skin Contact: Promptly flush the affected area with water (33, 172, 178) and remove any contaminated clothing (172, 178). If symptoms persist after washing, seek medical attention (33). Clean-up. Promptly clean up spills, using appropriate protective apparel and equipment and proper disposal (24, 33). See pp. 233-237 for specific clean-up recommendations.

(b)   Avoidance of "routine" exposure: Develop and encourage safe habits (23); avoid unnecessary exposure to chemicals by any route (23). Do not smell or taste chemicals (32). Vent apparatus which may discharge toxic chemicals (vacuum pumps, distillation columns, etc.) into local exhaust devices (199). Inspect gloves (157) and test glove boxes (208) before use. Do not allow release of toxic substances in cold rooms and warm rooms, since these have contained recirculated atmospheres (209).

(c)   Choice of chemicals: Use only those chemicals for which the quality of the available ventilation system is appropriate (13).

(d)   Eating, smoking, etc.: Avoid eating, drinking, smoking, gum chewing, or application of cosmetics in areas where laboratory chemicals are present (22, 24, 32, 40); wash hands before conducting these activities (23, 24). Avoid storage, handling, or consumption of food or beverages in storage areas, refrigerators, glassware or utensils which are also used for laboratory operations (23, 24, 226).

(e)   Equipment and glassware: Handle and store laboratory glassware with care to avoid damage; do not use damaged glassware (25). Use extra care with Dewar flasks and other evacuated glass apparatus; shield or wrap them to contain chemicals and fragments should implosion occur (25). Use equipment only for its designed purpose (23, 26).

(f)   Exiting: Wash areas of exposed skin well before leaving the laboratory (23).

(g)   Horseplay: Avoid practical jokes or other behavior which might confuse, startle or distract another worker (23).

(h)   Mouth suction: Do not use mouth suction for pipeting or starting a siphon (23, 32).

(i)   Personal apparel: Confine long hair and loose clothing (23, 158). Wear shoes at all times in the laboratory but do not wear sandals, perforated shoes, or sneakers (158).

(j)   Personal housekeeping: Keep the work area clean and uncluttered, with chemicals and equipment being properly labeled and stored; clean up the work area on completion of an operation or at the end of each day (24).

(k)   Personal protection: Assure that appropriate eye protection (154-156) is worn by all persons, including visitors, where chemicals are stored or handled (22, 23, 33, 154). Wear appropriate gloves when the potential for contact with toxic materials exists (157); inspect the gloves before each use, wash them before removal, and replace them periodically (157). (A table of resistance to chemicals of common glove materials is given on p.159). Use appropriate (164-168) respiratory equipment when air contaminant concentrations are not sufficiently restricted by engineering controls (164-5), inspecting the respirator before use (169). Use any other protective and emergency apparel and equipment as appropriate (22, 157-162). Avoid use of contact lenses in the laboratory unless necessary; if they are used, inform supervisor so special precautions can be taken (155). Remove laboratory coats immediately on significant contamination (161).

(l) Planning: Seek information and advice about hazards (7), plan appropriate protective procedures, and plan positioning of equipment before beginning any new operation (22, 23).

(m) Unattended operations: Leave lights on, place an appropriate sign on the door, and provide for containment of toxic substances in the event of failure of a utility service (such as cooling water) to an unattended operation (27, 128).

(n) Use of hood: Use the hood for operations which might result in release of toxic chemical vapors or dust (198-9). As a rule of thumb, use a hood or other local ventilation device when working with any appreciably volatile substance with a TLV of less than 50 ppm (13). Confirm adequate hood performance before use; keep hood closed at all times except when adjustments within the hood are being made (200); keep materials stored in hoods to a minimum and do not allow them to block vents or air flow (200). Leave the hood "on" when it is not in active use if toxic substances are stored in it or if it is uncertain whether adequate general laboratory ventilation will be maintained when it is "off" (200).

(o) Vigilance: Be alert to unsafe conditions and see that they are corrected when detected (22).

(p) Waste disposal: Assure that the plan for each laboratory operation includes plans and training for waste disposal (230). Deposit chemical waste in appropriately labeled receptacles and follow all other waste disposal procedures of the Chemical Hygiene Plan (22, 24). Do not discharge to the sewer concentrated acids or bases (231); highly toxic, malodorous, or lachrymatory substances (231); or any substances which might interfere with the biological activity of waste water treat-

ment plants, create fire or explosion hazards, cause structural damage or obstruct flow (242).

(q) Working alone: Avoid working alone in a building; do not work alone in a laboratory if the procedures being conducted are hazardous (28).

2. Working with Allergens and Embryotoxins

(a) Allergens (examples: diazomethane, isocyanates, bichromates): Wear suitable gloves to prevent hand contact with allergens or substances of unknown allergenic activity (35).

(b) Embryotoxins (34-5) (examples: organomercurials, lead compounds, formamide): If you are a woman of childbearing age, handle these substances only in a hood whose satisfactory performance has been confirmed, using appropriate protective apparel (especially gloves) to prevent skin contact. Review each use of these materials with the research supervisor and review continuing uses annually or whenever a procedural change is made. Store these substances, properly labeled, in an adequately ventilated area in an unbreakable secondary container. Notify supervisors of all incidents of exposure or spills; consult a qualified physician when appropriate.

3. Work with Chemicals of Moderate Chronic or High Acute Toxicity

Examples: diisopropylfluorophosphate (41), hydrofluoric acid (43), hydrogen cyanide (45). Supplemental rules to be followed in addition to those mentioned above (Procedure B of "Prudent Practices", pp. 39-41):

(a) Aim: To minimize exposure to these toxic substances by any route using all reasonable precautions (39).

(b) Applicability: These precautions are appropriate for substances with moderate chronic or high acute toxicity used in significant quantities (39).

(c) Location: Use and store these substances only in areas of restricted access with special warning signs (40, 229). Always use a hood (previously evaluated to confirm adequate performance with a face velocity of at least 60 linear feet per minute) (40) or other containment device for procedures which may result in the generation of aerosols or vapors containing the substance (39); trap released vapors to prevent their discharge with the hood exhaust (40).

(d) Personal protection: Always avoid skin contact by use of gloves and long sleeves (and other protective apparel as appropriate) (39). Always wash hands and arms immediately after working with these materials (40).

(e) Records: Maintain records of the amounts of

these materials on hand, amounts used, and the names of the workers involved (40, 229).

(f)   Prevention of spills and accidents: Be prepared for accidents and spills (41). Assure that at least 2 people are present at all times if a compound in use is highly toxic or of unknown toxicity (39). Store breakable containers of these substances in chemically resistant trays; also work and mount apparatus above such trays or cover work and storage surfaces with removable, absorbent, plastic backed paper (40). If a major spill occurs outside the hood, evacuate the area; assure that cleanup personnel wear suitable protective apparel and equipment (41).

(g)   Waste: Thoroughly decontaminate or incinerate contaminated clothing or shoes (41). If possible, chemically decontaminate by chemical conversion (40). Store contaminated waste in closed, suitably labeled, impervious containers (for liquids, in glass or plastic bottles half-filled with vermiculite) (40).

4.   Work with Chemicals of High Chronic Toxicity Examples: dimethylmercury and nickel carbonyl (48), benzo-a-pyrene (51), N-nitrosodiethylamine (54), other human carcinogens or substances with high carcinogenic potency in animals (38). Further supplemental rules to be followed, in addition to all those mentioned above, for work with substances of known high chronic toxicity (in quantities above a few milligrams to a few grams, depending on the substance) (47). (Procedure A of "Prudent Practices" pp. 47-50.)

(a)   Access: Conduct all transfers and work with these substances in a "controlled area": a restricted access hood, glove box, or portion of a lab, designated for use of highly toxic substances, for which all people with access are aware of the substances being used and necessary precautions (48).

(b)   Approvals: Prepare a plan for use and disposal of these materials and obtain the approval of the laboratory supervisor (48).

(c)   Non-contamination/Decontamination: Protect vacuum pumps against contamination by scrubbers or HEPA filters and vent them into the hood (49). Decontaminate vacuum pumps or other contaminated equipment, including glassware, in the hood before removing them from the controlled area (49, 50). Decontaminate the controlled area before normal work is resumed there (50).

(d)   Exiting: On leaving a controlled area, remove any protective apparel (placing it in an appropriate, labeled container) and thoroughly wash hands, forearms, face, and neck (49).

(e)   Housekeeping: Use a wet mop or a vacuum cleaner equipped with a HEPA filter instead of dry sweeping if the toxic substance was a dry powder (50).

(f)   Medical surveillance: If using toxicologically significant quantities of such a substance on a regular basis (e.g., 3 times per week), consult a qualified physician concerning desirability of regular medical surveillance (50).

(g)   Records: Keep accurate records of the amounts of these substances stored (229) and used, the dates of use, and names of users (48).

(h)   Signs and labels: Assure that the controlled area is conspicuously marked with warning and restricted access signs (49) and that all containers of these substances are appropriately labeled with identity and warning labels (48).

(i)   Spills: Assure that contingency plans, equipment, and materials to minimize exposures of people and property in case of accident are available (233-4).

(j)   Storage: Store containers of these chemicals only in a ventilated, limited access (48, 227, 229) area in appropriately labeled, unbreakable, chemically resistant, secondary containers (48, 229).

(k)   Glove boxes: For a negative pressure glove box, ventilation rate must be at least 2 volume changes/ hour and pressure at least 0.5 inches of water (48). For a positive pressure glove box, thoroughly check for leaks before each use (49). In either case, trap the exit gases or filter them through a HEPA filter and then release them into the hood (49).

(l)   Waste: Use chemical decontamination whenever possible; ensure that containers of contaminated waste (including washings from contaminated flasks) are transferred from the controlled area in a secondary container under the supervision of authorized personnel (49, 50, 233).

5.   Animal Work with Chemicals of High Chronic Toxicity

(a)   Access: For large scale studies, special facilities with restricted access are preferable (56).

(b)   Administration of the toxic substance: When possible, administer the substance by injection or gavage instead of in the diet. If administration is in the diet, use a caging system under negative pressure or under laminar air flow directed toward HEPA filters (56).

(c)   Aerosol suppression: Devise procedures which minimize formation and dispersal of contaminated aerosols, including those from food, urine, and feces (e.g., use HEPA filtered vacuum equipment for cleaning, moisten contaminated bedding before removal from the cage, mix diets in closed containers in a hood) (55, 56).

(d) Personal protection: When working in the animal room, wear plastic or rubber gloves, fully buttoned laboratory coat or jumpsuit and, if needed because of incomplete suppression of aerosols, other apparel and equipment (shoe and head coverings, respirator) (56).

(e) Waste disposal: Dispose of contaminated animal tissues and excreta by incineration if the available incinerator can convert the contaminant to non-toxic products (238); otherwise, package the waste appropriately for burial in an EPA-approved site (239).

F. Safety Recommendations

The above recommendations from "Prudent Practices" do not include those which are directed primarily toward prevention of physical injury rather than toxic exposure. However, failure of precautions against injury will often have the secondary effect of causing toxic exposures. Therefore, we list below page references for recommendations concerning some of the major categories of safety hazards which also have implications for chemical hygiene:

1. Corrosive agents: (35-6)
2. Electrically powered laboratory apparatus: (179-92)
3. Fires, explosions: (26, 57-74, 162-4, 174-5, 219-20, 226-7)
4. Low temperature procedures: (26, 88)
5. Pressurized and vacuum operations (including use of compressed gas cylinders): (27, 75-101)

G. Material Safety Data Sheets

Material safety data sheets are presented in "Prudent Practices" for the chemicals listed below. (Asterisks denote that comprehensive material safety data sheets are provided.)

*Acetyl peroxide (105)
*Acrolein (106)
*Acrylonitrile
Ammonia (anhydrous) (91)
*Aniline (109)
*Benzene (110)
*Benzo[a]pyrene (112)
*Bis(chloromethyl) ether (113)
Boron trichloride (91)
Boron trifluoride (92)
Bromine (114)

*Tert-butyl hydroperoxide (148)
*Carbon disulfide (116)
Carbon monoxide (92)
*Carbon tetrachloride (118)
*Chlorine (119)
Chlorine trifluoride (94)
*Chloroform (121)
Chloromethane (93)
*Diethyl ether (122)
Diisopropyl fluorophosphate (41)
Hydrogen chloride (98)
*Hydrogen cyanide (133)
*Hydrogen sulfide (135)
Mercury and compounds (52)
*Methanol (137)
*Morpholine (138)
*Nickel carbonyl (99)
*Nitrobenzene (139)
Nitrogen dioxide (100)
N-nitrosodiethylamine (54)
*Peracetic acid (141)
*Phenol (142)
*Phosgene (143)
*Pyridine (144)
*Sodium azide (145)
*Sodium cyanide (147)
Sulfur dioxide (101)
*Trichloroethylene (149)
*Vinyl chloride (150)

29 CFR 1910.1450 App. B References (Non-Mandatory)
Appendix B to 1910.1450—References (Non-Mandatory)

The following references are provided to assist the employer in the development of a Chemical Hygiene Plan. The materials listed below are offered as nonmandatory guidance. References listed here do not imply specific endorsement of a book, opinion, technique, policy or a specific solution for a safety or health problem. Other references not listed here may better meet the needs of a specific laboratory.

(a) MATERIALS FOR THE DEVELOPMENT OF THE CHEMICAL HYGIENE PLAN
1. American Chemical Society, Safety in Academic Chemistry Laboratories, 4th edition, 1985.
2. Fawcett, H.H. and W.S. Wood, Safety and Accident Prevention in Chemical Operations, 2nd edition, Wiley-Interscience, New York, 1982.
3. Flury, Patricia A., Environmental Health and Safety in the Hospital Labora-

tory, Charles C. Thomas Publisher, Springfield, IL, 1978.

4.   Green, Michael E. and Turk, Amos, Safety in Working with Chemicals, Macmillan Publishing Co., NY, 1978.

5.   Kaufman, James A., Laboratory Safety Guidelines, Dow Chemical Co., Box 1713, Midland, MI 48640, 1977.

6.   National Institutes of Health, NIH Guidelines for the Laboratory Use of Chemical Carcinogens, NIH Pub. No.812385, GPO, Washington, DC 20402, 1981.

7.   National Research Council, Prudent Practices for Disposal of Chemicals from Laboratories, National Academy Press, Washington, DC, 1983.

8.   National Research Council, Prudent Practices for Handling Hazardous Chemicals in Laboratories, National Academy Press, Washington, DC, 1981.

9.   Renfrew, Malcolm, Ed., Safety in the Chemical Laboratory, Vol. IV, J. Chem. Ed., American Chemical Society, Easlon, PA, 1981.

10.   Steere, Norman V., Ed., Safety in the Chemical Laboratory, J. Chem. Ed. American Chemical Society, Easlon, PA, 18042, Vol. I, 1967, Vol. II, 1971, Vol. III, 1974.

11.   Steere, Norman V., Handbook of Laboratory Safety, the Chemical Rubber Company, Cleveland, OH, 1971.

12.   Young, Jay A., Ed., Improving Safety in the Chemical Laboratory, John Wiley & Sons, Inc., New York, 1987.

(b) HAZARDOUS SUBSTANCES INFORMATION:

1.   American Conference of Governmental Industrial Hygienists, Threshold Limit Values for Chemical Substances and Physical Agents in the Workroom Environment with Intended Changes, 6500 Glenway Avenue, Bldg. D-7, Cincinnati, OH 45211-4438.

2.   Annual Report on Carcinogens, National Toxicology Program U.S. Department of Health and Human Services, Public Health Service, U.S. Government Printing Office, Washington, DC (latest edition).

3.   Best Company, Best Safety Directory, Vols. I and II, Oldwick, NJ, 1981.

4.   Bretherick, L., Handbook of Reactive Chemical Hazards, 2nd edition, Butterworths, London, 1979.

5.   Bretherick, L., Hazards in the Chemical Laboratory, 3rd edition, Royal Society of Chemistry, London, 1986.

6.   Code of Federal Regulations, 29 CFR part 1910 subpart Z. U.S. Govt. Printing Office, Washington, DC 20402 (latest edition).

7.   IARC Monographs on the Evaluation of the Carcinogenic Risk of Chemicals to Man, World Health Organization Publications Center, 49 Sheridan Avenue, Albany, New York 12210 (latest editions).

8.   NIOSH/OSHA Pocket Guide to Chemical Hazards. NIOSH Pub. No. 85-114, U.S. Government Printing Office, Washington, DC, 1985 (or latest edition).

9.   Occupational Health Guidelines, NIOSH/OSHA. NIOSH Pub. No. 81-123, U.S. Government Printing Office, Washington, DC, 1981.

10.   Patty, F.A., Industrial Hygiene and Toxicology, John Wiley & Sons, Inc., New York, NY (Five Volumes).

11.   Registry of Toxic Effects of Chemical Substances, U.S. Department of Health and Human Services, Public Health Service, Centers for Disease Control, National Institute for Occupational Safety and Health, Revised Annually, for sale from Superintendent of Documents, U.S. Govt. Printing Office, Washington, DC 20402.

12.   The Merck Index: An Encyclopedia of Chemicals and Drugs. Merck and Company Inc., Rahway, NJ, 1976 (or latest edition).

13.   Sax, N.I. Dangerous Properties of Industrial Materials, 5th edition, Van Nostrand Reinhold, NY, 1979.

14.   Sittig, Marshall, Handbook of Toxic and Hazardous Chemicals, Noyes Publications, Park Ridge, NJ, 1981.

(c) Information on Ventilation:

1.   American Conference of Governmental Industrial Hygienists. Industrial Ventilation (latest edition), 6500 Glenway Avenue, Bldg. D-7, Cincinnati, OH 45211-4438.

2.   American National Standards Institute, Inc. American National Standards Fundamentals Governing the Design and Operation of Local Exhaust Systems ANSI Z 9.2-1979, American National Standards Institute, NY 1979.

3.   Imad, A.P. and Watson, C.L. Ventilation Index: An Easy Way to Decide about Hazardous Liquids, Professional Safety, pp. 15-18, April 1980.

4.   National Fire Protection Association, Fire Protection for Laboratories Using Chemi-

cals NFPA-45, 1982. Safety Standard for Laboratories in Health Related Institutions, NFPA, 56c, 1980. Fire Protection Guide on Hazardous Materials, 7th edition, 1978. National Fire Protection Association, Batterymarch Park, Quincy, MA 02269.

5.   Scientific Apparatus Makers Association (SAMA), Standard for Laboratory Fume Hoods, SAMA LF7-1980, 1101 16th Street, NW, Washington, DC 20036.

(d) INFORMATION ON AVAILABILITY OF REFERENCED MATERIAL

1.   American National Standards Institute (ANSI), 1430 Broadway, New York, NY 10018.
2.   American Society for Testing and Materials (ASTM), 1916 Race Street, Philadelphia, PA 19103.

(Approved by the Office of Management and Budget under control number 1218-0131)
[55 FR 3327, Jan. 31, 1990]

Appendix B: Statement of Task

The National Academies, through its Board on Chemical Sciences and Technology, will review and update the 1995 publication, *Prudent Practices in the Laboratory: Handling and Disposal of Chemicals*. It will modify the existing content and add content as required to reflect new fields and developments that have occurred since the previous publication. Emphasis will be given throughout to the concept of a "culture of safety" and how that culture can be established and nurtured. Both laboratory operations and adverse impacts those operations might have on the surrounding environment and community will be considered.

Appendix C: Committee Member Biographies

Willliam F. Carroll (*Co-Chair*) is Vice President, Industry Issues of Occidental Chemical Corporation in Dallas, Texas, and an adjunct industrial professor of chemistry at Indiana University. He earned a B.A. from DePauw (1973), an M.S. from Tulane University (1975), and a Ph.D. from Indiana University (1978). He served as American Chemical Society (ACS) president in 2005 and as a member of the ACS Board of Directors from 2004 to 2006 and 2009 to 2011. Dr. Carroll has been an ACS member since 1974 and has served on and chaired a number of committees. He holds memberships in the Society of Plastics Engineers, the American Association for the Advancement of Science, the National Organization for the Professional Advancement of Black Chemists and Chemical Engineers, and the National Fire Protection Association, and was the recipient of an Indiana University Distinguished Alumni Service Award in 2009.

Barbara Foster (*Co-Chair*) is Safety Director for the C. Eugene Bennett Department of Chemistry at West Virginia University (WVU). She is also the safety program coordinator for the Eberly College of Arts and Sciences at WVU. In these capacities, Ms. Foster oversees all aspects of laboratory safety, including creation of Chemical Hygiene Plans, risk assessment and risk management, chemical inventory updates, ensuring adherence to safety regulations and building codes, and conducting laboratory inspections and audits. Ms. Foster has written over 20 safety and instructional publications, including the manual *Laboratory Safety and Management: A Handbook for Teaching Assistants*. She has presented over 60 talks on laboratory safety at professional meetings, workshops, and academic and governmental institutions across the nation. In 2007, Ms. Foster served as chair of the American Chemical Society Division of Chemical Health and Safety. She currently serves as a member of the Board of Editors of the *Journal of Chemical Health & Safety* and is an elected member of the National Fire Protection Association (NFPA) 45 Technical Committee. She holds a B.A. in biology from West Virginia University and is a National Registry of Certified Chemists Certified Chemical Hygiene Officer.

W. Emmett Barkley is the former director of laboratory safety at the Howard Hughes Medical Institute, and was a member of the last Prudent Practices committee. Dr. Barkley is President of Proven Practices, LLC, where he supports environmental health and safety programs at major academic research universities and government agencies. His prior experience includes 24 years at the National Institutes of Health (NIH) where he served as the founding director of the NIH Division of Safety. Dr. Barkley was a principal contributor to several authoritative guidelines in the fields of biological and chemical safety, including the NIH *Guidelines for the Laboratory Use of Chemical Carcinogens*, the NIH *Recombinant DNA Guidelines*, and the CDC/NIH publication *Biosafety in Microbiological and Biomedical Laboratories*. He received a bachelor of civil engineering degree from the University of Virginia and master of science and doctoral degrees in environmental health from the University of Minnesota.

Susan Cook is the biological safety officer for Washington University. She received her Ph.D. from the Johns Hopkins Bloomberg School of Public Health in the laboratory of Dr. Diane Griffin, where she studied viral pathogenesis. She then spent a year in the laboratory of Dr. Roy Curtiss III at Washington University in St. Louis studying vaccine design before accepting a biosafety fellowship at the Midwest Regional Center of Excellence, also located at Washington University. In October 2005, she earned the Certified Biological Safety Professional and Specialist Microbiologist in Biological Safety Microbiology certifications. In January 2006, following completion of the fellowship program, she was hired as the associate biological safety officer for Washington University in St. Louis. She became biological safety officer for Washington University on July 1, 2007. In addition, Dr. Cook oversees an NIH-sponsored biosafety postdoctoral fellowship program.

Kenneth P. Fivizzani retired from Nalco Company in 2009, after a 26-year career as a research scientist. He was the Chemical Hygiene Officer for Nalco's Naperville (IL) and Sugar Land (TX) Research Laboratories for 19 years. He received both B.S. and M.S. degrees in chemistry from Loyola University of Chicago and a Ph.D. in inorganic chemistry from the University of Wisconsin at Madison. He is a member of the American Chemical Society (ACS), the American Industrial Hygiene Association, the Industrial Research Institute's (IRI) Environmental Health and Safety Directors' Network, and Sigma Xi. He was the 2007 chair of the ACS Chicago Section. He was the 2000 chair for the ACS Division of Chemical Health and Safety. He is a past chair (2002–2004) and a current member of the ACS Committee on Chemical Safety (CCS). For 1999–2001, he served as chair of the IRI Environmental Health and Safety Directors' Network. In 1988–1989, he served as president of the Nalco Chapter of Sigma Xi, the Scientific Research Society. Dr. Fivizzani is certified as a Chemical Hygiene Officer by the National Registry of Certified Chemists. He serves on the Board of Editors of *Journal of Chemical Health & Safety* and writes columns for "The Last Word" in that journal. He has presented 30 papers at national ACS meetings to the Chemical Health and Safety, the Industrial and Engineering Chemistry, and the Chemical Education divisions. He has coauthored conference papers for the National Association of Corrosion Engineers, the Cooling Tower Institute, and the American Industrial Hygiene Conference & Exposition. He is an inventor or co-inventor on five U.S. patents. He has investigated corrosion inhibition and dispersion in boiler and cooling water systems. His current interests involve laboratory safety and recruiting of technical staff.

Robin Izzo is the associate director for laboratory safety in the Office of Environmental Health and Safety at Princeton University. She has more than 20 years' experience in the laboratory safety field, having held positions at the University of Vermont and Harvard University before her 16-year tenure at Princeton. Ms. Izzo is involved in a number of national and international organizations dedicated to health, safety, and environmental issues at colleges and universities. She was instrumental in working with the U.S. Environmental Protection Agency (EPA) to develop a proposed rulemaking to make compliance with chemical waste regulations more relevant to colleges and universities. Soon after the U.S. Department of Homeland Security (DHS) issued the Notice of Proposed Rulemaking for the Chemical Facility Anti-Terrorism Standards, Ms. Izzo was selected to serve on the University Working Group to advise DHS on how to revise and apply the new regulation for the laboratory and campus environment. Ms. Izzo is the chair of the Coordinating Committee for the EPA College and University Sector Strategy. She is a member of the Board of Directors for the Campus Safety Health and Environmental Management Association, the international association for college and university health and safety professionals. Ms. Izzo holds a B.S. in mathematics from the University of Vermont and an M.S. in environmental sciences from the New Jersey Institute of Technology.

Kenneth Jacobson is Chief of the Laboratory of Bioorganic Chemistry and the Molecular Recognition Section, at the National Institute of Diabetes and Digestive and Kidney Diseases (NIDDK) at the National Institutes of Health (NIH). He is a member of the NIH Senior Biomedical Research Service. Dr. Jacobson is a medicinal chemist with broad experience in organic synthesis and the specialized needs of pharmaceutical research in an academic/governmental laboratory setting. He is a "Highly Cited Researcher" in Pharmacology and Toxicology (Institute for Scientific Information). Since coming to NIH in 1983, he has trained >60 postdoctoral fellows, many of whom are synthetic chemists who now hold academic faculty positions. Dr. Jacobson served as chair of the Medicinal Chemistry Division (membership >10,000) of the American Chemical Society and has also been a consultant to the pharmaceutical industry. He was inducted into the ACS Medicinal Chemistry Hall of Fame in 2009.

Dr. Jacobson holds a B.A. in liberal arts from Reed College and a Ph.D. in chemistry from University of California, San Diego.

Karen Maupins is the team leader, Health, Safety, and Environmental in Lilly Drug Discovery Research. In this capacity, she has developed and implemented industrial hygiene plans for multiple pharmaceutical manufacturing and research and development (R&D) laboratory sites. She has also worked to develop corporate technical standards and audit protocols for the worldwide industrial hygiene program. She has developed, documented, and implemented a comprehensive health, safety, and environmental management system that outlines business processes for the management of health, safety, and environmental activities in all U.S.-based pharmaceutical and development operations. Ms. Maupins holds an M.S. in industrial safety and industrial hygiene from University of Duluth.

Kenneth G. Moloy is a Research Fellow for E. I. DuPont de Nemours & Company. He received a Ph.D. in inorganic chemistry from Northwestern University in 1984 and a B.S. in chemistry from Indiana University in 1980. Following graduate school, he joined Union Carbide Corp. in South Charleston, WV, working in long-range R&D. He then moved to DuPont (1995), joining their Central Research and Development Department in Wilmington, DE. Dr. Moloy's interest and expertise lie in the areas of organometallic chemistry, homogeneous and heterogeneous catalysis, organic chemistry, process chemistry, and material science. Although primarily a researcher, Dr. Moloy is also an active contributor to laboratory safety at DuPont and serves as an internal safety trainer, instructor, and consultant.

Randall B. Ogle has been employed with the Oak Ridge National Laboratory for over 20 years, serving as the environmental health and safety (EHS) lead for R&D advanced materials (and corporate industrial hygiene program manager for 2 years). For the past 5 years he has been the operations manager for the Center for Nanophase Materials Sciences, one of five U.S. Department of Energy (DOE) Nanoscale Science Research Centers. Mr. Ogle has been a lead in developing the DOE guidance on safety with nanomaterials, which has been used internationally. He has lectured and presented papers on related topics at national, international, and regional events. Mr. Ogle has over 30 years of experience in EHS and has been certified in industrial hygiene, industrial safety, and hazardous materials management. Mr. Ogle worked for the University of Alabama for 10 years and was a director of EHS consultation programs. He also worked for the U.S. Occupational Health and Safety Administration. He has degrees from the University of Virginia (1974), East Tennessee State University (1977 M.S. in environmental health), Texas A&M and University of Alabama (health sciences degrees). He has been a member of the American National Standards Institute Technical Advisory Group to the International Organization for Standardization (ISO) Technical Committee 229 on Nanotechnologies since its inception.

John Palassis combines 36 years of professional experience of which 6 were with private industry and 30 years were with the federal government in the National Institute for Occupational Safety and Health (NIOSH), a research agency under the Centers for Disease Control and Prevention. At NIOSH, Mr. Palassis developed industrial hygiene air sampling and analytical methods for hazardous chemicals found in the workplace and worked as a safety coordinator. He taught industrial hygiene and occupational safety courses and was course director for the NIOSH Laboratory Safety course. His current interests and projects involve chemical safety in schools, young worker safety, and safety management systems. He collaborated with federal and nongovernmental organizations to develop the newest NIOSH product, *School Chemistry Laboratory Safety Guide*. He has published more than 60 papers and chapters in refereed journals of industrial hygiene, safety, and analytical chemistry, the *NIOSH Manual of Analytical Methods*, NIOSH Health Hazard Evaluation Reports, NIOSH docu-

ments, and professional magazines and has presented at over 70 national and international conferences.

Mr. Palassis is a diplomate of the American Academy of Industrial Hygiene and a Certified Industrial Hygienist, a Certified Safety Professional, and a Certified Hazardous Materials Manager. He is a member of the American National Standards Institute ANSI Z490 and the ANSI Z10 Committees. Mr. Palassis is a member of the Management Committee and also of the Communications and Training Methods Committee of the American Industrial Hygiene Association. He is an active member of the organizing Safety Program Development committee of the Ohio Safety Congress, where he frequently presents papers. Mr. Palassis is a consultant to the Air Sampling Instruments Committee of the American Conference of Governmental Industrial Hygienists. Internationally, he collaborates with the World Health Organization and with the International Social Safety Association (worldwide federal agencies in occupational safety and health organization). Mr. Palassis is a member of the U.S. Environmental Protection Agency School Chemical Cleanout Campaign (SC3).

Russell Phifer is the Principal of WC Environmental, LLC in West Chester, PA. He has over 30 years of experience in the field of environmental health and safety. His background is varied, and includes management of health and safety on Superfund sites, chemical safety and waste management training, and consulting on environmental health and safety for lab and industrial facilities. He is immediate past chair of both the American Chemical Society Committee on Chemical Safety and the American Chemical Society Division of Chemical Health and Safety, as well as an elected ACS Fellow. Mr. Phifer is a 30-year member of the ACS Task Force on Laboratory Chemical and Waste Management, serving as Chair of that group for 6 years. He is an OSHA Authorized Trainer and holds various other professional certifications associated with hazardous materials management.

Peter A. Reinhardt is the director of the Office of Environmental Health and Safety at Yale University. EHS is responsible for the university's workplace safety, environmental affairs, industrial hygiene, biological safety, and radiation safety. Prior to moving to New Haven in 2007, he was the director of Environment, Health, and Safety at the University of North Carolina at Chapel Hill for 6 years, and worked for 21 years at the University of Wisconsin–Madison's Safety Department, last serving as their assistant director for chemical and environmental safety. He is a member of the American Chemical Society's Task Force on Laboratory Chemical and Waste Management, the Campus Safety, Health, and Environmental Management Association's Board of Directors, and the Council on Government Relations' Working Groups on Research Security and Export Control. He coauthored *Hazardous Waste Management at Educational Institutions*, *Infectious and Medical Waste Management*, and *The Environmental Compliance Guide for Colleges and Universities*. He co-edited *Pollution Prevention and Waste Minimization in Laboratories*. He previously served on two subcommittees of the committee that wrote the 1995 *Prudent Practices in the Laboratory*. He has a B.S. and an M.A. from the University of Wisconsin–Madison.

Levi T. Thompson is currently the Richard Balzhiser Professor of Chemical Engineering. Other honors and awards include the National Science Foundation Presidential Young Investigator Award, Union Carbide Innovation Recognition Award, Dow Chemical Good Teaching Award, College of Engineering Service Excellence Award, and Harold Johnson Diversity Award. He is cofounder, with his wife Maria, of T/J Technologies, a developer of nanomaterials for advanced batteries and fuel cells. He is also consulting editor for the *AIChE Journal*, and a member of the External Advisory Committee for the Center of Advanced Materials for Purification of Water with Systems (National Science Foundation Science and Technology Center at the University of Illinois) and AIChE Chemical Engineering Technology Operating Council. Professor Thompson earned his B.ChE. from the University of Delaware, and M.S.E. degrees in chemical engineering and nuclear engineering, and a Ph.D. in chemical engineering from the University of Michigan. Research in Professor Thompson's group focuses primarily on defining relationships between the structure, composition, and function of nanostructured catalytic and electrochemical materials. In addition, he has distinguished himself in the use of micromachining and self-assembly methods to fabricate microreactor, hydrogen production, and micro-fuel cell systems. Professor Thompson leads a large multidisciplinary team developing compact devices to convert gasoline and natural resources into hydrogen. Recently, he was appointed founding director of the Hydrogen Energy Technology Laboratory.

Leyte Winfield is a synthetic organic chemist employed as a chemistry professor at Spelman College. She has experience in academic, industrial, and military laboratories. From the combined experiences she has gained knowledge of diverse procedures relating to laboratory operation and safety. Her experimental background encompasses both instrumental and synthetic techniques. Dr. Winfield arrived at Spelman College in August 2003 following the completion of a postdoctoral position at Florida A&M University. While at Florida A&M, she focused on the design and synthesis of compounds targeted to the inhibition of cyclooxygenase-2 (COX-2). Prior to this training, she completed her doctorate at the University of New Orleans. Her tenure at the University of New Orleans afforded her the opportunity to design and synthesize dopamine reuptake inhibitor analogs for potential cocaine abuse therapeutics. As an assistant professor of chemistry, Dr. Winfield is currently involved in the development of an organic chemistry laboratory manual. In addition, her current research involves the development of novel COX-2 inhibitors specifically targeted at inducing apoptosis in cancer cells.

Index

A

Absorbents/absorbency
  Amberlite® polymeric resins, 204
  chemical hood filters, 196, 231
  gloves, 176
  spill containment, 28, 120, 125, 133, 134, 145, 177, 302
  waste management, 193, 203, 204
Absorption in the body (*see also* Exposure)
  gastrointestinal tract, 58
  prevention, 124, 179
  radiation, 81
  respiratory tract, 57, 124, 166, 179
  skin, 55, 58, 60, 62, 124, 139, 166, 179
Academic laboratories
  Chemical Hygiene Officer, 16
  chemical hygiene responsibilities, 16-17, 187
  culture of safety, 2, 3-4, 5, 6
  department chairperson or director, 16-17
  departmental safety committee, 17
  facilities, maintenance, and custodial personnel, 17
  high school laboratories, 3
  laboratory personnel, 17, 187
  laboratory supervisor, 17
  liability concerns, 6
  management of chemicals, 88, 93, 102, 296-297
  recordkeeping, 3
  regulations for, 194
  research laboratories, 4
  training program, 250
  undergraduate laboratories, 3-4
  ventilation design, 251
  waste management, 3, 186-187, 194, 196, 201

Access control, 20, 22, 28, 29, 92, 95, 101, 123-124, 126, 163, 206, 208, 214, 256, 257, 262, 263, 264, 301, 302
Accidents (*see also* Emergency; Fire; Injuries and illnesses; Safety and emergency equipment; Spills and releases)
  CHP component, 295, 299, 300, 302, 303
  costs of, 6
  equipment-related, 149, 150, 151, 153
  liability consequences, 6, 102, 116
  misuse of materials, 7
  MSDS information, 50
  plant-scale, 139
  preparation for, 107, 110, 125, 130, 150, 153, 158, 160, 299
  prevention, 16-17, 21, 86, 87, 95, 110, 113, 140, 141, 150, 151, 155, 216, 302
  regulations, 272, 281, 299
  reporting and record keeping, 5, 16, 19, 29, 30, 181, 281, 299
  reviews, 3, 17
  slips, trips, and falls, 77, 150, 164
  vehicular, 102, 193
  working alone and, 17
Acetaldehyde, 67, 72, CD
Acetic acid, 67, 68, 72, 96, 134, 136, 189, 202, 203, 204, CD
Acetone
  control banding, 64
  –dry ice coolant, 173
  dual-use hazard, 260
  flammability, 65, 67, 68, 120, 173
  inhalation, 57
  LCSS, CD
  spills, 48
  storage of chemical in, 135
  waste management, 85, 190

Acetyl peroxide, 303
Acetonitrile, 67, 69, 160, 204, CD
2-Acetylaminofluorene, 273
Acetylene and acetylenic compounds, 62, 69, 71, 72, 129, 135, 140, 166, 172, CD
Acetylides, 70, 71, 130
ACGIH (*see* American Conference of Governmental Industrial Hygienists)
Acid baths, 220
Acid inhibitors, 70
Acids (*see also specific compounds*)
  allergic response to, 61
  chemical hoods and exhaust systems, 139, 214, 224, 226, 233-234, 236, 240, 242, 254
  cleaning with, 73, 86, 138, 139-140
  corrosivity, 190, 245
  decontamination concerns, 254
  digestion bombs for microwaves, 159
  dilution/dilute, 17, 137, 138, 140, 189, 209
  drying with, 140
  emergency procedures, 118, 120, 180
  in environmental rooms, 245
  equipment for use with, 88
  explosion hazards, 70, 71, 96, 130, 131, 136, 139-140, 214, 224
  fire hazards, 67, 68, 69, 98, 99, 128-129
  gloves and protective clothing, 111, 113
  incompatibilities, 70, 73, 96, 98, 100, 128-129, 138, 139
  injuries, 58, 61, 118, 137, 138, 180, 214, 260
  labels/classification, 69
  neutralization of, 120, 137, 138, 195, 196, 209
  neutralization with, 121, 136, 208

oxidizers, 69, 73, 96, 98, 120, 128-129, 131, 138, 188

peroxidizables, 72, 134, 139

radioactive mixtures, 202, 203, 205

rules for working with, 301

scrubbers, 236

spills, 28, 120

storage, 20, 96, 97, 98, 99, 100, 125

testing with, 188, 189

waste management, 196, 202, 203, 204, 205, 209, 210, 301

Acquisition of chemicals

CHP component, 295, 297, 298

compressed gases, 89, 140, 164-165

computerized, 88, 89

controlled substances, 261

cost containment, 5, 92

donations and gifts, 87, 92

just-in-time deliveries, 43, 89

minimization, 5, 21, 84, 164, 186

monitoring, 16, 19

ordering, 21, 86, 88-89

receiving, 21, 79, 89-90, 263, 278, 299

in smaller containers, 86

standards for, 21, 22, 278, 299

Acrolein, 60, 70, 303, CD

Acrylamide, CD

Acrylonitrile, 112, 273, 303, CD

Action levels, 16, 123, 269, 289, 292, 293-294

Activated carbon, 73, 136, 204, 231, 236

Acute toxicity (*see also* Allergens, sensitizers, and allergic reactions; Corrosives and corrosivity; Irritants)

aquatic, 50

defined, 59

dose-response relationships, 54

employee protection standards, 274, 275, 290, 293, 295, 301-302

empty containers, 191

examples of highly toxic compounds, 56, 60, 275, 301

exposure factors, 60

information sources, 48, 50, 51, 59

lethal dose/lethal concentration, 54-55, 59, 60, 275

oven drying of samples, 156-157

"particularly hazardous substances," 55, 59, 274-275

risk assessment, 54, 55, 59-60

signs and symptoms, 19, 123, 137, 141

storage of chemicals, 97, 101

systemic effects, 59

waste regulation, 277

Administration and supervision (*see also* Chemical Hygiene Officers; Chemical Hygiene Plan; Documentation and record keeping; Inspections and audits)

chemical hygiene responsibilities, 15, 16, 17, 18, 22, 24, 297, 298-299

and culture of safety, 2, 3, 5, 7

emergency preparedness, 39, 42, 117, 178, 180, 281, 299

hazard controls, 14, 18, 19, 24, 108, 142

hazard evaluation and risk assessment, 47, 55, 60, 66, 75, 108, 123, 131, 301

laboratory supervisor, 17, 22, 24, 26, 75, 117, 177, 178, 180, 231, 297-298, 299

management of chemicals, 84, 123, 173

OSHA recommendations for, 296-300

reporting to supervisors, 15, 16, 17, 226, 301

security, 123, 145, 262

statutory requirements, 279, 280, 281, 295

waste management, 209, 302

Adverse reactions, 271, 279

Aerosols

biological hazards, 60, 126, 221

defined, 57

in environmental rooms, 245

filters and scrubbers, 124, 236, 301, 302-303

flammable, 50, 99, 290

housekeeping and, 114

inhalation, 57-58, 110, 221

monitoring, 144

radioactive, 204, 208, 221

risk assessment, 55, 60, 80, 221

suppression, 114, 124, 126, 208, 236, 240, 244, 301, 302-303

Air (*see also* Hot air; Respirators; Ventilation and environmental control systems)

autooxidation, 136

cleanliness in clean rooms, 244

compressed, 145, 165, 171, 180-181

dispersion modeling, 124

filters, 124, 126, 142, 143, 178, 179, 221, 236

liquefied, 69, 135, 138

motors, 151, 154

pollutants, 86, 234, 235-236

pressure control, 214, 219, 240

reactivity in, 70, 72, 87, 93, 125, 130, 134, 135, 136, 137, 139, 140, 141, 143, 160, 173

showers, 244

supply, 179, 180-181, 217, 222, 226, 227, 241

Airborne contaminants (*see also* Exhaust systems; Filters and filtration)

action levels, 269

in chemical hoods, 291

dry materials, 144

evaluating exposure, 209

explosive limits, 65, 66-67

lethal (LC) values, 56

monitoring, 60, 143, 144, 208, 228, 298

nanoparticles in, 135, 143, 144, 248

particulate materials, 57

pressure buildup and, 135

preventing exposure, 107, 178, 219, 296

protection against, 127, 144, 178-180

risk assessment, 219

skin exposure, 179

threshold limits, 60

vapor pressure and, 57

ventilation systems and, 219, 221, 241, 251, 296, 298

ALARA (as low as reasonably achievable) principle, 82, 219, 221

Alarms

audibility in environmental rooms, 245, 299

building code requirements, 218

in chemical hoods, 169, 226, 250

equipment, 216
  fire, 15, 27, 28, 43, 107, 117, 121,
    122, 127, 166, 176-177, 178,
    217, 245, 249, 299
  gas detectors, 137, 168, 169, 2
    49
  heat sensors and smoke
    detectors, 100, 178, 249
  low-oxygen, 163, 173
  security, 124, 256, 258, 264
  signage and postings for, 250
  spills and releases, 35, 82, 124
  testing/inspection, 43, 193,
    245, 249, 250
Alcohols (*see also specific*
    compounds)
  consumable, 95, 245
  denatured, 259
  disinfection with, 247
  flammability, 99
  incompatibilities, 73, 130, 139
  secondary, 72
  shipment of samples in, 101
  in thermometers, 88
Aldehydes, 210
Aliphatic hydrocarbons, 133
Alkali metals, 59, 70, 73, 128, 130-
    131, 140, 178, 190, 210
Alkalis and bases
  explosive hazards, 69, 70, 73,
    130-131, 190
  fire control, 140, 178
  incompatibilities, 73, 128
  injuries, 58
  neutralization, 120, 209, 210
  spills, 120
  waste management, 209, 210
Alkenes, 72, 135
Alkyl chloromethyl ethers, 63-64
Alkyl nitrites, 71
Alkyl perchlorates, 71
Alkylating agents, 64
Alkylbenzenes, 203
Alkylhydroperoxides, 71
Alkyllithium compounds, 96,
    135-136
Allergens, sensitizers, and allergic
    reactions, 16, 58, 59, 60-61,
    279, 295, 301
Allylic halides, 61, 72
Alumina, 136, 160, 236
Aluminum chloride, *see*
    Aluminum trichloride
Aluminum trichloride, 136, CD
Amberlite® polymeric resins, 204

American Biological Safety
    Association, 80
American Chemical Society, 17,
    53, 86, 109, 116, 218, 262,
    267, 274
American Chemistry Council, 267
American Conference of
    Governmental Industrial
    Hygienists (ACGIH), 267,
    297
  exposure limits, 48, 86
  ventilation manual, 239
American Heart Association, 180
American Industrial Hygiene
    Association, 267, 269
American Institute of Chemical
    Engineers, 267
American National Standards
    Institute, 48, 108, 162, 217,
    253, 269, 270
American Society of Heating,
    Refrigeration, and Air
    Conditioning Engineers,
    253, 273
American Society of Mechanical
    Engineers
  Boiler and Pressure Vessel
    Code, 170-171
Americans with Disabilities Act,
    6, 218, 272
Aminechromium
    peroxocomplexes, 71
Ammonia, 57, 61, 69, 121, 129,
    136, 141, 164, 166, 170, 260,
    270, 303, CD
Ammonium hydroxide, 71, 96,
    CD
Ammonium nitrate, 69, CD
Ammonium phosphate
    extinguishers, 178-179
Anaphylactic shock, 61, 112
Anhydrides, 61, 115, 210
Aniline, 303, CD
Animal and Plant Health
    Inspection Service, 205, 260,
    270, 273
Animals (*see* Laboratory animals;
    Service animals)
Annual Report on Carcinogens, 63,
    291, 304
Anosmia, 60
ANSI (*see* American National
    Standards Institute)
ANSI/ASHRAE standards, 221,
    222, 225, 227, 231, 249

Aprons, 15, 113, 138, 175
Aqueous solutions
  acids, 121
  bleach, 206
  corrosives, 61, 121
  hydrogen fluoride, 137
  hydroxides, 61, 139
  pH testing, 188, 189
  plastic equipment for, 168
  potassium iodide, 134, 189
  reducing agents for peroxides,
    134
  waste management, 107, 185,
    190, 192, 195, 196, 202, 203,
    204-205, 206, 207
Argon, 62, 130, 131, 134, 135, 159,
    160, 161, 172, 174, 234
Armored
  chemical hoods, 132
  mercury thermometers, 88
Aromatic hydrocarbons, 63, 133
Arsine, 60, 239, CD
Asbestos, 57, 141, 179, 273
Ascarite®, 136
ASHRAE (*see* American Society of
    Heating, Refrigeration, and
    Air Conditioning Engineers;
    ANSI/ASHRAE standards)
Asphyxiants and asphyxiation,
    62, 74, 90, 100, 140, 163, 168,
    172, 173, 178, 242, 245
Aspirators, 75, 110, 153, 174
Assembly point, emergency, 39,
    193
Atomic absorption
    spectrophotometers, 121,
    237, 254
Atomic Energy Act, 202, 271
Audits (*see* Inspections and
    audits)
Autoclaves and autoclaving, 171
  animal carcasses, 208
  canopy hood, 239
  contaminated labware, 206
  dedicated rooms or equipment,
    208, 215
  headspace, 166
  hydrogenation reactions, 135
  infectious waste, 206
  multihazardous waste, 205,
    206, 207, 208
  pressurization, 135, 166, 172
  testing for interior surface
    contamination, 208
  thermometer, 88

Autoignition, 65-66, 69, 73, 130, 135, 136, 139, 140, 141, 156
Autoxidation, 72, 73, 136
Azides, 60, 70-71, 96, 130, 136, 210, 260, 303, CD
Azo compounds, 72, 273

B

Back injuries, 77
Bar code labeling, 19, 91
Barriers and barricades, 74, 82, 108, 110, 120, 123, 125, 131, 132, 137, 151, 213, 223, 224, 229, 252, 256, 258
Bases (*see* Alkalis and bases)
Benchtop
    ventilated enclosures, 143, 220, 228, 240
Benzene, 57, 63, 65, 86, 96, 190, 273, 303, CD
Benzyl azides, 96
Benzyl halides, 61
Benzylamine, 96
Benzylic hydrogens, 72
Benzyltrimethylammonium hydroxide, 96
Biodegradation and biodegradable materials, 88, 196, 203
Biological materials and biohazards
    aerosols, 60, 126, 221
    animal tissues and carcasses, 205, 206, 207-208, 303
    biosafety levels, 80, 126, 126
    Chemical Hygiene Plan, 15
    containment and biosafety facilities, 75, 80, 126, 143, 220, 221, 231, 236, 243, 244, 245-248, 251
    disinfection and decontamination, 206, 254
    dual-use agents, 256
    exposure limits, 51
    filters, 236
    gloves and protective clothing, 111
    information sources, 51, 79, 80, 126, 189, 246, 260, 276
    procedures for working with, 126-127
    radioactive, 207-208
    regulation, 79, 189, 270-271, 272, 276, 278, 280-281

risk assessment, 13, 79, 80, 207, 221
    security, 256, 260, 262, 263, 264
    select agents, 79, 270
    shipping, export, and import, 101, 272, 278, 280-281
    storage, 97
    synthetic microorganisms, 79
    waste management, 87, 108, 127, 185, 187, 189, 195, 201, 202, 205-208, 278
Biosafety cabinets (BSCs), 75, 80, 126, 143, 220, 221, 231, 236, 244, 245-248, 251
Biosafety in Microbiological and Biomedical Laboratories, 79, 80, 126, 189, 246, 260, 276
Biosafety levels, 80, 126
Birth defects (*see* Developmental toxicity)
Bis(chloromethyl)ether, 54, 273, 303
Blankets, 36, 119, 121, 122, 180
Boiling chips, 73
Boiling eruptions, 73
Boiling points, 48, 65, 67, 68, 73, 86, 99, 139, 172
Boron halides, 136
Boron trichloride, 140, 303
Boron trifluoride, 140, 170, 303, CD
Bourdon tubes, 167
Bretherick's Handbook of Reactive Chemical Hazards, 52, 65, 304
Bromine, 61, 62, 303, CD
Building code requirements, 11, 22, 27, 95, 98, 100, 215, 217, 218, 242, 253, 272-273
Bulk materials, 77, 78, 87, 89, 94, 107, 133, 141, 142, 145, 195, 198, 236
Bunsen burners, 27, 69, 128
Bunsen tube, 172
Burns, 19, 28, 58, 59, 61, 66, 75, 113, 118, 119, 122, 136, 137, 138, 139, 141, 149, 161, 173, 214
tert-Butyl hydroperoxide, 303, CD
Butyllithiums, 135, 139, 140, CD

C

Cabinets (*see* Biosafety cabinets; Storage)
Cadmium, 273

Calcium, 130
Calcium gluconate gel, 138
Calcium oxide, 61, 70
Calorimetry, 131
Cancer, 51, 59, 63, 64, 82, 101 (*see also* Carcinogens)
Canopy hoods, 220, 237, 239, 247, 298
Capacitors, 76, 150, 152, 153
Carbon-14, 81
Carbon brushes, 151
Carbon dioxide, 62, 86, 132, 135, 138, 140, 170, 173, 177, 178, 205, 245, 252
Carbon disulfide, 56, 62, 66, 67, 69, 128, 129, 136, 303, CD
Carbon inventory, 251
Carbon monoxide, 56, 62, 63, 69, 129, 166, 179, 303, CD
Carbon powder, 73, 131, 136
Carbon steel, 174
Carbon tetrachloride, 137, 140, 178, 303, CD
Carbonyls, 166, 302, 303
Carcinogens
    clothing and protective apparel, 113
    combinations of compounds, 64
    defined, 63, 275, 291-292
    examples, 54, 63, 302
    experiment planning, 64
    exposure factors, 56, 64
    information resources, 48, 50, 52, 53, 63, 291, 304
    labeling, 101
    LCSS information, 63
    minimizing/avoiding exposure, 55, 64, 86, 274
    MSDS information, 48
    radiation, 75
    risk assessment, 54, 55, 63-64
    select carcinogens, 55, 59, 63-64, 86, 97, 274-275, 290, 291-292
    spills, 273
    standards for handling, 59, 63, 101, 273, 275, 290, 293, 302
    storage, 97, 101
Card files, 91, 92
Carts, 6, 23, 77, 89, 90, 114, 153, 168, 216
Casarett and Doull's Toxicology: The Basic Science of Poisons, 52
Catalog of Teratogenic Agents, 52, 62

Catalysts, 70, 72, 73, 74, 78, 130, 135, 136-137, 139, 160, 210
Cathode ray tubes, 163
Cathodes, 75
Caustics, 28, 58, 111, 120, 196 (*see also* Corrosives and corrosivity)
Celite, 84
Centers for Disease Control and Prevention, 77, 111, 205, 207, 260, 270, 273, 276
Centrifuges, 88, 161-162, 172, 216, 217
Characterization of waste
   identification responsibilities, 187
   for offsite management, 186-187
   test procedures for unknowns, 187-189
   unidentified materials (unknowns), 187
Checklists
   continuity of laboratory operations, 41
   emergency planning, 41
   inspection, 23, 25, 26, 27
   laboratory closeout, 253, CD
   laboratory hazard assessment, CD
   security vulnerability, 262
Chemical Abstracts Service (CAS), 52-53
   registry numbers, 87, 91, 94
Chemical Carcinogenesis Research Information System, 53
Chemical Facility Anti-Terrorism Standards (CFATS), 89, 259, 261, 270, 273
Chemical hazardous waste
   academic laboratories, 186-187, 194, 196, 201
   acids, 196, 202, 203, 204, 205, 209, 210, 301
   alkalis and bases, 209, 210
   assignment of tasks for handling, 194
   atmospheric disposal, 196
   central area for, 192-194
   characteristic waste, 189-190
   characterization, 186-189
   choosing transporter and disposal facility, 198
   collection and storage, 191-194

container size and fill level, 23
determining regulatory status, 190-191
empty containers, 191
gases, 198
at generation site, 191-192
generator types and definitions, 276
hazard reduction in-laboratory, 186
identification responsibilities, 187
incineration, 193, 195, 196, 197, 199
inspection, 26
labeling, 23
liability concerns, 198-199
limited waste, 190
manifesting, 199
minimization, 277
minimum requirements for generators, 277
monitoring offsite transport and management, 197-198
nonhazardous and nonregulated waste managed as, 194-195
offsite management, 186-187, 197-198
"on-site" definition, 276-277
preparation for shipment, 198
records and recordkeeping, 199-201
regulation, 189-191, 194, 271, 276-278
sanitary sewer disposal, 196
test procedures, 187-189
transportation of, 277-278
treatment and recycling, 196-197
unidentified materials (unknowns), 187-189
Chemical hoods (*see also* Exhaust systems)
   acid use in, 139, 214, 224, 226, 233-234, 236, 240, 242, 254
   adsorbents, 236
   airborne contaminants, 291
   airflow types, 229-232
   airfoils and baffles, 229
   alarms, 169, 226, 250
   alternatives to, 252
   armored, 132
   auxiliary air, 230-231

benchtop, 232
bypass, 230
California chemical fume hood, 233
in clean rooms, 234, 244
compressed gas cylinders in, 222
configurations, 232-234
constant air volume, 229-230
constant operation, 224-225
distillation (knee-high or low-boy), 229, 232, 233
ductless, 231-232
energy conservation, 251, 252
exhaust treatment and systems, 234-237, 241
explosion-proof, 248
exposure monitoring, 228
design and construction, 229-232
face velocities, 221-222, 226-227
filters, 196, 220, 231, 232, 235, 236
general design recommendations, 229
general rules, 15, 110
housekeeping, 223
instrumentation, 228
liquid scrubbers, 235-236
low-flow or high-performance, 231, 251
maximizing efficiency, 222-223
nonbypass, 229-230
perchloric acid, 233-234
performance-related factors, 222
proximity to windows and doors, 222
radioisotope hoods, 234
sashes, 223-224, 252
scrubbers and contaminant removal systems, 235
shielding, 131-132, 160, 177, 187
supply air diffuser proximity, 222
testing and verification, 225-228
thermal oxidizers and incinerators, 237
traffic proximity, 222
variable air volume, 230, 252
walk-in, 232-233
and waste disposal, 222

Chemical Hygiene Officers
(CHOs), 2-3, 15, 16, 18, 20,
21, 22, 23, 26, 27, 28 n.4, 47,
51, 63, 79, 114, 274, 289, 290,
295, 297
Chemical Hygiene Plan (CHP)
academic institutions, 16
access to, 19, 47, 293
compliance with, 17, 19, 23
defined, 15, 289
development and
implementation, 15, 16, 18,
19, 20, 289, 292, 293, 295,
296, 303-305
elements/topics, 3, 15, 47, 104,
117, 222, 274, 275, 292-293,
295, 296-305
general principles for work
with chemicals, 296-297
governmental laboratory, 20
guidance for development,
296-305
industry research facility, 18
information resources for
developing, 303-305
laboratory facility, 297-298
MSDSs, 303
OSHA Laboratory Standard, 3,
15, 47, 269, 274, 275, 292-293,
295, 296
responsibility for chemical
hygiene, 3, 51, 297
review and update, 15, 19, 293
rules and procedures for
working with chemicals, 61,
300-303
safety recommendations,
303
training and communication
on, 17, 18, 29
Chemical management program
(*see* Acquisition of
chemicals; Inventory and
tracking; Shipment and
transport; Storage; Waste
management)
Chemical properties (*see*
Laboratory Chemical
Safety Summaries; Material
Safety Data Sheets; *specific
properties*)
Chemicals of interest (COI), 23,
89, 259, 260, 261, 262, 270,
275

Children, 16
Chlorinated hydrocarbons, 63,
136
Chlorine, 60, 61, 69, 70, 136, 170,
260, 303, CD
Chlorine dioxide, 75, 130, 136
Chlorine trifluoride, 140-141, 303
Chloroform, 137, 202, 204, 205,
303, CD
1-Chloromethyl-4-fluoro-1,4-
diazoniabicyclo[2.2.2]octane
bis(tetrafluoroborate), 86
Chloromethyl methyl ether, 63-64,
273, CD
Chromates, 69, 73, 130, 210
Chromatography/
chromatographs, 24, 69, 86,
93, 113, 134, 192, 195, 216,
220, 237, 241
Chromic acid, 86, 98, 99
Chromium, 61, 73, 138, 174, 205
Chromium-51, 81
Chromium trioxide and other
chromium(VI) compounds,
71, 86, 98, 99, 273, CD
Chromosomal damage, 62, 275,
291
Chronic exposure, 19, 56, 87
Chronic toxicity and health
effects, 11, 18, 19, 48, 50,
51, 52, 56, 59, 62, 63, 87,
121, 123, 290, 295, 301,
302 (*see also* Carcinogens;
Neurotoxicity; Reproductive
toxins)
Circuit boards, 94
Circuit breakers and fuses, 28, 76,
120, 150, 151, 152, 153, 154,
155, 162, 218
Clean Air Act, 196, 278
Clean benches/laminar-flow
hoods, 143, 219, 221, 239,
248, 250
Clean rooms
casework, furnishings, and
fixtures, 216, 244
chemical hoods, 234, 244
classification, 243, 244
pressure control system, 219,
243
protocols, 243-244
Clean Water Act
Cleaning (*see* Housekeeping and
cleaning)

Clinical and Laboratory
Standards Institute, 269
Clinical laboratories, 79, 201, 202,
205, 269, 276, 281
Closure or loss of institution
or building (*see also*
Decommissioning)
alternative laboratory facilities,
42, 43
checklist for continuity of
operations, 41
laboratory closeout checklist,
CD
long-term, 42
short-term, 41-42
Clothing and protective apparel
(*see also* Gloves and hand
protection; Personal
protective equipment)
aprons, 15, 113, 138, 175
choosing, 175
contaminated, 118, 122, 126,
137, 144, 146, 175, 181, 300,
301, 302
disposal, 113, 118, 144, 302
entanglement in equipment,
153, 163
fires on, 122, 178, 180, 181
footware, 120, 144, 175-176,
244, 300, 302
jumpsuits, 175, 303
laboratory coats, 15, 19, 26, 109,
113, 126, 131, 132, 138, 144,
172, 214, 244, 247, 301, 303
laundering, 109, 113, 118, 144,
175
materials, 111-112, 132
for nanomaterials work, 143,
144
radiation protection, 80, 127,
162
static discharge from, 128
storage, 113, 175
survival kit, 36
Cobalt, 61, 136
Cold baths, 173
Cold burns, 66, 74, 119, 173
Cold rooms, 76, 97, 109, 152, 245,
298, 299, 300
Cold storage, 97-98
Cold traps, 75, 115, 151, 172,
173-174
Color coding for identification,
100, 165

Combustion (*see* Explosive and highly reactive hazards; Fire; Flammability and flammable chemicals; Ignition sources and causes)

Commingling of waste, 191, 193, 194, 198, 203

Communication during emergencies
contact list, 38
e-mail, 38
emergency contacts for individual staff, 39
internet and blogs, 38-39
media and community relations, 39
plan for, 38-39
telephone, 38
text messages, 38

Comprehensive Guide to the Hazardous Properties of Chemical Substances, 51, 59

Compressed air, 145, 165, 171, 180-181

Compressed Gas Association (CGA) standards, 168, 169

Compressed gas cylinders (*see also* Pressure vessels and reactions)
acquisition and returns, 89, 140, 164-165
in chemical hoods, 222
code requirements, 164, 168
fire extinguishers, 178
hazards, 19
inspection, testing, and records, 26, 27, 164, 165, 169, 170
labeling, 140, 165
leak prevention and control, 120-121, 140, 169, 170
lecture bottles, 89, 100, 165, 239
outlet connections, 168-169
outside location, 34
precautions for handling and use, 164-165, 168-170, 303
pressure-relief devices and regulators, 26, 164, 168, 169-170
securing and storing, 26, 35, 90, 96, 100, 111, 114, 154, 164, 166, 168, 170
transporting, 111, 114, 168
valves, 168, 169
venting, 168, 170

Compressed gases
chemical hazards, 74, 140-141
compatibility, 96
corrosive, 121
defined, 164, 290
in environmental rooms, 245
flammable, 69-70, 74, 114, 120-121, 165, 170, 190
inert, 165
information sources, 15
in-house gas systems, 100, 140
inventory control, 140, 164
list of hazardous chemicals, 140-141
monitors and alarms, 169
oxidizing, 120-121, 170
physical hazards, 74, 291
procedures for working with, 140-141
regulations, 114, 164, 270, 290, 291
toxic gases, 96, 121
transferring, 166

Computer simulation of experiments, 5

Computer systems and services
backup systems, 36, 259
cables and wiring, 216
checklist programs, 26
communication during emergencies, 38-39
EHS information services, 52-53
ergonomic considerations, 77, 216
face velocity testing, 227
facilities for, 213, 214, 216
fire extinguishers for, 177
information and data security, 258-259, 262
inventory and tracking, 22, 91
loss of data or systems, 36, 38
MSDSs, 47-48, 89
physical protection, 35
purchasing systems, 88, 89
viewing laser operations on, 75

Condensation, 69, 74, 92, 98, 133, 151, 152, 172, 173-174, 224, 225, 232

Condensers and condensing, 93, 115, 116, 130, 133, 135, 138, 149, 156, 172, 173, 174, 222, 232, 292

Confidential or Sensitive Information, 259

Connectors, 115, 139, 156, 166, 167, 169

Consumer Product Safety Commission, 49

Contact lenses, 36, 109, 127, 301

Containers and packages (*see also* Compressed gas cylinders; Glass and glassware; Labels and labeling)
accumulation containers, 209
on benchtops, 113
bulk and economy sizes, 21, 87, 114
caps and seals, 154, 159, 192
corrosives, 137, 192
cyanide, 22
damaged or deteriorating, 90, 92, 114
dry ice in, 136
empty, 92, 94, 137, 190, 191, 198
flammable and combustible substances, 22, 23, 58, 69, 88, 98-99, 128, 129
gases, 102, 198
hazardous, 137, 191
headspace, 23, 73
for hydrogen fluoride, 137
inspection of, 22, 135
inventory and tracking, 91
metal, 69, 90, 128
in microwaves, 75-76, 159
for nanomaterials, 103, 143, 145
opening, 77, 100, 139
overpacks, 97
for peroxide formers, 21, 69, 133, 134
plastic, 99, 129, 145
recycling, 93-94, 191
in refrigerators, 154
regulatory standards, 21, 94, 103, 191, 194, 298, 301
remote handling, 132
secondary containment, 23, 98, 113, 114, 125, 145, 298, 301
securing, 95, 97
sharps disposal, 58, 111, 114, 206-207
small, 86-87
for storing chemicals, 35, 77, 89, 95, 96, 97, 98-99, 113, 114, 125
transfers among, 23, 84, 87, 94, 104, 113, 123, 125, 240-241
transport, 6, 23, 89, 90, 114, 125, 298

unpacking and inspecting, 90
vented, 139-140
waste, 23, 58, 87, 93, 110, 111, 114, 120, 125, 127, 144, 145, 191, 192, 193, 194, 195, 198, 199, 206-207, 209
Contingency plans, 12, 15, 55, 66, 107, 125, 193, 197, 199, 302
Contractors, 4, 10, 12, 13, 14, 264, 281
Control banding, 64-65
Controlled substances. 95, 261
Convulsions, 119
Cooling baths, 76, 136, 149, 158, 170, 173
Copper, 85, 133, 137, 138, 160, 165-166, 167, 174, 189
Corrosives and corrosivity (*see also specific classes and chemicals*)
classes, 61
containers for, 137, 192
engineering controls, 220
equipment compatibility and damage, 50, 76, 88, 165-166, 167, 170, 171, 174
evacuation of systems, 153, 167
examples, 61, 73, 135, 137, 138, 139, 140-141
experiment planning, 61, 66
eye contact, 58, 61, 135, 141, 190
gases and gas leaks, 61, 121, 140-141, 165-166, 167, 170, 172, 174
gloves and protective apparel, 111, 113, 176
information sources, 52, 303
inhalation, 61
labeling, 69, 165, 192, 199
neutralization, 209
PPE, 171-172, 176
shipping and transport, 199
skin contact, 58, 61, 111, 135, 141, 190
storage, 97, 100, 166, 192, 221
wastes, 153, 174, 190, 191, 192, 199, 209
Cryogens
asphyxiation hazard, 163, 173
explosion hazard, 131
gases, 75, 90, 100, 173, 242
information sources, 52
liquids, 19, 27, 74, 75, 76, 119, 135, 163, 170, 172-174, 176

nonflammable, 74
short-term storage and conveyance, 175
transfer lines, 174
Culture of laboratory safety
academic laboratories, 3-4, 5
accessibility for scientists with disabilities, 6-7
administration and supervision, 2, 3, 5, 7
communication and, 15
continuous learning environment, 14
environmental impact, 5-6
and federal funding, 6
industrial and governmental laboratories, 4-5
legal and regulatory changes, 6
management commitment to, 10
operational excellence, 14
oversight, 14
responsibility and accountability, 2-3, 4, 14
security considerations, 7
technology and, 5, 6
tips for encouraging, 5
Custodial and maintenance personnel, 17, 25, 58, 110, 120 n.3, 142, 177 n.3, 216, 219, 245, 253, 268-269, 282
Cuts, 58, 76, 115, 118-119, 149, 164
Cutting and drilling tools and operations, 16-17, 77, 78, 140, 143, 149, 152, 164
Cyanides, 22, 54, 56, 60, 62, 63, 70, 125, 166, 180, 187, 189, 190, 210, 260, 275, 301, 303
Cyanogen bromide, CD
Cyanogen chloride, 260
Cylinders (*see* Compressed gas cylinders)

D

Deactivation, 86, 127, 139, 160, 209
Decay in storage, 203-204, 207, 208, 278
Decommissioning
ANSI standard, 253, 254, 269
checklist, 253, CD
cleaning and decontamination, 254
clearance, 254

donated materials from, 92
equipment, 163, 171
hazard assessment, 253-254
information sources, 253
laboratories, 92, 93, 94
removal of hazards, 92, 253, 254
Decomposition, 70, 76, 87, 93, 130, 133, 135, 136, 138, 139, 157, 166, 290, 292
Decontamination
of accident victims, 28, 29
authorized personnel, 29
biosafety cabinets, 247-248
in chemical hoods, 124
of chemical hoods, 234
devices, 75
documentation, 253
of equipment and glassware, 124-125, 208, 302
facilities for personnel, 143
of gloves and other PPE, 112, 124, 176
information sources on, 253, 293
of laboratories for decommissioning, 253, 254
radiological, 127, 205, 208, 234, 253
regulatory standards, 275, 293, 300, 302
respiratory protective equipment for, 178
verifying, 121, 254
of waste, 205, 206, 207, 208, 300, 302
of work surfaces and work areas, 27, 121, 124-125, 302
Dehydrating and drying agents, 61, 136, 138, 139, 140, 159, 160, 175
Delivery (*see also* Receiving rooms and loading areas; Shipment and transport)
service disruptions, 42
Department of Energy (DOE)
exposure limits, 78, 82, 143
Integrated Safety Management System, 13-14
nanomaterials guidelines, 77, 141, 142, 145
pollution prevention, 201
Department of Homeland Security (DHS), 23, 36-37, 89, 259, 261, 262, 263, 267, 275

Department of Transportation (DOT), 23, 49, 68, 90, 94, 98, 99, 101, 102, 103-104, 116, 164, 172, 186, 190, 193, 197, 198, 199, 267, 271, 277, 278, 279, 300
Desiccators, 115, 140, 175
Design of laboratory facilities
   access control, 214-215
   accommodations for individuals with disabilities, 6, 218
   adaptability, 215-216
   casework, 216
   closed or separate laboratory spaces, 214-215
   doors, 216
   equivalent linear feet of workspace, 215
   flooring, 216
   furnishings and fixtures, 216
   layout, 215-216
   noise and vibration issues, 216-217
   older facilities, 218-219
   open design, 213-214
   safety equipment, 217-219
   storage rooms, 22
   utilities, 217-218
   wet spaces relative to other spaces, 213
   windows and walls, 216
Designated areas, 22, 36, 39, 89, 101, 123, 125, 135, 145, 275, 290, 293
Detergents, 85, 86, 162, 191
Developmental toxicity, 52, 53, 55, 62
Dewar flasks, 74, 115, 133, 138, 148, 172, 173, 174-175, 300
Diacetylene, 71
Diacyl peroxides, 71
Dialkyl peroxides, 71
Diazomethane, 60, 61, 70, 72, 132, 136, 301, CD
Diborane, 60, 239, CD
1,2-Dibromo-3-chloropropane, 273
Dichloroacetylene, 140
Dichloromethane (methylene chloride), 57, 64, 136, 160, 190, 273, CD
Dichromates, 69, 73, 301
Dicyclohexylcarbodiimide, 61

Diethyl ether, 28, 65, 67, 68, 69, 72, 120, 128, 134, 136, 303, CD
Diethylamine, 67, 96, 302, 303
Diethylene glycol dimethyl ether (diglyme), 72
Diethylnitrosamine, 273, 302, 303, CD
Differential manometers, 88
Differential scanning calorimetry, 131
Differential thermal analysis, 131
Digital thermometers, 58
Diisopropyl ether, 136
Diisopropylfluorophosphate, 301, 303
Diisopropylnaphthalene, 203
Dilution of
   acids, 17, 137, 138, 140, 189, 209
   bases, 209
   bleach, 206
   chemical hood effluent, 241, 242
   flammable substances, 129, 162, 228, 240
   high-energy reagents, 115
   peroxides, 133, 134
   vapors and toxic gases, 129, 162, 240, 242-243
   waste chemicals, 134, 196, 204, 209
Dimethoxyethane, 120
Dimethyl sulfate, CD
Dimethyl sulfoxide (DMSO), 58, 67, 68, 136, CD
4-Dimethylaminoazobenzene, 273
Dimethylformamide, 204, CD
2,5-Dimethylhexane, 72
Dimethylmercury, 29, 60, 302
Dioxane, 72, 85, 134, 136, CD
Diphosgene (trichloromethyl) chloroformate, 85
Disabilities (*see* Scientists with disabilities)
Disinfection, 79, 126, 127, 206, 207, 208, 245, 247, 254
Disposal (*see* Waste management)
Distillation
   avoiding, 159
   chemical hoods and exhaust systems, 229, 232, 233, 237, 241
   column purification systems, 160-161
   commercial recyclers, 185

   containment, 128, 159, 174, 204
   explosive compounds, 109
   flammable/combustible compounds, 159
   heat guns and, 159
   peroxide hazards, 72, 134, 136
   PPE, 174
   radioactive solutions, 204
   rotary flash evaporation, 204
   shutoff device, 159
   solvent stills, 93, 159-160
   thermal, 160
   vacuum, 140, 153, 159, 174
   venting stills, 140, 153, 300
   waste, 195, 204
Distilled water, 77, 157, 189
Divinyl acetylene, 72
Documentation and record keeping
   academic laboratories, 3
   accidents, 5, 16, 19, 29, 30, 181, 281, 299
   compressed gas cylinders, 26, 27, 164, 165, 169, 170
   equipment, 42-43, 165, 167, 170-171
   exports and imports, 280
   waste management, 199-201
Donations and gifts of chemicals, 87, 92
Dose-response relationships, 54
Downdraft hoods or tables, 220, 237, 239
Drains, 16, 28, 34, 114, 154, 180, 217, 232, 233, 234, 235 (*see also* Sewer discharges)
Drills and exercises, 37, 43, 125, 259, 299
Dry benzoyl peroxide, 136
Dry fire extinguishers and sprinkler systems, 177, 178, 217
Dry ice, 40, 41, 77, 109, 133, 136, 140, 153, 164, 172, 173, 245
Dry nanoparticles, 78, 103, 142, 143, 144, 145, 248
Dry sand, 131
Dry sweeping, 114, 145, 302
Dry traps, 216
Dryboxes, 89, 132, 173
Drying
   with acids, 140
   agents, 61, 136, 138, 139, 140, 159, 160, 175

gases, 136
glassware, 17, 85, 156, 157, 159
ignition hazards, 128, 136, 156, 233
oven drying of samples, 151, 156-157
radioactive waste, 204, 208
solvents, 78, 140
and spontaneous combustion, 73
train, 139
Dual-use materials, 95, 256, 257, 258, 259-260, 261, 262, 263
Dusts (*see* Powders and dusts)

E

Earthquakes (*see* Seismic activity)
Eating and drinking, 26, 109, 175, 300
ECOTOX, 53
Education and training in safety practices
   academic laboratories, 3-4
   case studies, 53
   computer simulations, 5
   and culture of safety, 2, 3, 4
   EHS management system, 12
   emergencies and spills, 34
   first aid, 16, 29, 37, 137, 138, 299
   hands-on, 4, 5, 29, 53, 121, 146
   informal, 4
   new employees, 29
   radioisotope users, 13
   regulatory requirements, 4
   risk assessment and, 53
   student manuals, 4
   in teaching laboratories, 3-4
   topics, 29
Electrical equipment (*see also* Distillation; Heaters and heating equipment; Refrigerators and refrigeration; Vacuum systems and operations; *other types of equipment*)
   capacitors, 76, 150, 152, 153
   codes, 149, 150, 151, 152, 154
   electromagnetic radiation hazards, 162-164
   general principles, 149-153
   high-current or high voltage, 152-153

as ignition sources, 21, 69, 76, 128, 129, 149, 153, 154, 245
   inspection, 26
   instruments, 162
   magnetic field hazards, 163
   noise extremes, 164
   outlet receptacles, 150
   personal safety techniques, 152
   precautions, 151-152
   rotating equipment and moving parts, 163
   stirring and mixing devices, 154
   ultrasonicators, 161
   wiring, 26, 150-151
Electrical fires, 68, 76, 150, 177, 178
Electrical power interruption
   discontinuation of experiments, 40
   dry ice, 41
   environmental and storage conditions, 40
   generator power, 41
   laboratory procedures, 40
   long-term, 40
   potential effects, 40
   preplanning, 40-41
   security issues, 40
   short-term, 40
   UPS, 41
Electrochemical equipment, 149
Electrocution and electric shock, 76
Electromagnetic radiation hazards
   laser light sources, 75, 162
   radio frequency and microwave sources, 75-76, 162
   visible, ultraviolet, and infrared laser light sources, 75, 162
   X-rays and electron beams, 162-163
Elephant trunks, 220-221, 237, 241, 250
Elevators, 23, 39, 69, 90, 114, 298
Emergency action plan, 15, 22, 23, 27
Emergency equipment (*see* Safety and emergency equipment)
Emergency preparedness
   administration and supervision, 39, 42, 117, 178, 180, 281, 299

assembly point, 39
checklists, 41
communications, 38-39
community-wide emergencies, 42
data or computer systems, 36
decision makers, with succession, 37
delivery and service disruptions, 42
drills and exercises, 43
equipment or materials losses, 36, 43
essential knowledge and supplies, 36-37
essential personnel, 37-38
evacuations, 39
fire, 34, 42-43
flood, 34
general preparation, 117
highly toxic materials, 125
high-profile visitors, 35
high-value or difficult-to-replace equipment, 36
impact/occurrence mapping, 34
institutional or building closure, 41-42
intentional acts of violence or theft, 35-36
leadership, 37-38
loss of laboratory, 42-43
mission-critical equipment, 34, 36
outside responders and resources, 43
pandemic-related, 35
political or controversial researchers or research, 35
power loss, 40
preplanning, 33-37, 40-41, 43
priorities, 37
records for replacement of equipment, 42-43
regulations, 281
seismic activity, 35
shelter in place, 39-40
spills or releases, 35, 117, 125
staff shortages, 35, 42
survival kit, 36-37
training of laboratory personnel, 29, 37
vulnerability assessment, 33-36
weather-related, 34-35

Emergency response (*see also*
  Safety and emergency
  equipment)
  accident procedures, 29, 117
  acid spills, 118, 120, 180
  burns from heat, 119
  clothing contamination, 118
  cold burns, 119
  convulsions, 119
  cuts, 118-119
  eye splashes, 118
  fires, 27, 121-122
  gas cylinder leaks, 120-121
  general procedures, 181
  information sources, 48
  ingestion, 119
  notification of personnel, 117
  skin exposures, 118
  spills, 28-29, 118, 120
  training, 29
  treatment of contaminated/
    injured personnel, 117-119
  unconscious victims, 119
Engineering controls, 14, 220
    (*see also* Chemical hoods;
    Gloveboxes)
Environmental health and safety
    (EHS) management systems
    (*see also* Chemical Hygiene
    Plan; Chemical management
    program; Emergency
    . . .; Inspections and audits;
    Safety rules and policies)
  change management, 13
  elements, 10, 11
  employee safety training
    program, 29
  example, 13-14
  functions, 14
  implementation, 12
  information sources, 11, 14
  management commitment to
    performance, 3, 10
  performance measurement,
    12-13
  planning, 10-12
  policy and policy statement, 10
  principles, 13-14
  regulations, 10
  review by management, 13
  safety committees, 12, 16-17,
    23, 24-25, 64, 276
  staff and responsibilities, 3
Environmental health and safety
  policy, 10

Environmental Protection Agency
  (EPA), 49, 53, 56, 83, 94, 102,
  117, 145, 186, 187, 189, 190,
  191, 194, 195, 198, 199, 201,
  202, 205, 206, 207, 219, 253,
  261, 267, 271, 272, 273, 275,
  276, 277, 278, 279-280, 281
Environmental rooms (*see also*
  Cold rooms)
  acids in, 245
  aerosols in, 245
  alarm audibility in, 245, 299
  alternatives to, 245
  compressed gases in, 245
  design, 245
  guidelines for working in, 245
Equipment (*see also* Electrical
  equipment; Safety and
  emergency equipment;
  Vacuum systems and
  operations)
  accidents with, 149, 150, 151,
    153
  for acid use, 88
  alarms, 216
  cutting and drilling tools, 164
  glass components, 167-168,
    171-172
  high-termperature, 27
  incompatibility with
    corrosives, 50, 76, 88, 165-
    166, 167, 170, 171, 174
  inspections, 24
  low temperature, 172-174
  noise extremes, 164
  pressure extremes, 170, 172
  repair and maintenance,
    114-115
  rotating, 27, 160, 162, 163
  safety switches, 27
  water-cooled, 149
Ergonomics, 77, 149, 164, 216, 239
Ethane, 62
Ethanol (ethyl alcohol), 67, 196,
  206, 270, CD
Ethers, 72, 114, 136, 138 (*see also*
  specific compounds)
Ethidium bromide, CD
Ethyl acetate, 72, 190, CD
Ethyl benzene, 190
Ethyl chloromethyl ether, 63
Ethyl ether, 190
Ethylene dibromide (EDB), CD
Ethylene glycol ethers, 62, 72, 149
Ethylene oxide, 60, 137, 273, CD

Evacuation of systems, 153, 167
Evacuations, 27, 36, 37, 39, 82,
  117, 120, 127, 169, 178, 299
Excess material (*see* Unused and
  excess material)
Exhaust systems (*see also*
  Chemical hoods)
  benchtop ventilated
    enclosures, 143, 220, 228,
    240
  canopy hoods, 220, 237, 239,
    247, 298
  clean benches/laminar-flow
    hoods, 143, 219, 221, 239,
    248, 250
  downdraft hoods or tables,
    220, 237, 239
  elephant trunks, 220-221, 237,
    241, 250
  filters, 239, 243, 302
  flammable-liquid storage
    cabinets, 239-240
  fume extractors, 237, 239
  gas cabinets, 239
  hybrid, 242
  manifolded (common header),
    242
  room purge, 242
  slot hoods, 220, 237, 238
  snorkels, 143, 220-221, 226, 237,
    238, 240, 241, 250, 251, 298
  stacks, 242-243
  ventilated balance enclosures,
    220, 238
Exits and passageways, 27, 28, 39,
  95, 113, 117, 122, 123, 127,
  193, 213, 214, 264, 277, 299
Experiment planning (*see also*
  Hazard evaluation; Risk
  assessment)
  access control, 123-124
  change management, 13
  corrosives, 61, 66
  demonstrations and magic
    shows, 116
  designated areas, 123
  emergency response to
    accidents and spills, 125
  explosive and highly reactive
    substances, 130-140
  highly toxic substances, 64,
    122-122
  minimizing exposure, 124-125
  multihazardous materials, 125
  protocols, 123

scaled-up reactions, 13, 115-116, 130
storage and waste disposal, 125
unattended experiments, 116
working alone, 116
Expiration (*see* Shelf life and expiration dates)
Explosive and highly reactive hazards
acids, 70, 71, 96, 130, 131, 136, 139-140, 214, 224
airborne concentrations, 65, 66-67
alkalis and bases, 69, 70, 73, 130-131, 190
azo compounds, 72
boiling eruptions, 73
catalysts, 130
conducting reaction operations, 133
cryogens, 131
decomposition rates, 130
emergency planning, 130
experiment planning, 130-140
gases, 133, 135, 136, 140-141, 172, 174
glassware, 70, 72, 73-74, 109, 131-132, 136, 153, 157, 174, 175
hydrogenation reactions, 135
incompatibles, 130-131, 140
information sources
list of materials requiring special attention, 135-140
other oxidizers, 73
peroxides and peroxidizables, 19, 72-73, 133-135
personal protective apparel, 132
powders and dusts, 19, 73
protective shields and devices, 131-132
quantities of reactants, 132-133
reaction rates, 130
risk assessment, 70-74, 132
scaling up experiments, 130
shielding, 109, 110, 130, 131, 161, 174
Exports and imports
biological materials, 280-281
chemical exports, 23, 279-280
imports from R&D Labs, 280
R&D exemption, 279
record-keeping requirements, 280
TSCA requirements, 279-280

Exposure
airborne contaminants, 60, 107, 178, 219, 296
biohazards, 51
duration and frequency, 19, 56
engineering controls, 108
eye protection, 108-109
highly toxic substances, 56, 64, 86, 109-110, 124-125, 274
limits, 51, 60, 81 (*see also* Lethal dose/lethal concentration)
minimization, 18-19, 51, 56, 64, 86, 107, 108-113, 124-125, 178, 219, 274, 296
radioactive materials, 81
safety rules and policies, 18-19
risk assessment, 57-58, 209
routes, 18-19 (*see also* Eye contact and effects; Ingestion hazards; Inhalation hazards; Injection hazards; Skin contact and effects)
signs and symptoms, 60
Extension cords, 151-152, 159, 217
Eye contact and effects
corrosives, 58, 61, 135, 141, 190
emergency response, 118
protection against, 15, 26, 36, 61, 75, 108-109, 114, 124, 131, 132, 144, 162, 173, 174, 176
toxic chemicals, 58
Eyeglasses
contact lenses, 36, 109, 127, 301
prescription, 109, 164, 176
safety, 26, 36, 75, 108, 109, 124, 132, 144, 162, 176
Eyewashes and eyewash units, 15, 27, 28, 29, 34, 117, 118, 125, 128, 176, 180, 181, 217, 218, 273, 297, 299

F

Face shields, 26, 61, 108, 109, 114, 131, 132, 138, 144, 162, 173, 174, 176
Facilities (*see* Laboratory facilities)
Fail-safe devices, 40, 150, 156
Falls (*see* Slips and falls)
Federal Water Pollution Control Act, 271
Ferrous sulfate, 134, 136
Fiberglass, 57, 157-158, 233

Filters and filtration
activated alumina, 136, 160, 236
activated carbon, 73, 136, 204, 231, 236
for aerosols, 124, 236, 301, 302-303
air, 124, 126, 142, 143, 178, 179, 221, 236
aqueous-based wastes, 195
biosafety cabinets, 236, 246, 247, 248
in chemical hoods, 196, 220, 231, 232, 235, 236
clean rooms, 243
disposal of contaminated filters, 107, 191, 192
in exhaust systems, 239, 243, 302
flammable paper, 139
gas phase, 236
for gases, 153, 223, 235, 236, 237, 240, 247, 302
HEPA, 79, 124, 126, 143, 145, 221, 231, 234, 235, 236, 239, 243, 247, 248, 302-303
hydrogenation reactions, 130
molecular sieves, 236
nanomaterials, 79, 142, 143, 145, 221, 236, 240, 244, 248
oil filters, 179
radioactive particles, 236
replacing filters, 236
in respirators, 178-179
solvent use, 84
ultra-low penetration air (ULPA), 236, 248
vacuum cleaners, 121, 145, 302-303
Fire
code requirements, 11, 22, 27, 51, 89, 95, 98, 100, 113, 114, 128, 151, 154, 214, 216, 243, 270, 272-273
on clothing, 122, 178, 180, 181
control, 140, 178
electrical, 68, 76, 150, 177, 178
emergency response, 121-122
information resources, 51-52
vulnerability assessment, 34
Fire department inspections, 25
Fire Protection for Laboratories Using Chemicals, 51
Fire Protection Guide to Hazardous Materials, 52, 68

Fire safety equipment
  alarms, 15, 27, 28, 43, 107, 117, 121, 122, 127, 166, 176-177, 178, 217, 245, 249, 299
  automatic fire-extinguishing system, 21, 178, 244
  blankets, 15, 28
  fire extinguishers, 15, 16, 27-28, 29, 34, 37, 43, 68, 113, 117, 121-122, 125, 127, 128, 130, 138, 140, 156, 177-178, 180, 181, 193, 217, 234, 299
  fire hoses, 178
  heat sensors and smoke detectors, 34, 100, 178, 249
  training, 34
First aid
  equipment and supplies, 15, 28, 36, 125, 137, 180
  information, 27, 48, 49, 50, 51, 52, 66, 94, 299
  procedures, 137-138, 152, 181
  regulations, 272, 299
  training, 16, 29, 37, 137, 138, 299
Flammability and flammable chemicals (*see also* Fire; Flash . . .; Ignition sources and causes)
  acids, 67, 68, 69, 98, 99, 128-129
  aerosols, 50, 99, 290
  alcohols, 99
  basic precautions, 129
  catalyst ignition of, 130
  characteristics, 65-67
  classification systems, 67-69
  containers, 22, 23, 58, 69, 88, 98-99, 128, 129
  filter paper, 139
  gases, 69-70, 74, 114, 121, 129, 165, 170, 190
  information sources, 65
  LCSSs, CD
  limits, 66-67
  liquids, 129
  procedures for working with, 127-130
  risk assessment guide, 66
  storage hazards, 98-99
  substances, 65
Flash arrestors, 129, 170
Flash hazards, 128, 152
Flash lamps for lasers, 151
Flash points, 65, 67, 68, 69, 86, 99, 127, 128, 140, 158, 188, 190, 290

Floods, 34
Fluorides, 137, CD
Fluorination, 86
Fluorine, 61, 86, 137, 141, 170, CD
Formaldehyde, 54, 57, 61, 139, 207, 220, 273, CD
Formalin, 206, 273
Food and beverages, 109, 300
Footware, 120, 144, 175-176, 244, 300, 302
Freezing, 72, 119, 149, 169, 176, 192, 224, 234 (*see also* Refrigerators and refrigeration)
Friction, 70, 132, 133, 161, 190, 290
Friction tape, 172, 174-175
Frostbite, 141, 172, 173
F-TEDA-BF4, 86
Fume extractors, 237, 239
Furnaces, 75, 146, 154, 157, 158, 172

G

Gamma rays, 81, 203
Gas burners, 128
Gas cabinets, 239
Gas chromatographs, 24, 216, 220, 237, 241
Gas cylinders (*see* Compressed gas cylinders)
Gases (*see also* Compressed gases; *specific gases*)
  byproducts of reactions and fires, 73, 122, 130, 131, 153, 190
  corrosive, 61, 121, 140-141, 165-166, 167, 170, 172, 174
  cryogenic, 75, 90, 100, 173, 242
  detectors and alarms, 34, 137, 168, 169, 249
  drying, 136
  equipment composition and, 174
  explosive, 133, 135, 136, 140-141, 172, 174
  exposure monitoring, 228
  filters, scrubbers, and exhaust systems, 153, 223, 235, 236, 237, 240, 247, 302
  flammable, 50, 69-70, 74, 114, 121, 129, 135, 141, 165, 170, 172, 190
  hydrogenation reactions, 135, 171

  inert, 116, 128, 129, 135, 140, 145, 159, 160, 165, 168, 169, 173, 174
  inhalation, 55, 57, 141
  in-house systems, 100
  liquefied, 69-70, 114, 135, 138, 140-141, 164, 166, 172-174, 176, 242
  oxidizing, 50, 70, 141, 291
  packagings, 102, 198
  pressure reactions, 171
  reactive, 57, 70, 133, 135, 140, 141
  respirator protection, 179
  toxic, 57, 85, 122, 140, 172, 190, 261, 275
  tracer, 221
  waste, 198
Gastrointestinal tract, 58 (*see also* Ingestion hazards)
Gauges
  diaphragm, 167
  liquid-level, 167
  magnehelic, 249
  McLeod, 75
  pressure, 88, 167, 172, 180, 243, 249
  thermal conductivity, 75
  vacuum, 75
Genetic Toxicology Data Bank, 53
Gifts (*see* Donations and gifts of chemicals)
Gland joints, 166, 169
Glass and glassware (*see also* Cuts)
  airbaths and tube furnaces, 158
  amber, 73
  beads, 140
  breakage, 35, 74-75, 76, 114, 115, 119, 191, 258
  cleaning and decontamination, 70, 73, 85, 86, 93, 114, 124, 138, 214, 215, 302
  contaminated, 58, 114, 138, 144
  cooling, 138
  cutting, 16-17
  drying, 17, 85, 156, 157, 159
  equipment, 167-168, 171-172
  explosion hazards, 70, 72, 73-74, 109, 131-132, 136, 153, 157, 174, 175
  for food and beverages, 109, 300
  handling and storage of, 16-17, 76-77, 115, 164, 176, 223, 229, 300

heating elements, 155
high-temperature operations, 109
hose connections, 115
inspection, 26, 66, 76-77, 115, 174
pressure and vacuum equipment, 26, 74-75, 109, 115, 131-132, 135, 149, 153, 167, 170, 171-172, 174-175
protective apparel and equipment, 26, 36, 75, 108, 109, 124, 132, 172, 176, 214
reactivity with, 137, 141
recycling, 94
shielding materials, 177, 225, 229, 248
stirring and mixing devices and, 154
stoppers, 133, 154
storage in, 73, 94, 99, 100, 129, 133, 137, 154, 191, 192, 302
tubing, 76, 115, 164
windows, 132, 258
waste disposal, 77, 114, 115, 203, 205
Glassblowing, 109, 115, 167
Glasses (*see* Eyeglasses; Goggles; Personal protective equipment)
Globally Harmonized System (GHS) for Hazard Communication, 47 n.1, 49-50
Glovebags, 234
Gloveboxes, 19, 58, 110, 112, 123, 124, 126, 127, 145, 220, 221, 243, 244, 247, 248, 293, 298, 300, 302
Gloves and hand protection (*see also* Personal protective equipment), 19, 23, 26
absorbency, 176
butyl, 111
chemical-resistant, 111, 124
cleaning/decontamination, 111, 124, 176, 206
corrosion-resistant, 36, 111, 113
cut-resistant, 16-17, 77, 111, 115, 176
disposable, 112
disposal, 144, 203, 205, 208
double gloves, 111, 132, 144, 176
gauntlet-type, 144

heat-resistant, 158
inspection, 15
insulated, 152, 173, 176
Kevlar®, 111, 115, 176
latex, 111, 112, 176
leather, 111, 115, 132, 176
neoprene or rubber, 111, 112, 138, 139, 152, 233, 303
nitrile, 111, 112, 144, 176
polyethylene, 112
polyurethane, 112
poly(vinyl chloride), 112
radiation-resistant, 80
selection and use guidelines, 61, 109, 111-113, 118, 120, 124, 126, 127, 131, 144, 149, 162, 172, 173, 176, 247, 301
Silvershield® or 4H®, 112, 138
spill cleanup, 48, 120, 121
storage and replacement, 112, 144, 176
Viton®, 111, 112
Goggles (*see also* Personal protective equipment), 19, 26, 108-109, 120, 132, 138, 144, 162, 173, 176, 233
Governmental laboratories, 4-5, 20
Green chemistry
degradable products, 86
experiment planning and risk assessment, 87
less toxic reagents, 85
mercury replacements, 87-88
microscale work, 84-85
multihazardous waste minimization, 87
real-time controls, 86-87
safer solvents and other materials, 85-86
waste prevention, 84
wet chemistry elimination, 85
Ground-fault circuit interrupters, 150, 152, 159, 217-218
Grounding, 22, 66, 69, 76, 128, 129, 132, 150, 152, 153, 155, 158, 159, 168, 193

H

Hair
dryers, 159
facial, 178-179, 181
long, 163, 300
Halides, 14, 61, 70, 136, 210

Haloacetylene derivatives, 71
Halocarbons, 70
Halogens and halogenating agents, 62, 71, 73, 115, 130, 136, 137, 141, 187, 189, 190, 192, 193, 205
Halon, 128, 178
Hands (*see* Gloves and hand protection)
Hazard classifications, 26-27
Hazard evaluation (*see also* Risk assessment)
administration and supervision, 47, 55, 60, 66, 75, 108, 123, 131, 301
basic principles, 53-58
environmental, 49
explosivity, 70-74
flammability, 65-70
health hazards, 26, 48, 50
information sources, 47-53
nanomaterials, 49, 77-79
physical hazards, 48, 50, 74-77
reactivity, 70
toxicity, 53-65
Hazard reduction (*see* Waste management)
Hazardous Materials Transportation Uniform Safety Act, 277
Hazardous Substance Data Base, 53
Health hazards (*see also* Acute toxicity; Carcinogens; Chronic toxicity and health effects; Exposure)
information sources, 48
Heat of reaction, 131
Heat sensors and smoke detectors, 100, 178, 249
Heat transfer, 115, 130, 158
Heaters and heating equipment (*see also* Hot air; Hot plates; Ovens)
general precautions, 154-156
mantles and tapes, 157-158
oil, salt, or sand baths, 158
tube furnaces, 158
Heavy metals, 56, 70, 86, 132, 135, 136
Helium, 62, 76, 135, 143, 163, 169, 171, 172
HEPA filters, 79, 124, 126, 143, 145, 221, 231, 234, 235, 236, 239, 243, 247, 248, 302-303

Hexafluoropropylene, 96

Hexamethylphosphoramide, CD

Hexane, 62, 67, 86, 120, CD

High-performance liquid chromatography, 69, 134, 204, 216, 241

High-pressure reactions, 26, 74, 165 (*see also* Pressure vessels and reactions)

High school laboratories, 3, 218, 277

Horseplay, 300

Hoses, 115, 149, 156, 169, 170, 171, 178, 179, 217, 237 (*see also* Tubing)

Hot air
baths, 129, 133, 154, 156, 158
guns, 69, 154, 159

Hot plates, 66, 128, 136, 149, 154, 155, 156, 157, 158, 225

Housekeeping and cleaning (*see also* Laundry)
acids, 73, 86, 138, 139-140
and aerosols, 114
chemical hoods, 223
dry sweeping, 114, 145, 302
general practices, 19-20
glassware, 70, 73, 85, 86, 93, 114, 124, 138, 214, 215, 302
inspection of, 23, 24, 114, 297, 298
nanomaterials and, 144-145
personal, 300
rules, 113-114, 298
safety aspects, 19-20, 77, 113-114, 122, 163
security aspects, 20
vacuum cleaners, 121, 145, 302-303

Hydrazine, CD

Hydrides, 70, 90, 86, 122, 128, 136, 138, 177, 178, 210

Hydrobromic acid, CD

Hydrocarbons, 52, 63, 115, 133, 136, 140, 168, 177, 190, 236

Hydrochloric acid, 61, 96, 134, 188, 189, 205, 209, CD

Hydrocyanic acid (*see* Hydrogen cyanide)

Hydrogen, 34, 67, 69, 70, 72, 73, 81, 129, 130, 135, 136, 137, 139, 140, 141, 165-166, 172, 173, 174, 190, CD

Hydrogen chloride, 54, 57, 136, 170, 303

Hydrogen cyanide, 56, 60, 62, 70, 125, 166, 275, 301, 303, CD

Hydrogen fluoride (HF), 61, 96, 137, 141, CD

Hydrogen peroxide, 61, 69, 70, 96, 130, 138, 139, 260, 291, CD

Hydrogen phosphides, 141

Hydrogen selenide, 141

Hydrogen sulfide, 56, 60, 69, 96, 129, 141, 170, 189, 275, 303, CD

Hydrogenation reactions, 74, 130, 135, 171, 224

I

Ice, 74, 77, 113, 119, 154, 164, 216, 224 (*see also* Dry ice)

Ice baths, 172

Ice chests, 109

Ignition sources and causes
catalysts, 130
compressed or liquefied gases, 69-70, 121, 170
controlling/eliminating, 42, 65, 98, 99, 113, 128, 129, 135-136, 153, 154, 228, 240
electrical equipment, 21, 69, 76, 128, 129, 149, 153, 154, 245
gas burners and heating equipment, 69, 128, 129
heating elements and hot surfaces, 66, 128, 136, 156, 159, 160
nanomaterials, 142
oxidants other than oxygen, 69
pyrophorics, 70, 135-136
risk assessment, 34, 65, 66
shock-sensitive compounds, 72, 136
spontaneous combustion, 69, 73, 130, 135, 136, 139, 140, 141
static discharge from fabric, 113, 128

Ignition temperature, 65-66, 67, 68, 155

Incineration and incinerators
waste, 85, 87, 185, 193, 195, 196, 197, 199, 203, 205, 206, 207, 208, 235, 237, 278, 297, 300, 302, 303

Incompatibility of chemicals
acids, 70, 73, 96, 98, 100, 128-129, 138, 139
alcohols, 73, 130, 139
alkalis and bases, 73, 128
compressed gases, 96
with equipment materials and fittings, 166, 167, 169
explosive and highly reactive, 130-131, 140
information sources, 52
storage guidelines, 21, 22, 27, 96-97

Induction heaters, 75

Induction motors, 151, 154

Induction periods, 73, 74, 116, 130

Industrial hygienists, 3, 29, 47, 52, 60, 62, 108, 164, 228, 229, 231, 237, 249

Industrial laboratories, 2, 3, 4-5, 18-19, 48, 52, 65, 186, 194, 196, 242, 263

Inert gases, 116, 128, 129, 135, 140, 145, 159, 160, 165, 168, 169, 173, 174

Infectious agents (*see* Biological materials and biohazards)

Information sources (*see also* Documentation and recordkeeping; Education and training in safety practices; Labels and labeling; Laboratory Chemical Safety Summaries; Material Safety Data Sheets)
ADA compliance, 6-7
biohazards, 51, 79, 80, 126, 189, 246, 260, 276
carcinogens, 48, 50, 52, 53, 63, 291, 304
compressed gases, 15
computer services, 52-53
corrosives, 52, 303
cryogens, 52
fire protection, 51-52
first aid, 27, 48, 49, 50, 51, 52, 66, 94, 299
flammability, 51-52
GHS, 47 n.1, 49-50
incompatible chemicals, 52
informal forums, 53
LCSS preparation, 52, CD
reactivity of chemicals, 52
risk assessment, 47-53
toxicity, 48, 50, 51, 52, 59

Infrared radiation, 75, 162, 176

Ingestion hazards, 58, 109-110, 119

Inhalation hazards, 55, 57-58, 61, 110, 141, 221

Inhibitors, 70, 72, 73, 134

Injection hazards, 58, 111

Injuries and illnesses (*see also* Accidents)
 acid burns, 58, 61, 118, 137, 138, 180, 214, 260
 alkalis and bases, 58
 assisting recovery and return to work, 12
 cuts, 58, 76, 115, 118-119, 149, 164
 emergency response, 117-119
 regulations, 281

Inorganic compounds, 28, 52, 57, 61, 62, 73, 96, 97, 120, 125, 130, 134, 135, 136, 139, 189, 196, 210, 273, 279

Inorganic Syntheses, 209

Inspections and audits
 alarm systems, 43, 193, 245, 249, 250
 checklists, 23, 25, 26, 27
 compressed gas cylinders, 26, 27, 164, 165, 169, 170
 conducting, 26
 containers, 22, 135
 corrective actions, 26
 EHS management system performance, 24-25
 elements of, 25-26
 fire department, 25
 glassware, 26, 66, 76-77, 115, 174
 hazardous waste, 26
 items to include in, 26-27
 peer, 24
 PPE, 15, 180
 preparing for, 25
 pressure equipment, 165, 167, 170-171
 program audits, 24
 regulatory agencies, 25
 reports, 26
 routine, 24
 self-audits, 24
 storage, 27
 types, 24-25
 ventilation systems, 250

Insurance, 36, 42, 43, 95, 102, 116, 198, 267, 269

Integrated Risk Information System, 53

Integrated Safety Management system, 13-14

Interlocks, 40, 108, 124, 149, 155, 170

International Agency for Research on Cancer (IARC), 63, 275, 291

International Air Transport Association, 23, 101, 271, 279

International Association for Assessment and Accreditation of Laboratory Animal Care, 23

*International Chemical Safety Card*s, 51

International Civil Aviation Organization, 279

International Code Agency, 98

International Code Council, 273 n.1

International Commission on Non-ionizing Radiation, 76

International Conference of Building Officials, 273 n.1

International Electromagnetic Field Project, 76

International Labour Organization, 49

International Mechanical Codes, 242

International Organization for Standardization (ISO), 78 n.3, 243, 280

International Programme on Chemical Safety, 51

International Toxicity Estimates for Risk (ITER), 53

Inventory and tracking (*see also* Labels and labeling; Recycling)
 benefits, 22
 compressed gases, 140, 164
 computerized, 23, 91
 by container, 91
 and emergency response, 94
 exchanges between laboratories and stockrooms, 92-93
 general considerations, 90-92
 information in, 22-23
 regulations, 22
 safety issues when performing, 23
 security issues, 23

Iodine, 61, 81, 87, 136, 159, 203, CD

Ion exchange resins, 204

Ionizing radiation, 62, 80, 81, 162, 271

Irritants, 50, 55, 58, 59, 60, 61, 112, 137, 138, 139, 140, 141, 169, 179, 290

Isopropyl alcohol, 67, 172, 173

Isopropyl ether, 72

K

Kekulé, August, 1

L

Labels and labeling
 acids, 69
 bar code, 19, 91
 color coding, 100, 165
 compressed gas cylinders, 140, 165
 corrosives, 69, 165, 192, 199
 experimental materials, 94
 flammability and flash point, 65, 68
 GHS system, 47 n.1, 49-50
 hazard communication, 49-50, 51, 68, 94
 information to be included on, 94
 inspection of, 22, 23, 27
 and inventory and tracking, 94
 nanomaterials, 103, 145
 quality of information on, 51
 reading and heeding, 22, 28
 regulatory standards, 51, 92, 268, 270, 271, 272, 274, 277, 279, 294, 298
 replacement, 92
 by suppliers, 21, 51, 90, 94
 transfer and storage containers, 21, 51, 87, 94, 154, 277
 verification on receipt, 90, 298
 warning, 51, 101
 waste, 23, 187, 191, 192, 193, 194, 197, 198, 203, 206

Laboratory animals
 biohazards, 80, 205
 carcinogenicity testing and results, 63, 275, 291-292, 302
 disposal of tissues and carcasses, 205, 206, 207-208, 303
 facilities, 215, 239, 262

personal protection for working with, 303
security/protection of, 37, 260-261, 262
toxicity testing and results, 54-55, 56, 59, 60, 62, 78, 269, 295, 302-303
Laboratory Chemical Safety Summaries, 91, 108, 129, 140
access to, 91, 123
content, 60, 62, 63, 68
criteria for selection of, 50-51
limitations of, 50
preparation of, 51, 52, 55, 63, CD
and risk assessment, 55, 59, 60, 61
Laboratory coats, 15, 19, 26, 109, 113, 126, 131, 132, 138, 144, 172, 214, 244, 247, 301, 303
Laboratory Decommissioning Standard (ANSI), 253, 254, 269
Laboratory facilities (*see* Closure or loss of institution or building; Decommissioning; Design of laboratory facilities; Ventilation and environmental control systems)
computer laboratories, 213, 214, 216
Laboratory security (*see also* Security plans)
administration and supervision, 123, 145, 262
alarms, 124, 256, 258, 264
barriers, 258
basics, 256-259
biohazards, 256, 260, 262, 263, 264
checklists, 262
chemicals of interest, 23, 89, 259, 260, 261, 262, 270, 275
controlled substances, 261
controversial research and researchers and, 35
and culture of safety, 7
door locks, 257
dual-use materials, 259-260, 262
elevated (Level 2), 263-264
high (Level 3), 264
information and data, 23, 258-259, 262

lighting, 258
normal (Level 1), 263
operational, 258
physical and electronic, 256-258
protesters and civil disobedience, 35
radioactive materials and related equipment, 261
requirements, 260-261
research animals, 260-261
risks to, 7
systems integration, 259
terrorist attacks, 35-36
video surveillance system, 258
vulnerability assessment, 261-262
Laboratory supervisor, 17, 22, 24, 26, 75, 117, 177, 178, 180, 231, 297-298, 299
Laboratory Ventilation Standard (AIHA/ANSI), 249
Labpacks, 87, 186-187, 193, 194, 198, 199
Lamps, 75, 151, 162
Landfills, 185, 190, 193, 194, 195, 196, 197, 300
Lasers, 19, 29, 34, 75, 76, 109, 137, 149, 151, 162, 176, 214, 215, 216, 217
Laundry, clothing, 109, 113, 118, 144, 175
Lead and lead compounds, 62, 73, 81, 136, 166, 189, 193, 202, 203, 205, 206, 234, 273, 301, CD
Leadership, 2, 4, 37
Leaks
chemical hoods, 221, 224, 225, 226, 227, 229, 231
compressed gas cylinders, 120-121, 140, 166, 169, 170
corrosive gases, 121
detection, 100, 165, 171
flammable gases, 35, 69, 100, 120-121, 166, 170
gas cabinets, 239
gloveboxes, 243, 302
microwave, 159
packages, 21, 90
refrigerant, 245
reporting, 245
secondary containment and, 23, 98, 113, 114, 125, 127, 145, 298, 301

toxic gases, 121
waste containers, 185, 192, 197
water, 151
Lecture bottles, 89, 100, 165, 239
Lethal dose/lethal concentration, 54-55, 56, 59, 60, 275
Liability and litigation, 6, 102, 116
Life cycle analysis, 86, 251
Lifting injuries, 164
Light and lighting, 21, 41, 72, 100, 116, 154, 172, 216, 224, 245, 250, 252, 258, 262, 301
explosion hazards, 70, 130, 133, 248
fire hazards, 67, 136
Liquefied gases, 69-70, 114, 135, 138, 140-141, 164, 166, 172-174, 176, 242 (*see also* Compressed gases; Cryogens; *specific gases*)
Liquid scintillation counter, 85, 203
Liquid scintillation fluid, 202
Lithium, 70, 130, 178
Lithium aluminum hydride, 90, 96, 138, CD
Lithium hydroxide, 96
Lower explosive limit (LEL), 66-67, 68, 228, 240
Low-valent metal salts, 70

M

Magnesium, 73, 122, 130, 131, 136, 141, 159, 178
Magnesium hydroxide, 209
Magnetic fields, 76, 163
Maintenance and repair, 4, 17, 24, 41, 76, 102, 110, 114-115, 143, 149, 150, 162, 166, 167, 171, 180, 218, 221-222, 231, 232, 241, 248-249, 250, 252, 257, 295, 297, 298-299
Manometers, 75, 87, 88, 249
Material Safety Data Sheets (MSDSs)
access to, 27, 47-48, 89, 91, 268
audience for, 48-49, 78-79
content, 22, 48, 50, 62
for emergency responders, 117, 118, 119, 138, 181
for experimental materials, 101, 280, 281
guidelines on use of, 13, 15, 21, 22, 51, 59, 97, 108, 110, 118, 123, 127-128, 140

information sources for
        preparing, 52-53
    limitations, 49, 91
    for nanomaterials, 103, 104
    regulatory requirement, 47,
        270, 272, 274, 279, 280
    and risk assessment, 47-49, 51,
        55, 59, 61, 62
    spills and, 28, 29
    from suppliers, 72, 89
McLeod gauges, 75
Mechanical shock, 70, 130, 131,
        174-175
Melting points, 48, 116, 158
Mercaptans, 210, 240
Mercury, CD
    bubblers, 170
    decontamination, 254
    incompatibilities, 166
    in meters and gauges, 75, 87-
        88, 157
    MSDSs, 303
    properties, 57
    reclamation, 93
    replacement, 87-88
    spills, 121, 157, 254
    toxicity, 62, 75
    waste, 174, 195, 196, 205, 206
Metal acetylides, 71
Metal alkyls, 128, 178
Metal azides, 210
Metal fires and explosions, 132,
        140, 141, 160, 177-178, 254
Metal fulminates, 71
Metal halides, 70, 210
Metal hydrides, 70, 122, 128, 136,
        178, 210
Metal hydroxides, 61
Metal nitrides, 70
Metal oxides, 57, 70, 86, 141
Metal perchlorates, 254
Metal peroxides, 71, 73, 130-131
Methane, 62, 70, 129, 172
Methanol, 57, 65, 67, 69, 96, 190,
        196, 202, 203, 204, 214, 303,
        CD
Methyl acetate, 204
Methyl acetylene, 72
Methyl chloride, 141
Methyl chloroform, 278
Methyl cyclopentane, 72
Methyl ethyl ketone (MEK), 67,
        CD
Methyl fluorosulfonate, 60
Methyl iodide (*see* CD)

Methyl-isobutyl ketone, 72, 190
Methyl lithium, 96
Methyl methacrylate, 72, 177
1-Methyl-2-pyrrolidinone, 72
Methylene chloride, 57, 64, 160,
        190, 273
Methylene diphenyl diisocyanate,
        61
Methylenedianiline, 273
Methylmercury, 275
Met-L-X, 122, 130, 178
Microscale and miniaturized
        operations, 5, 49, 70, 74, 84-
        85, 131, 203
Microwave ovens, 141
    acid digestion bombs, 159
    containers in, 75-76, 159
Microwaves, 75-76, 162
Molecular sieves, 204, 236
Molecular weights, 48
Monitoring (*see also* Laboratory
        security)
    aerosols, 144
    airborne concentrations, 60,
        143, 144, 208, 228, 298
    chemical hoods, 17, 23, 124,
        223, 225, 226, 231, 236,
        297-298
    cold storage, 97-98
    exposure, 16, 81, 228, 269, 289,
        292, 293-294, 295
    heating equipment, 157, 158
    nanomaterials, 79, 144, 248
    pressure equipment, 135, 141,
        165, 166, 243
    radiation, 81, 82, 127, 132
    regulatory compliance, 20, 123,
        272, 273, 294
    spills and leaks, 35, 79, 140,
        168, 245, 249
    ventilation systems, 252, 298
    wastes, 197-198
Multihazardous waste
    animal tissues and carcasses,
        206, 207-208
    chemical–biological waste,
        205-207
    chemical–radioactive–
        biological waste, 208
    chemical–radioactive waste,
        202-205
    commercial disposal services,
        205
    hazard reduction (treatment),
        204-205

    incineration, 203, 205, 206, 207,
        208
    labware, 206, 208
    medical waste and sharps,
        206-207
    minimization, 202, 203, 207
    offsite management, 207
    radioactive–biological
        laboratory waste, 207-208
    risk assessment, 202
    sewer disposal, 206
    storage, 203-204
Multimeters, 152

N

Nanomaterials
    airborne concentrations, 135,
        143, 144, 248
    in biosafety cabinets, 248
    in chemical hoods, 248
    clothing and PPE, 143, 144
    containers, 103, 143, 145
    controls for R&D laboratories,
        141-146
    determining appropriate
        controls, 142-143
    engineering controls, 143-146,
        248
    facilities, 248
    filters 79, 142, 143, 145, 221,
        236, 240, 244, 248
    hazards, 77-79
    housekeeping, 144-145
    information sources, 142
    marking, labeling, and signage,
        145
    monitoring and
        characterization, 111
    offsite transport and shipment,
        103
    onsite transfer and transport,
        103-104
    personnel competency,
        145-146
    planning and hazard
        assessment, 142
    regulations, 280
    ventilation systems, 143, 248
    waste disposal, 145
    work area design, 143
    work practices, 145
NARM (naturally-occurring
        or accelerator-produced
        radioactive material), 80, 202

National Electrical Code (NEC), 149, 152

National Emission Standards for Hazardous Air Pollutants, 270

National Fire Protection Association
  chemical hood standards, 224, 231
  electrical code, 149, 152
  fire code, 22, 51, 67, 68, 95, 98, 224, 231, 243, 270, 273
  fire protection guidelines, 52, 68
  hazard classification, 68, 69
  storage regulations, 22, 95, 98, 273

National Institute for Occupational Safety and Health (NIOSH), 48, 51, 52, 53, 59, 60, 64, 66, 77, 78-79, 109, 111, 112, 141, 142, 144, 164, 178, 179, 268, 269

National Institutes of Health, 77, 79, 126, 201, 203, 207, 260

National Library of Medicine, 52, 53, 63

National Pollutant Discharge Elimination Systems (NPDES), 271

National Toxicology Program (NTP), 63, 275, 291

Near-infrared radiation, 75

Needle valves, 168

Needles and syringes, 30, 58, 111, 114, 131, 164, 206-207, 208, 278

Neurotoxicity, 54, 55, 56, 59, 62, 290

Neutralization and neutralizers
  acids, 120, 121, 136, 137, 138, 195, 196, 208, 209
  alkalis and bases, 120, 209, 210

Nickel, 61

Nickel alloys, 166, 174

Nickel carbonyl, 60, 166, 302, 303, CD

Nitrates, 69, 70, 71, 96, 130, 138, 139, 190

Nitric acid, 69, 73, 96, 120, 131, 138, 139, 214, CD

Nitrites, 69, 71, 140

Nitrogen, 34, 62, 69, 73, 74, 75, 76, 77, 111, 129, 130, 131, 133, 134, 135, 138, 140, 153, 154, 158, 159, 160, 163, 164, 165, 169, 170, 171, 172, 173, 174, 239, 243, 245

Nitrogen compounds, 71

Nitrogen dioxide, 56, 57, 60, 61, 275, 303, CD

Nitrogen mustard, 260

Nitrogen oxides, 138

Nitrogen triiodide, 70, 136

Nitrosamines (*see* CD)

N-Nitroso compounds, 70, 71, 138-139, 273, 302, 303

Noise extremes, 77, 132, 149, 164, 214, 216-217, 229, 250, 272

Notifying personnel of emergencies, 117

Nuclear Regulatory Commission (USNRC), 79, 81, 82, 127, 189, 201, 202, 204, 205, 207, 208, 253, 261, 271, 276

Nucleating agents, 73, 129

O

Occupational and Educational Eye and Face Protection, 108

Occupational Safety and Health Act, 268, 270, 281

Occupational Safety and Health Administration (*see also* OSHA Standards)
  GHS implementation, 49
  information resources, 53
  state laws, 268

Odor and odor thresholds, 60, 92, 95, 99, 137, 139, 141, 179, 187, 216, 237, 240, 293, 301

Odorants, 70

Oil
  baths, 88, 93, 128, 129, 133, 154, 155, 156, 158
  bubblers, 170
  filters, 179
  fires, 174
  PCBs, 275-276
  reactivity, 139, 140, 165, 167, 169, 170
  vacuum pump, 93, 133, 153, 173, 174
  waste, 174, 277

Oil Dri, 120

Oily rags, 178

Olfactory fatigue, 60

Online services (*see* Computer systems and services; *specific databases*)

Open houses, 25

Ordering chemicals (*see* Acquisition of chemicals)

Organic compounds, 52, 58, 61, 62
  acetylenic, 135
  adsorbents, 236
  air emissions, 278
  azides, 70, 71, 72, 136
  compatible, 98
  drying, 157
  explosion hazards, 138, 139-140, 141, 173
  flammability/combustion, 65, 69, 159, 190, 205
  halides, 136, 205
  mercury, 62
  nitrates, 70, 130
  oxidizable, 73, 208
  peroxides, 50, 70, 72, 73, 100-101, 130-131, 133-135, 189, 210, 290, 291
  solvents, 62, 70, 76, 93, 111, 112, 136, 138, 139, 140, 151, 153, 157, 159, 177, 188, 190, 196
  waste, 186-187, 188, 189, 190, 196, 203, 204-205, 208, 210
  water solublity, 188, 192, 196

Organisation for Economic Co-operation and Development, 49

Organolithium compounds, 136

Organomercurial compounds, 301

Organometallic compounds, 70, 122, 130, 139, 170

Organophosphates, 62

OSHA standards
  chemical-specific, 273
  Control of Hazardous Energy Standard, 150
  enforcement, 268
  exposure limits, 269, 273
  Hazard Communication Standard, 47, 48, 49, 51, 101, 268-269, 270, 275
  Laboratory Standard, 2, 6, 13 n.2, 14-15, 17, 18, 20, 21, 22, 48, 59, 62, 63, 101, 175, 268-269, 270, 272, 273-275, 289-305

Occupational Exposure to
Bloodborne Pathogens
Standard, 79, 206
Personal Protective Equipment
Standard, 111, 112, 175
Respiratory Protection
Standard, 110, 178, 179, 180,
270
Osmium tetroxide, 60, 260, CD
Ovens
drying samples, 156-157
general precautions, 156-157
microwave, 141, 159
Oxidants and oxidation (*see also
specific compounds*)
acids, 69, 73, 96, 98, 120, 128-
129, 131, 138, 188
adsorbents, 236
corrosive, 61, 121, 138
fire hazard, 34, 69, 70, 128-129,
141
gases, 50, 69, 70, 120-121, 135,
137, 141, 169, 170, 291
incompatibilities, 70, 73, 139,
140
liquids, 69
metal oxidants, 57, 70, 86, 141
nonoxygen examples, 69
reactivity, 70, 130, 131, 139, 140
solids, 69
spills and leaks, 120-121
storage, 96, 97, 98, 99, 100, 140,
170
waste, 187, 188, 195, 204, 205,
208
Oxygen, 62, 65, 66, 67, 135, CD
cylinders/tanks, 139, 163, 169
depletion, 163, 173, 179, 245
gas/atmospheric, 70, 72, 73, 78,
129, 134, 135
isotopes, 203
leak testing, 169
liquefied, 69, 74, 135, 172, 173
monitors, 163, 173, 245
oil/lubricant reactivity, 139,
140, 165, 167, 169, 170
scavenging/scavengers, 160
Ozone, 57, 60, 69, 76, 139, 151,
271, 278, CD

P

Palladium on carbon, 139, 130,
CD
Pandemic planning, 35

Paracelsus, 18
Particularly hazardous substances
(PHSs), 5, 13, 14, 18, 23, 55,
59, 60, 62, 63, 64, 80, 118,
216, 220, 274-275, 278, 293,
296
Particulates, 49, 57, 79-80, 144,
176, 179, 181, 221, 231, 235,
236, 237, 239, 243-244, 254
(*see also* Nanomaterials;
Powders and dusts)
Pentanes, 67, 72, 120
Peracetic acid, 303, CD
Perchlorates, 69, 70, 71, 74, 98, 99,
130, 131, 139, 254
Perchloric acid, 69, 71, 73, 96, 100,
129, 131, 139, 224, 226, 233-
234, 240, 242, 254, CD
Permanganates, 69, 73, 98, 99, 130,
139, 210, 236
Permissible exposure limits
(PELs), 60, 268, 269, 273, 274,
292, 294, 295, 297
Peroxides and peroxidizables (*see
also specific compounds*)
acids, 72, 134, 139
containers, 21, 69, 133, 134
detection tests, 134
dilution, 133, 134
distillation, 72, 134, 136
explosives, 19, 72-73, 133-135
metal peroxides, 71, 73, 130-131
organic peroxides, 50, 70, 72,
73, 100-101, 130-131, 133-
135, 189, 210, 290, 291
peroxidizable compounds, 134
precautions for handling,
133-134
reducing agents for, 134
storage, 21, 72, 100, 133-134
waste disposal, 134-135, 187,
188-189, 190, 210
Personal protective equipment
(*see also* Clothing and
protective apparel; Gloves
and hand protection;
Goggles; Respirators)
airborne contaminants, 127,
144, 178-180
corrosives, 171-172, 176
eye protection, 26, 36, 75, 108,
109, 124, 132, 144, 162, 176
face protection, 26, 61, 108, 109,
114, 131, 132, 138, 144, 162,
173, 174, 176

familiarity with, 22
foot protection, 175-176
glass hazards, 26, 36, 75, 108,
109, 124, 132, 172, 176, 214
PHSs and, 13
Petroleum ether, 120
Pets, 16
Phenol, 58, 61, 139, 202, 204, 303,
CD
Phorbol esters, 64
Phosgene, 57, 60, 85-86, 239, 260,
303, CD
Phosphate buffer, 54
Phosphine, 139, 141
Phosphorus, 61, 69, 81, 90, 96, 139,
203, CD
Phosphorus pentachloride, 70
Phosphorus pentoxide, 61, 136
Phosphorus trichloride, 139
Physical hazards (*see also specific
hazards*)
compressed gases, 74, 291
cryogens (nonflammable), 74
electrocution, 76
ergonomic, 77
high-pressure reactions, 74
information sources, 48
magnetic fields, 76
radio frequency and
microwaves, 75-76
sharp edges, 76-77
slips, trips, and falls, 77
ultraviolet, visible, and
infrared radiation, 75
vacuum work, 74-75
Pipets and pipetting, 77, 110, 126,
127, 206, 252
Planning (*see* Chemical Hygiene
Plans; Emergency action
plan; Emergency response;
Experiment planning)
Plastic equipment and containers,
76, 88, 94, 96, 97, 98, 99, 112,
113, 114, 115, 120, 128, 129,
130, 144, 145, 149, 150, 156,
168, 169, 172, 173, 174, 175,
191, 203, 224, 233, 234, 236,
240, 243, 244, 245, 302, 303
Platinum, 61, 130, 135, 139
Polychlorinated biphenyls
(PCBs), 187, 206, 270,
275-276
Polymerization, 70, 72, 73, 116,
130, 292
Polyols, 139

Potassium, 69, 70, 75, 130, 134, 140, CD
Potassium amide, 72
Potassium chloride, 195, 196
Potassium cyanide, CD
Potassium hydride, 122, CD
Potassium hydroxide, 120, 140, 207, CD
Potassium iodide, 134, 188, 189
Potassium metals, 72, 122, 134, 178, 180
Potassium permanganate, 73, 236
Potassium peroxides, 139
Powders and dusts, 55, 57, 60, 73, 78, 112, 124, 130, 131, 136, 145, 161, 177, 178, 179, 190, 222, 231, 237, 244, 248, 254, 302
Preparation (*see* Emergency preparedness; Experiment planning; Procedures for working with materials)
Pressure control systems/devices
air, 214, 219, 240
clean rooms, 219, 243
compressed gas cylinders, 26, 164, 168, 169-170
gauges, 88, 167, 172, 180, 243, 249
precautions, 166-167
Pressure vessels and reactions (*see also* Autoclaves and autoclaving; Compressed gases; Vacuum systems and operations)
and airborne contamination, 135
assembly and operation, 165-168
cleaning, 165
closed-system vessels, 166
compatibility of equipment materials and fittings with chemicals, 166, 167, 169
gas monitors and alarms, 168
glass equipment, 26, 74-75, 109, 115, 131-132, 135, 149, 153, 167, 170, 171-172, 174-175
headspace, 166
for highly reactive materials, 166
maintenance, 167
maximum allowable working pressure, 167

piping, tubing, and fittings, 164, 166, 167, 168, 169
plastic equipment, 168
pressure gauges, 167
records, inspection, and testing, 165, 167, 170-171
relief devices and regulators, 166-167
shielding, 166, 170, 172, 174, 175, 300
stuffing boxes and gland joints, 166
Teflon tape applications, 165, 168
valves, 166, 167, 168
venting, 167
warning signs, 166
Principal investigators and project managers, 3, 4, 6, 19, 38, 251, 262 (*see also* Laboratory supervisor)
Procedures for working with materials (*see also* Experiment planning)
biohazards, 126-127, 247-248
compressed gases, 140-141
equipment and glassware maintenance, 114-115
explosive and highly reactive materials, 130-140
flammable chemicals, 127-130
highly toxic substances, 122-125
housekeeping, 113-114
information sources, 48
minimizing exposure, 108-113, 124-125
nanomaterials, 141-146
personal behavior, 108
radioactive materials, 127
storage of chemicals, 114
transport of chemicals, 114
Procurement (*see* Acquisition of chemicals)
Publicly owned treatment works (POTW), 196, 202, 204, 278
Pull stations, 121, 127
Putrescible waste, 204, 206, 207, 208
Pyridine, 72, 303, CD
Pyrolysis, 73, 136
Pyrophorics, 19, 34, 50, 69, 70, 74, 96, 100, 113, 128-129, 135-136, 141, 145, 159, 233, 291

R

Radiation and radioactive materials (*see also* Multihazardous waste)
absorbed dose, 81-82
acids, 202, 203, 205
aerosols, 204, 208, 221
biological materials, 207-208
clothing and PPE, 80, 127, 162
decay in storage, 203-204, 207, 208, 278
disposal, 202-205, 207-208
filters, 236
hazards, 75, 79-82
procedures for working with, 127
shielding, 81, 82, 127, 203, 234
training requirements, 13
Radio frequency hazards, 75
Raney nickel, 73, 130, 136, 139
Reaction hazard index, 65
Reactivity (*see also* Explosive and highly reactive hazards; Incompatibility of chemicals; Pyrophorics; Water reactivity; *specific compounds*)
in air, 70, 72, 87, 93, 125, 130, 134, 135, 136, 137, 139, 140, 141, 143, 160, 173
gases 57, 70, 133, 135, 140, 141
glass, 137, 141
hazards, 70
information sources, 52, 65, 304
pressure reactions, 166
storage hazards, 100-101
Receiving chemicals, 21, 79, 89-90, 263, 278, 299
Receiving rooms and loading areas, 89-90, 240, 256
Recycling
characterization of wastes for, 187, 190
containers, packaging, and labware, 93-94, 191
energy reclamation, 195-196, 198
general considerations, 92, 93, 192, 194
information sources, 201
manufacturer and supplier role
multihazardous waste, 201, 202, 204, 205
nonhazardous solid waste, 192

off-site, 93, 185, 186, 198
regulations for chemical
   hazardous waste, 193-194,
   195-196, 277, 300
solvents, 5-6, 84, 93, 185, 192
Reducing agents, 69, 70, 73, 100,
   130, 133, 138, 139-140, 170
Refrigerators and refrigeration
   access control, 123, 216
   alarms on, 216
   azo and peroxide storage, 72,
      100, 133-134
   containers, 154
   decommissioning, 253, 254
   defrosting and cleaning, 21,
      154
   explosion-proof, 21, 153-154
   flammability hazards, 21,
      153-154
   food or beverages, 21, 153
   freezer space for samples, 37
   labeling of samples, 154
   power loss, 40, 41, 76
   safety precautions, 153
   sharing between research
      groups, 216
   spark-proof, 154
   storage management, 97-98,
      154
   waste storage in, 207
Registry of Toxic Effects of
   Chemical Substances
   (RTECS), 59
Regulations and legislation (*see
   also* Occupational Safety
   and Health Administration;
   OSHA standards; *individual
   statutes*)
   accident reporting and record
      keeping, 272, 281, 299
   acquisition of chemicals, 21, 22,
      278, 299
   acutely toxic substances, 274,
      275, 290, 293, 295, 301-302
   administration and supervision
      of laboratories, 279, 280, 281,
      295
   biohazards, 79, 189, 270-271,
      272, 276, 278, 280-281
   building codes, 11, 22, 27, 95,
      98, 100, 215, 217, 218, 242,
      253, 272-273
   carcinogens, 59, 63, 101, 273,
      275, 290, 293, 302

characteristic waste, 189-190
chemical hazardous waste,
   189-191
Chemical Hygiene Plan, 3, 15,
   47, 269, 274, 275, 292-293,
   295, 296
chemicals used in laboratories,
   270, 273
compliance monitoring, 3
compressed gases, 114, 164,
   270, 290, 291
containers, 21, 94, 103, 191, 194,
   298, 301
and culture of safety, 6
determining regulatory status
   of waste, 190-191
electrical codes, 149, 150, 151,
   152, 154
emergency preparedness and
   response, 6, 272, 299
employee information and
   training, 6
empty containers, 191
and experiment planning,
   281-282
federal processes, 267
fire codes, 11, 22, 27, 51, 89,
   95, 98, 100, 113, 114, 128,
   151, 154, 214, 216, 243, 270,
   272-273
limited waste, 190
participation in regulatory
   processes, 267
PCBs, 275-276
penalties for violations, 6, 23,
   267-268
physical hazards, 6
radioactive materials, 79, 127,
   270, 276, 278
rationale for, 267
relationships between
   regulations and standards,
   268-269
risk-based performance
   standards, 89, 259, 261, 270,
   273
shipping, export, and import,
   23, 271-272, 278-281
state, 268
storage, 95
training in, 23
waste management, 6, 271,
   276-278
Reporting
   academic laboratories, 3

accidents, 5, 16, 19, 29, 30, 181,
   281, 299
supervisors, 15, 16, 17, 226, 301
TSCA requirements, 279
Reproductive toxins, 50, 52, 53,
   55, 56, 59, 62-63, 86, 97, 101,
   274-275, 290, 291, 293
Resource Conservation and
   Recovery Act (RCRA), 93,
   189, 190, 193, 194, 196, 201,
   202, 209, 267, 271, 272, 276,
   277-278
Respirators
   inspections, 180
   procedures and training, 180
   types, 179
Respiratory tract, 57, 124, 166, 179
   (*see also* Inhalation hazards)
Risk assessment (*see also* Hazard
   evaluation)
   aerosols, 55, 60, 80, 221
   airborne contaminants, 219
   biohazards, 13, 79, 80, 207, 221
   carcinogens, 54, 55, 63-64
   control banding, 64-65
   existing control systems, 11
   explosivity, 70-74, 132
   flammability, 65-70
   information sources, 47-49
   review of, 12
   toxicity, 54, 55, 58-60
Risk-Based Performance Standard
   (DHS), 262

S

Safety and emergency equipment
   (*see also* Fire safety
   equipment)
   access to, 15, 27
   automatic external
      defibrillators, 180
   eyewash units and safety
      showers, 15, 27, 28, 29, 34,
      113, 117, 118, 121, 122, 125,
      127, 128, 137, 176, 178, 180,
      181, 217, 218, 273, 297, 299
   first aid, 15, 28, 36, 125, 137,
      180
   respirators, 178-180
   shields, 177
   spill control kits and cleanup,
      177
   storage and inspection, 180-181

Safety committees, 12, 16-17, 23, 24-25, 64, 276
Safety rules and policies
  general guidelines, 15-17
  housekeeping practices, 19-20
  working alone, 17-18
Safety shields, 74, 108, 172, 176, 177, 187
Scientists with disabilities, 6-7
Scrubbers, 236
Security (*see* Laboratory security)
Security plans
  elements, 262-263
  levels of security, 263-264
  managing security, 264
  training, 264
Seismic activity, 35, 36, 70, 95, 96, 97, 100, 168, 199
Select agents, 79, 270
Select carcinogens, 55, 59, 63-64, 86, 97, 274-275, 290, 291-292
Sensitizers (*see* Allergens, sensitizers, and allergic reactions)
Service animals, 16, 218
Sewer discharges, 16, 35, 69, 85, 134, 145, 149, 153, 185, 186, 192, 193, 195, 196, 202, 203, 204, 206, 207, 208, 209, 210, 222, 232, 276, 278, 300, 301
Sharps
  disposal, 58, 111, 114, 206-207
  hazards, 76-77
Shelf life and expiration dates, 21, 22, 92, 100, 101, 112, 125, 132, 134, 138, 261
Shielding
  bench shields, 132
  blast/explosion, 109, 110, 130, 131, 161, 174
  chemical hoods, 131-132, 160, 177, 187
  full-face, 26, 61, 108, 109, 114, 131, 132, 138, 144, 162, 173, 174, 176
  glass materials, 177, 225, 229, 248
  for magnetic fields, 76
  microwave or radio frequency emissions, 162
  pressure/vacuum reactions, 166, 170, 172, 174, 175, 300
  radiation, 81, 82, 127, 203, 234
  rotating equipment, 160, 162, 163

safety shields, 74, 108, 172, 176, 177, 187 (*see also* Face shields)
Shipment and transport (*see also* Exports and imports)
  in alcohol, 101
  biological materials, 101, 272, 278, 280-281
  carts, 23
  chemicals, within U.S., 280
  containment, 6, 23, 89, 90, 114, 125, 298
  corrosives, 199
  gas cylinders 111, 114, 168
  information sources, 48
  "materials of trade" exemption, 102
  nanomaterials, 102-104, 280
  regulations, 23, 271-272, 278-281
  training, 23
  waste, 87, 186-187, 193, 194, 198, 199
Signage and postings, 39, 250, 298
Short-term exposure limits (STELs), 51, 60, 269
Showers
  air, 244
  safety, 15, 27, 28, 29, 34, 113, 117, 118, 121, 122, 125, 127, 137, 176, 178, 180, 181, 217, 218, 273, 297, 299
Silver and silver compounds, 166, 167, CD
Skin contact and effects
  absorption of toxins, 55, 58, 60, 62, 124, 139, 166, 179
  airborne contaminants, 179
  cold and cryogenic substances, 119
  corrosives, 58, 61, 111, 135, 141, 190
  emergency response, 118
  minimizing, 111-113
  toxic chemicals, 58
Slips and falls, 77, 164
Slot hoods, 220, 237, 238
Snorkels, 143, 220-221, 226, 237, 238, 240, 241, 250, 251, 298
Sodium, 69, 70, 75, 122, 130, 137, 139, 140, 159, 178, CD
Sodium amide, 72, 140
Sodium azide, 61, 136, 260, 303, CD
Sodium bicarbonate, 120, 177

Sodium bisulfate, 120, 136
Sodium bisulfite, 134
Sodium borohydride, 96
Sodium carbonate, 120, 195
Sodium cyanide, 54, 56, 60, 260, 303, CD
Sodium hydride, CD
Sodium hydrogen sulfide, 96
Sodium hydroxide, 61, 96, 136, 140, 190, 209, CD
Sodium hypochlorate, 96
Sodium nitrite, 140
Sodium peroxide, 90
Solvents (*see also specific chemicals*)
  drying, 78, 140
  filtration, 84
  flammable, 28, 120
  green chemistry, 85-86
  organic compounds, 62, 70, 76, 93, 111, 112, 136, 138, 139, 140, 151, 153, 157, 159, 177, 188, 190, 196
  recycling, 5-6, 84, 93, 185, 192
  spills, 28, 120
  stills, 93, 159-160
Southern Building Code Congress International, Inc., 273 n.1
Spills and releases
  absorbents, 28, 120, 125, 133, 134, 145, 177, 302
  acids, 28, 120
  alarms, 35, 82, 124
  alkalis and bases, 120
  cleanup, 28-29, 120, 280-281
  containment, 120
  debris management, 120
  disposal, 28
  emergency planning and response, 28-29, 35, 118, 120, 121, 280-281
  environmental releases, 28
  flammable solvents, 28, 120
  highly toxic substances, 29, 120, 273
  information sources, 28
  low-flammability and low-toxicity materials, 28
  mercury, 121
  policy, 28-29
  regulations, 281-282
  reporting, 28, 29, 281
Standards Completion Program, 52

Standards for Protection Against
  Radiation (USNRC), 79, 127,
  270, 276
Static electric discharges, 22, 23,
  69, 113, 128, 129, 132, 168,
  170, 176
Stoppers, 133, 154
Storage (*see also* Refrigerators and
  refrigeration)
  access control, 21, 22
  acids, 20, 96, 97, 98, 99, 100,
    125
  biosafety cabinets, 75, 80, 126,
    143, 220, 221, 231, 236, 244,
    245-248, 251
  carcinogens, 97, 101
  chemical storage cabinets, 22
  clothing and PPE, 113, 175
  cold storage, 21, 97-98
  compatibility, 21, 22, 27, 96-97
  compressed gas cylinders, 26,
    35, 90, 96, 100, 111, 114, 154,
    164, 166, 168, 170
  containers, 35, 73, 77, 89, 94, 95,
    96, 97, 98-99, 100, 113, 114,
    125, 129, 133, 137, 154, 191,
    192, 302
  controlled substances. 95
  corrosives, 97, 100, 166, 192,
    221
  cyanides, 22
  dating, 21
  experiment planning, 125
  flammable and combustible
    liquids, 22, 98-99
  gas cylinders, 100
  general considerations, 95-96,
    114
  in glass, 73, 94, 99, 100, 129,
    133, 137, 154, 191, 192, 302
  highly toxic substances, 22, 97,
    101, 125
  inappropriate places, 21
  information sources, 48
  inspection, 27
  labeling, 21, 51, 94, 154
  odiferous materials, 22
  peroxides, 21, 72, 100, 133-134
  reactive substances, 100-101
  regulations, 95
  rooms, 22
  shelving, 21
  size of containers, 99
  sprinkler head distance, 21
  ventilation, 22

Sulfur, 73, 81, 121, 131, 136, 203,
  209
Sulfur dioxide, 96, 170, 303, CD
Sulfuric acid, 61, 69, 73, 96, 120,
  139-140, CD
Suppression of aerosols, 114, 124,
  126, 208, 236, 240, 244, 301,
  302-303
Survival kit, 36

T

Teachers and instructors, 3-4
Teaching laboratories (*see*
  Academic laboratories)
Technology advances, and culture
  of safety, 5
Temperature sensors and smoke
  detectors, 100, 178, 249
Teratogens, 52, 62, 275, 291 (*see
  also* Developmental toxicity;
  Reproductive toxins)
Testing
  with acids, 188, 189
  alarm systems, 43, 193, 245,
    249, 250
Tetrahydrofuran, 28, 67, 72, 96,
  120, 134, 136, 138, 160, CD
Thermometers, 17, 87-88, 157
Titanium, 73, 136
Titanium dioxide, 49, 78-79, 141
Toluene, 67, 85, 86, 93, 96, 133,
  140, 202, 203, CD
Toluene diisocyanate, 61, CD
Toxic Substances Control Act
  (TSCA), 279-280
Toxins and toxicity (*see also*
  Acute toxicity; Allergens,
  sensitizers, and allegic
  reactions; Asphyxiants and
  asphyxiation; Carcinogens;
  Chronic toxicity and health
  affects; Corrosives and
  corrosivity; Developmental
  toxicity; Exposure; Irritants;
  Neurotoxicity; Reproductive
  toxins)
  access control, 123-124
  compressed gases, 96, 121
  designated areas, 123
  dose-response relationship,
    54-56
  duration and frequency of
    exposure, 56
  emergency planning, 125

  exposure routes, 57-58
  gases, 57, 85, 122, 140, 172, 190,
    261, 275
  hazard evaluation, 53-65
  information sources, 48, 52
  minimizing exposure, 124-125
  planning experiments, 122-123
  protocols, 123
  risk assessment, 55, 58-60
  storage, 101, 125
  target organs, 63
  types, 60-65
  waste disposal, 125
TOXLINE, 53
TOXNET, 52, 53, 63
Tracer gases, 221
Training (*see* Education and
  training in safety practices)
Transfer and exchange of
  chemicals
  between containers, 23, 84,
    87, 94, 104, 113, 123, 125,
    240-241
  compressed gases, 166
  nanomaterials, 103-104
Transport (*see* Shipment and
  transport)
Trifluoroacetic acid, CD
Trimethylaluminum, CD
Trimethyltin chloride, CD
Tubing, 17-18, 76, 115, 128, 137,
  149, 154, 156, 164, 165-166,
  167, 168, 170, 171, 172, 175,
  176, 237, 239 (*see also* Hoses)

U

ULPA (ultra-low penetration air)
  filters, 236, 248
Ultrasonic equipment, 85, 86, 149,
  161
Ultraviolet peroxidation, 203,
  208
Ultraviolet radiation, 72, 75, 76,
  109, 139, 151, 162, 176
Unattended experiments, 34, 39,
  40, 41, 42, 116, 124, 128, 149,
  151, 154, 155, 156, 158, 160,
  161, 170, 301
Unconscious victims, 119
Uniform Fire Code, 270
Uniform Hazardous Waste
  Manifest, 199, 200
United Nations
  Environment Programme, 51

Recommendations on the Transport of Dangerous Goods, 49
Sub-Committee of Experts on the Transport of Dangerous Goods, 49
Unused and excess material, 73, 86, 87, 114, 133, 185, 192

V

Vacuum cleaners, 121, 145, 302-303
Vacuum systems and operations
  assembly of apparatus, 175
  desiccators, 175
  Dewar flasks, 174-175
  glass vessels, 26, 74-75, 109, 115, 131-132, 135, 149, 153, 167, 170, 171-172, 174-175
  hazards, 74-75
  pumps, 153
  rotary evaporators, 175
  spill cleanup, 121
Valves, gas cylinders, 168, 169
Vapor pressure, 56, 57, 65, 122, 128, 164, 219, 220, 290
Variable autotransformers, 151, 154, 155, 156, 157, 158, 170
Vehicular accidents, 102, 193
Ventilated balance enclosures, 220, 238
Ventilation and environmental control systems (*see also* Chemical hoods; Exhaust systems)
  and airborne contaminants, 219, 221, 241, 251, 296, 298
  code requirements, 253
  choosing, 253
  commissioning, 251
  constant air volume systems, 241
  containers, 139-140
  design criteria, 249-250
  engineering controls for personal protection, 220
  exhaust systems, 241-243
  goals performance measurement, 250-251
  information sources, 253
  inspection, 250
  maintenance, 248-249, 250
  management program, 249-251
  monitoring systems, 252
  retrocommissioning, 252
  risk assessment, 219-221
  supply systems, 179, 180-181, 217, 222, 226, 227, 241
  sustainability considerations, 251-253
  training program, 250
Venting (*see also* Exhaust systems)
  gas cylinders, 168, 170
Vinyl acetylene, 71
Visible radiation, 75
Visitors, 15, 16, 35

W

Waste management (*see also* Characterization of waste; Chemical hazardous waste; Multihazardous waste)
  absorbents, 193, 203, 204
  academic laboratories, 3
  administration and supervision, 209, 302
  air emissions, 278
  biohazards, 87, 108, 127, 185, 187, 189, 195, 201, 202, 205-208, 278
  commingling of waste, 191, 193, 194, 198, 203
  containers, 23, 58, 87, 93, 110, 111, 114, 120, 125, 127, 144, 145, 191, 192, 193, 194, 195, 198, 199, 206-207, 209
  corrosives, 113, 118, 144, 302
  wastes, 153, 174, 190, 191, 192, 199, 209
  cost-effectiveness, 5-6
  and culture of safety, 5-6
  experiment planning, 125
  filter disposal, 107, 191, 192
  highly toxic materials, 125
  information sources, 48
  landfill disposal, 185, 190, 193, 194, 195, 196, 197, 300
  packaging and shipment, 87, 186-187, 193, 194, 198, 199
  radioactive waste, 202-205, 208, 278
  reduction/minimization, 5, 6, 85, 86, 108, 195, 201, 202, 203, 207, 277
  regulations, 271, 276-278
  sewer discharges, 16, 35, 69, 85, 134, 145, 149, 153, 185, 186, 192, 193, 195, 196, 202, 203, 204, 206, 207, 208, 209, 210, 222, 232, 276, 278, 300, 301
  sharps, 77, 114, 115, 203, 205
  source reduction, 5, 122, 185, 195, 201, 202
  treatment (laboratory-scale) of surplus and waste chemicals, 209-210
Water reactivity
  hazardous materials, 70
Weather, severe, 34-35
Windows, 132, 258
Working alone, 15, 17-18, 116, 120, 122, 301
World Health Organization, 51, 76

X

Xylene, 62, 67, 85, 93, 190, 203

Z

Zirconium, 73, 136